Physical Geography Today
A Portrait of a Planet

THIRD EDITION

THIRD EDITION

Physical Geography Today
A Portrait of a Planet

Robert A. Muller
Louisiana State University

Theodore M. Oberlander
University of California, Berkeley

 Random House New York

Third Edition
987654321
Copyright © 1974, 1978, 1984 by Random House, Inc.

Library of Congress Cataloging in Publication Data

Muller, Robert A.
  Physical geography today.

  Includes bibliographies and index.
  1. Physical geography.   I. Oberlander, Theodore.
II. Title.
GB55.M84 1984   910'.02   83-23122
ISBN: 0-394-33264-4

Manufactured in the United States of America

Cover Photo: *Twilight in the Wilderness*, 1860. Frederic Edwin Church. Oil on canvas, 101.6 × 162.5 cm. Courtesy, The Cleveland Museum of Art, Purchase, Mr. and Mrs. William H. Marlatt Fund.

Cover and text design: Leon Bolognese

Preface

The diversity of the planet we inhabit is so great that no single point of view or scientific discipline is adequate to describe or explain the earth fully. Artists, physicists, economists, historians, and geologists all have their unique and valuable perspectives, and each group is likely to see the world differently. The discipline of geography has its own view of the earth—an especially comprehensive one that focuses on the patterns of natural and human phenomena on the earth's surface, the causes of these phenomena, and the interrelationships between them.

This book focuses its attention on physical geography, which is mainly concerned with the natural world as it is seen from the human perspective. Physical geography is the geography of the human environment, and is more than a mere composite of other physical sciences such as meteorology, climatology, biology, pedology, and geology. Physical geography not only describes the natural phenomena at the earth's surface but also, and more importantly, seeks explanations of how and why physical processes act as they do, and how they are significant to life on our planet. The subject matter of physical geography extends from the atmosphere through the hydrosphere and biosphere to the upper portion of the lithosphere, but centers upon the surface region where land, sea, and air meet and interact, and where life flourishes. Human impacts on the natural systems operating at this interface are an important consideration in physical geography, and are touched upon in almost every chapter of this text.

The processes that influence surface phenomena involve both energy and materials. Energy continuously cascades through physical systems ranging in scale from the global biosphere to single microbes. Materials such as carbon, water, and oxygen flow through these systems in never-ending cycles that involve inputs, storages, outputs, and quasiequilibrium states. The study of physical geography emphasizes the ways in which the various physical systems interact, constantly exchanging energy and materials. While it is our desire to organize the complexity of our planet by using the concepts of energy systems, it is not our intention to permit these concepts to diminish the fascination derived through consideration of the individual wonders of the real world. Despite our use of generalizing theory, our focus is on the world itself, in all its wondrous but comprehensible complexity.

Our aim has not been to introduce and define every term used by physical geographers in their professional work, nor to touch on every phenomenon that enters into the realm of physical geography. Rather than lightly skimming a very large surface, we have sought to develop an understanding of the most essential facts and relationships by extended treatment of the topics that seem to be of paramount importance. Some of these have not previously been explored in beginning textbooks in physical geography.

Based largely on instructor experiences with the second edition of this text and with the subsequent *Essentials of Physical Geography Today*, a reorganized and more concise treatment, we have restructured the third edition of *Physical Geography Today*. We have adopted the organizational sequence of *Essentials of Physical Geography Today*, which has been well received by instructors and users of the latter text.

Specifically, the discussions of the forces that drive the atmospheric and oceanic circulations have been extracted from Chapter 5 of the second edition and have become Chapter 4 of the present text; in this new position the circulation patterns immediately follow the global energy and temperature distributions that drive the atmospheric and oceanic circulations. The secondary atmospheric circulations in the tropics and midlatitudes become the focus of Chapter 6, which follows the chapter on atmospheric moisture and precipitation.

The chapter entitled "Ecologic Energetics" in the second edition has been redeveloped to give more emphasis to the concerns of biogeography. It now follows the material concerning climatic classification and precedes the coverage of vegetation in relation to climate.

Following the chapter on soils, the discussion of the lithosphere and its materials has been separated from the treatment of basic geomorphic concepts, as in the *Essentials* text. Fluvial and aeolian processes and associated landforms are treated in Chapter 14, followed by chapters dealing with structural and climatic influences on landforms, and glacial and marine processes. Case Studies, which provide close-up views of interesting phenomena like El Niño, climatic change, and species extinctions, now follow *every* chapter, along with annotated References, Review Questions, and suggested Applications, new problems that require students to think creatively about the information presented in the chapter.

The text is designed to be appropriate for an academic quarter or semester, and, with the addition of selected readings, this edition could serve as the basis for a two-semester sequence. Much attention has been devoted to continued improvement of the book's graphics and their captions, which have been popular features of past editions of *Physical Geography Today* and *Essentials*. Many new illustrations have been introduced in this third edition in the hope of enhancing student understanding of processes and interactions, and of the overall perception of physical landscapes.

Acknowledgments

We are grateful for the assistance of the reviewers whose comments and suggestions helped us shape the third edition of *Physical Geography Today*. A preliminary draft of the manuscript was reviewed by Robert B. Batchelder, Boston University; Vernon Meentemeyer, University of Georgia; Laurence S. Kalkstein, University of Delaware; Wayne N. Engstrom, California State University, Fullerton; Patricia F. McDowell, University of Oregon; Patrick J. Bartlein, University of Oregon; the late Jack R. Villmow, Northern Illinois University; Marlon L. Shelton, University of California, Davis; Kenneth L. White, Texas A. & M. University; and Jay R. Harman, Michigan State University. Detailed comments on each chapter were provided by Robert B. Batchelder, Boston University; John J. Alford, Western Illinois University; David McArthur, San Diego State University; Carl L. Johannessen, University of Oregon; John Lier, State University of California at Hayward; and John Street, University of Hawaii. We are also grateful for the work of Harry Spector, who collated reviewers' comments and edited the text. We gratefully acknowledge the assistance of many people at Random House in the preparation of this edition: especially Barry Fetterolf, Suzanne Thibodeau, Sylvia Shepard, Susan Israel, Brian Hogley, Leon Bolognese, Charlotte Green, Cindy Spiegel, David Rothberg, and David Saylor. We appreciate their encouragement as well as all of their contributions. It would be neither fair nor wise to overlook the many ways in which the assistance of our wives, Jeanne and Lucille, helped bring this undertaking to fruition. And, of course, we must acknowledge the contributions of our students, our teaching assistants, and the countless persons whose paths we have crossed to our benefit in the preparation of this book—as well as those whose needs have had to wait while project-related deadlines and emergencies were being met.

Overview

Contents

Physical Geography Today *A Portrait of a Planet*

THIRD EDITION

The Mystery, by T. M. Oberlander, 1966

The origin of the universe and our solar system with its nine planets remains a matter of scientific speculation. What is clear is that the planet earth has changed astonishingly since its formation about 5 billion years ago. Physical geography is the study of the processes that continue to modify our planetary environment.

In every country of the world there are individuals who identify themselves as geographers. What is their function? Everyone — everywhere — that geography is concerned with *where things are*, but beyond this most people are unsure about what the study of geography involves. Some have encountered a type of "geography" that merely describes the earth and the natural processes and human activities occurring on its surface, and others are aware that geography is a field that seems to have no clear boundaries. Everything in the universe has some spatial properties. Once these are described, what more remains to be done? The answer is that description is only the beginning. Having determined the spatial pattern of any phenomenon, one naturally asks further questions: how was the pattern achieved? why did it develop in such a way? what are its current tendencies? how does it influence other phenomena? what is its future? These questions are the essence of geographical inquiry, and have been for more than 2,000 years.

The two broadest subdivisions within the field of geography are *physical geography,* which focuses on the natural processes that create physical diversity on the earth, and *human geography,* which is concerned with human activity on our planet. However, human activities clearly alter natural processes and the physical environment, and natural processes and features exert many influences on human activities. Thus physical and human geography are intimately interwoven. While they cannot be entirely separated, it is possible to focus upon either the human phenomena played out on the physical stage or the physical stage on which humans must perform. This book does the latter.

The earth's surface is a constantly changing arena in which many forms of both energy and matter continually interact. Here energy from the sun and from the earth's interior meet air, rock, soil, water, and a host of living organisms. All of these are themselves intricate systems linked by and to the physical processes that shape our natural surroundings. Because physical geography analyzes the complex natural processes that determine the human environ-

1
Evolution of the Earth

Formation of the Earth

Birth of the Universe
Formation of the Planets
Structure of the Earth
The Early Atmosphere
The Early Landscape

Life on the Earth

The Fossil Record
Organic Evolution
Interaction of Life with
 the Environment

Dating the Earth

Figure 1.1 Physical geography draws upon the specialized knowledge of many disciplines. The unique contribution of physical geography is its focus on the interactions of the varying phenomena that combine to give every place its particular character. (Tom Lewis)

ment, its subject matter must be extremely diverse, including that of most of the natural sciences (Figure 1.1). In this analysis we concentrate always on interactions, such as the effect of solar energy on atmospheric motion, the role of water in the development of soil, or the influence of vegetation on erosion processes. Similarly, human geography studies the interplay of physical, cultural, historical, and economic influences and their effect on human activities throughout the world. Geography's emphasis on the interactions among various physical and human systems provides a unique point of view, one that no other single field of science offers.

Physical geography, which is the subject of this book, is the original environmental science—traditionally concerned with the interaction between human beings and the physical environment. The relentless intensification of our extraction of water, food, fuel, and raw materials from the earth is increasingly affecting our environment and the natural processes that maintain it. In many cases human activities have unintentionally triggered changes in natural systems, affecting even the earth's capacity to support life. Knowledge of the subtle linkages within and between environmental systems can help us reduce the danger of negative side effects resulting from our activities on the surface of the earth.

As this chapter stresses, the present is part of a continuum of change on the earth that goes back some 5 billion years. The changes that have occurred have left a variety of traces. These include the deposits made by geological processes, which speak to us of changing climates, upheavals of the land, and rises and falls in the level of the seas; fossilized plants and animals, which record the history of life on our planet; datable materials of various types, which enable us to develop a reliable chronology of events; and even a record of the earth's changing magnetic field, which has contributed critical information about long-term movements of the earth's crust. Such clues have led to a generally accepted reconstruction of the principal events in the history of the earth. This chapter outlines those events, which have created the major physical systems that are the subjects of later chapters.

FORMATION OF THE EARTH

Birth of the Universe

Before the earth could be formed, the universe had to come into being. Scientific evidence suggests that the universe did not always exist, but originated at a definite point in time. Most scientists believe that before this event all the matter and energy in the universe had been squeezed into a single nucleus, or "cosmic egg." Between 10 and 20 billion years ago, its internal energy caused this nucleus to explode in what is called the "Big Bang," which threw matter outward in all directions. During this expansion, the elements were formed by the fusion of atomic protons, electrons, and neutrons. The first stars did not form until millions of years later, when they condensed out of vast clouds of gases and dust that were drawn together by the gravitational attraction of neighboring matter. During this consolidation process, the squeezed clouds of atoms became so heated that self-sustaining nuclear reactions commenced, emitting energy and causing them to glow. In this way they became stars—like our sun. The stars cluster by the billions in enormous rotating galaxies, many of them resembling pinwheels (Figure 1.2). The galaxies are so far apart that their light takes millions of years to pass across the voids between them.

The Big-Bang theory is supported scientifically by the discovery in 1965 that the universe seems to be pervaded by a faint "echo" of the original explosion. This is in the form of a much diminished background radiation from what appears, 10 or more billion years later, to be a very low temperature (3° Kelvin) source. Even earlier, astronomers concluded that the weak, red-tinged light arriving here from distant galaxies indicates that they are rushing away from one another as the universe continues to expand like a balloon.

The present apparent rate of expansion of the universe suggests that the Big Bang occurred some 10 to 20 billion years ago. To realize how long ago that was in relation to later events, think of 10 billion years as a 24-hour day, with the universe originating at midnight, 24 "hours" ago. Each second of such a day would represent 100,000 actual years! Our sun's

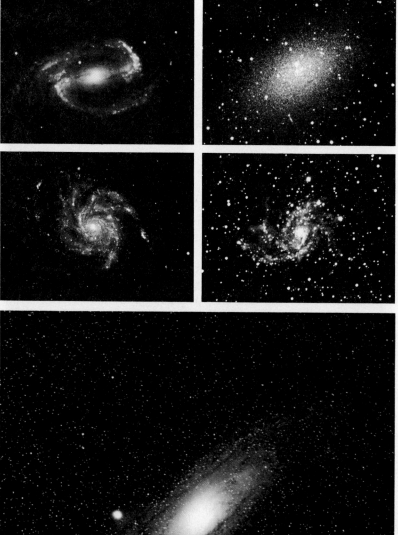

Figure 1.2 Stars cluster by the billions in galaxies separated by vast expanses of empty space. Galaxies take many forms, the more common of which are illustrated in these telescope photographs. At the top are a barred spiral galaxy (left) and an elliptical galaxy (right). Below these are a well-structured spiral galaxy (left), much like our own Milky Way, and an irregular spiral galaxy (right). The origins of the different forms remains conjectural, although an evolutionary sequence may be involved.

The lower photograph shows our own galaxy's nearest neighbor in space, the great spiral galaxy M31, which is visible to the naked eye as a hazy patch of light in the constellation Andromeda. The Andromeda galaxy M31 is virtually a twin to our Milky Way galaxy, but the distance between the two is so great that it takes more than 2 million years for the light from one to reach the other.

The separate stars visible in the photographs are those of our own galaxy, through which we view all more remote objects in space. (Hale Observatories, California Institute of Technology and Carnegie Institution of Washington, and U.S. Naval Observatory)

energy output suggests that it became active some 6 billion years ago. Using the most conservative estimate of the age of the universe, this would be almost 10 hours after the beginning of the 24-hour "day" initiated by the Big Bang.

Formation of the Planets

Although there are billions of stars, the only planets we can detect are the nine that are circling our own sun. Since our sun seems to be an average star, it is assumed that many stars have similar planetary systems. However, they are invisible, even through the most powerful telescopes, because planets give off only reflected light, which is extremely faint compared with light emitted by the furnace-like stars.

The most widely accepted explanation of the birth of our planetary system is that of Harold Urey, an American chemist and winner of a Nobel Prize. Urey proposed that our solar system began as a rotating, disk-shaped cloud of gases and dust. As the center of the cloud condensed to form the star we call the sun, the outer portions broke into separate eddies that themselves condensed into spinning swarms of solid matter in the form of *planetesimals*. Gravitational attraction caused the separate swarms of planetesimals to condense individually to form the existing planets. The moons of the planets are regarded as left-over planetesimal masses. Several types of evidence indicate that the earth formed in this manner about 4.6 billion years ago, or about 1 P.M. (one hour after noon in our imaginary 24-hour day).

Urey's theory is supported by two simple facts. First, all the planets circle the sun in the same direction, presumably the direction of spin of the original cloud of matter. Second, the orbits of all the planets lie in approximately the same plane, thought to be the plane of the disklike cloud (Figure 1.3).

The rotation imparted to the earth at its creation is gradually slowing, and is what produces our cycle of daylight and darkness. The earth's rotation is also a primary cause of atmospheric motion and thereby of weather and climate. The fact that the earth's axis of rotation is oblique to the plane of its orbit around the sun is the basis for our seasons, as will be seen in Chapter 3.

Structure of the Earth

According to Urey's model, the earth was formed by the condensation of planetesimals at moderate temperatures. Our planet's hot interior is thought to have developed later by heating resulting from the *radioactive decay* of elements in the earth's interior. This process continues today; however, the heat that is generated is transported outward and is lost at the surface. The temperature of the earth's outer layers seems to have been stable for at least 2 billion years. Radioactive decay affects the atoms of certain unstable heavy elements (such as uranium) that change by casting off protons from their atomic nuclei. This releases energy that heats the surrounding material. As the early earth's interior was strongly heated by radioactive decay of unstable elements, solid material began to melt and move. Over millions of years light material slowly rose toward the surface, while heavier material gradually sank

toward the center. Eventually three concentric shells evolved (Figure 1.4). These can be detected by observing the behavior of earthquake shock waves as they pass through the earth.

Extending about half way to the surface from the earth's center is the *core*, which has a radius of 3,400 kilometers (km), or 2,100 miles. The core has such a high density that it is thought to be dominated by a mixture of iron and nickel. The physical state of this material is uncertain, since the combination of high pressure and high temperature at the center of the earth cannot be duplicated in laboratories.

Surrounding the core is the *mantle*, which is some 2,900 km (1,800 miles) thick. It is composed of high-density rock material dominated by silica, magnesium, and iron. Temperatures in the mantle are high enough to melt this material, but pressures are so great at these depths that most of the mantle is rigid. However, in the upper mantle there is a zone, extending from about 70 to 240 km (40 to 150 miles) beneath the earth's surface, in which molten mantle material can move at a very slow rate. This zone is known as the *asthenosphere*. Movements of mantle material in the asthenosphere have an important influence on the earth's surface, as we shall see in Chapter 12.

Above the asthenosphere is the outer zone of solid rock termed the *lithosphere*. This includes both the uppermost portion of the mantle and the chemically distinct outermost shell of the earth, which is called the *crust*. The crust varies in thickness from some 10 to 70 km (6 to 40 miles) and is the only portion of the earth to which humans have had access. It is composed of the solid rock that forms the continents and ocean floors. As we shall see in Chapter 12, the continents are raft-like slabs of low-density rock, solidly embedded in a continuous layer of higher-density rock that underlies them as well as the ocean floors. The rock of the continental areas is diverse in origin and in many places has been severely disturbed by vertical and horizontal crustal movements. The rock underlying the deep sea floors is mainly solidified lava produced by submarine volcanic eruptions. The origin of the rocks of the continents and sea floors is discussed in detail in Chapter 12.

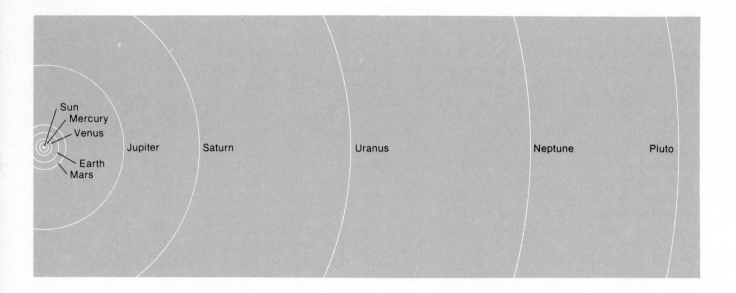

The Early Atmosphere

Since the earth and sun are thought to have formed from the same cloud of matter, they should initially have had similar compositions. But the gaseous element neon, which is abundant in the sun and other stars, is now absent from the film of gases, known as the *atmosphere,* which surrounds the earth. This indicates that the earth lost its original gaseous atmosphere. The loss may have been caused by heating when the planetesimals condensed, even before the planets were formed.

The absence of an atmosphere was only temporary, for volcanic eruptions on the land began feeding new gases to the surface, gradually producing a new atmosphere. At first its composition would have been similar to that of volcanic gases, being dominated by carbon dioxide, with some water vapor, nitrogen, and sulfurous gases, and only traces of free oxygen. The planet Venus has such an atmosphere at present—one that is quite incapable of supporting life. This inhospitable atmosphere was eventually altered, as we shall see shortly.

The Early Landscape

The earth's surface 3 or 4 billion years ago must have resembled the volcanic landscape in Figure 1.5. Flows of molten rock and blankets of volcanic ash covered the land. The ground was completely barren, and no living organisms were present. Water (condensed from clouds of volcanic steam) fell as rain, helping to decay rock by chemical processes. Large and small rock fragments and mineral grains littered the ground. There was no soil, and rainwater streamed off, cutting deep gullies and carrying rock particles and dissolved minerals to the growing seas. Between rains, windstorms swept sand and dust across the naked landscape.

With no vegetative protection, erosional forms in rainy regions must have been spectacular. Hillslopes were steep, rocky, rutted with gullies, and bordered by piles of rock debris. Sand dunes were no doubt common. The combination of much rainfall and no covering vegetation and soil gave these early landscapes an appearance seen nowhere on earth today. The naked scenery lacked the warm colors that characterize our present deserts, for there was little or no free oxygen to combine with iron-bearing substances. Iron oxides color both rocks and soils the rusty hues to which we are accustomed.

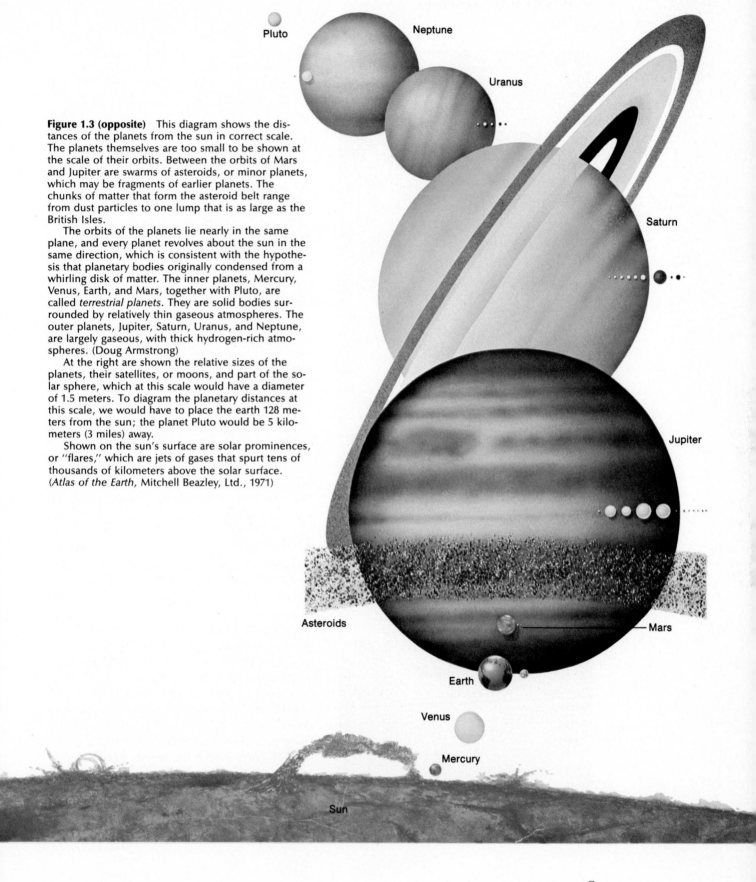

Figure 1.3 (opposite) This diagram shows the distances of the planets from the sun in correct scale. The planets themselves are too small to be shown at the scale of their orbits. Between the orbits of Mars and Jupiter are swarms of asteroids, or minor planets, which may be fragments of earlier planets. The chunks of matter that form the asteroid belt range from dust particles to one lump that is as large as the British Isles.

The orbits of the planets lie nearly in the same plane, and every planet revolves about the sun in the same direction, which is consistent with the hypothesis that planetary bodies originally condensed from a whirling disk of matter. The inner planets, Mercury, Venus, Earth, and Mars, together with Pluto, are called *terrestrial planets*. They are solid bodies surrounded by relatively thin gaseous atmospheres. The outer planets, Jupiter, Saturn, Uranus, and Neptune, are largely gaseous, with thick hydrogen-rich atmospheres. (Doug Armstrong)

At the right are shown the relative sizes of the planets, their satellites, or moons, and part of the solar sphere, which at this scale would have a diameter of 1.5 meters. To diagram the planetary distances at this scale, we would have to place the earth 128 meters from the sun; the planet Pluto would be 5 kilometers (3 miles) away.

Shown on the sun's surface are solar prominences, or "flares," which are jets of gases that spurt tens of thousands of kilometers above the solar surface. (*Atlas of the Earth,* Mitchell Beazley, Ltd., 1971)

Pluto

Neptune

Uranus

Saturn

Jupiter

Asteroids

Mars

Earth

Venus

Mercury

Sun

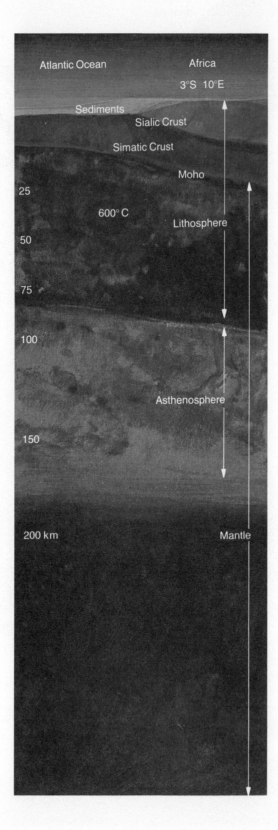

Figure 1.4 The earth in cross section. The earth may be visualized as a series of concentric shells having different properties. The *crust* varies in thickness from about 10 km (6 miles) under the oceans to 40 km (25 miles) or more under the continents. The second shell, or *mantle,* extends to a depth of about 2,900 km (1,800 miles), where the *outer core* begins. At 5,000 km (3,000 miles) below the surface is the *inner core,* believed to be an alloy of iron and nickel (possibly including some silicon or sulfur).

The boundary between the crust and upper mantle is referred to as the Mohorovičić discontinuity (after its discoverer) or *Moho.* The outer 200 km (125 miles) of the earth may be separated into two classes of material: the zone of low strength and low seismic velocity is called the *asthenosphere;* the more rigid layers that overlie it, which include both the crust and the upper mantle, are called the *lithosphere. Lithospheric plates* may be visualized as rigid slabs that are free to move on the low-strength asthenosphere. It is important to note that lithospheric plates include both crust and mantle and that a single plate may include both oceanic and continental crust. These concepts are discussed further in Chapter 12.

Our knowledge of the inner properties of the earth has come primarily from a study of the shock waves generated by earthquakes. These waves, called *seismic waves,* travel through the earth at speeds that vary according to the properties of the material through which they pass. (Robert Kinyon/Millsap & Kinyon after R. Phinney)

LIFE ON THE EARTH

About 3.5 billion years ago—with only 8 or 9 hours left in our imaginary 24-hour day—the history of the earth took a unique turn. Life appeared! How this occurred remains a puzzle. Experiments show that when mixtures of the simple volcanic gases present in the earth's primitive atmosphere are subjected to electric discharges, complex molecules are formed. These include amino acids, which are the building blocks of the proteins that are found in every living cell. In the early atmosphere, lightning and ultraviolet radiation from the sun could have provided the energy to break apart and recombine molecules. These new, more complex molecules would have been transported to the oceans during rainfalls. In time the oceans may have become an "organic broth," rich with complex molecules that were not "living" but had the potential to be organic building blocks. The discovery of previously unknown amino acids in meteorites has suggested the additional possibility of an extraterrestrial source for some primitive organic compounds.

Whatever the source, an unusual combination of molecules eventually appeared that was able to absorb energy from its surroundings and to reproduce itself. This combination of molecules was the first "living" organism.

The Fossil Record

We can only guess about the origin of life on earth, for there are no known traces of the first life forms. Our understanding of life in the remote past is based on the *fossil record*. *Fossils* are mineral replacements of the remains of plants and animals that were buried by mud, silt, or sand, before they could decompose (Figure 1.6). As the covering deposits accumulated to great depths, they slowly became compacted and cemented to form layers of rock. Such a process requires hundreds of thousands of years. Over such spans of time the remains of plants and animals within the deposits are replaced by more resistant minerals that become hard casts of the original forms, thus preserving them indefinitely. In undisturbed sedimentary rocks, the oldest life forms are of course found as fossils in the lowest layers of rock. The fossils of more recent life forms are in higher layers. Thus an erosional trench through a thick mass of sedimentary rock, as at the Grand Canyon of the Colorado River, gives us a catalogue of the life forms that have inhabited the region through tens to hundreds of millions of years.

The catalogue is not complete, however, for most fossils represent only hard bones, shells, and woody plant parts. The soft tissues of organisms usually disappear before burial and fossilization. Furthermore, if sediment is not accumulating at a location, there is no possibility of natural burial, and there is a gap in the fossil record for organisms of that region. Finally, much of the fossil record at a locality may have been removed by natural erosion of fossil-bearing rocks.

Figure 1.5 Because the earth's early landscapes lacked any form of vegetative protection, areas of erodible materials would have been deeply gashed by the effects of water running off the land surface, as is this area in California's Death Valley. The rocks seen here are volcanic lavas and sediments composed largely of volcanic ash. (T. M. O.)

Organic Evolution

The earliest living organisms preserved in the fossil record had already evolved much beyond the earth's initial life forms. The earliest fossilized organisms resemble the blue-green algae seen on the surfaces of ponds today. These early plants obtained energy by *photosynthesis*, a process that uses the visible light of solar radiation as an energy source to produce carbohydrates, such as sugars and starches, from carbon dioxide and water, with oxygen given off as a by-product (see Chapter 9). This brings us to an important question: how did more complex plants and animals arise from the earliest simple forms of life?

An answer came from the British naturalists Charles Darwin and Alfred Russel Wallace, both of whom published theories of *organic evolution* in 1859. According to the various theories of evolution, most organisms do not reproduce exact copies of themselves. In fact, the offspring of any one pair of parents may vary considerably. In a given environment some variations have survival value. They may increase an organism's ability to gather food, escape pre-

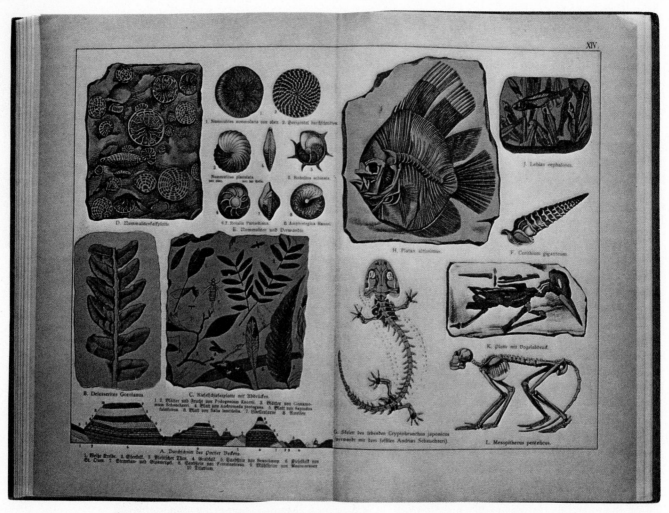

Figure 1.6 This nineteenth-century German scientific treatise illustrates the diagnostic fossils of rocks of different ages found in the Paris Basin of France, which is portrayed in a geological cross section at the lower left. The wide range of fossil types present in the rocks of this region is clearly evident. (Photo by Werner Kalber, from the John Sinkankas Collection)

dators that would use *it* for food, or withstand environmental stress. Such a favored organism has an above-average chance to survive and transmit the helpful variation to its offspring. This affects the evolution of the species as a whole, which thus slowly changes by the process of *natural selection.*

If an organism finds an environment to which it is extremely well adapted, additional changes offer no new benefits and the species may not evolve further. An example is the opossum, which is virtually unchanged over the past 60 million years. On the other hand, natural selection may result in alteration in a species in just a few decades. A widely cited example is the English peppered moth. Once a light-colored moth with dark speckles, it became predominantly dark in color between 1850 and 1900 as industrialization altered its physical environment. The darker coloration apparently helped the moth hide from predatory birds in the newly soot-covered towns and nearby woodlands.

In time, the changes in species resulting from the process of natural selection may produce entirely new species—groups of mutually fertile individuals that differ from similar groups in some constant way. According to

Darwin and Wallace, and most other biologists, this is the process that has produced the amazing number of plant and animal species that populate the earth today. All of these are genetically related to the first primitive organisms that formed in the earth's ancient seas.

Interaction of Life with the Environment

The appearance of microscopic green plants in the oceans 3.5 billion years ago—at 3:36 P.M. in our 24-hour day—set in motion new processes that changed the face of the earth in several ways. For a long time plants were confined to the sea, where water shielded the organisms from the sun's cell-destroying ultraviolet radiation. But the simple marine plants had to remain in the sunlit upper portions of the seas to carry on the vital photosynthesis process. During photosynthesis water molecules are split into hydrogen and oxygen atoms. Plant cells use some of the oxygen and release the rest. Little of the oxygen released by the first plants remained in the atmosphere. Most of it combined with iron-bearing minerals in decomposed rock, coloring the rock various shades of red, brown, or yellow. It was only after the reactive minerals exposed at the earth's surface had been oxidized, that the free oxygen released by photosynthesis began to accumulate in the atmosphere. The oxidation process appears to have required some 2 billion years—about 5 hours in our 24-hour day. Then, at last, the atmosphere began to absorb free oxygen generated by biotic activity. Today's atmosphere contains more than 20 percent free oxygen, most of it the special gift of the green plants that have lived and died through the ages.

At 10:30 P.M. in our 24-hour day the fossil record reveals the sudden appearance of complex, multicellular marine animals with shells. How they developed from preceding plant life is still unknown. What is clear is that these animals evolved the ability to make use of the oxygen that was being released by marine plants, as well as the ability to use plant material for food. Once these adaptations appeared, genetic variation and the natural selection pro-

cess produced explosive expansion in the diversity of oxygen-breathing animal species.

Soon marine animals and plants were working together to alter the earth's environment. Aquatic plants extracted carbon from CO_2 dissolved in water to produce carbohydrates, as part of the photosynthesis process. Tiny aquatic animals that fed on these plants built hard shells by combining the carbon with calcium weathered from rocks and washed from the land. When these animals died, their remains drifted to the ocean floor to form a growing deposit of calcium carbonate sediment, which was gradually transformed into limestone rock (Figure 1.7). Limestone, one of the world's most abundant rock types, is the chief storehouse of the carbon that once dominated the earth's atmosphere. In many places, carbon from the remains of marine organisms combined with hydrogen to form the hydrocarbons: petroleum and natural gas.

The removal of carbon from seawater by marine organisms gave the oceans greater capacity to gain carbon from other sources—mainly from atmospheric carbon dioxide. Thus while biological processes were adding oxygen to the atmosphere, they were also removing carbon dioxide from it. This changed the atmosphere very significantly, reversing the initial proportions of oxygen and carbon dioxide. Today carbon dioxide is present in the atmosphere in only minor amounts, about 335 parts per million (ppm).

The growing quantities of oxygen in the earth's atmosphere led to the formation of *ozone*, a molecule composed of 3 atoms of oxygen, as compared to the oxygen molecule's 2 atoms. Ozone is created in the upper atmosphere by high-energy solar radiation that splits apart oxygen molecules, the atoms of which recombine as ozone molecules. Ozone is of crucial importance to life on the earth because it absorbs much of the biologically destructive ultraviolet radiation streaming from the sun. The development of an ozone layer, high in the atmosphere, that could screen out lethal radiation, was required before the earth's land surfaces could become habitable to living organisms. Not until about 400 million years ago—only one hour before midnight in our 24-

(a)

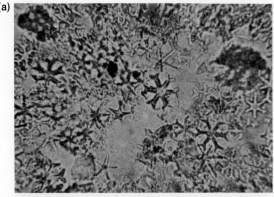

(b)

hour day—were plants able to begin colonizing the land surfaces. Small invertebrate animals, such as insects and scorpion-like arachnids, appeared soon afterward, with larger invertebrates moving onto the land some 50 million years after the arrival of the first plants.

When plants appeared on the land, the process of rock decay, erosion, and movement of sediment all were altered. The chemical action of plant acids helped decay rock surfaces. Later, the prying action of the roots of more highly evolved plants assisted in the breakup of rock masses. The activities of terrestrial plant and animal organisms began to create the first true soils, which thickened under vegetative protection, softening the previously angular outlines of landscapes. The soil cover absorbed moisture and retained it, improving the opportunity for further vegetative colonization.

The widespread growth of land vegetation had the effect of removing carbon from the atmosphere on a massive scale. Here and there vigorous plant growth in swampy coastal areas led to the accumulation of thick masses of undecomposed plant remains. These became buried by sediment and were eventually transformed into coal. An enormous amount of carbon that was taken from the atmosphere is now fixed in mineral form in the world's coal fields.

While plant and animal life has completely transformed our planet, the nonliving environment has at times taken its toll on the living

Figure 1.7 (a) Photomicrographs of biologically produced (biogenic) deep-sea sediment. At the top is an accumulation of the calcareous (calcium-rich) skeletons of star-shaped *discoasters*, which are now extinct. Their disappearance from the seas is thought to be an indication of the onset of the Ice Ages. Below is an accumulation of *foraminifera* fragments, which are an important component of present marine sediments. The specific organisms forming biogenic marine limestones indicate both the age of the sediment and the temperature of the seas inhabited by the organisms. (Lamont-Doherty Geological Observatory/CORE Laboratory)

(b) The summit of Mount Everest, on the border of Nepal and Tibet, is the highest point on our planet. This giant peak in the Himalaya Range rises 8,848 meters (29,028 ft) above sea level, and is composed of limestone that originated on the sea floor about 100 million years ago. (Barbara Brower)

world. Climatic changes due to astronomical or other causes have affected the evolutionary development of plants and animals. So also has such geological activity as the uplift of mountain ranges and changes in the size and position of continents and ocean basins (see Chapter 12).

The early small amphibians evolved into giant reptiles known as dinosaurs (Greek: "terrible lizards"), which dominated the earth some 130 million years ago (Figure 1.8). These huge creatures suddenly disappeared, along with many other life forms, about 65 million years ago—just 10 minutes before midnight in our 24-hour day. Various causes have been proposed. One hypothesis involves a major asteroid impact that shrouded the earth in dust for some months, reducing the penetration of sunlight and causing the collapse of plant-dependent food chains. Chemical evidence for such an event has recently been uncovered in rocks dating from the time of the mass extinctions of species some 65 million years ago. Others have proposed climatic change as the reason for the mass extinctions (see the Case Study following Chapter 9).

The evidence is controversial in either case. The giant reptiles were replaced by the ancestors of modern mammals. Not until about 4 million years ago—only half a minute before midnight—did the first human-like creatures appear, presumably our own ancestors. Although we are newcomers on the earth, we are certainly not intruders here. Like all our fellow organisms we have evolved with the earth and are adapted to it. Our respiration depends on the gases in the current atmosphere, and our eyes respond to those wavelengths of solar radiation that reach the earth's surface in greatest abundance. We are a product of the earth's environment as a whole, and our interactions with its complex individual environments will continue to affect our progress as a species.

DATING THE EARTH

Before the year 1800 it was generally believed that the earth's landscape had been formed by a series of violent catastrophes that had lasted only a few thousand years. Toward the end of the eighteenth century, James Hutton, a Scottish naturalist, took a fresh look at the mountains and streams of his native land and interpreted what he saw as the products of erosion and mountain uplift. He theorized that these slow processes, given enough time, could have produced the landscapes of the British Isles, and concluded that the eras of geologic time must be much longer than previously thought. Not long afterward, Charles Darwin concluded that evolution required time spans on the order of hundreds of millions of years.

To check the ideas of Hutton and Darwin, scientists in the nineteenth century attempted to develop an accurate way to measure geologic time. The fossil record gave the succession of the geologic eras, but it did not indicate their actual ages. Therefore scientists measured the current rates of sediment deposition to calculate how long it would have taken to build up the known thicknesses of sedimentary rock layers. The accuracy of this method was uncertain because rates of sedimentation have varied with time and because the record contains many gaps.

The accurate natural timekeeper that scientists sought was found early in the twentieth century when it was discovered that the passage of time could be measured by using the decay rates of *radioactive isotopes.* An isotope is one of a number of different forms of a single chemical element, such as the isotopes uranium 234 and uranium 238 (the number representing the mass of each of its atoms). A radioactive isotope decays spontaneously into another element, at a certain predictable rate that is unaffected by changes in temperature, pressure, or other external factors. The rate of decay is specified in terms of the isotope's *half-life,* which is the time required for half of an initial amount of radioactive atoms to decay.

Suppose a rock initially contained a certain amount of uranium 238 when it solidified from the molten state. Uranium 238 decays with a half-life of 4.5×10^9 years (see Appendix 1) through a chain of steps to the stable end product, lead 206. Each uranium atom that decays eventually becomes one atom of lead. Since the

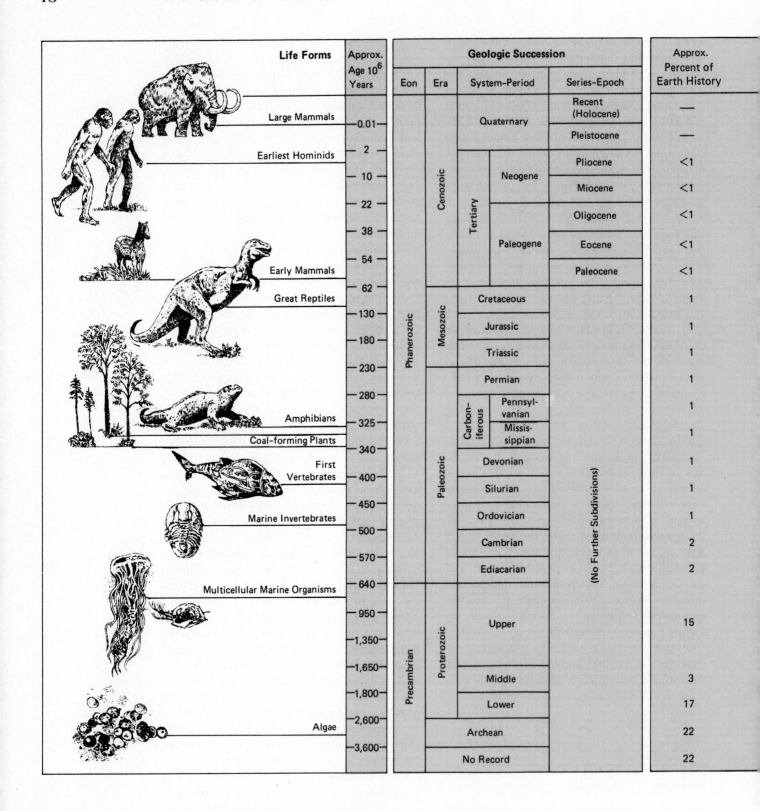

Life Forms	Approx. Age 10^6 Years	Geologic Succession				Approx. Percent of Earth History
		Eon	Era	System-Period	Series-Epoch	
				Quaternary	Recent (Holocene)	—
Large Mammals	0.01				Pleistocene	—
Earliest Hominids	2		Cenozoic	Tertiary — Neogene	Pliocene	<1
	10				Miocene	<1
	22			Tertiary — Paleogene	Oligocene	<1
	38	Phanerozoic			Eocene	<1
Early Mammals	54				Paleocene	<1
	62			Cretaceous		1
Great Reptiles	130		Mesozoic	Jurassic		1
	180			Triassic		1
	230			Permian		1
	280			Carboniferous — Pennsylvanian		1
Amphibians	325			Carboniferous — Mississippian	(No Further Subdivisions)	1
Coal-forming Plants	340		Paleozoic	Devonian		1
First Vertebrates	400			Silurian		1
	450			Ordovician		1
Marine Invertebrates	500			Cambrian		2
	570			Ediacarian		2
Multicellular Marine Organisms	640					
	950			Upper		15
	1,350	Precambrian	Proterozoic			
	1,650			Middle		3
	1,800			Lower		17
Algae	2,600			Archean		22
	3,600			No Record		22

Figure 1.8 (opposite) The geologic time scale. To the left are shown selected life forms and the approximate times when they first appeared or were dominant. A general trend from less complex to more complex life forms reflects the processes of organic evolution. Note that the time intervals in the diagram are not evenly spaced but rather emphasize the most recent 600 million years. (John Dawson after *Adventures in Earth History,* edited by Preston Cloud, W. H. Freeman Company. Copyright © 1970)

Table 1.1   Half-Lives of Radioactive Elements

Radiometric Clock	Half-Life (Years)
Rubidium 87/Strontium 87	5.0×10^{10}
Uranium 238/Lead 206	4.5×10^{9}
Potassium 40/Argon 40	1.3×10^{9}
Uranium 235/Lead 207	0.7×10^{9}
Uranium 234/Thorium 230	248×10^{3}
Carbon 14	5730

Source: Gray, Dwight E. (ed.). 1972. *American Institute of Physics Handbook.* 3rd ed. New York: McGraw-Hill.

rate of decay is known, measurement of the relative numbers of uranium 238 and lead 206 atoms in the rock allows us to compute the elapsed time. For example, if the rock contains equal amounts of uranium 238 and lead 206, the elapsed time since the rock solidified must have been one half-life, or 4.5×10^{9} years.

If the rock initially contained some lead 206, or is younger than about 60 million years, the uranium 238/lead 206 method will not be accurate. It is therefore sometimes necessary to use more than one radioactive element to establish a date. The most useful radiometric clocks for geologic time are listed in Table 1.1. The decay of potassium 40 to argon 40 has been particularly useful as an age indicator, because potassium occurs in several of the minerals composing common rock types formed by volcanic activity. The age of the volcanic rock sets a minimum age for the deposit found under it, and a maximum age for the deposit above it. Perhaps the most widely noted uses of potassium/argon age determinations have been the mapping of the ages of ocean floors (Chapter 12) and the dating of fossils of early human-like creatures (hominids) discovered in the volcanic regions of East Africa (see the Case Study following this chapter).

Using radiometric dating methods, it is possible to make a fairly accurate determination of the age of the earth. The oldest rocks presently identified on the earth are in Greenland. Their radiometric age is about 3.8×10^{9} years, meaning that the earth is at least 3.8 billion years old. However, since erosion may have destroyed still earlier rocks, it is probable that the earth is even older. (Rocks from the moon have been dated at 4.2 billion years.) According to currently accepted models of the creation of the solar system, meteorites were formed at the same time as the earth. Radiometric dating of meteorites should therefore be an accurate method of determining the earth's true age. This method results in an age of 4.6 billion years. Direct measurements of the total uranium and lead content of the earth's crust also point to an age of 4.6 billion years.

Carbon 14 is a radioactive isotope that has been important for dating events occurring within the past 40,000 years or so. Carbon is a useful substance for age determination because it is a constituent of virtually all plant and animal matter. Carbon 14—a radioactive form of carbon that is continuously formed in the earth's atmosphere by the action of cosmic rays on nitrogen 14—makes up a definite proportion of all the carbon ingested by plants and animals. When an organism dies, it ceases to take in carbon. The amount of carbon 14 in its cells steadily declines due to radioactive decay, but the amount of ordinary carbon remains constant. Therefore, measuring the ratio of radioactive carbon to ordinary carbon allows the age of the sample to be calculated. Carbon 14 dating has provided highly accurate data about relatively recent events in the earth's history. For example, by dating wood from trees that were killed by the advance of glacial ice, it has been possible to fix the date of the ice's advance through a particular area. Carbon 14 dating can also be applied to charcoal, shells, bone, and the calcium carbonate in soils. Thus it is of great assistance in archaeological work.

Another radiometric dating method uses the decay of uranium 234 to thorium 230. Since uranium 234 has a half-life of 248,000 years, the

Figure 1.9 Above is an increment borer used to re-move thin cores from living trees. The rings visible in the core are counted to determine the age of the tree. **(Left)** The upper portion of this roadcut, adja-cent to California's Sierra Nevada, exposes volcanic *tuff* having a radiometric age of about 700,000 years. Under the tuff is a deposit of glacially transported rock debris, including large boulders. The relation-ships depicted establish that the Sierra Nevada was glaciated before 700,000 years ago. (T. M. Oberlander)

uranium/thorium method is applicable to older materials than is the carbon 14 method. It has been particularly useful in dating underground cavern development and limy deposits of var-ious types, including shells.

A number of other dating techniques are useful for measuring relatively short time spans of thousands of years. One method uses *varves,* which are annual sediment layers deposited in lakes and on the seafloor (see Figure 1.9). Coarse sediment is deposited seasonally during periods of stream flow, with fine sediment set-tling out during the dry season or when water surfaces are frozen. In Scandinavia varve anal-ysis has been extended more than 10,000 years into the past, and correlation of varve patterns from different localities has enabled scientists to reconstruct the history of glacial recession in northern Europe.

Tree-ring analysis is similar to varve analysis and has been used both in age determination and to reconstruct climatic conditions in the recent geologic past. It involves counting the annual growth rings of trees and measuring their relative widths to estimate how moisture and temperature have fluctuated through time. In this way, trees more than 4,000 years old have been found among the bristlecone pines of California and Nevada. Logs used in the con-struction of ancient Indian pueblos have been dated by correlation of their rings with ring sequences of known age. The ages of many kinds of topographic surfaces of recent origin can be determined by counting the rings of the largest trees growing on them.

Other dating techniques are applicable in special circumstances, and the number of chronological indicators will surely increase with further research.

SUMMARY

Physical geography is concerned with the in-teracting processes that have created the phys-ical environment we inhabit. Our environment is the result of continuous change since the formation of the earth almost 5 billion years ago. The universe itself appears to have origi-nated explosively some 10 to 20 billion years ago. The planets, including earth, were formed by the condensation of matter in orbit around the sun. The earth's internal structure consists of a core, a surrounding mantle, and a thin outer crust that contains the continents and ocean basins.

The earth has changed greatly from its original condition. The earth's early atmosphere was dominated by carbon dioxide vented during volcanic eruptions. Ancient land surfaces were barren and rutted by water erosion. The advent of life in ancient seas about 3.5 billion years ago initiated the changes resulting in our present environment. Life originated in the sea, where plant photosynthesis began to extract carbon dioxide from the atmosphere and to replace it with free oxygen. This made animal life possible. It also led to the formation of an ozone shield against ultraviolet radiation, which permitted life to exist on the land.

Fossils preserved in sedimentary rock record the evolution of organisms on the earth. The carbon removed from the atmosphere by biological activity was stored in the form of limestone, coal, oil, and natural gas. In the course of biological evolution, many organisms have disappeared after a period of dominance. Mammals, which are the dominant vertebrates today, are relatively recent arrivals on the scene. Although the human species emerged but a moment ago in geologic time, it has already developed the capability to change the environments of all other organisms.

Our understanding of the earth's history depends on the development of accurate techniques for age determination that cover the span of geologic time. Several such techniques are based on the decay rates of unstable radioactive isotopes. Two of the most useful are the uranuim/lead and the potassium/argon methods, both of which can be applied to rocks formed by crystallization hundreds of thousands to billions of years ago. The uranium/thorium method is applicable to limy materials that are too old for carbon 14 dating, which is itself accurate over the past 40,000 years. Other chronological indicators for the recent geologic past include varve analysis and tree-ring analysis.

REVIEW QUESTIONS

1. In what way does the field of geography differ from all other sciences?
2. How has the age of the earth been determined?
3. What is the source of the earth's internal energy?
4. How was the atmosphere of the early earth different from the earth's present atmosphere?
5. During what proportion of the earth's history were the surfaces of the continents devoid of any form of life?
6. How did the appearance of life in the seas eventually alter the global environment?
7. Indicate two reasons for the appearance of plant life, which produces energy by photosynthesis, before animal life could develop.
8. What two factors are thought to be the most important influences on organic evolution and the development of new species?
9. What has happened to the carbon that was once a dominant component of the earth's atmosphere?
10. What is the physical basis of radiometric age determination techniques?

APPLICATIONS

1. Where within North America would you look for landscapes most similar to those of 500 million years ago? To those of 3 billion or more years ago?
2. What are the ages of the rocks present in the area of your campus? Around your home? Do these rocks contain fossils? If so, what life forms are represented? Where in your

area is a good collection of local fossils on display?

3. For what purely physical reasons would life not be likely on either Venus or Mars?

4. What are some clear examples of "natural selection," by which specific life forms have become adapted to particular environments? Is the human species, *Homo sapiens*, adapted to a particular environment?

FURTHER READING

Abell, George O. *Exploration of the Universe*, 3rd ed. New York: Holt, Rinehart and Winston (1975), 738 pp. This introductory text often used in astronomy courses is especially useful for recent ideas about the formation of the universe and solar system.

Cloud, Preston. *Cosmos, Earth, and Man.* New Haven: Yale University Press (1978), 372 pp. A lucid account of the history of the universe and planet earth, emphasizing the major changes that have occurred due to the origin and evolution of life, by a leading earth scientist.

Dott, Robert H., Jr. and **Roger L. Batten.** *Evolution of the Earth.* New York: McGraw-Hill (1971), 649 pp. This introductory text traces the evolution of crustal features, flora, and fauna from a geological perspective.

Eicher, Donald L. *Geologic Time.* Englewood Cliffs, N.J.: Prentice-Hall (1968), 150 pp. This brief monograph, part of the Foundations of Earth Science series, surveys each of the methods used in dating crustal materials and features.

Poirier, Frank E. *Fossil Evidence: The Human Evolutionary Journey,* 3rd ed. St Louis: C. V. Mosby (1981), 428 pp. A full account of evolutionary systematics and the hominid fossil remains that may represent the ancestors of the human race. It is well illustrated and highly informative.

Sagan, Carl. *Cosmos.* New York: Random House (1980), 365 pp. A brilliantly written and copiously illustrated anecdotal account of cosmic evolution, from the birth of the universe to the present. The book is based on the 13-part television series produced by the author, a well-known astronomer.

The Solar System, A Scientific American Book. San Francisco: W. H. Freeman (1975), 145 pp. One of a series of reprints of original articles in *Scientific American.* A good overview of the planets including the earth and our moon, with recent ideas on their origin and evolution.

Dating the Earliest Humans

The most interesting of all problems of age determination is that of dating the early ancestors of the human race. Since the end of the Ice Age about 10,000 years ago, all ice-free portions of the earth have been inhabited by modern humans—*Homo sapiens sapiens*—who first appeared in Eurasia and Africa about 40,000 years ago. Earlier, between about 40,000 and 100,000 years ago, a somewhat different human form was dominant. These were the *Neanderthals,* who used fire, built shelters, practiced religious rituals, and possessed sophisticated technologies for fabricating well-made tools from stone, bone, wood, fiber, and animal skins. Although differing physically from modern races in ways that would give them a primitive appearance today, most Neanderthals would have fallen within the range of variation of modern individuals. Their brains were even larger than the present human average. Whether they were ancestral to the existing human race, or reflect a separate evolutionary track, remains a matter of discussion among anthropologists.

The skeletal remains of Neanderthals, regarded as the earliest *Homo sapiens,* are widely dispersed throughout Europe, Africa, and Asia, and have been dated by the decay of the unstable isotope of carbon (C-14) in bone and associated charcoal, peat, and cave deposits, as well as by stratigraphic methods involving the study of successive layers of sediment deposited over tens of thousands of years in lakes, caves, and stream valleys.

The Neanderthals (named after the Neander Valley, in Germany, where they were first described) are now known to have been preceded in history by still more primitive fire-using and tool-fabricating hominids, which have been designated *Homo erectus* (upright-walking humans). Fossils of this smaller-brained species have been found in Europe, Africa, Asia, and Indonesia, and range in age from about 1.5 million years to less than 200,000 years. Dating is principally by stratigraphy and analysis of fossil pollen, which permits climatic oscillations to be recognized by related changes in vegetation types.

The oldest *H. erectus* finds have come from East Africa, and were first made at Olduvai gorge in Tanzania by the famous English prehistorian Louis Leakey after he had worked in the area for 30 years. The ages of Leakey's *H. erectus* finds in the Olduvai area have been established by application of the radiometric dating technique based on the decay of the unstable isotope potassium 40 to argon 40. The potassium/argon (K/Ar) age determination technique is applicable to most volcanic rocks, which generally contain potassium, with the decay process beginning after crystallization of the rocks from a molten condition. The fossil hominid finds at Olduvai occur in geological formations that contain many layers of volcanic *tuff* (volcanic ash that has turned to rock). Subsequent finds have also come from several other East African locations that also expose volcanic rocks, offering excellent opportunities to establish precise dates using the K/Ar method. The archaeological material may be within the volcanic material itself, or in sediments above or below datable beds of tuff. Thus objects found between tuff layers I (lower) and II (higher) are younger than the K/Ar age of tuff I and older than the K/Ar age of tuff II. In many cases the ages of the bracketing tuffs establish the age of fossils and stone artifacts quite precisely. The volcanic environment of East Africa millions of years ago, as at present, was an excellent one for human occupation, as the volcanic activity blocked streams and created shallow lakes and marshes that were lures to a host of animals that could be hunted as a food source.

The human remains in East Africa, consisting of fossilized fragments of skulls and skeletons, are associated with vast numbers of stone tools. Fairly sophisticated tools accompany *H. erectus* remains in deposits as old as 1.5 million years. These take the form of handaxes and cleavers formed by careful, well-controlled flaking, some of it by wood, bone, or antler hammers. This industry, which persisted for more than a million years, is known as *Acheulian,* after the village of St. Acheul in France, where such primitive Paleolithic (Old Stone Age) tools were first studied.

Still older artifacts at Olduvai consist of stream cobbles of tennis-ball size, modified into crude chopping tools by the controlled removal of a few flakes by stone-on-stone percussion. These simple pebble tools of the so-called "Oldowan" culture apparently were fabricated by beings even more remote in time than *H. erectus.* They come from deposits as

CASE
STUDY

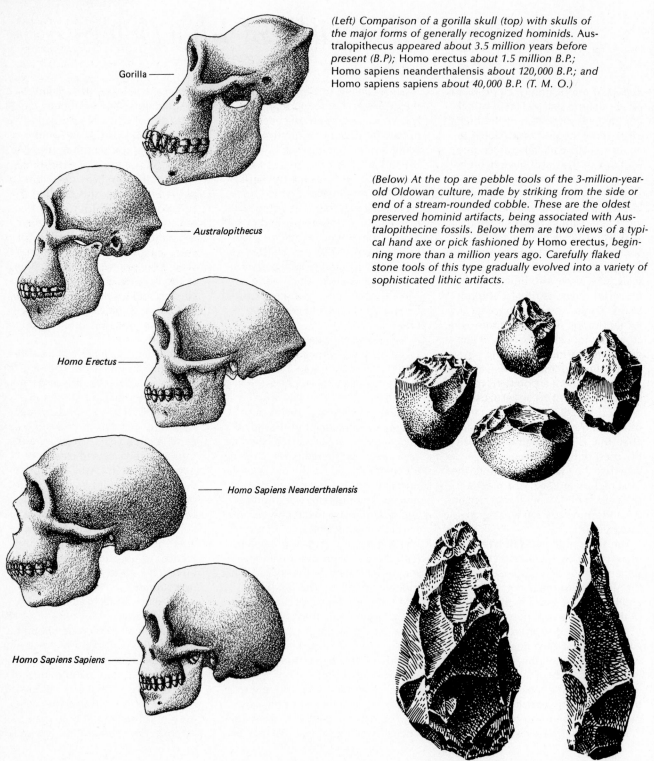

(Left) Comparison of a gorilla skull (top) with skulls of the major forms of generally recognized hominids. *Australopithecus* appeared about 3.5 million years before present (B.P); Homo erectus about 1.5 million B.P.; Homo sapiens neanderthalensis about 120,000 B.P.; and Homo sapiens sapiens about 40,000 B.P. (T. M. O.)

(Below) At the top are pebble tools of the 3-million-year-old Oldowan culture, made by striking from the side or end of a stream-rounded cobble. These are the oldest preserved hominid artifacts, being associated with Australopithecine fossils. Below them are two views of a typical hand axe or pick fashioned by Homo erectus, beginning more than a million years ago. Carefully flaked stone tools of this type gradually evolved into a variety of sophisticated lithic artifacts.

Gorilla

Australopithecus

Homo Erectus

Homo Sapiens Neanderthalensis

Homo Sapiens Sapiens

old as 3.5 million years, according to K/Ar age determinations on the geological formations in which they occur.

Hominid fossils are associated with the Oldowan tools. They are an extremely primitive type that has been given the name *Australopithecus*. Although the details of their bone structures indicate that they walked erectly, the australopithecines had massive jaws, protruding muzzles, and brains much smaller than those of *H. erectus*. Nevertheless the australopithecines were distinctly different from any known apes, and are regarded by many anthropologists as a direct human ancestor.

Louis Leakey himself did not believe the australopithecines to be the first tool users. He recognized another form, contemporary with them, which he called *Homo habilis*—a less robust but larger-brained species that he preferred to the australopithecines as a human ancestor. Other influential anthropologists disagree, regarding Leakey's *H. habilis* as merely advanced, gracile (slender) australopithecines—perhaps females of the species.

Later finds by other scientists working in East Africa have complicated the picture even more. Among these finds are discoveries by Louis Leakey's son, Richard, and wife, Mary. Richard Leakey has found gracile forms that initially appeared older than the australopithecines, providing a more reasonable link to modern humanity, but subsequent K/Ar dating has suggested that these were, in fact, later in time, and were therefore possibly themselves derived from australopithecines. Mary Leakey has unearthed fossils of gracile australopithecines, and even their surprisingly modern-appearing footprints, petrified in rock that has been K/Ar-dated at 3.5 million years. She argues that the australopithecines should themselves be included in the genus *Homo*. But other scientists disagree vigorously on the basis of their own discoveries.

Thus the perceived course of human evolution is argued on the basis of the latest age determinations on volcanic tuff from East Africa, where both datable volcanic rocks and the remains of early hominids are uniquely abundant and in very fortunate association. New fossil finds and new K/Ar determinations constantly alter previous theories, bringing new scientists into the public eye and sending yesterday's champions back to the field to dig for more evidence in support of their own hypotheses.

The Starry Night by Vincent Van Gogh, 1889. (Collection, the Museum of Modern Art, New York; acquired through the Lillie P. Bliss bequest)

Ceaseless flows of energy cause continuous change on the earth. Energy streaming from the sun and from the planet's interior powers the earth's dynamic physical systems, which constantly interact to maintain the environments we inhabit.

E nergy is involved in every process and is the source of all change. But energy cannot produce change unless there is something that can react to it. The moon receives energy from the sun, but having no atmosphere, no water, and no life—systems that can react to energy—the moon has changed little over billions of years. The earth, however, is a complex of active physical systems—the atmosphere, the oceans, landforms, soils, and the plant and animal realms—all changing restlessly in response to energy in various forms. The ways in which these systems respond to energy and use it to interact with and thus influence one another, are major themes of physical geography. To set the stage for later discussions, this chapter deals with energy—its sources and the forms it takes, and its general effects on the earth's physical systems.

2
Energy and the Earth's Systems

SOURCES OF ENERGY

The most important source of energy for the earth is the sun. The sun's radiation provides the power to set in motion most of the physical processes important to life, including the movements of the atmosphere that distribute energy, produce weather, and drive the oceanic circulation. Solar energy also causes the growth of plants, which, in turn, sustain animal life on our planet.

Two other sources of energy reside within the earth itself. One is gravitational force, which exerts a toward-the-center-of-the-earth pull on all objects, near and far. The pull is strongest on or near the earth's surface. This causes flows of air and water and falls of earth and rock. The second internal source of energy is the heat generated within the earth by radioactive decay of unstable atomic isotopes. The sudden shock of an earthquake or the eruption of a volcano reminds us of unseen forces below the earth's surface. These forces work over long spans of time to lift mountain ranges, to open new oceans and close old ones, and to change the positions of the continents.

FORMS OF ENERGY

The total amount of energy in the universe remains constant. Energy cannot be created, nor can it be destroyed. It may seem that matter is being converted to energy in a nuclear reaction, but such reactions merely release the energy binding atomic nuclei, energy that is required for matter to exist. More important, energy can be converted from one form to another. While atomic energy can produce titanic effects and can even be harnessed for human use, the forms of energy most significant to natural processes on the earth's surface are solar radiant energy, heat energy, gravitational energy, kinetic energy, and chemical energy (Figure 2.1).

Radiant energy from the sun heats the atmosphere and the earth's surface. It does this both directly and indirectly, as we shall see in Chapter 3. Solar radiation reaches the earth in about 8⅓ minutes, traveling at a speed of about 300,000 km (186,000 miles) per second. The largest portion of the sun's energy output is in the wavelengths of visible light (Chapter 3), which our eyes are adapted to sense. The invisible longer-wavelength thermal radiation is felt as heat, and the invisible shorter-wavelength ultraviolet radiation is what causes sunburn. On the earth, solar radiation varies in intensity from place to place and from time to time. It is the seasonal geographical variation of solar radiation that creates the earth's weather systems and climates (Chapter 3).

Heat energy is the energy resulting from the random motion of the atoms and molecules of substances. The hotter a substance is, the more vigorous is the motion of its atoms. This motion must be generated by an input of energy, often in another form, such as solar radiation.

There is a distinction between temperature, which is a measure of average molecular motion in a substance, and heat energy, which refers to the total energy in the entire volume of a substance. A cup of hot coffee at 50°C (122°F) has a higher temperature than a bathtub of warm water at 35°C (95°F), but there is actually much more heat energy in the bathtub water because of its larger volume.

As we shall see later in this chapter, there is also an important difference between sensible heat and latent heat. *Sensible heat* is the ordinary heat, created by molecular motion, which we can feel and which can be measured directly with a thermometer. *Latent heat* is a type of stored energy that cannot be measured directly. It becomes sensible heat only when released by a change of state of a substance, as when water vapor condenses to liquid water or liquid water freezes to ice.

Gravitational energy is the attractive force exerted by the mass of an object, and is proportional to the mass and inversely proportional to the square of the distance from the center of the object. On or near the surface of the earth gravitational energy is the potential energy an object has due to its elevation, or distance from the earth's center. A large boulder on a hillside or a parcel of air aloft in the atmosphere has the potential to move downward if its support, or the force holding its aloft, is lost. Gravitational energy is proportional to altitude and mass. A large boulder near the top of a mountain has more potential energy—and will make a bigger splash in a lake below it—than a smaller rock at the same elevation or a somewhat larger boulder much farther downhill.

Kinetic energy is energy of motion. The higher the speed at which an object is moving, the more energy it has. The size of the splash produced by a boulder that falls into a lake is actually a consequence of the kinetic energy it gains as it loses its potential energy. For a given velocity, kinetic energy is proportional to the mass of the object. If equal volumes of air and water are moving at the same speed, the volume of water will have much greater kinetic energy because of its vastly greater mass. This is most apparent along coasts, where the destructive force of wind-driven ocean waves is obvious. The winds that generate storm waves blow with much greater velocity than water waves can attain. But the wind can barely displace coarse sand, while the impact of a wave can move masses of rock weighing many thousands of kilograms (tens of tons).

Chemical energy is energy stored in the electrical bonds that bind together the molecules or atoms of substances. When substances react chemically, energy is released, absorbed, or converted to other forms of energy. During combustion, as when one strikes a match, chemical energy is transformed into heat and light. In plant photosynthesis (Chapter 10) solar radiation is used to generate carbohydrates. These are stored in plant tissues and may later be consumed by animals and subsequently oxidized (combined with oxygen), releasing the chemical energy animals require. Chemical energy is not the same as atomic energy, which results from the separation and recombination of the protons and neutrons in atomic nuclei. The energy produced by a nuclear reactor is a result of the splitting of heavy uranium-235 atoms, which produces isotopes of lighter elements plus an explosive release of the energy that previously bound the uranium nucleus together. Stars, including the sun, are natural nuclear reactors in which hydrogen is being converted to helium with energy as a by-product.

Energy Transformations

Standing in the sunlight, you feel warm because solar radiant energy excites the molecules of your skin, creating heat energy there. Running uses chemical energy stored in the cells of your body to produce energy of motion. Coasting downhill on a bicycle, your speed increases as your initial potential energy is converted to kinetic energy.

As noted earlier, energy is never destroyed but is constantly being converted from one form to another. Some of these forms are of little use. When a snowball hits a wall, its kinetic energy is abruptly transformed. Its atoms, and some of those of the wall, are violently jostled, producing heat energy that diffuses into the air. In fact, some heat is generated in every process that involves the transfer of energy. This heat—or random motion of molecules—merely warms the surrounding environment and is not otherwise usable. Since energy transfers always generate a certain amount of heat that is quickly dissipated, usable energy can never be transferred with 100 percent efficiency.

Energy and Work

All processes in the physical world can be viewed in terms of energy and work. Energy is the capacity for doing work, which generally means producing change. Heat energy within the earth creates volcanoes, warps the earth's crust, lifts mountains, and moves the continents and ocean floors. Chemical energy stored in plants causes their growth, supports animal life, and helps to form soils. The energy in some vegetation is stored as fossil fuel, the main source of power for industrial society.

Radiant energy from the sun is what directly or indirectly produces most change on the earth. In one important sequence of events, it causes molecules of water to move from the earth to the atmosphere, where they have potential energy. When this water falls as rain, it trades potential energy for kinetic energy, which on impact with the ground causes soil particles to break up or move to new locations. Some of the water falling on the land runs into streams. As a stream flows toward lower altitudes, its kinetic energy is used in the work of altering the landscape—eroding the land and transporting and depositing sediment. In cold climates and at high altitudes, precipitation may take the form of snow. Accumulated snow can become the moving ice of a glacier that also alters the landscape by erosion and the deposition of sediment.

By producing temperature differences on the earth's surface, radiant energy also causes motion of the atmosphere. The resulting wind, in turn, transfers energy to the ocean, giving rise to ocean currents and waves. The kinetic energy of waves modifies shorelines by eroding rocks and transporting sand to and from beaches. More important still, water in both the seas and the atmosphere continually transports energy from warmer to cooler regions of the earth, helping to equalize the earth's energy distribution. Solar radiant energy is also important in many other ways that will be apparent in later chapters.

WATER: AN ENERGY CONVERTER

Water plays a critical role in converting energy from one form to another, in moving energy

Figure 2.1 This painting interprets some of the forms of energy important in physical geography, and illustrates ways in which energy interacts with systems on the earth.

Most change on the earth is caused by two kinds of energy: solar radiation (1) and the earth's internal heat (14). Solar energy powers the motion of the atmosphere (2) by causing temperature differences on the earth's surface. The wind in turn transfers energy to the ocean, producing ocean currents and waves (3). Ocean currents and the moving atmosphere help distribute energy from the hot equatorial regions of the earth to the cool polar regions. The energy of waves actively shapes coastlines (4) by eroding rocks and by transporting sand to and from beaches.

On the seas and on the land, energy from the sun frees water molecules to enter the atmosphere by evaporation (5). Water vapor is transported by the moving atmosphere and returns to the earth's surface as precipitation (6). Some of the water falling on the land runs into streams (7). As a stream flows toward lower altitudes, its gravitational, or potential, energy causes erosion of the land and transportation and deposition of sediment. In cold climates or at high altitudes, winter precipitation takes the form of snow (8). Accumulated snow produces glaciers (9), which are moving masses of ice that erode the land beneath.

Vegetation (10) absorbs solar radiant energy and transforms it to stored chemical energy by the process of photosynthesis. Some of the stored energy is passed on to animals (11), which ultimately depend on green plants as their source of food energy. Vegetation, moisture, and rock materials interact to form soils (12). The chemical energy in some vegetation and tiny marine animals becomes stored as fossil fuels, which constitute the main power source for modern industrial society (13).

The energy released by radioactive decay (15) creates heat within the earth. This geothermal energy can be tapped and may become an important source of power. Uneven heating below the earth's crust causes motion within the asthenosphere (Figure 1.4). At "hot spots" (17) currents rise and spread. This moves the continents and ocean floors, causing the crust to be wrinkled in some places and broken by faults (16) in others. Where large segments of the crust are forced together (18 and white arrows), the heavier segment, usually an oceanic plate, descends into the mantle and melts. The molten material rises back toward the surface to create volcanic eruptions (19). (John Dawson)

from place to place, and in energy storage. To do all this water must have some unique characteristics. Let us now look at the two properties of water that enable it to play such an important part in planetary processes.

Ability to Change Phase

All substances can take three physical forms—solid, liquid, and gaseous. These are called *phases* of the substance. The phase in which a particular substance occurs depends on its temperature and pressure environment. For example, we ordinarily see iron as a metallic solid. But we know that foundries melt iron to a liquid that can be poured to make castings. Subjected to even more energy, such as a high-intensity beam of atomic particles, iron can be vaporized. Water, however, is the only common chemical compound that occurs in all three phases under the natural conditions prevailing at the earth's surface. Morever, our planet is the only one in the solar system that has the proper range of temperatures to permit water to appear in all three phases at the planetary surface.

For a substance to change phase, heat energy must be either absorbed or released by its molecules or atoms. A change to a phase having tighter binding of molecules or atoms releases latent energy. This results in sensible heat that can be measured. Such phase changes can be seen when water vapor condenses to liquid water, when liquid water freezes to ice, and when water vapor freezes directly to ice in the form of frost. A change to a phase that has looser binding of molecules, as when water or

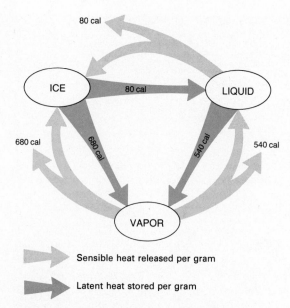

Figure 2.2 Significant energy changes occur when water changes from one physical state to another. Energy must be added to change water from a state in which the water molecules are tightly bound to a state in which they are more loosely bound. To change 1 gram of water from solid ice directly into gaseous water vapor requires adding approximately 680 calories of energy, called the *latent heat of sublimation*. The change from solid to liquid requires 80 calories per gram, called the *latent heat of fusion*. The energy required to change 1 gram of liquid water to gaseous vapor, the *latent heat of vaporization*, varies slightly with the temperature of the water. For water at 15°C the latent heat of vaporization is 590 calories per gram, and for water at 100°C it is 540 calories per gram.

Energy is released when a substance changes from a state in which its molecules are loosely bound to a state in which they are more tightly bound. The amount of energy that must be released is numerically equal to the latent heats described above; for example, 590 calories are released from 1 gram of water vapor at 15°C to form 1 gram of liquid water at the same temperature.

ice evaporates or ice thaws, absorbs energy, which becomes latent heat in the resulting substance.

Energy is commonly measured in units called *gram calories*, or simply *calories*. One gram calorie (with a small "c") is the amount of energy required to raise the temperature of a gram of water from 14.5°C to 15.5°C. (The Calories, spelled with a capital "C", used in relation to food energy in diets are kilocalories, equal to 1,000 gram calories.) Table 2.1 and Figure 2.2

Table 2.1 Heat Transfers Associated with Phase Changes of Water

Phase Change	Heat Transfer	Type of Heat
Liquid water to water vapor	540–590 calories absorbed	Latent heat of vaporization
Ice to liquid water	80 calories absorbed	Latent heat of fusion
Ice to water vapor	680 calories absorbed	Latent heat of sublimation
Water vapor to liquid water	540–590 calories released	Latent heat of condensation
Liquid water to ice	80 calories released	Latent heat of fusion
Water vapor to ice	680 calories released	Latent heat of condensation

summarize the caloric values of the energy absorbed or released by the phase changes of water.

Because phase changes in water occur at temperatures that are commonly found on the earth, latent heat is an important factor in the earth's temperature balance. In moist climates and over warm seas, much of the incoming solar energy is utilized for evaporation rather than heating, resulting in significant local cooling. The atmospheric circulation transports this energy in its latent form in water vapor to other regions, where it is released by the condensation that creates clouds. Where there is little water to be evaporated, as in desert regions or on city streets, most incoming solar radiation is converted to sensible heat, causing local temperatures to be higher than those in moist areas.

High Heat Capacity

In addition to its readiness to change phase, water also has an unusual capacity to act as a reservoir of heat energy. Water requires a large input of energy to become warmed, but once warmed it cools more slowly than most other substances. The amount of heat energy required to raise the temperature of one gram of a substance by 1°C at normal atmospheric pressure is known as the substance's *specific heat*. The specific heat of water (1 calorie per gram of water per degree Celsius) is about five times that of soil or air. This means that one gram of water must absorb five times more heat energy than one gram of soil or air to increase in temperature an equal amount.

The capacity of a substance to absorb heat in relation to its volume is its *heat capacity*, defined as the amount of heat required to raise the temperature of the entire volume of the substance by 1°C. Therefore, the heat capacity of a substance is equal to its specific heat multiplied by its mass. Because of the very high density of water relative to air, the heat capacity of water is many times that of air, and is two to three times that of dry soil. Its high heat capacity means that water can absorb and store large amounts of energy without changing tem-

perature greatly. By contrast, small amounts of energy will produce large changes in the temperatures of land surfaces or air. This helps explain why temperatures vary much less in the oceans than on the land.

THE ATMOSPHERE: AN ENERGY TRANSPORTER

The earth's atmosphere is a major topic in physical geography because it is the chief means of transporting energy and moisture over the earth's surface. To understand how the mixture of gases forming in the atmosphere behaves, it is necessary to know something about the behavior of gases in general.

Unlike the rigid structures of solids and the loosely bound molecules of liquids, the molecules of gases are free to move independently of one another. Gas molecules frequently collide with solid surfaces and with each other. Each collision exerts a push on the molecule or surface that is struck. The total force that molecular collisions exert on a given area of surface in contact with the gas is called the *pressure* of the gas.

The pressure exerted by the earth's atmosphere decreases rapidly as altitude increases. At an altitude of 5 to 6 km (3 to 3.6 miles) the atmospheric pressure is only half the pressure at sea level, as shown in Figure 2.3. A *barometer* measures atmospheric pressure in terms of the height of a mercury column that exerts the same downward pressure as the atmosphere. Under average conditions at sea level, this is 760 millimeters (mm), or 29.9 inches of mercury. Atmospheric pressure is expressed also in *millibars* (mb)—a unit of pressure equal to 1,000 dynes per square centimeter. (A *dyne* is the force required to cause a mass of one gram to accelerate by 1 centimeter per second in each second of time.) Standard atmospheric pressure at sea level is about 1,000 millibars.

The pressure, temperature, and volume of a given amount of gas are dependent on one another. A change in one always causes a change in either or both of the others (Figure 2.4 a and

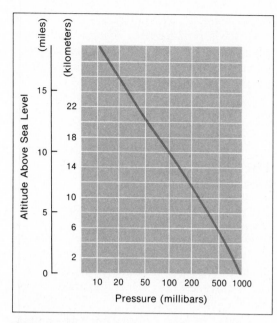

Figure 2.3 The pressure exerted by air in the atmosphere decreases rapidly with increased altitude. As the graph indicates, average pressure at sea level is approximately 1,000 millibars, but at an altitude of 5 km (3 miles) it falls to about 540 millibars. Note that the pressure is shown on a *logarithmic scale,* which is useful when graphing a quantity that varies over a wide range. The pressure at sea level can be thought of as the weight per unit area of a column of the earth's atmosphere. The graph shows the standard value of atmospheric pressure at each altitude; actual atmospheric pressure at a particular location on a particular day can be a few percent higher or lower than the standard value. If the air is denser than usual, the pressure is higher than normal, and if the air is less dense, the pressure is lower than normal. (Doug Armstrong adapted from *Handbook of Geophysics and Space Environment,* edited by Shea L. Valley, Air Force Cambridge Research Laboratories, U.S. Air Force, 1965)

b). A rise in temperature will increase the volume of an unconfined gas. But if a gas is sealed in a container so that it cannot expand, a rise in temperature will cause the pressure of the gas to rise.

If a volume of unconfined air close to the earth's surface is heated, it will expand without changing its pressure. This expanded air is less dense than the surrounding cooler air because it has fewer molecules per unit of space. The heated air is pushed upward by a buoyant force that exceeds the downward gravitational force on the air, and the warm air rises like a hot-air balloon (Figure 2.4c). Differences in the density of air thus cause sensible heat to be transferred upward in the lower atmosphere. This vertical transfer of air and of heat energy in turn sets air in motion horizontally, producing wind.

The general effect of the earth's wind systems is to transfer warm air from regions that receive greater amounts of solar radiation to regions that receive lesser amounts. Winds also carry the latent heat present in water vapor from regions where evaporation has occurred to far-off places, where the vapor condenses back to water or ice. The circulation of the atmosphere and its significance in energy transfer are the subjects of Chapter 4.

THE EARTH'S PHYSICAL SYSTEMS

A system is any collection of interacting objects. The earth as a whole can be thought of as a single physical system. However, the earth contains so many phenomena interacting in so many ways that to understand them it is necessary to subdivide the world into smaller systems. None of these is truly independent; all interact to some degree.

The environment at the earth's surface is divisible into four major systems. These are the *atmosphere,* which is the envelope of gases surrounding the solid surface of the earth; the *hydrosphere,* comprising all the waters of the earth; the *lithosphere,* composed of the solid material forming the outer shell of the earth; and the *biosphere,* which encompasses all life forms present on our planet.

All physical systems have some common features. All have energy sources and contain materials of some type. Most are *open systems,* with *inputs* of energy and materials from outside the system, *transformations* of energy and materials within the system, *storage* capacity, and *outputs* of energy and materials that become the inputs for other systems (Figure 2.5). In a *closed system* there would be no external inputs and no outputs to other systems. Closed systems would be entirely self-contained, like a battery-powered wristwatch. When its internal

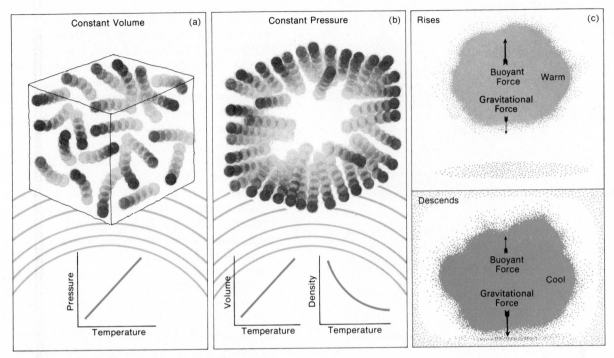

Figure 2.4 (a) The pressure, volume, and temperature of a fixed amount of gas are interdependent; a change in one of the three quantities is always accompanied by a change in one or both of the others. If a gas is confined in a sealed box so that its volume is constant, changing the temperature of the gas changes the pressure the gas exerts on the walls of the box. The gas molecules move more rapidly at higher temperatures and therefore collide more forcefully and more frequently with the walls. Therefore, pressure increases with increased temperature if the volume is held constant.

(b) A parcel of air in the atmosphere is unconfined. If the air is heated, it must expand its volume to maintain a constant pressure. The volume of the parcel therefore increases with increased temperature. The greater the volume occupied by the gas molecules in the parcel, the lower the density of the gas. Density therefore decreases with increased temperature for an expanding gas at constant pressure.

(c) The transfer of energy to and from parcels of air in the atmosphere can produce motion. A parcel of air that is hotter than the surrounding atmosphere is less dense than the atmosphere; in such a situation, the upward buoyant force on the parcel exceeds the downward gravitational force, and the parcel rises. Conversely, if a parcel of air is cooler than the surrounding atmosphere, it is denser. The downward gravitational force in this case exceeds the upward buoyant force, and the parcel descends. Upward motion causes a horizontal inflow of the surrounding air at the surface, and downward motion causes horizontal outflow at the surface. (Tom Lewis)

energy source exhausts itself, a closed system would cease to function. In actuality it is hard to conceive of a completely closed system, since some of its energy would be lost as heat during every energy transformation within it.

When we look at the inputs and outputs of any system, we become aware of the interdependence of different systems. Think for a moment of waves breaking on a beach. The inputs to the beach system are wave energy and sand particles. The outputs are relocated sand and the form of the beach itself.

Daily changes in sand flow and beach form are the direct result of wave action against the shore, as will be discussed in Chapter 18. However, by tracing the inputs and outputs farther, we can see that the beach system is part of a much larger picture. The kinetic energy of the waves was originally produced by the force of wind against the water surface hundreds or thousands of kilometers out to sea. The wind resulted from geographical variations in atmospheric pressure. The pressure variations are a consequence of place-to-place variations in solar energy input to the atmosphere. Solar energy itself is an output of atomic reactions

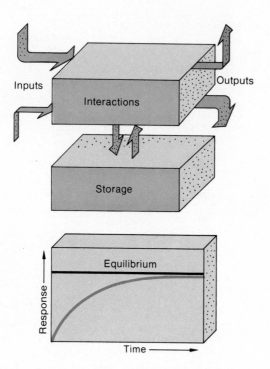

Figure 2.5 A system is any collection of interacting objects. Physical geography is concerned with a wide variety of systems, including landforms, soils, the biosphere, and the atmosphere. Thinking about the world in terms of systems is useful because systems have many features in common, some of which are illustrated in this schematic diagram. The inputs to a system represent energy and material received from outside the system's boundaries. The inputs are transformed by system interactions to new forms that are either placed in storage for a time or transferred to other systems as outputs of material and energy. Consider a field of corn as an example of a system: the inputs include radiant energy from the sun, carbon dioxide from the atmosphere, and moisture and nutrients from the soil. Part of the input is stored as chemical energy in the process of photosynthesis. Outputs include the water vapor returned to the atmosphere by plant transpiration, the free oxygen released in photosynthesis, and the chemical energy that is eventually used as food.

The lower diagram illustrates schematically one of the ways that systems respond in time to an input. In this model of systems behavior, the output does not reach full strength immediately upon application of an input. Time is required for a steady condition to be achieved. Some systems, such as the atmosphere, respond rapidly to new inputs; other systems, such as soils, may require hundreds of thousands of years to attain equilibrium. (Tom Lewis)

within the sun. The material input of the beach system—sand—is an output of a complex erosional system on the land, and was probably delivered to the shore by a river. Finally, one output of the beach system—the changing form of the sand deposit—has important effects on the living organisms that inhabit the beach. This is but one simple example of the complexity of almost any open system.

Systems can store energy or materials for varying amounts of time. Rocks that are being stretched or squeezed in the earth's crust are storing energy. They are behaving like a compressed or stretched spring. Eventually, when their deformation, or *strain*, exceeds their strength, they snap. This releases the stored energy in the form of an earthquake. The slow bending of rocks in western California has stored so much energy that a major destructive earthquake is expected sometime in the 1980s. Hurricanes are another example of the release of stored energy. In this case, the summer build-up of both sensible and latent heat in the tropics finds release in powerful tropical storms that vent this energy into higher latitudes.

System Equilibrium

Many systems tend toward a stable condition in which continued inputs of energy and materials produce no long-term changes in the system or its outputs. Such a system is said to be in *equilibrium*. Short-term changes may occur, yet the long-term behavior is constant. For example, a flood may cause a stream channel to be enlarged and deepened by erosion of its bed and banks. But when the flood subsides, the channel is restored to its previous state by the deposition of sediments that fill in the portions enlarged by erosion. Although there are short-term fluctuations, most stream channels remain about the same from year to year.

The maintenance of equilibrium in a physical system requires that inputs of energy and materials maintain an average balance with outputs over the period of time considered. When the balance is upset, the system reacts by changing in such a way that a new balance can be achieved.

In general, systems do not react to every

small change in inputs; rather they incorporate a certain amount of inertia. They resist change until the degree of imbalance produced by changed inputs is too great to withstand. This inertia minimizes work—the expenditure of energy—in the system. The point at which a system becomes so unbalanced that it begins to change to restore equilibrium is known as a system *threshold*.

Think again of a boulder, with all its stored potential energy, resting on a steep slope. Imagine scratching away some of the soil at the foot of the boulder on its downhill side. At first nothing happens. But no sensible person would continue digging incautiously at the downhill base of the boulder. Even without thinking of potential energy or thresholds, anyone would realize that at some point the digging would undermine the boulder, which would topple over and go crashing down the hillside. The digging would steepen the slope under the boulder to the threshold value at which the boulder would be moved by the force of gravity.

There are different types of thresholds in most systems. Some are externally controlled, involving changes in inputs of energy or materials; others are internal, having to do with storage of materials or energy within the system. Systems change when certain parameters either rise above or fall below threshold values. A current of water begins to pick up and carry solid particles of a certain size when its velocity increases to a certain value. Similarly, a current of water that is carrying particles begins to drop them when its velocity decreases to a certain point. Thresholds can be reached by either gradual or sudden changes, which may be either natural in origin or a direct result of human activities.

The time required for an equilibrium system to begin to change as a result of a change in system parameters is the *response time*. The time required for a system to reestablish equilibrium after a change has begun is the *relaxation time*. These times vary greatly in different systems. A river channel responds in hours to a change in inputs and reestablishes equilibrium quickly, within days. Other systems, such as hillslopes, have response and relaxation times of years, centuries, or even longer.

When a system threshold is reached, the resulting change may be violent, as in the case of an earthquake or a hurricane, each of which releases stored energy as suddenly as the toppling boulder described previously. Other threshold phenomena can cause a system to assume a new form. Damming a river reduces the load of sediment it carries to the sea. This reduces the supply of sand to coastal beaches. Within a few years or decades after a dam is built, coastal beaches may shrink or may disappear altogether. In this case the supply of sand has dropped below the threshold value required to sustain the beaches.

Cycles of Materials

As long as the earth and its atmosphere do not experience a long-term change in temperature, we can assume that the earth's input and output of energy are in balance. But this energy is used and transformed in many ways between the time solar radiation, the major energy input, arrives on our planet and is returned to space in a different form. These intervening energy exchanges are the driving force for most physical systems on the earth.

The materials composing these systems are fixed in amount. Except for rare meteorite falls and the escape of some gaseous molecules, the earth is composed of the same matter that came together when the planet formed 4.6 billion years ago. Since the earth's materials are finite, they must be constantly recylced to permit the earth's systems to continue functioning.

The rate at which materials are recycled varies greatly from system to system. Water can evaporate from the ocean surface and enter the atmosphere as vapor, then recycle to the ocean as precipitation, all in a few hours. However, if the water were to fall as snow in a polar region, it could become part of a glacier, where it might be held as ice for hundreds or even thousands of years before returning to the sea. Some of the ice in Antarctic glaciers is more than 120,000 years old.

Other materials participate in even longer cycles. Oceanic salts have been stored for hundreds of millions of years in beds of rock salt

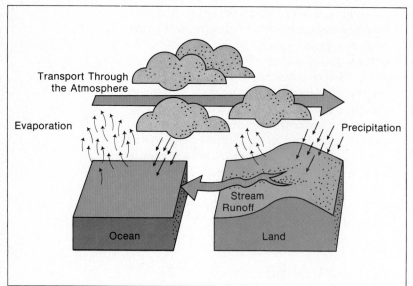

Transport Through
the Atmosphere

Evaporation

Precipitation

Stream
Runoff

Ocean   Land

Figure 2.6 The hydrologic cycle is a key element in physical geography because water has important interactions with systems such as vegetation and landforms. The processes involved in the cycling of water include evaporation of water from the oceans, evaporation and transpiration from the continents, transport of water vapor through the atmosphere, and the return of water from the atmosphere to the surface as precipitation. The annual evaporation loss from the oceans exceeds the amount gained from precipitation, but the deficit is made up by streams flowing from the continents to the oceans. (Tom Lewis)

extending from New York to Ohio and Michigan—the bed of a shallow sea 400 million years ago. Eventually the slow erosion of the North American continent will return this salt to the oceans. The earth's crust is itself being recycled ever so slowly by geological processes, as we shall see in Chapter 12.

The earth's hydrosphere is a dynamic physical system that is dependent on the *hydrologic cycle*, illustrated in Figure 2.6. Streams carry water endlessly from the continents to the oceans. If there were no "return flow" back to the continents, the land masses would eventually become waterless. The hydrologic cycle in its simplest form begins with evaporation of seawater, which enters the atmosphere as water vapor. The atmospheric circulation transports much of this water vapor to the continents. Over the land, water vapor condenses into clouds composed of water droplets. The droplets coalesce and fall as rain or snow. Some of the rainwater or melted snow runs off into streams that carry the water back to the oceans to complete the cycle.

Actually this is an oversimplification. Much precipitation also falls on the oceans themselves, producing a short subcycle within the larger cycle. And evaporation takes place on the continents—from various bodies of water and from plant leaves, which transpire vast amounts of moisture. Most rainfall does not run off directly into streams but enters the soil and becomes soil moisture. Some of it percolates deeper to become groundwater, which nourishes springs and wells and sustains streamflow between rains. Even though it is a simplified model, the basic concept of the cycle is correct— it involves circulation and changes of state of material, with periods of *residence* in different forms.

In the hydrologic cycle, water has several residence forms (Figure 7.1, p. 156). About 97 percent of the earth's free water is salt water residing in the oceans. The next largest reservoir of water is the ice in glaciers, particularly the Antarctic ice sheet. Most of the remaining water is stored in porous rock or loose deposits below the land surface, forming the groundwater that is tapped by wells. Lakes, ponds, swamps, and rivers contain less than 1 percent of the total supply of free water. The atmosphere holds even less. If all the water present in the atmosphere at any moment were to descend to the earth, it would form a layer only about 2.5 centimeters (cm), or 1 inch, in depth.

Another material cycle of great importance

to life on the earth is the *carbon cycle*, depicted in Figure 2.7. The most critical substance in the carbon cycle is gaseous carbon dioxide (CO_2). As noted in Chapter 1, green plants require atmospheric carbon dioxide, along with water and solar energy, to manufacture carbohydrates in the process of photosynthesis. Although the carbon dioxide in the atmosphere is continually being depleted by plant activity, it is also continually released at the surface by plant respiration and decay, as well as by animal respiration, which expels CO_2 in exhaled breath as a waste product of the process by which animals oxidize nutrients to obtain energy.

Only a small percentage of the earth's carbon is present as CO_2 in the atmosphere. Far larger amounts of carbon compounds are dissolved in the seas and stored in vegetation, limestone bedrock, and deposits of fossil fuels (coal, oil, and natural gas). However, since the beginning of the Industrial Revolution in the nineteenth century, the burning of fossil fuels has increased the CO_2 content of the atmosphere by 10 to 15 percent, as shown in Figure 2.8. This is a matter of concern because CO_2 is a strong absorber of the thermal energy that is reradiated from the earth's surface (see Chapter 3).

The increase in atmospheric CO_2 over the past 100 years already appears to have produced an 0.3°C rise in the temperature of the lower atmosphere. It is estimated that the level of CO_2 in the atmosphere will double in 50 to 70 years, causing a 3°C global temperature increase. Continuation of the current tendency toward global warming would eventually reduce the size of the polar icecaps in Antarctica and Greenland. The change of phase removing water from glacial storage would gradually raise the level of the seas, causing heavily populated coastal regions to become submerged, as discussed further in Chapters 17 and 18.

Budgets of Energy and Materials

Inputs, outputs, storage, and balance can be treated as parts of the *budget* of a system. As the term suggests, the budget concept is similar to a financial account that balances income and savings against expenses.

The example of an ordinary concrete swimming pool can serve to illustrate the budget concept. The pool gains water from rainfall and the city water supply. It loses water by evaporation and the splashing out of water by pool users. But to account for all the water passing through the pool, we must understand all interactions in the system. If there are cracks in the concrete, leaks must be included in the output side of the budget. So must the films of water clinging to persons leaving the pool. Similarly, any runoff of rainwater into the pool must be added to the input.

We could also develop an energy budget for the pool, in which energy inputs and outputs are reflected in the changing temperature of the pool water. Temperature changes indicate changes in the energy budget in much the same way that water-level changes reflect the material budget.

The accounting system for energy and materials sometimes reveals that things are occurring in the system that we do not understand or have failed to measure. For example, the global CO_2 budget shows that the increase in atmospheric CO_2 produced by the burning of fossil fuels over the past century is only about half the expected amount. The missing CO_2 is believed to have gone into solution in the oceans or to have been absorbed by vegetation on the land. The CO_2 budget shows that either ocean water or land vegetation has a higher CO_2 storage capacity than expected, somewhat reducing the harmful effect of fossil fuel use.

One of the tasks of physical geography is to identify the significant similarities and differences among the systems at work in the various regions of the earth. Systems that are subject to similar inputs make similar responses. Thus the problem met, the lessons learned, and the solutions developed in one area can be applied to others of a similar type. This can help us avoid the costly errors that have already led to environmental degradation in many parts of our planetary home.

Figure 2.7 This diagram illustrates the large number of systems through which carbon can pass during its cycle. The numbers are estimated values in billions of tons of carbon released or absorbed annually in each process (red) or the total amount stored in each reservoir (black).

Transfer of CO_2 to and from the atmosphere is an essential part of the carbon cycle. CO_2 enters the atmosphere primarily from the respiration and decay of organisms, from the burning of fossil fuels, and from the CO_2 dissolved in the oceans. Photosynthesizing green plants and the oceans absorb CO_2 from the atmosphere. Some of the carbon absorbed by plants is locked into long-term storage as coal and peat. The largest reservoir of carbon is limestone bedrock. Much of the calcium carbonate ($CaCO_3$) composing limestone is precipitated chemically from seawater, but some is composed of the shells and skeletal parts of tiny marine animals. The remains of marine organisms are also the source of petroleum and natural gas.

The transfer of CO_2 to and from organisms is believed to be nearly in balance over the year. The CO_2 content of the atmosphere, however, is increasing by several percent each decade. This increase is the result of human industrial activities; it would be even greater if it were not for the CO_2 taken up by the oceans. (John Dawson after Gilbert and Plass, "Carbon Dioxide and Climate," *Scientific American*, 1959)

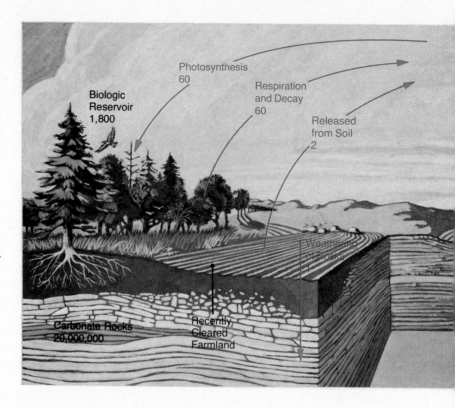

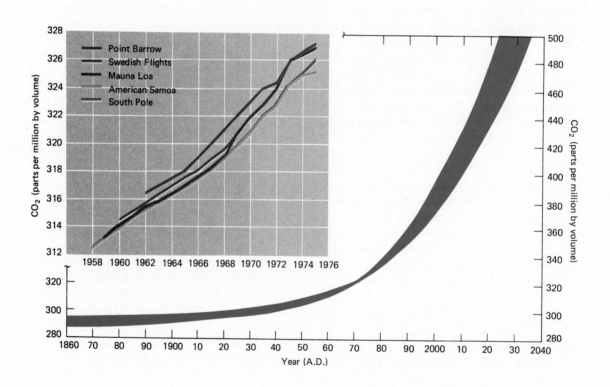

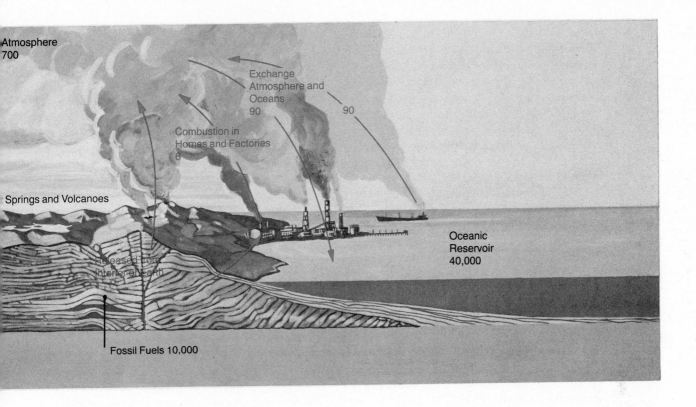

Atmosphere
700

Exchange
Atmosphere and
Oceans
90   90

Combustion in
Homes and Factories
6

Springs and Volcanoes

Released from
Interior of earth

Oceanic
Reservoir
40,000

Fossil Fuels 10,000

Figure 2.8 (opposite) Estimates of the global increase in atmospheric CO_2 since the Industrial Revolution and continuing into the twenty-first century. The colored envelope includes the varying data of five different observers. All estimates take into account uptake of CO_2 by the oceans and the biosphere, and assume a continued exponential increase in the burning of fossil fuels (especially coal), with about half the released CO_2 remaining in the atmosphere, where it contributes to global warming.

The inset shows the increase in atmospheric CO_2 measured over a seventeen-year period at five widely separated locations far from major sources of CO_2 emissions. (After W. W. Kellogg and R. S. Schware, *Climate Change and Society,* Westview Press, 1981)

SUMMARY

Energy is the capacity to do work, meaning to cause change in the state or position of materials. The principal sources of energy for physical processes on the earth are solar radiation, gravitational force, and the earth's internal energy resulting from radioactive decay of unstable elements. Although energy cannot be created or destroyed, it can take many forms.

Of most significance to environmental processes are solar radiant energy, heat energy, potential energy due to gravity, kinetic energy, and chemical energy. The earth's physical processes involve constant transformation of energy from one form to another. Some energy is lost as heat in every energy transformation.

Water is extremely important to the earth's

physical systems because of its role in energy conversion, storage, and circulation. All changes of state of water between gaseous, liquid, and solid forms either absorb energy as latent heat or release energy as sensible heat. The atmospheric circulation transports energy globally as measurable sensible heat or as latent heat that is released when water vapor condenses to liquid water. The atmospheric circulation is caused by pressure differences resulting from differential solar heating of the earth's surface. Water's unique properties allow it to store and release vast quantities of heat energy without undergoing large temperature changes. Thus water has a moderating effect on temperature changes at the earth's surface and in the atmosphere. The kinetic energy of water moving in streams, water waves, and flows of glacial ice also enables water to do work in the form of erosion and sediment transport, which produce landscape change.

The earth is a complex of interacting physical systems. The largest-scale systems are the atmosphere, hydrosphere, lithosphere, and biosphere. Most earth systems are open systems that receive inputs of energy and materials from other systems and have outputs to other systems. Most systems tend to move toward an equilibrium condition in which processes act and interact without causing long-term changes in the system itself. When changes in system inputs and internal characteristics reach certain threshold values, the system reacts by changing in such a way that a new balance is achieved. Physical systems can store energy or materials for varying amounts of time. When system thresholds are surpassed, the release of stored energy may cause violent events, such as earthquakes and hurricanes, but these are necessary to restore system equilibrium.

The continued activities of the earth's physical systems require energy and materials to be used over and over again, constantly changing their form. These cycles of energy and materials, involving inputs, transformations, storage, and outputs, can be evaluated in terms of budgets that must balance out over a period of time. The budget concept helps us to better understand environmental processes and to avoid errors in the management of the earth, our home.

REVIEW QUESTIONS

1. What are the principal energy sources that produce change on the face of the earth?
2. Why is energy due to the force of gravity known as *potential* energy?
3. How is energy related to the existence of matter?
4. What energy transformations are involved in the initial formation and later human utilization of coal and petroleum?
5. How are energy transfers at the earth's surface and in the atmosphere related to changes in phase by water?
6. What is the difference between specific heat and heat capacity?
7. Considering the heat capacity of water, how might the climate of a coastal region be expected to differ from that of a continental interior?
8. What is the difference in meaning between the response time and the relaxation time of a physical system?
9. What are the similarities and differences between the inputs, forms of residence, and outputs of the hydrological cycle and the carbon cycle?
10. The system budget for a swimming pool is analogous to that of a natural lake. What are some other systems for which energy and material budgets can be formulated?

APPLICATIONS

1. Diversion of "wasted" solar energy to controlled uses will cause decreases in solar energy input to other systems. Will this have noticeable consequences?
2. Latent heat is a type of energy that might be compared to gravitational energy. What are the similarities and differences between latent heat and gravitational energy?
3. How is national policy for energy production of concern to climatologists?
4. What is the source of the water you use each day? What artificial systems have been installed to bring this water to you? How has the simple hydrologic cycle described in the text been altered or "shortcircuited" in your area? Has this produced any unexpected consequences?
5. What is the largest scale closed system you can think of? What is the smallest scale open system imaginable?
6. What sort of artificial disturbances of natural systems have occurred and are occurring in your area? What are the response times of these disturbances? Are the relaxation times known?

FURTHER READING

Asimov, Isaac. *Life and Energy.* New York: Bantam (1965), 378 pp. An account of energy principles and their applications to living organisms by one of the most adept of all writers on scientific topics.

Chorley, Richard J., ed. *Water, Earth, and Man.* London: Methuen (1969), 588 pp. This is a most useful collection of 38 selections, mostly by British authors, dealing with various aspects of the hydrologic cycle. The selections are at introductory and intermediate levels and complement a number of chapters in this text.

Energy and Power, A Scientific American Book. San Francisco: W. H. Freeman (1971), 144 pp. A collection of articles on energy printed originally in the periodical, *Scientific American.* Energy sources, flows, and transformations are all treated in a lively, highly readable manner, with many excellent illustrations.

Odum, Howard T. *Environment, Power, and Society.* New York: Wiley-Interscience (1971), 331 pp. This book attempts to show how natural and engineered systems are related in terms of energy flows and basic system structures. Odum uses the principles of ecology to offer solutions to current problems of global importance. A particular contribution is the symbolic language used to depict energy flows and transformations.

Weyl, Peter K. *Oceanography: An Introduction to the Marine Environment.* New York: John Wiley (1970), 535 pp. This book complements *Physical Geography Today* very well. The chapters on oceanic salts and geochemical cycles are especially pertinent to Chapter 2 of this text.

CASE STUDY

Understanding Nature

The physical world around us is a complex mosaic that is capable of generating pleasure and delight, awe, and even fear. The relationships among the skies, lands, and waters have aroused curiosity from the very beginnings of human life. What are the global patterns of the units of the mosaic, and how do the natural processes interact to produce both environmental stability and ongoing change, and occasionally cataclysmic events?

In our attempts to sort through the earth's myriad phenomena, we tend to construct mental *models* to help us understand things and events and how they are interconnected. With models we find a way of perceiving order in nature, of gaining deeper knowledge of why and how things happen as they do. Consider a model for "tree." One of us might think of a pine tree; another, an elm or a centuries-old redwood. Even though these are different trees, they nevertheless have enough similarities to be understood by all as belonging to the same category, and we share an understanding of that model.

Ecologists and physical geographers use different models. The ecologist views a "tree" as a marvelous microenvironment where different creatures occupy different niches from roots to treetop, where intricate energy exchanges take place between the inhabitants and the surrounding environment. His model explains much more because he has probed a little deeper into a seemingly simple part of the world and discovered its larger dimensions. The view of the physical geographer, on the other hand, often focuses on the regional and global patterns of the trees, and especially on how they are related to climates, the availability of water, soils, and landforms.

All fields of inquiry claim models that are useful in explaining the world. For example, there are innumerable ways to look at and understand the nature of one of the earth's most important and most common substances, water. Water covers three-fourths of the earth's surface; it is also beneath the surface, permeating the soil and rock, and in the atmosphere, circulating in the form of vapor.

Different types of models describe water's different properties. A chemist uses a molecular model to explain water's fundamental makeup of two hydrogen atoms and one oxygen atom. A physical scientist is interested in water's unique ability to form three states of matter. A physical geographer, however, is concerned with water as a transporter of heat energy through the atmosphere and as an agent of change on the earth's surface. So models have been developed that describe how water flows across the earth in streams, rivers, and glaciers, how it circulates in the atmosphere and oceans, how it affects climate, vegetation, and soil formation, and how it shapes the landscape.

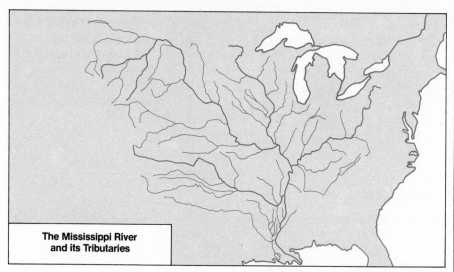

The Mississippi River and its Tributaries

Wherever rainfall runs off the land, it erodes its own drainage system. This map, which is an example of a graphic model, shows the area drained by the Mississippi River system (see also Chapter 14). Graphs, photographs, maps, and diagrams are types of graphic models used to aid our conceptualization of the physical world. (Doug Armstrong)

Physical models are replicas of environmental processes or systems. The photograph shows a section of the largest physical model of a drainage basin ever built in the United States. The model was built by the Corps of Engineers in the late 1940s near Jackson, Mississippi, to represent the main channel of the Mississippi River below Cairo, Illinois. During the flood along the lower Mississippi River during April, 1973, the physical model was operated each day to "model" estimates of changing conditions for the remainder of the flood event; an entire flood season could be "run through the model" in a day! The model was used to help with real-time decisions about diversions of Mississippi River water into the Atchafalaya Basin floodways upstream from New Orleans (see the Case Study for Chapter 14).

The concepts of "cycle" and "system" and "budget" are important components of models in physical geography. The hydrologic cycle mentioned earlier and described more fully in later chapters is a model to explain how all the earth's original supply of water is continually recycled. The same water is transported time and time again from the land and oceans into the air, and then back to earth again. The local water budget is a model to describe the availability of water for various processes and purposes in the environment. Water shapes the earth's surface features, and several of the models useful for analyzing flowing water are illustrated here.

Keep in mind that the models described in this book are the physical geographer's perceptions of the world—a world we can come to know much better through educated eyes.

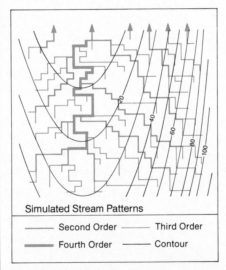

Simulated Stream Patterns

——— Second Order ——— Third Order
——— Fourth Order ——— Contour

Drainage systems and their properties can also be simulated by computers, as this output in graphical form from a computer model shows. This model was used to study the development of stream patterns and to predict changes in underground water levels. In the illustration, the curved lines represent contours of elevation and the blue lines show the development of a drainage basin pattern on homogeneous materials. The water flows down the slopes.

Computer models are used extensively now to simulate water levels and volumes in streams and rivers. The model inputs are similar to but much more complex than the water-budget models introduced in Chapter 7. For example, vital forecasts of flood stages along the Pearl River in Mississippi and Louisiana were issued twice each day by the Lower Mississippi River Forecast Office of NOAA during the April, 1983, floods along the river. The computer models not only estimate water levels quite effectively, but the models are also capable of routing the flood crest down the river day-by-day. (Doug Armstrong after A. Armstrong and John Thornes, The Geographical Magazine, © IPC Magazines, Ltd., 1972 by permission of New Science Publications)

Number 8 by Mark Rothko, 1952. (From the collection of Mr. and Mrs. Burton Tremaine, Meriden, Connecticut)

Solar radiation is the principal source of energy for the natural processes that give rise to diversity and change on the earth. However, if the earth continually received energy from the sun without returning an equal amount to space, the oceans would boil and the land would be scorched. Since the average temperature of the earth remains much the same from year to year, the earth must be returning as much energy to space as it receives from the sun.

Of course, not all locations on earth have equal energy gains and losses. Each year, tropical regions receive a greater amount of energy than they radiate back into space. Polar regions, on the other hand, annually lose more energy to space than they receive from the sun. We know that the tropical regions are not progressively heating up, nor are the polar regions cooling off. This means that there must be a flow, or *flux*, of energy from areas of excess to areas of deficiency. The atmosphere and oceans circulate the energy that the earth receives, transporting warm air and water from the tropics to the poles in exchange for cool air and water that move back toward the equator. In this chapter we shall examine the energy gains and losses that maintain the earth's temperature balance.

3
Energy and Temperature

SOLAR RADIATION

Solar energy is transferred through space as *electromagnetic radiation*. Such radiation travels in waves and can be classified according to wavelength—the distance between similar points on successive waves. Radiation wavelengths are measured in *micrometers;* one micrometer is equal to one millionth of a meter.

All life processes are driven by complex exchanges of visible and invisible forms of radiant energy from the sun. The heat of the tropics and the cold of the polar zones are consequences of the earth-sun relation in space and of energy transfers on the earth.

Figure 3.1 presents the *electromagnetic spectrum*, in which various types of electromagnetic radiation are identified according to wavelength. Visible light includes wavelengths from 0.4 to 0.7 micrometers. Our eyes sense the various wavelengths of visible light as different colors. When visible sunlight is bent, or refracted, and then reflected by water droplets in the atmosphere, we see the various wavelengths as a rainbow. Electromagnetic radiation also includes wavelengths both longer and shorter than those of visible light. Ultraviolet radiation, X-rays, gamma rays, and cosmic rays are all emitted at wavelengths shorter than those of visible light, while infrared radiation and radio waves have longer wavelengths than radiation that is visible.

Every solid, liquid, and gas, whether warm or cold, emits electromagnetic radiation as a result of the motion of its molecules. The temperature of the radiating substance determines the wavelengths of emission. The hotter the object, the shorter the wavelengths of the radiation emitted. When we turn on the heating element of an electric stove, the coil remains dull black while warming up. During this time it emits infrared radiation, which we can feel but not see. As it grows hotter, the wavelength of maximum radiative intensity becomes shorter, shifting over into the visible portion of the spectrum. The coil glows dull red, then bright red. If it could be heated further without melting, it would eventually glow yellow-white, like the sun.

At the earth's surface and in the lower atmosphere, temperatures normally range between −40° and +40°C (−40° and 104°F). Substances at such temperatures emit electromagnetic radiation with wavelengths of 4 micrometers or more. The surface of the sun, on the other hand, has a temperature of more than 6,000°C (10,800°F). At this temperature, most radiation is emitted at wavelengths of less than 4 micrometers. Hence we can think of most solar radiation as *shortwave radiation*, and all radiation emitted by terrestrial sources (the earth's surface and atmosphere) as *longwave radiation* (Figure 3.2).

One of the laws of physics is that a surface emits an amount of radiation per unit of time that is in direct proportion to the surface tem-

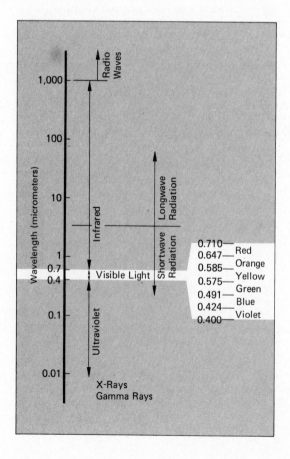

Figure 3.1 The electromagnetic spectrum is conventionally divided on the basis of wavelength. The divisions most important in physical geography are infrared radiation, visible light, and ultraviolet radiation. Visible light consists of wavelengths between approximately 0.7 and 0.4 micrometers. Infrared radiation consists of wavelengths intermediate between the wavelengths of radio waves and those of visible light, and ultraviolet light corresponds to wavelengths shorter than those of visible light. At extremely short wavelengths, the ultraviolet portion of the spectrum merges into the X-ray and gamma ray divisions.

Radiation with a wavelength longer than 4 micrometers is called longwave radiation; radiation in wavelengths shorter than 4 micrometers is called shortwave radiation. The distinction is useful in physical geography because solar radiant energy input to the earth is mostly at wavelengths shorter than 4 micrometers, while the energy radiated from the earth is at wavelengths longer than 4 micrometers (see Figure 3.2). (Doug Armstrong)

perature. A very hot object, such as the sun, not only emits shorter wavelength radiation than a cooler object like the earth, but also emits the radiation at a far greater rate. Accordingly, solar radiation is shortwave radiation emitted at a high intensity, and any radiation emitted by the earth and its atmosphere must be longwave radiation emitted at a much lower rate.

SOLAR ENERGY INPUT TO THE EARTH

The sun radiates energy equally in all directions. The earth, which in relation to the sun is like a grain of sand to a football 100 yds (91.5 m) away, intercepts only a tiny fraction of the total radiation emitted by the sun. But this small fraction is an enormous quantity of energy, amounting to 2.6×10^{18} calories per minute. The solar energy intercepted by the earth in one minute is about equal to the total electrical energy artificially generated on earth in one year.

Not all this radiant energy reaches the earth's surface because, as we shall see later in this section, the earth's atmosphere modifies the solar radiation that strikes it. Nor do equal amounts of radiant energy strike all parts of the upper atmosphere. This is because the distribution of solar radiation reaching the top of the atmosphere is controlled by the length of the daylight period and the elevation of the sun above the horizon. Both of these factors vary with latitude and are determined by various earth-sun relationships.

Earth-Sun Relationships

Like every planet in the solar system, the earth follows a fixed path, or *orbit*, around the sun. The orbit is slightly elliptical, so that the earth's distance from the sun varies by a small amount during the year it takes for the earth to circle the sun. However, this distance never varies more than 1.8 percent from the average distance of 149.5 million km (92.9 million miles). As a result, the monthly income of solar energy over the entire earth varies from the 12-month

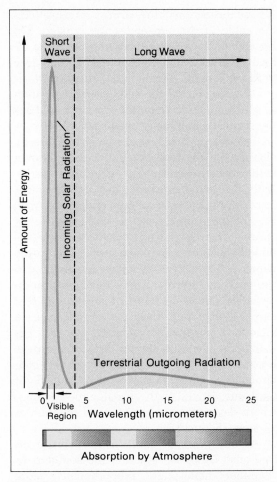

Figure 3.2 Because of the high surface temperature of the sun, energy is emitted primarily at wavelengths shorter than 4 micrometers, with much of the energy concentrated in the visible region of the spectrum. In contrast, the longwave radiation emitted by the earth is confined to wavelengths longer than 4 micrometers and has a broad peak at about 10 micrometers, because of the earth's comparatively low average surface temperature.

The lower portion of the figure indicates the degree to which atmospheric gases, primarily carbon dioxide and water vapor, absorb electromagnetic energy near the earth's surface. Wavelength bands of strong absorption are shown in red, and bands of relative transparency are indicated by yellow. The lower atmosphere is relatively opaque to longwave radiation, so that much of the radiant energy emitted by the earth's surface is absorbed by gases there. The atmosphere is relatively transparent to electromagnetic radiation in the band from 8.5 to 11 micrometers, and radiant energy in this band can escape to space if the sky is clear. (Doug Armstrong after G. M. Dobson, *Exploring the Atmosphere*, 1963, by permission of the Clarendon Press, Oxford)

average by no more than 7 percent. The earth intercepts the greatest amount of solar radiation in early January, when it is closest to the sun (*perihelion*), and receives the least in July, when it is farthest from the sun (*aphelion*). Later we shall see why seasons in the northern hemisphere are not in phase with the total delivery of solar radiation to the earth and its atmosphere.

In addition to orbiting the sun, the earth rotates on its axis; each rotation period (one day) has been divided artificially into 24 hours. The rotation introduces an additional daily cycle of solar energy income related to the times of sunrise and sunset. Because the earth is essentially spherical, the sun illuminates only half of the globe at any time, neglecting twilight, as shown in Figure 3.3. The boundary line between the light and dark halves is the *circle of illumination*, or in space age parlance, the *terminator*.

The Earth's Tilted Axis

Figure 3.3 shows that as the earth circles around the sun, the earth's axis always remains tilted in the same direction. This phenomenon is the cause of the seasons and the variation in daylight periods from the equator to each of the poles. As the earth rotates on its axis, the circle of illumination appears to sweep around the earth from east to west over a period of 24 hours. Of course, the circle actually remains fixed, facing the sun, as the earth turns through it from west to east. If the earth's axis of rotation were perpendicular to the plane of its orbit, every place on the earth would have 12 hours of daylight and 12 hours of darkness every day of the year. In the middle and high latitudes, the seasonal change from long daylight periods in summer to short daylight periods in winter would not occur. But the earth's axis tilts at an angle of about 23½° from the perpendicular. This causes the duration of daylight and darkness to vary seasonally everywhere except at the equator, which is always cut exactly in half by the circle of illumination.

The inclination of the earth's axis also affects the angle of the sun above the horizon. On December 21 or 22, the sun appears to be directly overhead at noon at latitude 23½°S, close to Rio de Janeiro. This latitude, called the *tropic of Capricorn*, marks the southernmost position at which an observer on the ground would see the midday sun directly overhead.

The day on which the noon sun is directly overhead at the tropic of Capricorn is known in the northern hemisphere as the *winter solstice*. In the northern hemisphere this day, December 21 or 22, has the fewest hours of daylight, with the sun appearing low in the southern sky even at noon. In areas north of the Arctic Circle (latitude 66½°N), the sun is not visible at all at the winter solstice. Regions within the Arctic Circle, such as Greenland and northern Scandinavia, experience 24 hours of darkness on that date. Moving south from the Arctic Circle, an increasing proportion of each parallel lies within the circle of illumination, so the days become longer. Still, the period of darkness continues to be longer than the period of daylight until one arrives at the equator (0° latitude), where day and night are of equal duration at all times (Figure 3.3).

As one proceeds south of the equator on December 21 or 22, entering the southern or "summer" hemisphere, proportionately more of each parallel lies within the circle of illumination, so the daylight period becomes progressively longer than the period of darkness. At the Antarctic Circle (latitude 66½°S) and beyond it to the South Pole, the sun remains above the horizon for 24 hours, so there is no night at all. Of course, while it is the winter solstice in the northern hemisphere, it is the summer solstice in the southern hemisphere. Table 3.1 indicates the varying lengths of the daylight period at different latitudes at the time of the northern hemisphere winter solstice.

As the earth circles the sun, the tilt of the earth's axis causes the noon sun to be directly overhead at different locations at different times of the year (Figure 3.3). Six months after the northern hemisphere winter solstice, the noon sun is directly overhead north of the equator at latitude 23½°N (the approximate latitude of Ha-

Table 3.1 Length of Daylight During Winter Solstice*

Latitude		Daylight
90°N	North Pole	0
80°N		0
70°N		0
60°N		5 hr 52 min
50°N		8 hr 4 min
40°N		9 hr 20 min
30°N		10 hr 12 min
20°N		10 hr 55 min
10°N		11 hr 32 min
0°	Equator	12 hr 7 min
10°S		12 hr 28 min
20°S		13 hr 5 min
30°S		13 hr 48 min
40°S		14 hr 40 min
50°S		15 hr 56 min
60°S		18 hr 8 min
70°S		24 hr
80°S		24 hr
90°S	South Pole	24 hr

* Not counting twilight. Daylight includes time when at least the upper edge of the sun's disk is above the horizon.

Source: Adapted from List, Robert J. (ed.) 1951. *Smithsonian Meteorological Tables*, 6th rev. ed. Washington, D.C.: Smithsonian Institution Press.

vana, Cuba), known as the *tropic of Cancer*. The moment the vertical rays of the sun fall on the tropic of Cancer, about June 21, is the northern hemisphere *summer solstice*. At the summer solstice, the regions north of the Arctic Circle have 24 hours of daylight, while in latitudes south of the Antarctic Circle the sun does not appear above the horizon.

Midway between the solstices—about March 21 and September 21—the noon sun is directly overhead at the equator. At these two times the circle of illumination cuts through the poles and coincides with meridians of longitude. This causes the periods of daylight and darkness to be of equal duration everywhere. Thus these dates are known as the *equinoxes*—the *vernal equinox* in March and the *autumnal equinox* in September, from the perspective of the northern hemisphere. The exact dates of the

solstices and equinoxes vary slightly from year to year because the astronomical relationships between the earth and sun do not coincide exactly with the days determined by the earth's rotation.

The tilt of the earth's axis not only produces solstices and equinoxes, it also affects the intensity of solar radiation. Imagine holding a flashlight close to a square of cardboard representing the earth and its atmosphere. With the flashlight perpendicular to the cardboard, the circle of light on the cardboard is small but very bright. If the cardboard is tilted to make an angle with the beam of light, the illuminated area of the cardboard becomes larger but also dimmer. The same amount of light is emitted by the flashlight, but the light intensity on each unit of area of the cardboard has decreased.

The geometric relationship between the elevation of the sun above the horizon (beam angle) and the intensity of the solar beam at the earth's surface is shown in Figure 3.3 (bottom left). When the sun is directly overhead, the solar elevation is at the maximum of 90° and the intensity of the solar beam per unit of surface area is greatest. As the solar angle above the horizon decreases from the maximum, the intensity of the solar beam per unit of surface area decreases, reaching zero at sunset, when the solar elevation is 0°.

The amount of solar energy received per unit of surface area thus depends primarily on the angle at which the incoming radiation strikes the surface. Winter sunbathers in Miami recognize this, propping themselves in lounge chairs in order to be perpendicular to the winter sun's rays. Vineyards are planted on steep south-facing slopes along the Rhine and Mosel rivers in western Germany for the same reason. When the sun is 60° above the horizon, a field on a 30° slope facing toward the sun receives nearly 15 percent more solar energy than a horizontal field of the same size. It is the constant tilt of the earth's axis that helps produce summer in the northern hemisphere when the earth is farthest from the sun in its orbit. Solar elevations are higher in summer, and the northern hemisphere receives more solar radiation per unit area then.

Figure 3.3 **(top)** The amount of solar energy that reaches a given location at the top of the atmosphere depends on the distance between the earth and the sun and on the orientation of the earth. The top half of the diagram shows the earth at twelve different times during the year. The maximum distance of the earth from the sun is 152 million km (94.4 million miles) and occurs early in July. The minimum distance is 147 million km (91.4 million miles) and occurs early in January. Because solar energy input to the earth varies with the earth-sun distance, the amount of solar radiation the earth receives is 1.07 times greater in early January than in July.

The earth's axis of rotation is inclined 23½° from the perpendicular to the plane of the earth's orbit. As the earth circles the sun, the orientation of the axis remains the same. At the winter solstice, the sun is overhead on the tropic of Capricorn in the southern hemisphere, and at the summer solstice, it is overhead on the tropic of Cancer in the northern hemisphere. At the vernal and autumnal equinoxes, the sun is overhead at the equator.

(bottom left) The amount of solar radiant energy that a unit area on the earth's surface intercepts depends on the angle between the sun's rays and the plane of the surface. As the sketch shows, the solar beam is spread over a wider area where it meets the surface obliquely, reducing the energy received per unit of area. A surface intercepts the greatest amount of radiant energy when the surface is perpendicular to solar rays.

(bottom center) The solar radiant energy input to a location on the earth during a given 24-hour period depends partly on the duration of daylight. At any single moment, one-half of the earth is illuminated by the sun. The circle of illumination is the boundary between the light and dark regions of the earth. Because of the tilt of the earth's axis, the duration of daylight is different at different locations. The diagram illustrates the situation at winter solstice in the northern hemisphere.

(bottom right) The surface of the earth has been divided into two sets of intersecting grid lines to enable locations on the earth to be specified. The *parallels of latitude* are circles parallel to the plane of the equator. These circles connect points having the same angular distance north or south of the plane of the equator, the angle being formed at the center of the earth. The North Pole is at latitude 90°N.

The second set of grid lines, the *meridians of longitude*, are circles drawn with the earth's axis as a diameter. One meridian, chosen to be the *prime meridian*, is designated as 0° longitude. In the United States, and in many other countries, the prime meridian is taken to be the meridian on which the astronomical observatory in Greenwich, England, is situated. The other meridians connect points having the same angular distance east or west of the plane of the prime meridian. Longitudes vary from 0° to 180°.

For precision, a degree can be divided into 60 minutes of angular measure, and a minute can be divided into 60 seconds. The latitude and longitude of Washington, D.C., for example, can be written 38 degrees 54 minutes north, 77 degrees 2 minutes west or in abbreviated form as 38°54'N, 77°02'W. (John Dawson)

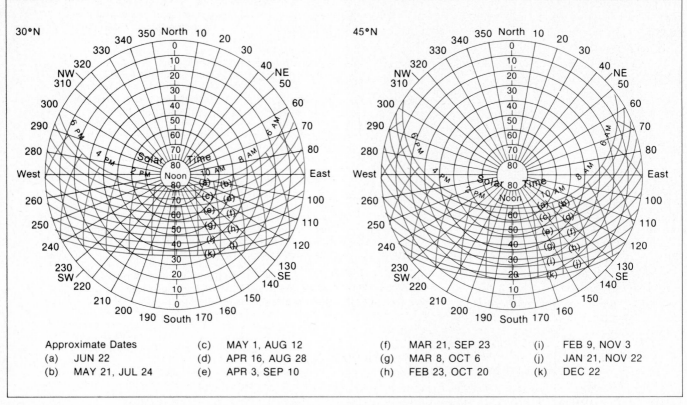

Approximate Dates

(a)	JUN 22	(c)	MAY 1, AUG 12	(f)	MAR 21, SEP 23	(i)	FEB 9, NOV 3
(b)	MAY 21, JUL 24	(d)	APR 16, AUG 28	(g)	MAR 8, OCT 6	(j)	JAN 21, NOV 22
		(e)	APR 3, SEP 10	(h)	FEB 23, OCT 20	(k)	DEC 22

Figure 3.4 The solar altitude and azimuth chart on the left is for latitude 30°N, the latitude of New Orleans, Louisiana, and the chart on the right is for latitude 45°N, the latitude of Minneapolis, Minnesota. The charts represent the dome of the sky as it would appear to an observer standing at the center of the circle. The numbers around the rim of the circle stand for azimuth, or degrees from north. The vertical line of numbers gives the altitude of the sun in degrees, starting with 0° at the horizon. To find the sun's position in the sky at a given time for these two latitudes, first select the red grid line corresponding to the date, and then find the point of intersection with the red line representing the time of day. Intermediate dates and times can be estimated by interpolation. The sun's altitude can be read from the scale of concentric circles, and the sun's azimuth can be read from the scale of radial lines. On March 21 at latitude 30°N, for example, the sun's altitude at 11:00 A.M. is 57° above the horizon, and its azimuth is approximately 150° from true north.

The duration of daylight can be estimated from solar position charts by finding the times on a given day when the sun's altitude is greater than 0°. On December 22 at latitude 45°N, the sun has an altitude greater than 0° between 7:45 A.M. and 4:15 P.M. Because these charts show the geometric position of the center of the sun, the duration of sunlight is a few minutes longer than these times suggest. (After List, 1951)

At any location, the angle of the solar beam at any time can be specified by taking into account the hour of the day, the position of the earth in its orbit, the tilt of the earth's axis, and the earth's spherical shape. The results of such determinations can be charted for each latitude. The two solar position charts in Figure 3.4 give the sun's position in terms of its *altitude* and *azimuth*, a system for specifying location in the sky that is analogous to the latitude and longitude location system used on the spherical surface of the earth.

To determine the solar altitude and azimuth, imagine that you are facing north with your arm outstretched and pointing north; then turn your body clockwise toward the sun and raise your arm until it points straight at the sun. The angle through which you turn your body is the solar azimuth, and the angle through which you raise your arm to point at the sun is the solar altitude. Azimuth is specified as an angle between 0° and 360°, measured in a clockwise direction from north. Altitude is an angle between 0° and 90°, measured upward from the level horizon. Thus an azimuth of 90° and an altitude of 45° indicate that the sun is due east and halfway between the horizon and the point directly overhead.

One of the things indicated on a solar altitude and azimuth chart is the maximum elevation of the sun above the horizon at any latitude. Using the charts in Figure 3.4 you can verify that at the summer solstice the maximum altitude of the sun is about 83° at New Orleans, Louisiana, and about 68° at Minneapolis, Minnesota. At the winter solstice, the maximum solar altitude is about 36° at New Orleans and only about 21° at Minneapolis.

Solar Radiation at the Top of the Atmosphere

The rate at which perpendicular rays of solar radiation strike the top of the earth's atmosphere is known as the *solar constant*. The average value over the year, based on satellite, rocket, and high mountain data, is about 1.94 calories per square centimeter (sq cm) per minute. Scientists express amounts of solar radiation by a unit known as the *langley*. One langley is equal to 1 calorie per sq cm. The solar constant, therefore, can be expressed as 1.94 langleys per minute.

The amount of solar radiation striking an area at the top of the atmosphere during one day depends on the value of the solar constant, the true distance between the earth and the sun, the duration of daylight, and minute-by-minute changes in the angle of the solar beam. Geographic variations in solar radiation received at the top of the atmosphere result from the last two factors, which are related to latitude.

The total daily solar radiation input to the top of the atmosphere at different latitudes is shown in Figure 3.5. Each curve indicates solar radiation input in langleys per 24 hours. For example, the energy input at the latitude of Philadelphia or Denver (about 40°N) on September 1 is about 800 langleys. This decreases to about 325 langleys at the winter solstice. The shaded areas represent periods of continuous darkness in the polar regions.

Figure 3.5 shows that through the year the solar radiation input varies much more at the poles than at the equator. At the equator, the variation is from about 780 langleys per day at the summer solstice to about 880 during late February and early November. The hours of daylight and darkness are always about equal at the equator, and the sun's altitude at noon is always high. The most extreme annual variation is at the South Pole, ranging from more than 1,100 langleys in mid-December, when the earth is closer to perihelion, to 0 from mid-March to early September. This is because the South Pole receives continuous sunlight between September and March, when the earth is near perihelion, whereas between March and September the sun is always below the horizon. Averaged over a full year, however, locations at the equator receive nearly two and a half times as much solar radiation as the South Pole.

Solar Radiation and the Atmosphere

Before solar radiation can reach the surface of the earth, it must pass through the earth's atmosphere, which is composed of gases, particles, and clouds. All of these respond differently to solar radiation at various wavelengths. The atmosphere therefore exerts a strong influence on the amounts and types of solar radiation that reach the earth's surface.

The chemical makeup of the atmosphere is given in Table 3.2. In Chapter 1 we saw how events during the early history of the earth caused nitrogen, oxygen, and carbon dioxide to be introduced into the atmosphere. The inert gases neon, helium, and krypton are probably remnants from the earth's original atmosphere. Argon, a more abundant inert gas, seems to have been produced largely by the decay of radioactive potassium in the earth's crust.

Nitrogen, oxygen, and the inert gases are present in the same relative proportions everywhere in the atmosphere. The amounts of other gases, such as carbon dioxide, water vapor, and ozone, vary with time, location, and altitude. Water vapor, which enters the atmosphere from the earth's surface through evaporation and plant transpiration, rarely appears above an al-

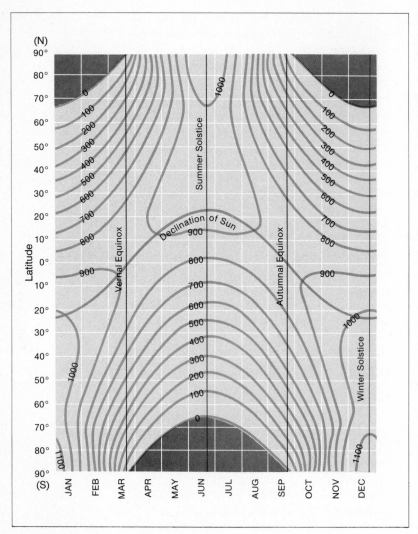

Figure 3.5 This graph shows solar radiant energy input to the top of the atmosphere. The horizontal scale is marked off in months of the year. The vertical scale lists latitude north and south of the equator. The curved contour lines give the solar radiant energy input to the top of the atmosphere in units of langleys per 24 hours, based on a value of 1.94 langleys per minute for the solar constant. The shaded areas of the diagram poleward of latitude 66½° represent times of continuous darkness when there is no solar radiant energy input.

The energy input into the northern and southern hemispheres is not symmetrical. At summer solstice in the northern hemisphere (June 21), locations at latitude 15°N receive about 900 langleys per 24 hours, but at summer solstice in the southern hemisphere (December 21), locations at latitude 15°S receive close to 1,000 langleys per 24 hours. The reason is that the earth is nearer the sun in January than in July, and the energy input to the earth is 7 percent higher in January than in July.

The patterns on the graph are related to the seasonal regimes of solar altitudes and day length at each latitude, used in determining how much solar radiation reaches the top of the atmosphere each day. At latitude 40° in the northern hemisphere, for example (close to Philadelphia or Madrid), the low values in January (about 350 langleys/day) are due to relatively low solar altitudes at noon and less than 10 hours of daylight. The high values in June (about 940 langleys/day) are associated with much higher solar altitudes at noon and more than 14 hours of daylight. The orange curve in the graph, the declination of the sun, represents the latitude where the solar elevation at noon is 90°. (Doug Armstrong after *Smithsonian Meteorological Tables*, edited by Robert J. List, 6th ed., 1971, by permission of the Smithsonian Institution)

Table 3.2   Principal Gases in the Earth's Lower Atmosphere

Gas	Molecular Formula	Number of Molecules of Gas Per Million Molecules of Air	Proportion (Percent)
Nitrogen	N_2	7.809×10^5	78.09
Oxygen	O_2	2.095×10^5	20.95
Water Vapor	H_2O	variable	variable
Argon	Ar	9.3×10^3	0.93
Carbon Dioxide	CO_2	330.0	0.03
Neon	Ne	18.0	1.8×10^{-3}
Helium	He	5.0	5.0×10^{-4}
Krypton	Kr	1.0	1.0×10^{-4}

Source: Adapted from List, Robert J. (ed.) 1951. *Smithsonian Meteorological Tables*, 6th rev. ed. Washington, D.C.: Smithsonian Institution Press.

titude of 10 km (6 miles). Ozone occurs primarily at an altitude of about 25 km (15 miles), where it is produced by reactions between solar radiation and oxygen molecules.

One can think of the atmosphere as being divided into "spheres" related to how temperature changes with altitude. In the *troposphere*—the portion of the atmosphere nearest the earth's surface—the temperature decreases with altitude at an average rate of 6.5°C per kilometer (3.6°F per 1,000 ft). The level at which temperatures stop decreasing with altitude is known as the *tropopause*. In the *stratosphere*, on the other hand, temperatures increase with al-

Figure 3.6 The atmosphere is conventionally divided into layers according to the variation of temperature with altitude. The *troposphere* decreases in temperature with increased altitude at the average rate of 6.5°C per km (3.6°F per 1,000 ft). The troposphere is warmest at the earth's surface, on the average, because it is heated primarily from below by the transfer of heat energy from the surface. The troposphere contains water vapor, and most clouds and weather phenomena are confined to this layer. Temperature increases with increased altitude in the *stratosphere,* largely because atmospheric gases such as ozone absorb a portion of the radiant energy incident from the sun. The stratosphere is nearly devoid of water vapor, so clouds seldom form there. Above the stratosphere, molecules of the gases forming the atmosphere are very widely dispersed and atmospheric pressure is negligible. The *thermosphere* is composed of scattered free electrons and other subatomic particles.

The temperature curve in the diagram corresponds to the values assumed for the United States standard atmosphere. A standard atmosphere is meant to represent the average state of the atmosphere, but the actual temperatures and pressures on a given day may differ from the standard values. (Doug Armstrong after the U.S. Air Force, 1965)

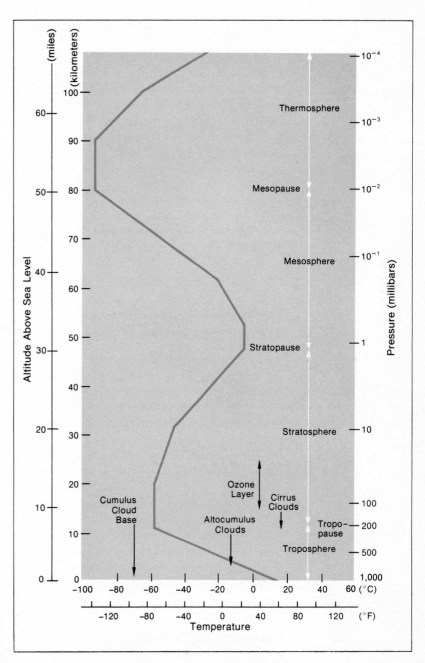

titude. Figure 3.6 shows these temperature/altitude relationships, as well as those in the still higher *mesosphere* and *thermosphere,* which are composed of thinly scattered gaseous molecules and subatomic particles.

Solar radiation interacts with gas molecules in the atmosphere through the processes of *absorption* and *scattering*. In the process of absorption, a molecule takes up radiant energy and converts it to heat. Scattering refers to the action in which gas molecules, dust particles, and water droplets deflect incoming solar radiation from its original path. Because of scattering, the earth's surface receives solar radiation from all parts of the sky, as well as directly from the sun. Shorter wavelengths, especially the blue portion of visible light, are scattered more effectively than longer wavelengths; that is what causes the sky to appear blue. The water droplets that form clouds, in addition to absorbing and scattering incoming radiation, reflect some solar radiation back out to space. In fact, more incoming radiation is reflected back to space by clouds than is absorbed by the atmosphere.

Gas molecules absorb most strongly at wavelengths in the longwave portion of the spectrum. Since most solar radiation is in the range of visible light (shortwave), the greater part of the radiation passes through the atmosphere. However, as we shall see later in this chapter, nearly all the radiation absorbed by the lower atmosphere is longwave radiation from the earth's surface, rather than from incoming solar radiation.

Ultra-shortwave radiation, such as X-rays, gamma rays, and ultraviolet radiation, requires special mention because its energy can destroy the complex molecules required for life processes. Fortunately, little of this radiation actually reaches the earth's surface; most of it is absorbed in the upper atmosphere. Molecular oxygen in the upper layers of the earth's atmosphere, and the ozone layer in the stratosphere, strongly absorb ultra-shortwave radiation, causing these portions of the atmosphere to be heated to temperatures close to those at the earth's surface (Figure 3.6).

THE ENERGY BALANCE OF THE EARTH-ATMOSPHERE SYSTEM

Because the earth is neither heating up nor cooling down, the amount of energy it receives from the sun and the amount it radiates back into space must be in balance. To understand this we must examine the ways in which the earth and its atmosphere lose energy and thereby maintain an energy balance. It is important to keep in mind that there are both seasonal and geographical variations in energy exchanges within the earth-atmosphere system, but our discussion will be in terms of averages for the entire globe over a year. For this purpose, we shall assume that the total solar radiation intercepted by the earth and its atmosphere over a year is equivalent to 100 units of energy.

Reflection and Albedo

Airline passengers are often startled by the dazzling brightness when their plane emerges from thick clouds after taking off under a heavy overcast. Thick clouds are very effective reflectors of shortwave radiation, throwing back as much as 90 percent of the solar energy that falls on their upper surfaces. This reflected radiation combines with the incoming solar beam to produce the extremely bright zone just above the clouds.

Table 3.3   Albedo of Various Surfaces

Surface	Albedo (Percent)
Fresh Snow	80–95
Dense Stratus Clouds	55–80
Ocean (sun near horizon)	40
Ocean (sun halfway up sky)	5
Bare Dark Soil	5–15
Bare Sandy Soil	25–45
Desert	25–30
Dry Steppe	20–30
Meadow	15–25
Tundra	15–20
Green Deciduous Forest	15–20
Green Fields of Crops	10–25
Coniferous Forest	10–15

Sources: List, Robert J. (ed.) 1951. *Smithsonian Meteorological Tables*, 6th rev. ed. Washington, D.C.: Smithsonian Institution Press. Budyko, M.I. 1974. *Climate and Life.* D. H. Miller, tr. New York: Academic Press, pp. 54–55.

The proportion of incoming solar radiation that an object reflects is called its *albedo*. The albedo of a good mirror approaches 100 percent. Clouds and fresh snow reflect 55 to 95 percent of the incoming solar beam. Most land surfaces, except those covered by snow, have albedos between 10 and 30 percent. The albedo at a given location varies from season to season as snow appears and disappears, and as bare fields become covered with crops (Table 3.3). It is particularly important to take albedo into account in analyzing the radiation balance because in the process of reflection very little radiation may be absorbed; it is simply redirected—generally outward to space. The reflected energy is lost to the earth-atmosphere system.

Assuming that 100 units of solar radiation energy are intercepted by the earth and atmosphere in a year, Figure 3.7 shows that the albedo of the whole earth-atmosphere system is about 33 percent. About 25 percent of the incoming radiation is reflected by clouds, and 8 percent by the earth's surface. Absorption by atmospheric gases and dust accounts for another 22 percent. Of this, 3 percent is ultraviolet radiation absorbed by ozone in the stratosphere. Ultimately, 45 percent of the solar ra-

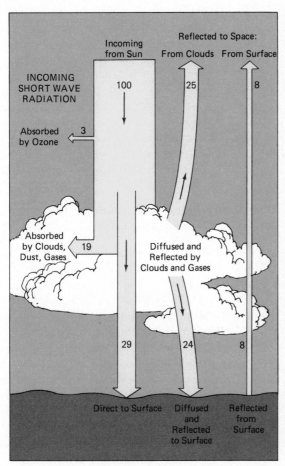

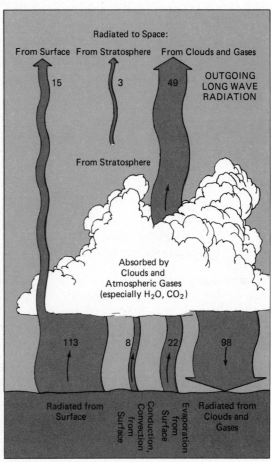

Figure 3.7 This diagram traces 100 units of shortwave solar radiation that arrive at the top of the earth's atmosphere, and the interactions of this radiation with the atmosphere and the surface. The numbers represent average global values. The interactions of incoming shortwave radiation are shown on the left, and the interactions of outgoing longwave radiation are shown on the right. All the indicated interactions occur at a given location during daylight hours, but between sunset and sunrise the shortwave radiation is, of course, zero. At night, conduction, convection, and evapotranspiration are normally very small.

The earth as a whole is neither gaining nor losing radiant energy; for every 100 units of solar radiant energy received, an average of 100 units (33 + 67) are returned to space. Similarly, energy is in balance at the earth's surface, on the average. The surface receives 143 units of energy (45 + 98) and loses 143 units (113 + 30). Can you use the diagram to show that the average gain and loss of energy by the atmosphere are also in balance?

Note that the 98 units of longwave radiant energy returned to the surface by the atmosphere are important components in the energy balance at the surface. If it were not for the energy contributed by the atmosphere, the surface would cool to the temperature where the radiated energy was in balance with the absorbed energy. In the absence of an atmosphere, the average temperature of the earth would be approximately −20°C (−4°F). (Tom Lewis after Herbert Riehl, *Introduction to the Atmosphere*, © 1972 by McGraw-Hill Book Company)

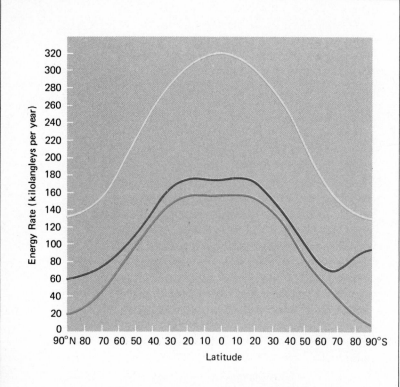

Figure 3.8 This graph shows the average annual distribution of solar radiation in the earth-atmosphere system by latitudes. The yellow curve represents incoming solar radiation at the top of the atmosphere, the red curve the incoming solar radiation just above the earth's surface, and the green curve the net solar radiation gain of the surface. In polar regions the combined effects of low solar altitudes, clouds, snow cover, and icecaps are evident. In the tropics seasonal cloudiness reduces the radiation gain at the surface. The highest gains are in the cloud-free subtropical desert regions. (Vantage Art, Inc., after William D. Sellers, *Physical Climatology*, © 1965, The University of Chicago Press)

diation is absorbed by the land and water areas of the earth. Figure 3.8 summarizes the average annual latitudinal distribution of solar radiation at the top of the atmosphere, contrasted with absorption at the surface.

Clouds have an important effect on the geographical distribution of solar radiation at the earth's surface. Figure 3.9 shows the average radiation input at the surface during winter and summer for the United States and for the entire

earth. At the top of the atmosphere the amount of solar energy input varies only with day length and the angle of the sun's rays. But at the earth's surface, differences in cloudiness can cause irregularities in the pattern of energy input. The southwestern deserts in the United States are farther north than Florida but receive more energy because their skies are generally clear. In North America, the lowest energy inputs at the surface are in the east and northwest, where fog and clouds are common. The world maps in Figure 3.9 show strong north-south differences in energy received at the surface in the winter hemisphere. The highest values occur in deserts in the summer hemisphere.

Longwave Radiation: The Greenhouse Effect

With an average temperature of about 15°C (59°F), the earth's surface is warm relative to its atmosphere and the space beyond it. Therefore it emits longwave radiation upward to the atmosphere and space. If the earth had no atmosphere, all this longwave radiation would be lost to space, and the surface would cool very rapidly after each sunset. This does not occur, however, because the atmosphere acts like a greenhouse, allowing shortwave solar radiation to pass through to the surface, but absorbing the longwave radiation emitted by the earth itself. Thus the atmosphere acts much like a blanket that keeps the earth warm.

Water vapor and carbon dioxide molecules are primarily responsible for the absorption of longwave radiation in the lower atmosphere. The absorbed radiation heats the lower atmosphere, especially in humid regions where there is abundant water vapor. The atmosphere, in turn, reradiates longwave radiation. Some of this is directed upward and is lost to space, but much more is returned downward to the surface of the earth because the lower altitudes are normally warmer and radiate at higher intensities. On the average, such longwave radiation exchanges result in a net loss of longwave radiation at the surface and also in the troposphere. Hence both the surface and the troposphere are cooled by longwave radiation exchanges.

Surface cooling by longwave radiation is dramatically reduced by the presence of clouds. Clouds absorb most of the longwave radiation emitted from the surface, significantly reducing longwave radiation losses to space. Like the atmospheric gases, clouds also emit longwave radiation, much of which is received by the surface. That is why, in general, cloudy nights tend to be much warmer than clear nights with starry skies.

Radiation Exchanges in the Earth-Atmosphere System

The right-hand side of Figure 3.7 traces the exchanges of longwave radiation within the energy balance of the earth-atmosphere system. The figure shows that an average of 113 units of longwave radiation are emitted by the earth's surface. Most of these (98 units) are absorbed by clouds, water vapor, and carbon dioxide in the troposphere, and only 15 units escape to space. The troposphere, in turn, radiates longwave radiation (98 units) back to the surface, and an additional 49 units upward to space. The warm stratosphere also loses some longwave radiation (3 units) to space.

The longwave radiation emitted by the troposphere to space, and back downward to the surface, exceeds the longwave radiation received from the surface. This is because it is supplemented by 19 units of solar radiation absorbed by clouds, dust, and gases, and converted to longwave radiation in the troposphere. It is also supplemented by 30 units of energy transferred from the surface by evapotranspiration, conduction, and convection, which are explained further in the following section.

Figure 3.7 shows that the troposphere radiates 49 units of longwave radiational energy to space. With the radiation from the surface that escapes (15 units) and the loss from the warm stratosphere (3 units), the total longwave loss to space equals 67 units, shown at the top of the diagram.

We have considered the total solar energy intercepted by the earth and atmosphere as 100 units of shortwave radiation. As Figure 3.7 shows, returned to space are 33 units of reflected shortwave radiation and 67 units of longwave radiation emitted from the surface, the troposphere, and the stratosphere. Therefore, outgoing shortwave and longwave radiation ($33 + 67 = 100$ units) is equal to the incoming solar radiation (100 units), and the earth-atmosphere system as a whole is in balance.

Considered separately, however, the atmosphere and the earth's surface do not show balanced radiation budgets. Figure 3.7 shows that radiative losses in the troposphere exceed gains by 30 units. Similarly, the earth's surface gains 143 units and radiates away only 113 units. Consequently, the surface experiences an annual net gain of 30 units—an amount just equal to the net loss in the troposphere. We know that the surface is not becoming progressively warmer, nor is the troposphere becoming cooler. Therefore, other energy exchanges must compensate for the imbalance in the troposphere and at the earth's surface. These energy exchanges occur through conduction, convection, and evapotranspiration.

Conduction, Convection, and Evapotranspiration

When two objects at different temperatures are in contact, heat flows from the warmer to the cooler object by the process of *conduction*. The same process plays a role in the transfer of heat energy at the surface of the earth. By midday, solar radiation has caused the top layer of the ground to become warmer than the deeper layers. As a result, heat is transferred deeper into the ground by conduction. At night, when radiation losses have cooled the surface layer, heat flows from the deeper layers back to the surface.

Heat is also conducted upward from the earth's surface to the atmosphere when land and water surfaces are warmer than the air in contact with them. This helps reduce some of the net radiation gain of the surface and it offsets some of the net radiation loss of the troposphere.

The efficiency of the conduction process is

Figure 3.9 The average radiation received at the surface in the United States during January and July and globally during December and June is shown in units of langleys per day.

During December and January, when the sun is overhead in the southern hemisphere, the solar radiation received in the northern hemisphere decreases rapidly with increased latitude. Southern Florida receives more than 300 langleys per day during January, on the average, but northern Minnesota receives only 150 langleys per day. Solar radiation input is small at high latitudes during the winter because of the short duration of daylight and the low altitude of the sun in the sky. The solar radiation input to a given location is larger when skies are clear than when skies are cloudy; for this reason the arid southwestern United States receives more solar radiation than cloudy locations at the same latitude.

During the summer months (June to September in the northern hemisphere, December to March in the southern hemisphere), the solar radiation input exhibits little variation with latitude. Locations in northern Canada receive the same input as locations in Texas. The increased duration of daylight with increased latitude compensates for the lower altitudes of the sun toward the pole. The variation in solar radiation input between different locations during the summer is caused primarily by differences in cloudiness. Where the isolines (lines connecting points of equal value) are dashed, the data are missing or incomplete. Data for Greenland are unavailable for June. (*The National Atlas of the United States of America*, 1970; and Löf, Duffie, and Smith, *World Distribution of Solar Radiation*, Report No. 21, University of Wisconsin, 1966)

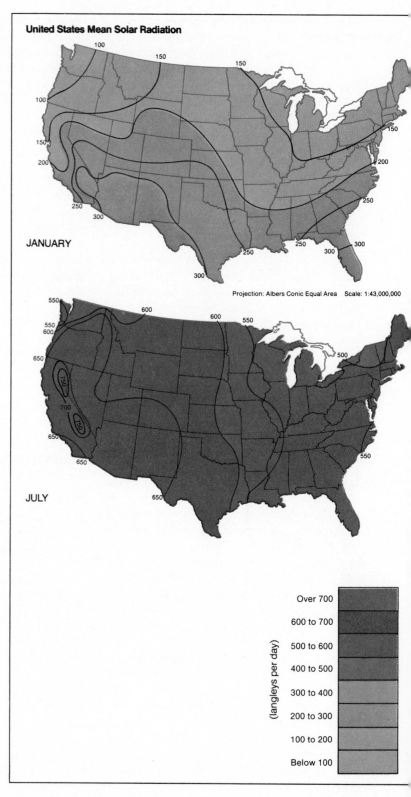

United States Mean Solar Radiation

JANUARY

Projection: Albers Conic Equal Area   Scale: 1:43,000,000

JULY

(langleys per day)

Over 700
600 to 700
500 to 600
400 to 500
300 to 400
200 to 300
100 to 200
Below 100

Global Solar Radiation

Projection: Flat Polar Quartic Interrupted and Condensed

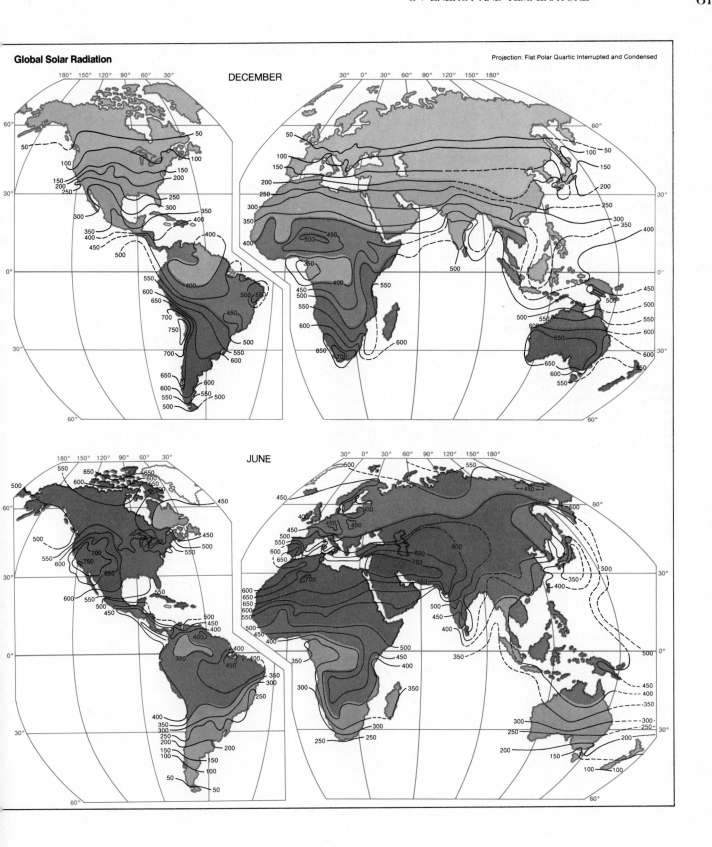

increased by the process of *convection*—the vertical rise of parcels of lower-density warm air. Convection replaces warmer surface air with cooler air from aloft, thus allowing heat flow by conduction to continue. However, when warm air moves over cooler land or water surfaces, heat flows from the atmosphere downward to the surface by conduction. This also occurs at night when the earth's surface cools by longwave radiation. Conduction in this direction chills the lower atmosphere and removes any possibility of upward convection. As shown in Figure 3.7, a net flow of 8 units of energy is transferred from the surface upward to the atmosphere by conduction and convection.

A most important energy exchange between the earth's surface and the troposphere involves the change of liquid water and solid ice to water vapor. Recall from Chapter 2 that every gram of water that is vaporized under average conditions transfers heat energy from the surface to the atmosphere. Evaporation from water surfaces of all types is a constant process that transfers heat to the atmosphere. Less obvious is the process by which plants absorb liquid water from the soil and emit water vapor to the atmosphere through pores in their leaves. This process is called *transpiration*. Since transpiration and evaporation involve similar energy exchanges and are difficult to measure separately, they are often considered jointly as *evapotranspiration*.

Figure 3.7 shows that, on the average, 22 units of energy are transferred from the earth's surface to the troposphere by evapotranspiration. This energy is stored in water vapor in the form of latent heat. It becomes sensible heat, capable of warming the surrounding air, only when the water vapor condenses back into water droplets or ice crystals elsewhere in the atmosphere.

Heating of the atmosphere by evapotranspiration, conduction, and convection equalizes the net radiation gain of the earth's surface and the net radiation loss of the atmosphere in Figure 3.7, balancing the global energy budget. A final component, plant photosynthesis, uses solar energy, but it utilizes too small a fraction to be of concern in an analysis of the energy balance of the earth-atmosphere system.

Latitudinal Differences

The energy balance of the earth-atmosphere system shown in Figure 3.7 represents an annual average for the system as a whole. If we look at specific locations on the earth, however, we discover that energy exchanges are not in balance. Figure 3.10 shows that equatorial and tropical regions receive much more solar energy than they return to space through reflection and longwave radiation. Polar regions, on the other hand, lose more energy through reflection and longwave radiation than they gain from solar radiation.

These latitudinal imbalances in energy gains and losses are the cause of the atmospheric and oceanic circulations discussed in Chapter 4. Flows in the atmosphere and oceans transport warm air, latent heat in water vapor, and warm water into the higher latitudes, and they transport colder air and water toward the equator. As Figure 3.11 indicates, there is a strong seasonal character to the poleward flow of energy.

Figure 3.10   The graph shows average annual global values of absorbed solar radiant energy and emitted longwave radiation by latitudes in the northern hemisphere for the earth's surface and the atmosphere together. Between the equator and about latitude 38°N, absorption exceeds emission, and there is a net input of energy to the earth. Poleward of 38°N, emission exceeds absorption, and there is a net loss of energy. The temperature in the lower latitudes therefore rises until the rate of heat flow toward the poles is sufficient to carry away the excess energy. (Doug Armstrong after F. K. Hare, *The Restless Atmosphere*, 1966, Hutchinson Publishing Group, Ltd.)

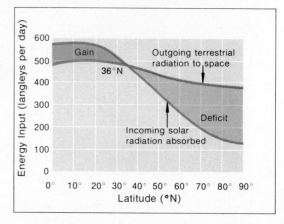

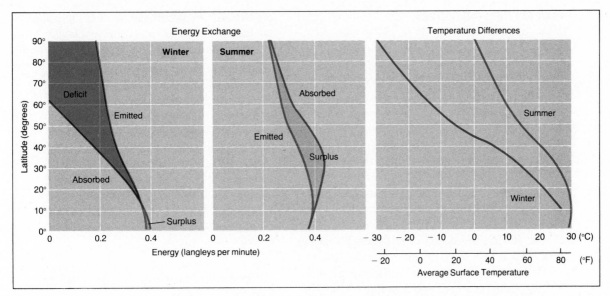

Figure 3.11 Energy exchange and temperature in the northern hemisphere. The diagram on the left shows the average rates at which the earth and atmosphere absorb solar radiant energy and emit longwave radiation to space. During the winter absorption exceeds emission only at latitudes near the equator, but during the summer the rate at which energy is absorbed exceeds the rate of emission at every latitude. Much of the excess absorbed energy warms the oceans and is stored there.

The diagram on the right shows the average temperature of the atmosphere at a height of approximately 2 meters (6.6 feet) above the surface. The temperature difference between low and high latitudes is greater during the winter than during the summer. The temperature differential helps to power atmospheric motion and a poleward flow of energy from the tropics. (Doug Armstrong after Herbert Riehl, *Introduction to the Atmosphere*, © 1972 by McGraw-Hill Book Company)

In the winter hemisphere only the zone between 20° latitude and the equator receives more energy from radiation than it loses, but in the summer hemisphere all regions gain more radiational energy than they lose. Because temperature differences between low and high latitudes are greater during the winter, the atmospheric circulation tends to be most vigorous then.

ENERGY BALANCES AT THE EARTH'S SURFACE

The large-scale energy exchanges between the sun, the earth-atmosphere system, and space have been measured with increasing accuracy in recent years by a number of specialized satellites. Our knowledge of local energy balances at the earth's surface is based on measurements for research at various experimental stations. By interpreting these data in accordance with the principles of energy exchange discussed earlier in this chapter, scientists are gaining a better understanding of how the energy balances at particular locations are affected by such factors as cloud forms, vegetation cover, local water bodies, and human activities.

A study of surface energy balances usually includes an evaluation of the radiation budget— the income and outgo of shortwave and longwave radiation at the surface. The net gain or loss of radiative energy at the earth's surface is generally called the *net radiation*, which is normally positive during the daytime (heating) and negative at night (cooling). The all-inclusive energy balance contains, in addition to the radiation budget, the other energy balance components: latent heat (evapotranspiration) and sensible heat (measurable heating of the air and soil).

Local Energy Balances

Figure 3.12 provides examples of hourly patterns of energy balance factors for two locations. In Figure 3.12a, Hancock, Wisconsin, was selected to be representative of humid midlatitude locations. During the sunny daytime hours, more shortwave and longwave radiation is received at the surface than is reflected or radiated back to the atmosphere or into space. Following sunrise, the rate of incoming solar radiation rises rapidly, resulting in a net radiation gain that peaks near solar noon, and returns to zero a few minutes before sunset. Thus, during the day, the radiation budget at Hancock is said to be positive. The daytime net radiation gain is used for evapotranspiration, atmospheric heating, and heating of the soil. The portion used in plant photosynthesis is negligible.

At night, however, somewhat more longwave radiation is lost to the atmosphere than is received from it, so the radiation budget of the cooling surface is negative. Small amounts of energy are conducted downward from the warmer atmosphere and upward from the warmer subsoil to the cooling surface. Thus the lower atmosphere is heated by the surface during the day, when the surface radiation budget is positive, and cooled by the surface at night, when the surface radiation budget is negative.

In Figure 3.12a, evapotranspiration is also seen to be closely controlled by the net radiation. This is evident in the curve for latent heat transfer due to vaporization of water. Evapotranspiration is primarily a daytime phenomenon; there is little or no vaporization of water at night. In a dry climatic region (Figure 3.12b), little soil moisture is available for evapotranspiration, and the net radiation gain goes almost entirely into sensible heat—the heating of the air and soil. On the average, equal energy inputs will cause dry regions to be warmer during daylight hours than regions with high rates of evapotranspiration.

Figure 3.13 presents annual energy balances for four contrasting locations within the continental United States. Significant differences may be observed among these locations on the basis of latitude, yet it is apparent that the balances also vary according to other environmental factors, such as moisture availability. The greatest contrast is between Madison, Wisconsin, and Yuma, Arizona. At Yuma, a desert location, very little of the net radiation gain can be used up by evapotranspiration, so nearly all is available for heating the air and soil. As a consequence, Yuma is well known as one of the hottest year-round locations in the United States. By contrast, at Madison and West Palm Beach, Florida, abundant moisture in the summer permits high evapotranspiration rates, so less energy is available to heat the soil and air. Astoria, Oregon, is intermediate between these extremes.

Differential Heating and Cooling of Land and Water

Land surfaces and water bodies respond differently to the same energy inputs. This is because the heat capacities of land and water are different. As we saw in Chapter 2, water can absorb much energy without increasing its temperature greatly, but land areas heat rapidly as they receive energy. Whereas the heat capacity of water is about 1.0, the heat capacity of most land surfaces is between 0.2 and 0.4. This means that for a given energy input a volume of ground will increase in temperature three to five times more than an equal volume of water.

On a summer day, the temperature of a soil surface exposed to the sun can easily reach 40°C (104°F). Dry beach sand and asphalt pavement may become too hot for bare feet. Most of the energy gain is concentrated near the surface, because conduction to greater depths is relatively slow. At night, the soil surface cools rapidly by longwave radiation and, in turn, cools the atmosphere immediately above by conduction. If there is little or no wind to mix the lowest layers of air, the air near the ground will become several degrees cooler than the air a few meters higher. These differences can be seen in Figure 3.14, the vertical profiles of air and ground temperatures for a fair winter day in southern New Jersey.

Surface temperatures of large bodies of water change much less during the year than do land surface temperatures. This is due not

Figure 3.12 (a) The local energy balance and its principal components during a 24-hour period are shown for Hancock, Wisconsin, a moist midlatitude location. The net radiation is the difference between the rate at which solar radiant energy is absorbed at the ground and the net loss of longwave radiation in the exchange between the surface and the atmosphere. For net radiation, the sign convention is positive (+) for a radiational gain and negative (−) for a radiational loss. During most of the day the net radiation is positive. During the night the radiation budget is negative because the net flow of radiant energy is away from the surface.

Energy cannot accumulate at the surface, and the excess or deficiency of radiant energy can be accounted for as the sum of three components: the latent heat removed by evapotranspiration of water; the sensible heat flowing from the surface into the air by conduction and convection; and sensible heat that flows into the soil from the surface by conduction. For these three components, however, the sign convention is just the opposite of that for radiation. Energy flows away from the surface are treated as positive (+), and energy flows to the surface as negative (−).

At Hancock at midday, latent heat transfer is positive because of evaporation and transpiration from plants. The small negative value of latent heat transfer before sunrise indicates condensation of water vapor on the ground as dew. Heat transfer from the ground to the air is positive at midday, which indicates that surface heat is warming the air. Overnight, the conduction of heat from the soil to the surface is negative, indicating that heat is flowing by conduction from the warmer soil to the cooler surface.

(b) A similar energy balance for a dry lake bed in southern California shows that when there is little or no soil moisture available for evapotranspiration, the net radiation is utilized for heating the air and ground. This results in greater temperature ranges between daytime and nighttime than occur in humid regions. (Doug Armstrong after William D. Sellers, *Physical Climatology*, © 1965, The University of Chicago Press)

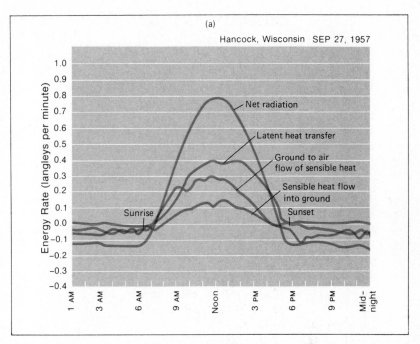

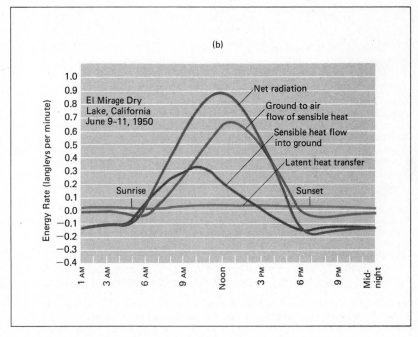

only to the greater heat capacity of water, but also to the fact that the energy gain of a water surface is distributed to much greater depths. Solar radiation penetrates water to a depth of tens of meters, and waves and currents also distribute the heat energy downward.

Local Temperature Variations

The physical principles involved in energy balances can help explain temperature variations on a local scale. Figure 3.15 shows air temperatures for a period of several days at Baton Rouge, Louisiana. Air temperature near the

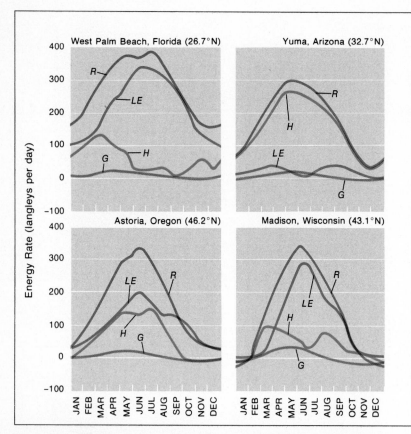

Figure 3.13 The graphs show the annual regime of the local energy balances at four places in the United States. Madison, Wisconsin, is representative of humid midlatitude regions with low solar radiation income during winter, West Palm Beach, Florida, of humid subtropical regions with moderate solar radiation income during winter, Astoria, Oregon, of humid and cloudy maritime climates with much winter cloudiness, and Yuma, Arizona, of hot desert regions with little cloudiness.

R is the radiation budget (solar energy received vs. longwave energy emitted). *LE* is the energy removed by evapotranspiration (latent heat transfer). *H* is the energy exchanged between the ground and the air by conduction and convection. *G* is energy conducted into the ground from the surface. The values of *R* are similar at all four locations during summer, but *R* is much smaller at Astoria and Madison during winter because of fewer hours of daylight and low solar altitudes; during winter in Madison, *R* is slightly negative.

LE is usually the largest component of *R* during the growing season. *LE* is zero at Madison during winter, when low temperatures inhibit evapotranspiration, and there is no net radiation gain. *LE* is low throughout the year at Yuma because of the absence of water for evapotranspiration, and most of the net radiation gain goes into heating the air. In each location, energy tends to be conducted from the surface into the soil during spring and back to the surface during fall. (Doug Armstrong after William D. Sellers, *Physical Climatology,* © 1965, The University of Chicago Press)

ground is an indication of surface temperature. In the first three days, clear weather permitted ground and air temperatures to rise markedly in the morning and to fall sharply in the afternoon and night. The lowest temperatures occurred just after sunrise, and the highest followed in the early afternoon. Although the solar radiation input peaks near noon and then declines, cooling begins some time in the afternoon, when radiation losses begin to exceed gains. During cloudy weather, the temperature changes much less throughout the day. Clouds reduce the incoming solar radiation during the day, and at night they trap outgoing longwave radiation and return much of it to the surface, reducing the daily temperature variation.

A large body of water also tends to reduce daily temperature ranges over adjacent land areas, because the temperatures of large water bodies change little from day to day. Onshore flows of marine air help maintain almost uniformly mild temperatures at coastal cities such as San Francisco and Seattle, in comparison with cities in continental interiors such as Kansas City.

Monthly temperature variations reflect much the same influences as daily variations. Figure 3.16 depicts winter and summer temperatures at San Diego, California, a coastal city; Elko, Nevada, an interior city in an arid region; and Cleveland, Ohio, a midcontinent city with frequent cloud cover. The graphs show the *distribution* of temperatures for January and July, which is found by measuring the temperature every hour and plotting the number of hours each temperature occurred.

Because of the reduced energy input in winter, the January temperature distributions in all three cities fall well below the distributions for July. The January and July distributions for San Diego show the most overlap and the least separation between peaks because coastal temperatures tend to be relatively uniform throughout the year. The monthly temperature distributions at Cleveland and Elko cover a greater range than those at San Diego because of the greater daily range of temperatures in the continental interior. The wide range of the July temperatures for Elko reflects the city's high elevation (1,500 meters; 5,000 ft) and desert lo-

Figure 3.14 This figure illustrates steep temperature gradients immediately above and below the ground surface during fair days with little wind. The data were taken at an experimental site near Seabrook, New Jersey, during December 5, 1956. Between 9:00 A.M. and 1:00 P.M., the warmest temperatures were within the first 5 cm above the surface, but during late afternoon, overnight, and early morning, the coldest temperatures were immediately above the surface, which was losing heat by longwave radiational cooling.

The profiles should also be interpreted as a time sequence beginning with 5:00 A.M. when the temperature profile in the air, at the surface, and down into the soil, resulted from overnight radiational cooling of the surface. By 9:00 A.M. the surface and the air above are warming, but little heat has penetrated the soil. By 1:00 P.M. the surface and the air have warmed a great deal, and significant warming by conduction has penetrated as far as 20 cm (8 in.) into the soil. Late afternoon and overnight cooling in association with a surface net radiation loss can be followed through 5:00 and 9:00 P.M. to 1:00 A.M. and, assuming no change in heat storage for the entire diurnal cycle, back to 5:00 A.M. again. (Vantage Art, Inc. after J. R. Mather, *Climatology: Fundamentals and Applications,* © 1974 by McGraw-Hill Book Company)

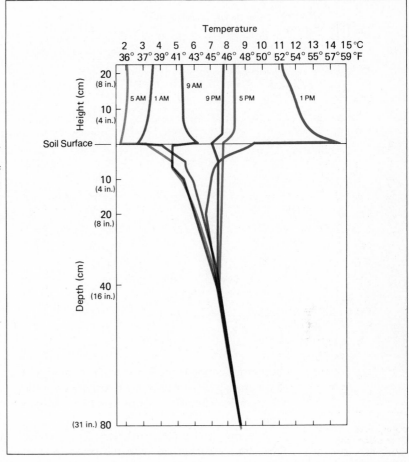

cation with minimal cloud cover, which are conducive to hot days and cool nights.

Average global temperatures for January and July are shown in Figure 3.17, and seasonal temperatures for the United States are depicted in Figure 3.18. The global maps in Figure 3.17 illustrate seasonal displacements of the regions of highest temperatures into the northern and southern hemispheres. Subtropical dry regions in the interiors of the continents are warmer than equatorial regions regardless of whether the surfaces are land or water. In the winter hemisphere, the coldest temperatures are always found over continental regions; on the map this is especially evident for January over continental regions of the northern hemisphere.

The four maps of the 48 states of the continental United States in Figure 3.18 illustrate the seasonal march of isotherms northward in spring and equatorward in fall. The isotherms also reflect the geographic patterns of elevation and proximity to persistent airflow from coterminus or oceanic sources. Good examples are the low temperatures in January in North Dakota compared to the mild temperatures over coastal Washington at the same latitudes.

Modifications of Energy Balances

Changes in any of a number of factors can alter local energy balances. Cloudiness, atmospheric pollution, surface albedo, irrigation, vegetation—all of these influence the utilization of radiant energy. Results of human activities,

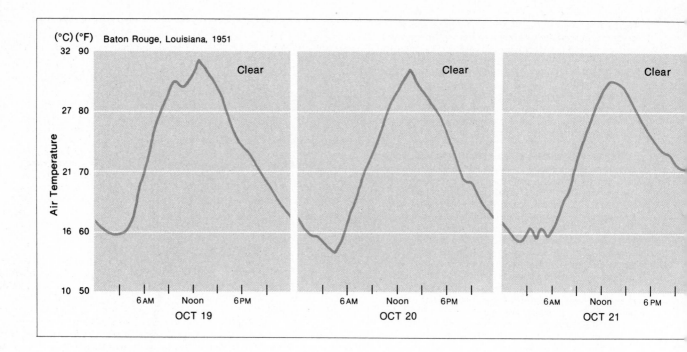

both unintentional and planned, have modified local energy balances, giving rise to measurable changes in climate in both rural and urban areas.

On the local scale, gardeners and orchard growers are concerned with the prevention of frost damage to plants. On clear nights the ground rapidly loses heat to the atmosphere by longwave radiation, and return radiation from the atmosphere is minimal. During early spring and late fall, such conditions can cause temperatures near the ground to drop below the freezing point. Orchard growers often use large fans or even helicopters to disturb the cold air near the ground and mix it with warmer air above. These devices have largely replaced orchard heaters, or smudge pots, which produce a blanket of smoke to absorb and reradiate longwave radiation back to the ground. Radiant heaters are sometimes used to protect plants in courtyards and around patios. Often ornamental plants and small citrus trees are covered with plastic sheeting to reduce longwave radiation losses that lead to frost damage.

Large urban areas produce their own distinctive climates by modifying local energy balances. Especially in winter, cities can be as much as 7° to 8°C (12° to 14°F) warmer than the surrounding countryside. Towering buildings and canyon-like streets trap above-average amounts of radiant energy, even when the sun is low in the sky. Asphalt, concrete, and bricks store sensible heat that would be used for evapotranspiration in a natural or agricultural environment. The burning of fossil fuels in homes, factories, and automobiles adds to the radiant heat input; and the dust, smoke particles, and pollutants in the air trap outgoing longwave radiation and return it to the surface. These effects combine to produce the phenomenon of the "urban heat island" (Figure 3.19).

Human modification of energy balances did not begin with the Industrial Revolution. Ancient agricultural practices, such as tropical slash-and-burn agriculture (see Case Study following Chapter 10), midlatitude forest cutting, and overgrazing in subtropical latitudes, have altered surface albedo, the water-holding capacity of soils, and evapotranspiration rates over vast areas. Removal of natural vegetation for agricultural purposes has allowed the wind to lift soil particles high into the troposphere, causing dust storms that darken the sky over areas of thousands of square kilometers.

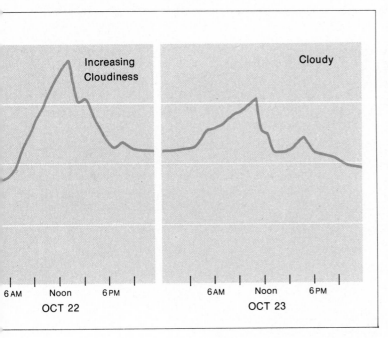

Figure 3.15 The temperature records at Baton Rouge, Louisiana, for October 19 to 23, 1951, illustrate the daily variation in air temperature as the ground heats and cools. The temperature rises in the morning, when the radiation budget becomes positive, and reaches a peak in the early afternoon. The radiation gain declines in the afternoon, and net radiation becomes negative at night, causing the temperature to drop.

On October 19 to 21, skies were clear, but on October 22 and 23, skies were cloudy. On a cloudy day, the daily variation in temperature is less extreme than on a clear day because clouds reduce solar energy input during the day and at night reduce surface heat loss caused by longwave radiation. (Doug Armstrong)

Figure 3.16 The average temperature distributions during January and July are shown for three cities in different climatic regions over the five-year period 1935 through 1939. The temperature distribution curves show the number of hours during the given month that the specified temperature was recorded; in San Diego, California, in July, for example, a temperature of 20°C (68°F) was recorded in 57 separate hours.

A narrow temperature distribution, which San Diego, a coastal city, shows, means that the temperature remains comparatively constant. At Elko, Nevada, in July, in contrast, the temperature distribution is broad, reflecting the hot days and cool nights of the cloudless desert. The distributions are moderately broad for Cleveland, Ohio, a midcontinent city with a moist climate. (Doug Armstrong after Arnold Court, *Journal of Meteorology, 8,* 1951, American Meteorological Society)

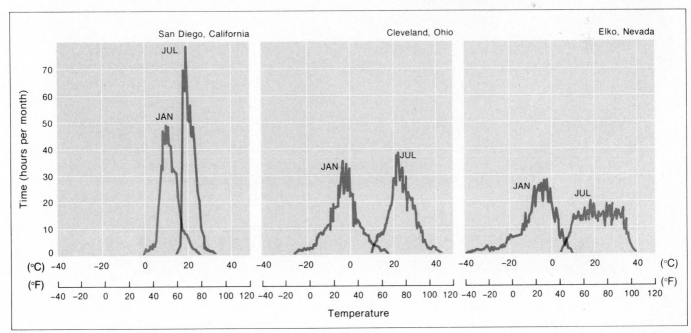

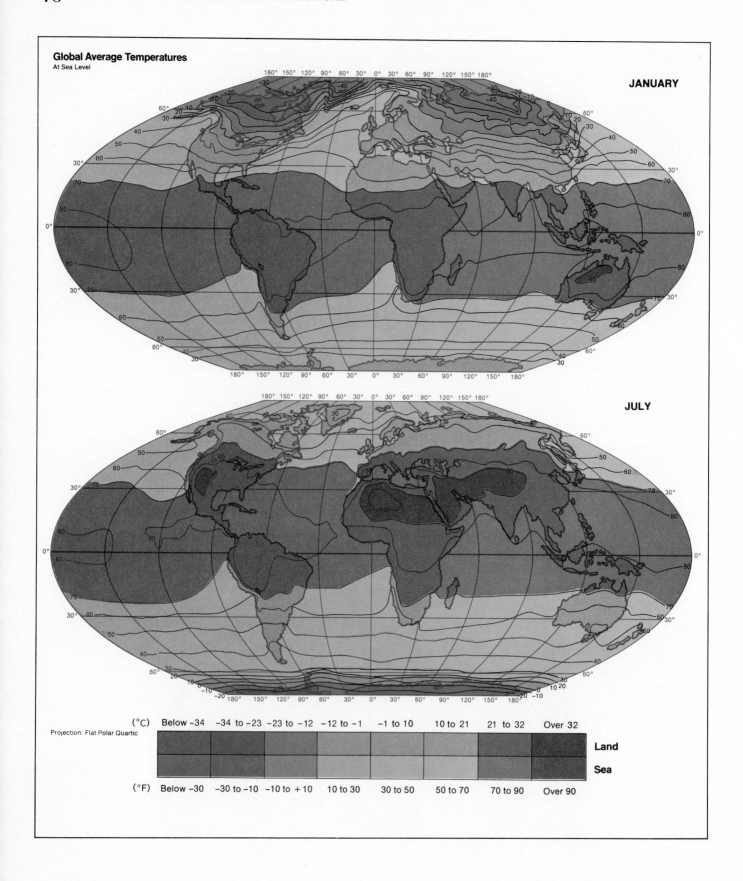

Global Average Temperatures
At Sea Level

JANUARY

JULY

Projection: Flat Polar Quartic

(°C)	Below −34	−34 to −23	−23 to −12	−12 to −1	−1 to 10	10 to 21	21 to 32	Over 32	
									Land
									Sea
(°F)	Below −30	−30 to −10	−10 to +10	10 to 30	30 to 50	50 to 70	70 to 90	Over 90	

Figure 3.17 (opposite) Global temperature distributions are shown for January and July. Temperatures have been reduced to approximate sea-level values by applying a correction for the decrease in temperature with increased altitude. In winter (January in the northern hemisphere, July in the southern hemisphere), temperatures generally depend on latitude and decrease regularly from the equator toward the poles. Note, however, the continental and maritime effects on temperature over North America and the North Atlantic during January. In summer, temperatures depend less strongly on latitude and more on land and water contrasts (see Figure 3.18).

Note that blue tones on this map indicate below-freezing temperatures, and "warmer" colors indicate above-freezing temperatures. The extreme poles are not shown on the flat polar quartic map projections used here. (Andy Lucas and Laurie Curran after Glenn Trewartha, *Introduction to Climate*, © 1968, McGraw-Hill Book Company and *Goode's World Atlas* © 1970, Rand McNally & Company)

The modern industrial age has intensified the human impact on energy balances. Jet aircraft leave exhaust products and condensation trails high in the troposphere and even in the lower stratosphere. Jet condensation trails resemble some natural clouds and may significantly reduce shortwave radiation in regions of heavy air traffic. Perhaps more disturbing are findings that various synthetic chemicals, such as the chlorofluorocarbons in spray-can products, and the methyl chloroforms in solvents, may have the potential to destroy the ozone layer in the stratosphere that prevents dangerous amounts of ultraviolet radiation from reaching the earth's surface.

At the surface artificial ponds and reservoirs, paved areas, forest removal, and the drainage of marshes all have altered albedo and evapotranspiration patterns. Data from satellite surveys of albedo, land use, and plant cover, and of the degree of cloudiness over the entire surface of the earth, are required to assess the effects of these changes. The launching of the Earth Resources Technology Satellite, ERTS-1, in the summer of 1972, was an important first step in this program, which has continued through data transmissions from a series of orbiting satellites, such as LANDSAT 1, 2, and 3, and SEASAT. An even more sophisticated satellite, LANDSAT 4, was launched in July 1982. The thematic matter (TM) on LANDSAT 4 will permit even more practical evaluations of environmental problems such as relationships of land use change to increased flooding in urban regions.

SUMMARY

The sun provides the energy for most of the natural processes that affect the earth's surface. Solar energy is transmitted through space in the form of electromagnetic radiation, which has a range of wavelengths. Solar radiation consists principally of ultraviolet, visible, and infrared radiation. Most solar radiation has a wavelength of less than 4 micrometers and is considered to be shortwave radiation.

The intensity of the solar beam at the top of the atmosphere is expressed as the solar constant. Daily amounts of solar radiation received at the top of the atmosphere depend on the number of daylight hours and the angle at which the sun's rays strike the atmosphere throughout the day. These factors are determined by earth-sun relationships. The most important aspects of these relationships are the distance of the earth from the sun and the inclination of the earth's axis of rotation with respect to the plane of its path around the sun. The inclination of the earth's axis is the cause of the seasons and controls the angle of the solar beam and length of day at various latitudes. Almost half of the solar radiation striking the atmosphere penetrates to the earth's surface. The rest is reflected back to space or is absorbed by the earth's atmosphere. Absorbed shortwave radiation is later re-emitted in the form of longwave radiation.

For global energy income and outgo to remain in balance, the earth-atmosphere system must return the same amount of energy to space as it receives from the sun. This outgoing

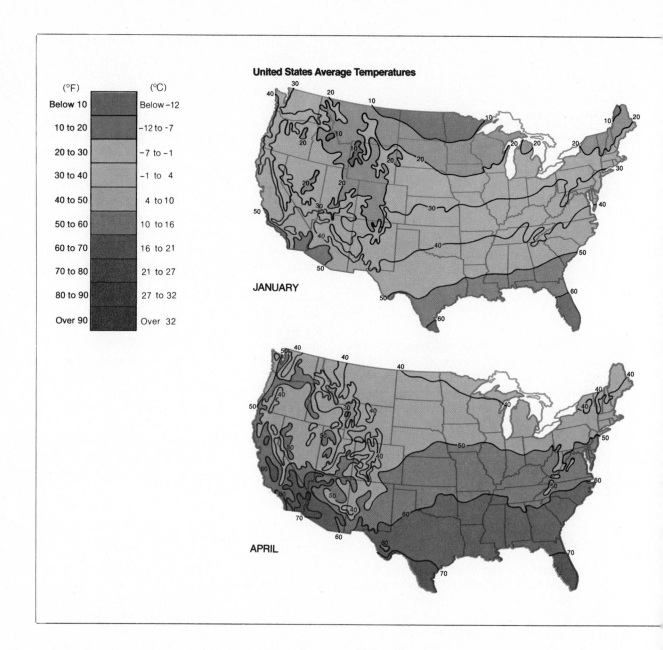

United States Average Temperatures

(°F)	(°C)
Below 10	Below −12
10 to 20	−12 to −7
20 to 30	−7 to −1
30 to 40	−1 to 4
40 to 50	4 to 10
50 to 60	10 to 16
60 to 70	16 to 21
70 to 80	21 to 27
80 to 90	27 to 32
Over 90	Over 32

JANUARY

APRIL

radiation consists of reflected shortwave radiation and longwave radiation from the earth and its atmosphere. The ratio of reflected to received solar radiation represents the earth's albedo. Water vapor and carbon dioxide molecules in the lower atmosphere absorb much of the longwave radiation emitted by the earth, heating the atmosphere from below. Longwave radiation re-emitted by the atmosphere is a substantial energy input to the earth's surface, preventing steep falls in overnight temperatures and maintaining high average surface temperatures. The earth's surface, therefore, gains energy not only directly from the sun, but also indirectly from its own atmosphere.

The earth's surface loses energy by longwave radiation, by evapotranspiration and latent heat loss, by conduction and convection of heat into the atmosphere, and by conduction of heat from the surface down into water bodies

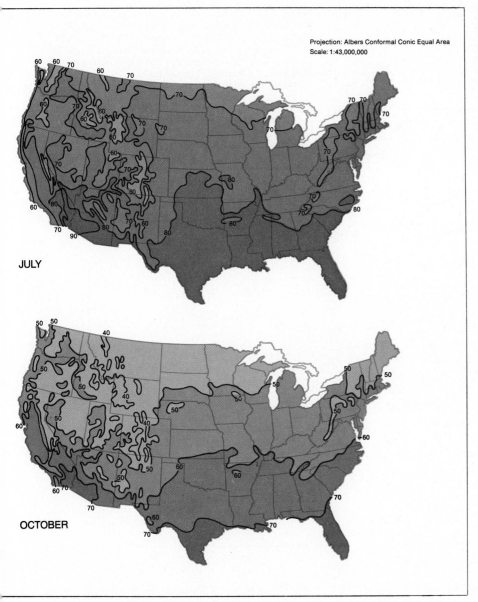

Projection: Albers Conformal Conic Equal Area
Scale: 1:43,000,000

JULY

OCTOBER

Figure 3.18   United States temperature distributions (in °F) are shown for selected months. In January, temperatures decrease regularly with increased latitude. In July, the variation with latitude is less marked because the input of solar radiant energy is relatively independent of latitude during the summer (see Figure 3.9). (Andy Lucas after *The National Atlas of the United States of America*, 1970)

and the soil. The daily variation of air temperature near the surface can be interpreted in terms of the local energy balance and the differential heating of land and water. Local energy balances have been modified significantly by the results of human activities, both deliberate and unintentional. These activities and their consequences are now being monitored by orbiting satellites such as LANDSAT and SEASAT.

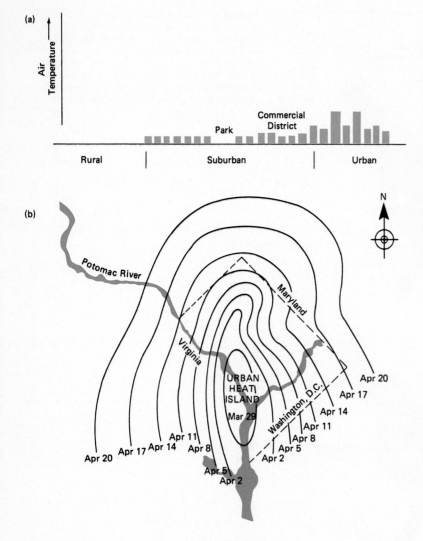

(a)

Figure 3.19 All metropolitan regions have urban heat islands where temperatures are significantly higher than over surrounding rural regions. Heat islands are due to complex interactions between local energy balances and the concentrated release of heat associated with the burning of fossil fuels.

(a) An idealized temperature cross section over an urban region. (After T. R. Oke, *Boundary Layer Climates*, 1978, Methuen)

(b) Average dates of the latest freezing temperature in spring in the Washington, D.C., metropolitan region. The longer freeze-free season in the core of the urban heat island is due to average winter minimum temperatures 3°C (7°F) warmer than in the northern suburbs. (After Clarence A. Woollum, *Weatherwise*, vol. 17, 1964)

REVIEW QUESTIONS

1. Describe the electromagnetic spectrum. What is the relation of temperature to wavelength?

2. What factors determine the amount of solar radiation reaching the top of the earth's atmosphere?

3. Describe the variation of day length from pole to pole at the winter solstice.

4. Describe how the atmosphere intercepts and interacts with incoming solar radiation.

5. Discuss the various ways the surface of the earth loses energy.

6. What is the role of the atmosphere in heating the earth?

7. What local environmental conditions modify a region's energy budget?

8. In energy-budget terms, why does the maximum daily air temperature tend to occur between 2:00 and 6:00 P.M.?

9. Compare the utilization of net radiation in humid and in dry climates.

10. Explain the effects of landmasses and oceans on the seasonal regimes of air temperatures.

11. What are various ways in which the energy balance of the earth-atmosphere system might be modified by human action?

APPLICATIONS

1. Using Figure 3.4, plot the annual regime of solar radiation at the top of the atmosphere at the latitude of your campus. Compare this with the regimes at latitudes 15 degrees to the north of your location and 15 degrees to the south of your location. Explain the differences in the three regimes.

2. Estimate how the landuse and landcover of your campus area have changed from pre-settlement days to the present. How would these landscape changes have affected the albedo of the campus area as a whole?

3. Over which regions of the earth would you expect the greatest seasonal changes in albedo? Explain.

4. Local daily temperature highs and lows are usually given on radio and TV and in newspapers: highs referring to late afternoon, and lows occurring in early morning hours. Keep track of the daily temperature range and the degree of cloud cover over a month-long period. How much do varying degrees of cloudiness affect the daily temperature range? Are other factors involved?

5. On a clear evening with no breeze, record the variations in temperature in different areas of your campus. You will need a thermometer that reponds quickly to temperature changes (check with your instructor). Explain the pattern of temperature variations you measure.

6. On both a clear day and a cloudy day, make a record of hourly temperatures at an unshaded ground surface and compare these readings with hourly air temperatures 5 feet above the ground. Make the same experiment measuring temperatures on bare ground and grassy surfaces. Explain your results.

FURTHER READING

Gedzelman, Stanley David. *The Science and Wonders of the Atmosphere.* New York: John Wiley (1980), 535 pp. This beautifully illustrated book emphasizes introductory level explanations of physical processes. Chapters 3, 4, and 8 include very helpful examples of earth-sun relationships, radiation laws, and atmospheric optics.

Geiger, Rudolf. *The Climate Near the Ground.* (Tr. of 4th German ed.) Cambridge, Mass.: Harvard University Press (1965), 611 pp. This classic stresses the results of field studies on each of the continents, with particular emphasis on radiation and temperature. Geiger is sometimes called the "father of microclimatology."

Mather, John R. *Climatology: Fundamentals and Applications.* New York: McGraw-Hill (1974), 412 pp. Chapter 2 includes basic discussions of radiation and temperature, instrumentation and data, as well as their applications and limitations.

Miller, David H. "A Survey Course: The Energy and Mass Budget at the Surface of the Earth." Publ. No. 7, Comm. on College Geog., Assoc. of American Geog. (1968), 142 pp. This very useful monograph is organized into study units that proceed from energy exchange processes through local energy budgets to regional synthesis. Emphasis is placed on professional papers from almost every region of the globe.

Oke, T. R. *Boundary Layer Climates.* London: Methuen (1978), 372 pp. This recent text is a more advanced analysis of surface energy exchanges. Emphasis is placed on local and geographical differences, and the text is an excellent primer for students who want to do special studies.

Sellers, William D. *Physical Climatology.* Chicago: University of Chicago Press (1965), 272 pp. This work focuses almost entirely on fundamental analyses, in both descriptive and mathematical forms, of the radiation and energy budgets at the earth's surface.

Trewartha, Glenn T., and **Lyle H. Horn.** *An Introduction to Climate.* 5th ed. New York: McGraw-Hill (1980), 416 pp. Focus on geographical distributions of solar radiation and temperature is found in Chapters 2 and 8-12.

CASE STUDY

Solar Energy

Suppose you leave your garden hose sprawling across a sun-baked patio or driveway on a hot summer day. When you come home at 5 in the afternoon, you rush out and pick up the hose to water some wilting plants. But you may do more harm than good. The water in the hose has been heated to more than 50°C (122°F) by absorption of solar radiation, and the hot water may be lethal to your tender plants.

At the same time, the soles of your feet will burn if you step bare-footed out on the blacktop pavement of the street. Black-top pavement and dark automobile tops can heat to more than 60°C (140°F). People have long been aware of these straightforward examples of solar heating, but, with few exceptions, there has been little attempt to harness this energy source. Only since the rapid rise in energy costs have people begun to pay serious attention to the potential uses of solar energy.

Solar systems for heating water and buildings represent the simplest and least expensive adaptations to solar energy. On a cold day the family cat will almost always take advantage of winter sun streaming through the glass panes of a window. The glass allows most of the solar radiation to enter the room, where it is absorbed by carpet and cat alike. On the other hand, most longwave radiation emitted from within the warm room is absorbed by the glass and reradiated back into the room—a simple application of the greenhouse effect discussed in Chapter 3.

Commercial systems for heating water by solar energy have been developed and are becoming common in Florida, California, and some foreign regions. These systems usually consist of a water-filled network of tubing coated with black paint and set into a sloping roof facing toward the south. The tubing is usually covered by glass and is set over dark metal sheets heavily insulated from the attic. The greenhouse effect is maximized. The heated water is then pumped to a storage tank, where it may be used directly, or it is circulated through a large volume of gravel or stone surrounding the tank, which serves for heat storage overnight and during cloudy days. This technology has also been adapted to heat air circulated through homes or buildings.

Direct solar heating normally supplements conventional heating systems. Initial costs are considerable, but operating costs are low. As the technology improves, costs should decrease, making solar heating increasingly competitive with conventional systems. Solar radiation is, of course, a continuously renewable source of energy, unlike gas, oil, or coal, and it is nonpolluting. It is especially adapted to Florida and the Southwest, where there is an abundance of solar radiation and an absence of large trees. In the more humid Northeast and the lower Mississippi River valley, the fossil energy saved by direct solar heating in the winter may be offset during hot humid summers by undesired solar heating of the structure. The trade-off will have to be compared against the much lower air-conditioning costs of one-story houses nestled within the shade of large trees. To use solar energy to the greatest advantage, the trees should be deci-

This experimental solar house at the University of Delaware has been designed so that solar heating and photovoltaic cells in the panels on the roof provide more than two-thirds of the energy needs of a family. (R. A. Muller)

duous—giving shade in the summer, but losing their leaves during the winter—so that the winter sun may be absorbed by the building.

By use of multiple reflectors and mirrors, it is also possible to focus the solar radiation striking a number of receiving surfaces onto a single point, producing very high temperatures in solar furnaces. Probably the best known is the one at Odeillo in the French Pyrenees. This furnace can achieve temperatures higher than 3300°C (6000°F), and it is used for experiments requiring high temperatures in a few moments.

The ultimate objective of solar energy utilization is the conversion of sunlight into electricity that can be fed into regional power grids. The light meter of a camera includes photovoltaic cells that convert solar radiation into electrical current; the deflection of the meter needle is related to light impinging on the cells, resulting in an electrical current. Larger amounts of direct-current electricity can be generated by multiple cells of semiconductor materials, such as silicon cells, connected in series. This technology was originally developed for the satellite and space programs, where the solar cells have proven to be reliable and long lasting, but relatively expensive and inefficient. At the present time, they can convert to electrical current only about 10 to 13 percent of the solar energy they receive.

A number of more ambitious schemes to utilize solar energy have been proposed in recent years. One such scheme involves construction of giant power-station satellites consisting of banks of solar cells that would intercept solar radiation in space. The converted solar energy would then be beamed to receiving stations on the earth by means of microwave transmission. Another proposal calls for solar farms in the deserts of the American Southwest. The solar farms would cover thousands of acres of desert land, producing large amounts of electricity—with the environment impact probably less than mining and drilling for fossil fuels. One such facility already is operating near Barstow, California.

Incidentally, most other energy sources are directly or indirectly related to solar energy. Wind and wave energy and hydroelectric power are all direct products of the earth's climate—all driven by variations in solar radiation over the surface of the earth. Wood products are the result of the photosynthetic conversion of solar radiation, carbon dioxide, and water; and the fossil fuels such as gas, oil, and coal also result from photosynthesis and represent solar radiation stored for millions of years within the earth's crust.

Direct solar heating and the conversion of solar radiation to electricity are especially appealing because total global energy consumption is far less than the solar energy received each day by the earth-atmosphere system. Pollution is minimal and the solar energy is a continuously renewable resource. But because solar radiation intensities are so low, the technologies of collection, concentration, and storage are more costly than fossil fuel utilization at the present time. However, there is no doubt that solar energy will be a growth industry in future years.

Snow Storm—Steam Boat off a Harbour's Mouth Making Signals in Shallow Water and Going by the Land by William Turner. (The Tate Gallery, London)

A furious storm at sea emphasizes the complex inter-actions of the atmospheric and oceanic systems. Winds drive the surface ocean currents, and the ocean temperatures influence the circulation patterns of the atmosphere.

Balloonists can control only their vertical rise or descent. But in 1979 and 1980 balloon trips were completed across the widths of both the Atlantic Ocean and North America. How can balloonists, with no control over their horizontal direction, attempt to cross a continent or an ocean? The answer is easy: the motion of the atmosphere decides the route. In the middle latitudes all long-distance balloon trips go from west to east, because the general motion of the atmosphere is in that direction. It is as impossible to go the other way as it is for a raft to drift up a river instead of down it.

Ocean waters too have directional characteristics. A capped bottle that is thrown into the sea off the Florida coast would probably come ashore in Iceland, Norway, or Ireland; it would never reach Brazil, or even nearby Cuba. To send a message to Cuba in a floating bottle, the best starting point would be Morocco; and objects floating ashore in Brazil would come from South Africa or Angola.

On a global scale, both the air and the seas move in directions that vary somewhat from season to season but are dependable from year to year. The large-scale semipermanent pattern of atmospheric flow, which drives the oceanic circulation, is one of the more important facts of physical geography and is known as the *general circulation.*

The general circulation of the atmosphere and the oceanic circulation that results from it are the principal mechanisms by which energy is transferred from equatorial and tropical regions having net solar energy gains to high-latitude regions having net energy losses to space. These flows of air, carrying both latent and sensible heat, along with the ocean water, carrying sensible heat, maintain the earth's thermal balance at all latitudes. The general circulation also delivers to the continents much of the water that evaporates from the oceans. It determines local weather and is the foundation for the entire pattern of global climates. This chapter explains the general circulation of the atmosphere and the resulting oceanic circulation.

4
General Atmospheric and Oceanic Circulation

Forces Causing Atmospheric Motion
The Pressure Gradient Force
The Coriolis Force and the Geostrophic Wind
The Force of Friction

The General Circulation of the Atmosphere
Pressure Patterns and Winds on a
    Uniform Earth
Observed Pressure and Wind Systems
The Upper Atmosphere and Jet Streams

The Oceans
The Oceans: Energy Banks
Surface Currents
The Deep Circulation and Upwelling

FORCES CAUSING ATMOSPHERIC MOTION

The atmosphere is in ceaseless motion, whether churning with furious storms or drifting so slowly that hardly a breeze is felt. Although the details of atmospheric circulation are complex, the general features can be understood by looking at the forces involved.

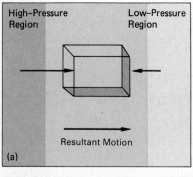

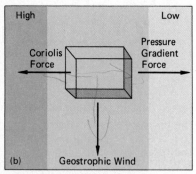

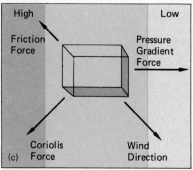

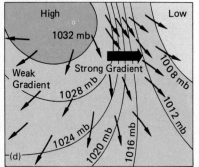

Figure 4.1   (a) The pressure gradient force pushes a parcel of air away from high pressure toward low pressure. If no other forces were present, the parcel would move toward the low-pressure region.

(b) A parcel of air aloft moving horizontally on the rotating earth experiences a Coriolis force and a pressure gradient force. The Coriolis force acts at right angles to the direction of motion of the parcel, and will continue deflecting the parcel of air until it is moving parallel to the isobars. Then the pressure gradient force and the Coriolis force are equal, and no further deflection or change of speed occurs. Air motion parallel to isobars is known as *geostrophic wind;* the speed of the geostrophic wind is highest where the pressure decreases most rapidly with distance. This diagram is drawn for the northern hemisphere. How would it appear for the southern hemisphere?

(c) Friction acts on air parcels moving near the ground, reducing wind speeds. Because the Coriolis force is proportional to the wind speed, the deflection is less than for the geostrophic winds above. Surface winds, therefore, flow at angles across the isobars indirectly from higher pressure toward lower pressure.

(d) Isobars are lines on weather maps connecting places with equal atmospheric pressure (adjusted to sea level). In this diagram pressures are shown in millibars (mb). Surface wind speeds and directions are determined by the interactions of the pressure gradient force, the Coriolis force, and the friction force, with wind speeds strongest where isobars are packed most closely together. (Vantage Art, Inc.)

Any force may be thought of as a push or a pull. According to the fundamental laws of motion developed in the seventeenth century by the English physicist Isaac Newton, a moving object's speed and direction of motion do not change unless a force is exerted on the object. Horizontal motion is influenced only by forces pushing or pulling in a horizontal direction, and vertical motion is influenced only by forces acting in the vertical direction. Although the force of gravity pulls every parcel of air downward toward the earth, gravity has no direct effect on the horizontal motion of air. The forces that act on a parcel of air moving horizontally in the atmosphere are the pressure gradient force, the Coriolis force, and friction.

The Pressure Gradient Force

Recall from Chapter 2 that differential surface heating on the earth results in small horizontal differences of temperature, air density, and atmospheric pressure. Pressure is force per unit area. If the pressure on one side of a parcel of air is greater than the pressure on the opposite side, the parcel will be pushed toward the region of lower pressure. The greater the difference in pressure, the greater the net push and the more rapid the resulting atmospheric motion, or wind. This "push," caused by the horizontal difference in pressures, is called the *pressure gradient force* (Figure 4.1a).

On a weather map, meteorologists represent atmospheric pressure patterns by lines called *isobars.* Isobars connect locations on the map that have equal atmospheric pressures (see Figure 4.1b). The pressures shown do not correspond to measured values; they have been corrected to sea-level values to compensate for the lower pressures measured at weather stations at elevations above sea level. The pressure gradient force is at right angles to the isobars and is strongest where the isobars are most closely spaced—that is, where the pressure changes most rapidly.

The Coriolis Force and the Geostrophic Wind

If it were not for the rotation of the earth, winds would simply follow the pressure gradients. But the earth's rotation complicates the motion of the atmosphere. Except over the equator, air moving down the pressure gradient is deflected, or turned, from straight paths when viewed from the surface of the earth.

Figure 4.2a shows the initial trajectories of rockets launched to the north and to the south from the central United States. As the rockets proceed toward the north and south as viewed from space, the surface of the earth rotates and turns underneath them; in other words, the east-west and north-south system of parallels

Figure 4.2 The Coriolis force is associated with the rotation of the spherical grid or parallels and meridians as the earth rotates on its axis from west to east.

(a) Rockets are launched from the central United States to either the north or south. The diagram on the right, for some time later, shows how the grid of parallels and meridians rotates under the rockets. From the earth's surface, it would appear as if the rockets had been deflected into curving trajectories to the right of the original north and south trajectories.

(b) These diagrams illustrate similar deflections to the right when the rockets are launched into east or west trajectories.

The Coriolis force affects the paths of both moving air and water over the surface of the earth. When looking downwind, for example, the deflection is to the right in the northern hemisphere and to the left in the southern hemisphere. (From Joe R. Eagleman, *Meteorology: The Atmosphere in Action.* New York: D. Van Nostrand Co., 1980)

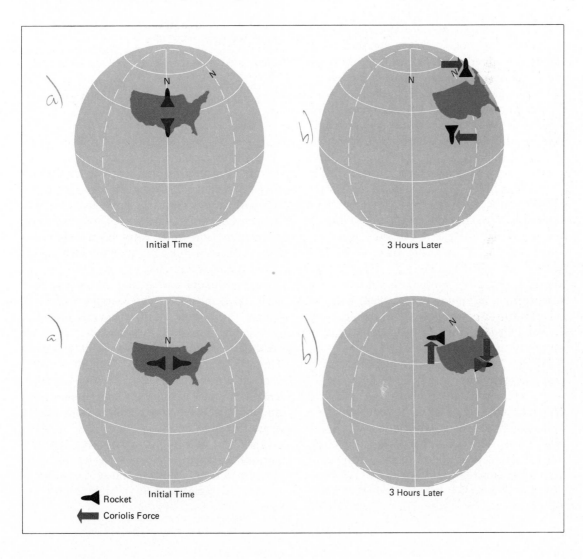

and meridians turns underneath the rockets, or *the moving air.* The right-hand diagram in the figure shows that from the earth's surface, the rockets, or the moving air, appear to be deflected to the right, as you look downwind. Figure 4.2b shows that there is a similar apparent deflection to the right when the moving air flows initially toward the east or west.

This deflection is known as the *Coriolis force* after Gaspard Coriolis, the nineteenth-century French engineer who first explained the phenomenon mathematically. Even though no real force is operating, the apparent deflection resulting from the earth's rotation affects all objects, such as air and ocean water, moving freely over the earth's surface. A correction for the Coriolis force must be added when aiming missiles and even long-range artillery.

The Coriolis force acts at right angles to the direction of the motion. In the northern hemisphere the apparent deflection is to the right and in the southern hemisphere to the left, so that the pattern of deflection in the two hemispheres is symmetrical. It is important to remember that deflection must always be thought of in terms of looking downwind in the direction toward which the air is moving. Air flowing toward the equator is turned to the west in both hemispheres, and air moving toward the poles is turned to the east. Likewise, air moving to the west is turned toward the poles, and air moving toward the east is turned toward the equator. Figure 4.3 indicates that the Coriolis force is zero at the equator. The deflection becomes significant in subtropical latitudes, and increases to a maximum at the poles. The Coriolis force is also proportional to the speed of the moving air, so that the deflection is greater with stronger winds.

Aloft the Coriolis force is equal to the pressure gradient force, causing air to flow at right angles to the pressure gradient force and parallel to isobars (Figure 4.1c). Meteorologists refer to airflow that is parallel to isobars as *geostrophic wind.* The circulation of upper-level air, above the frictional effects of the earth's surface, approximates geostrophic flow. Technically, the geostrophic wind is limited to situations with straight isobars, but the airflow aloft

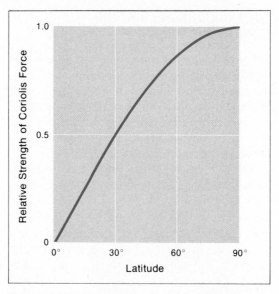

Figure 4.3 This diagram shows that the relative strength of the Coriolis force increases from zero at the equator to a maximum at the poles. Between 10° latitude and the equator the Coriolis force is of little meteorological significance, but by 30° of latitude, the deflection is 50 percent of the maximum at the poles for a given wind speed. (Doug Armstrong)

The geostrophic winds of the upper atmosphere are especially important to meteorologists concerned with weather forecasting, for they influence surface air movements, as will be seen later in this chapter. Upper air measurements of pressure, wind speed, and wind does maintain a direction approximately parallel to curved isobars also.

In the middle and higher latitudes of the northern hemisphere, the Coriolis deflection causes air to flow in a clockwise direction around high-pressure areas and in a counterclockwise direction around low-pressure areas (Figure 4.1b and Figure 4.4). Using these facts, the Dutch meteorologist Buys Ballot in 1857 described the relationship between pressure systems and the direction of air flow aloft in simple terms. Buys Ballot's Law states that if one's back is to the geostrophic wind in the northern hemisphere, low pressure is to the left and high pressure is to the right. In the southern hemisphere, low pressure is to the right, high pressure to the left.

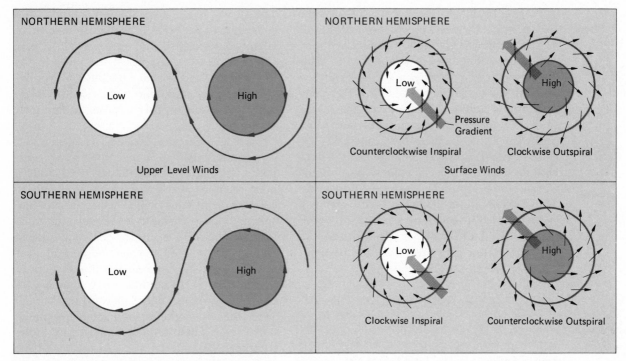

Figure 4.4 This figure shows the relationships between upper and surface winds for both northern and southern hemispheres. The Coriolis deflection is to the right in the northern hemisphere and to the left in the southern hemisphere, producing symmetrical patterns of deflection in the two hemispheres. In the northern hemisphere, the resultant geostrophic wind flows parallel to the isobars, with low pressure to the left, high pressure to the right. The surface flow below converges in a counterclockwise flow toward low pressure and diverges away in a clockwise flow from high pressure because of frictional effects. (Vantage Art, Inc. after Arthur N. Strahler, *Introduction to Physical Geography*, 1st ed., 1951, John Wiley & Sons)

direction, as well as temperature and humidity, are taken twice a day at more than 100 weather stations across the continental United States. From these measurements, meteorologists construct weather maps that estimate pressure and wind at various levels of the atmosphere over North America. These projections are the basis for predicting weather as much as 72 hours in advance.

The Force of Friction

Do you remember when you last played shuffleboard? The object of the game is to push wooden disks toward numbered scoring sections on the deck of the court. Friction between the disk and the court surface slows the disk; hopefully the right combination of push and friction allows the disk to come to rest in the section with the highest score. If the energy sources that drive the atmospheric circulation were suddenly to disappear, friction between the atmosphere and the earth's surface, and even within the atmosphere itself, would cause all atmospheric motion to slow down and eventually cease. Scientists regard friction as a "force" that opposes the motion of objects on the earth or in the atmosphere itself. For a parcel of air to maintain its movement, the driving force must overcome frictional resistance.

The effect of friction is at its maximum at the earth's surface and decreases upward with altitude. Recall that the Coriolis force is proportional to the speed of the air. It follows that near the ground, where surface winds are slowed by friction with the land surface, buildings, and trees, the Coriolis effect is reduced, but the pressure gradient force is not affected by the slower air speed. Consequently, instead of flowing parallel to the isobars, air near the surface responds to the pressure gradient force by flowing from higher to lower pressure at an angle across the isobars (Figure 4.1d). This causes surface air to *spiral into* centers of low pressure and to *spiral out of* centers of high pressure. In the northern hemisphere, the inward spiral is counterclockwise, the outward spiral clockwise. In the southern hemisphere, the reverse is true (Figure 4.4). Relating Buys Ballot's Law to surface winds, if your back is to the wind in the northern hemisphere, and you rotate 45° to the right, the low pressure area will be on your left, the high pressure area on your right. In the southern hemisphere, rotate 45° to the left, and low pressure will be on the right, high pressure on the left (Figure 4.4).

THE GENERAL CIRCULATION OF THE ATMOSPHERE

The pressure gradient force, the Coriolis force, and the force of friction are the basic factors that determine atmospheric motion. But other large-scale mechanisms are also involved in the atmospheric circulation. The following sections describe a generalized model of the global circulation of the atmosphere. This model represents a grand average for one year; the actual circulation at any moment will be somewhat different.

Pressure Patterns and Winds on a Uniform Earth

To understand the general circulation of the atmosphere, it is helpful to think first of the atmospheric circulation that would exist if the earth's surface were a uniform body of water. Let us start by also assuming that there is no rotation, and therefore no Coriolis force. Under such conditions, a low-pressure belt would develop over the equatorial region, where there is excessive radiational heating. High-pressure centers would develop over both polar regions, where there is excessive radiational cooling.

The air would rise over the low latitudes and subside over the poles. The atmospheric circulation would consist of a surface flow of air from the high-pressure regions at the poles to the low-pressure zone at the equator, and to complete the circulation loop there would be a return flow of air toward the poles in the upper atmosphere.

Such a vertical convective cell is called a *Hadley cell*, after the English meteorologist who proposed this model of circulation in 1735. Modern interpretations of the mechanics of general circulation restrict the Hadley cells in each hemisphere to the lower latitudes between the equator and about 30° latitude. The right side of Figure 4.5 shows an exaggerated vertical cross section of the Hadley cells north and south of the equator. As the meridians of longitude converge poleward, so does poleward moving air. Thus convergence aloft and longwave radiational cooling help to induce subsidence in the subtropical latitudes.

If we add rotation and the Coriolis force to this circulation model, the circulation pattern is altered considerably. Excessive radiational heating at the equator still produces rising air that flows out aloft toward the poles. Once away from the equator, however, the poleward flowing upper air in the northern hemisphere converges and is turned to the right by the Coriolis force, producing a westerly (west-to-east) ''jet stream,'' consisting of the subtropical jets shown in Figure 4.5. Leftward deflection of poleward moving upper-level air in the southern hemisphere creates a westerly jet stream there as well. The air aloft cools by longwave radiation to space, and a portion sinks into the lower atmosphere at about 30° latitude in each hemisphere. This produces high-pressure belts in the subtropics. Much of this subsiding air flows back down the surface pressure gradient toward the equator. This surface flow is deflected by the Coriolis force, again toward the right in the northern hemisphere and toward the left in the southern hemisphere (Figure 4.5). In both hemispheres the deflected flow is turned toward the west, so this surface flow is from northeast to southwest in the northern hemisphere, and from southeast to northwest in the southern hemisphere. These are the

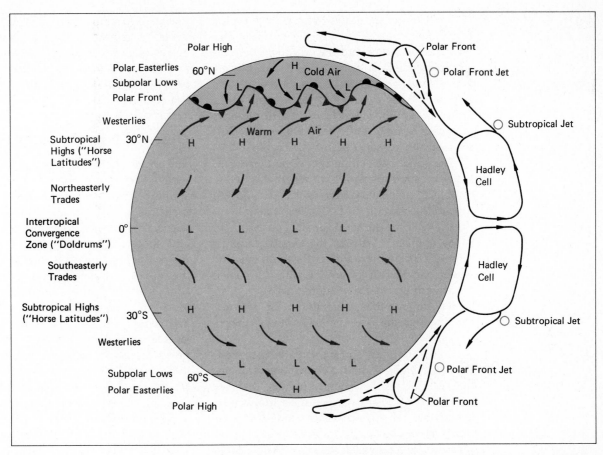

Figure 4.5 This sketch is a highly diagrammatic representation of the general circulation over a homogeneous earth with a water surface. The Coriolis effect is included. Semipermanent pressure and wind belts are shown on the surface of the sphere, and a much enlarged vertical cross section is shown on the right.

A detailed description of the pressure and wind systems is in the body of the text, and the text and figure should be studied carefully together. Note especially that winds are named for the direction from which the air is moving; for example, the air in the northeasterly trades moves from the northeast towards the southwest. For clarity, the polar front is shown only for the northern hemisphere. It is discussed in detail in later sections of this chapter and in Chapters 5 and 6. The red circles in the vertical cross sections represent the average location of jet streams; they are also discussed later in the chapter. (Vantage Art, Inc.)

"northeasterly trades" and "southeasterly trades" shown in Figure 4.5.

The surface pressure and wind systems we have described exist on the average and have traditional names dating from the era of wind-dependent sailing vessels. The belts of descending warm dry air, fair weather, and weak surface winds in the subtropics were known as the *horse latitudes*. Although other interpretations of this term have been offered, it is generally believed that when sailing ships were becalmed there, and supplies of food and drinking water dwindled, horses were the first passengers to go overboard.

Winds are traditionally named according to the direction *from which* they blow. Between the horse latitudes and the equator in the northern and southern hemispheres lie zones characterized, as noted previously, by the northeasterly and southeasterly *trade winds*. These "easterly

trades" are among the earth's steadiest and most persistent winds; they provided sea traders in sailing vessels with reliable westward routes across the oceans. Near the equator the trade winds converge into a low-pressure zone of generally light, variable winds and calms traditionally known as the *doldrums*. The doldrum belt is what present-day meteorologists refer to as the *intertropical convergence* (ITC).

Some of the subsiding air of the subtropical highs flows poleward near the surface. This flow is deflected toward the east to become the southwesterlies of the northern hemisphere and the northwesterlies of the southern hemisphere. Usually these winds are simply called *westerlies*, as they are in Figure 4.5 In the days of sailing vessels, these were the winds used for eastward passages across the oceans.

The surface air within the westerlies is normally of subtropical origin. When this warm "tropical" air meets colder "polar" air moving toward the equator, the lighter tropical air flows up and over the denser polar air. The *polar front* is the boundary between the two types of air (Figure 4.5). As the warm air flows up over the wedge of colder air toward the pole, it is chilled by expansion and radiational cooling. This cooled air subsides over the polar regions, forming high-pressure centers known as the *polar highs*.

Cold surface air spreads equatorward from the polar highs in both hemispheres and undergoes Coriolis deflection to become the *polar easterlies*. The polar easterlies meet the westerlies in the convergence zones along the polar fronts of both hemispheres; these belts of lower pressure are often called the *subpolar lows*. From time to time, particularly during winter, the polar front bulges toward the equator, allowing polar air to penetrate to subtropical latitudes. These bulges, known as *polar outbreaks*, are indicated by the dashed arrows in the vertical cross section in Figure 4.5.

Surface pressure and wind patterns on a uniform, rotating earth can be summarized as follows. There are three zones of low pressure and atmospheric convergence at the surface—the region of the polar front in each hemisphere and the intertropical convergence zone astride the equator. There are also four zones of high pressure and atmospheric divergence—the polar and subtropical highs in each hemisphere. In general, zones of low pressure and convergence are associated with clouds and precipitation, and fair weather prevails in zones of high pressure and divergence. Due to seasonal changes of solar radiation income at each latitude (discussed in Chapter 3), these zones shift poleward in summer and toward the equator in winter. The heat storage in oceans, however, delays poleward migration of the belts by one or two months from the solstices, so that these belts do not reach their highest latitudes until late summer.

Observed Pressure and Wind Systems

The actual patterns of pressure and wind at the earth's surface are more complicated than the simple belts associated with the uniform surface model. The presence of continents, mountain barriers, and oceans considerably influences the atmospheric circulation. The specific heat of rock or soil is much less than that of water, so the continents poleward of tropical latitudes become warmer than the oceans in summer and much cooler in winter. This differential heating and cooling of land and water masses affects surface pressure and winds, as thermal low-pressure cells develop over the land in summer to be replaced by high-pressure cells in winter. Mountain ranges also disturb the flow of the atmosphere, even at its upper levels.

The observed pattern of pressure distribution, shown in Figure 4.6, consists of separate cells rather than continuous belts of high and low pressure. Nevertheless, these cells tend to be distributed along the same bands of latitude as the idealized pressure zones on a uniform earth. The average direction of surface winds is closely related to the pressure distribution. Surface winds spiral outward from high-pressure regions and inward toward low-pressure regions, indicating geostrophic flow modified by friction with the earth's surface.

In the southern hemisphere, there is almost continuous ocean between latitudes 45° and 65°S, so the observed circulation, particularly in winter (July), tends to be similar to the model for the homogeneous all-water surface. In the northern hemisphere, the presence of large landmasses produces well-defined pressure and wind cells. These cells migrate north during the summer and south during the winter, reflecting changes in the receipt of solar radiation.

In winter, the northern regions of North America and Eurasia become very cold in comparison with the adjacent oceans. This causes the polar high to be displaced in the direction of the equator, forming two high-pressure centers—the massive Siberian high over north-central Asia and the much smaller Yukon high over eastern Alaska and the Canadian Arctic. Two subpolar lows, the Icelandic and Aleutian lows, are situated over the warmer oceans at similar latitudes. The strong pressure gradients associated with the low-pressure centers over the oceans produce frequent periods of windy and stormy weather at these latitudes.

During summer, when the continents become warmer than the oceans, the subtropical highs strengthen and expand over the cooler oceans. The subtropical highs over the oceans of the northern hemisphere are known as the Pacific or Hawaiian high and the Azores or Bermuda high; these nearly stationary high-pressure cells dominate the weather over the subtropical oceans and adjacent continental areas all summer long. Meanwhile, thermal low-pressure centers develop over the hot deserts of southwestern United States and southern Asia (Figure 4.6).

The seasonal shifts in pressure patterns generate seasonal shifts in winds. The inset in Figure 4.6 shows how the average position of the intertropical convergence—the meeting place of the trade winds—shifts from January to July. Note that there is relatively little change in the position of the ITC over the Atlantic and eastern Pacific oceans. However, over land areas, especially those bordering the Indian Ocean, there is a large seasonal displacement. Associated with the extreme shift in the position of the ITC over eastern and southern Asia is a reversal in wind direction between winter and summer. This seasonal change in the direction of surface winds produces very strong seasonal changes in the prevailing weather. During the winter season the wind tends to blow from the land to the sea; during the summer season the air tends to flow from the sea to the land. A regional wind system that reverses direction seasonally is known as a *monsoon*. The monsoon phenomenon is discussed at greater length in Chapter 6 in connection with tropical weather types.

The Upper Atmosphere and Jet Streams

The circulation pattern at the top of the troposphere is much simpler than that near the surface. Poleward of the subtropical highs, the upper atmosphere flows mostly from west to east in a vast circumpolar vortex. Since friction here is at a minimum, the Coriolis and pressure gradient forces are in balance. Therefore, airflow in the upper atmosphere approximates geostrophic conditions, with low pressure to the left and high pressure to the right in the northern hemisphere (Figure 4.7). The strongest flows are concentrated in the relatively narrow *jet streams*.

These high-velocity streams of air were first observed in the upper troposphere in the 1940s with the development of military aircraft that could fly at altitudes exceeding 10 km (6 miles). These so-called jet streams were found to occur at altitudes of 10 to 15 km. They are hundreds of kilometers wide and several kilometers thick. Wind speeds along the core of a jet stream can exceed 300 km (200 miles) per hour for a distance of 1,600 km (1,000 miles) or more. Figure 4.5 shows that, on the average, there are two jet streams in each hemisphere: the subtropical jets associated with the subtropical highs and the polar front jets that are normally above the polar fronts. The jet streams are utilized by eastward-bound airliners and are avoided as far as possible on westward flights. Assistance from the jet stream is the reason that east-bound flights across the full width of the United States require about an hour less than west-bound flights.

The jet streams generally follow a sinuous path, as shown in Figure 4.8a. There are typically three to six curves, or "waves," in each complete circuit of the jet stream around the earth. The waves usually remain positioned over certain geographical regions, but at times they undergo severe oscillations (Figure 4.8, b and c). When the waves push outward for several weeks, the weather under displaced waves becomes warmer or colder and wetter or drier than normal, depending on which way the air stream is curving. Areas of low pressure, known as *upper-level troughs*, are present where

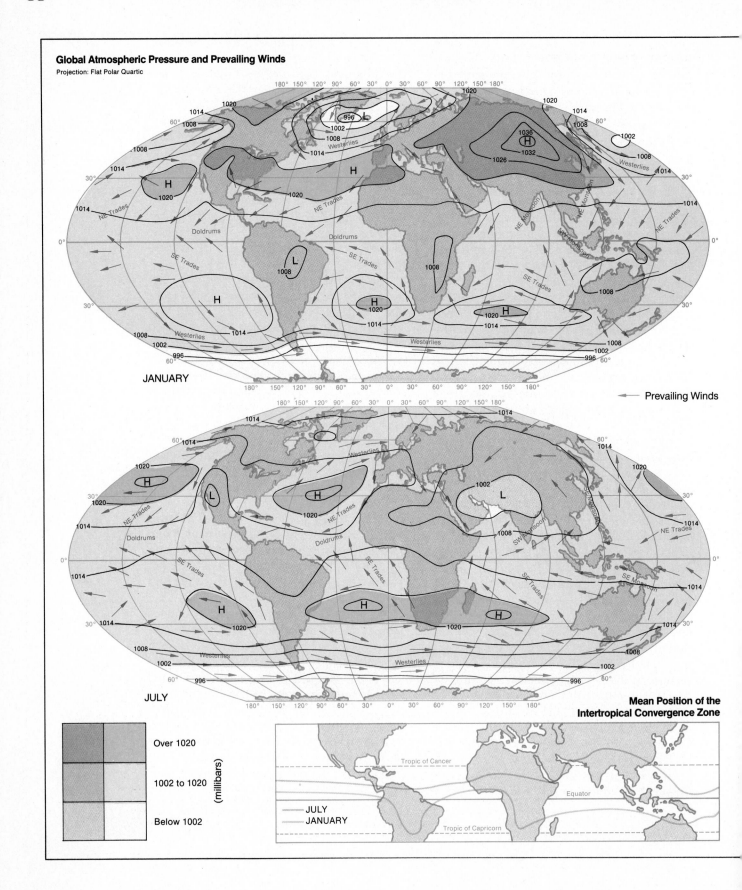

Global Atmospheric Pressure and Prevailing Winds
Projection: Flat Polar Quartic

JANUARY

JULY

Prevailing Winds

Over 1020

1002 to 1020

Below 1002

(millibars)

Mean Position of the
Intertropical Convergence Zone

Tropic of Cancer

Equator

Tropic of Capricorn

JULY
JANUARY

Atmospheric Pressure in Polar Regions
Projection: Stereographic

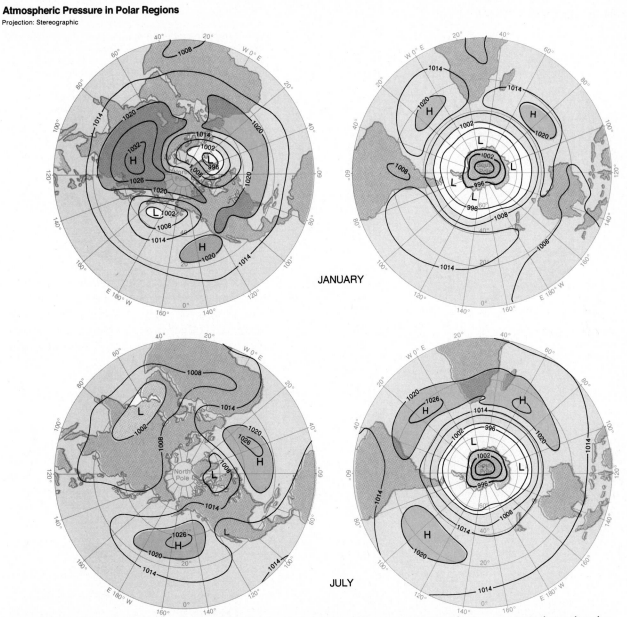

Figure 4.6 Global maps show the average direction of surface winds and the average atmospheric pressure at the surface for January and July. Pressure readings are in millibars. Near the Antarctic Circle, where there is open ocean entirely around the globe, the isobars form a continuous belt. Elsewhere, the presence of landmasses causes the regions of high and low pressure belts to be broken into individual pressure cells. The cells tend to be distributed along bands of latitude that are analogous to the pressure belts that would form on a uniform earth. The formation of pressure cells is affected by temperature differences between the oceans and the continents.

The average distribution of surface winds closely follows the pressure patterns, with winds tending to flow outward from high-pressure regions and inward toward low-pressure regions. In many areas, the winds approach geostrophic flow modified by friction with the earth's surface.

In both the northern and southern hemispheres, the distribution of pressure cells, especially the subtropical highs, generally shifts to the north in July and to the south in January. Over the poles, the high-pressure cells are more diffuse and less numerous in summer than in winter. The resulting seasonal shift in wind patterns generates a seasonal variation in the weather of many regions. The winds on the east coast of central Africa, for example, change from the northeast trades in January to the southeast trades in July. The inset map shows how the mean position of the intertropical convergence zone, which is the meeting place of the trade winds, shifts between January and July.

Upper Air Winds
Projection: Stereographic

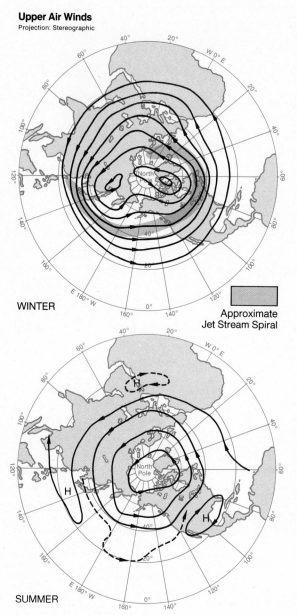

WINTER

Approximate
Jet Stream Spiral

SUMMER

Figure 4.7 The maps show winter and summer wind patterns for the northern hemisphere in the upper atmosphere near an altitude of 5.5 km (3.4 miles), where the atmospheric pressure is approximately one-half the pressure at sea level. Wind directions and speeds in the upper atmosphere are usually determined by tracking freely drifting radio-equipped balloons that transmit meteorological data to the ground. In winter the upper air winds are strong and form a well-defined pattern of circulation with several undulations. The shaded region indicates the average location of fast-moving jet streams. Many winter storms in the northern latitudes appear to be generated with the help of jet streams. The flow of upper air winds tends to be weaker during summer than during winter. Dashed lines indicate particularly weak flows. (Andy Lucas after Herbert Riehl, *Introduction to the Atmosphere,* © 1972 by McGraw-Hill Book Company)

so extreme that waves are "pinched off" from the main stream, forming detached upper-level cells of low or high pressure (Figure 4.8d). These cells affect surface weather, then slowly die away, leaving the pattern to resemble again that shown in Figure 4.8a. A wave cycle in which the path of the jet stream changes from gently sinuous to highly irregular to gently sinuous again generally requires one to two months.

The general circulation related to global patterns of pressure sets in motion smaller *secondary circulations* that produce local weather conditions. In Chapter 6 we shall examine these secondary circulations, which directly influence our daily activities.

THE OCEANS

The phenomena of the oceans and atmosphere are so closely linked that many scientists prefer to speak of "the ocean-atmosphere system." Air moving over the sea surface sets water in motion, producing ocean currents. These currents (like air currents) help redistribute energy by carrying warm waters toward the poles and cool waters toward the tropics. The oceans, in turn, affect the atmospheric circulation by releasing water and latent heat into the atmo-

the jet stream bulges toward the equator. Near the surface, cold air flows toward the tropics on the western edge of each trough, and warm air flows poleward on the eastern edge. When no undulations are present, there is little exchange of energy between the tropics and the poles. The undulating pattern is necessary to maintain a net poleward flow of energy from low to high latitudes. Periodically, the oscillations become

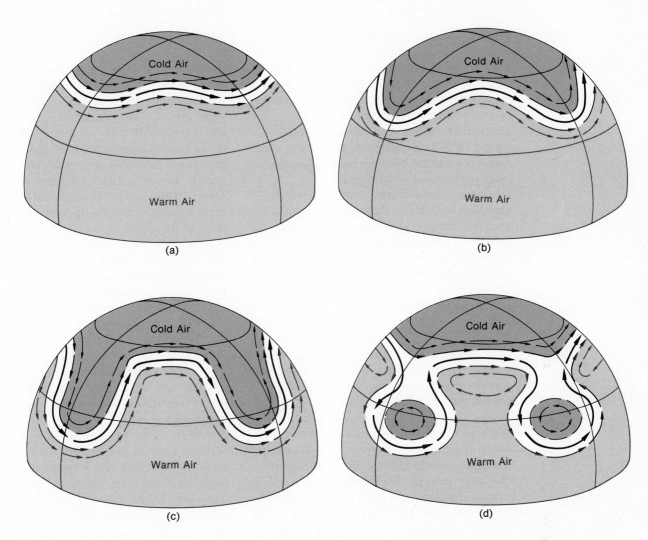

Figure 4.8 Circumpolar jet streams circle the earth at an altitude of 10 to 15 km (6 to 9 miles) and gain much of their energy from the temperature difference between warm tropical and cold polar air. (a) The jet streams appear in conjunction with large, slow-moving undulations, or waves, in the flow of the upper atmosphere. There are typically three to six waves in a complete pattern around the earth. The waves may undergo increasingly severe oscillations (b) (c), until cells of rotating warm and cold air are formed (d). The cells then die away, and the pattern once more resembles that shown in (a); the entire cycle is generally completed in four to eight weeks.

This mechanism results in the movement of warm air toward the poles and cold air toward the tropics and helps to maintain the poleward flow of energy from equatorial regions. Jet streams and upper air waves also appear to generate surface weather systems by causing near-surface convergence and ascent of air in cyclonic storms. (Doug Armstrong after Jerome Namias, "The Jet Stream," *Scientific American*, copyright © 1952 by Scientific American, Inc. All rights reserved.)

sphere through evaporation, and by functioning as reservoirs of heat that drive the atmospheric circulation and provide much of the energy for the hydrologic cycle described initially in Chapter 2.

The Oceans: Energy Banks

We have seen that the heat capacity of ocean water is two to three times that of continental surfaces and many times greater than that of the air. The great heat capacity of the oceans makes them reservoirs of the energy that powers global weather and climates. This energy is released into the atmosphere by conduction, convection, and latent heat transfer. The transfer of latent heat takes place during the evaporation process that moves enormous amounts of water vapor from warm sea surfaces to the atmosphere. While this water provides the earth's rainfall, it also helps power the atmospheric circulation by latent heat release during condensation and cloud formation.

Ocean temperatures are "conservative," changing very slowly through the seasons. But even small changes in ocean temperatures seem capable of producing large changes in weather patterns. Oceanographers have discovered vast and persistent surface "pools" of ocean water that are 1° to 2°C warmer or cooler than the water around them. These ocean temperature abnormalities, or "anomalies," seem to produce changes in the average atmospheric circulation. For example, unusually warm water off the east coast of the United States during 1971 and 1972 was associated with very wet weather in the coastal states, and it probably helped provide energy for Hurricane Agnes, a tropical cyclone that was especially destructive in the area between Virginia and Pennsylvania. Because of complex interactions with the upper air flow, the same anomaly may have played a role in freezes, droughts, and grain crop failures in the Soviet Union in 1972. A pool of colder than normal water in the central Pacific in 1976–1977 is believed to have caused changes in the general circulation, producing severe drought in the western United States along with the coldest winter in 60 years in the East, followed by a summer drought in the Midwest that equaled those of the Dust Bowl years of the 1930s.

Scientists have begun using water temperature patterns to predict temperature and precipitation patterns across the United States for one- to three-month periods, and the results are now being tested experimentally against the actual weather that occurs after the forecast. These climatic forecasts have proven useful, but much more needs to be learned about the very complex thermal interactions of the oceans and atmosphere, as the Case Study following this chapter illustrates.

Surface Currents

Wind blowing over the surface of the ocean exerts a push on the water, but the resulting motion of the water is only a fraction of the wind speed. This is due to the greater density and internal friction of water. Wind-driven ocean currents have speeds ranging from kilometers per day to kilometers per hour. Because friction causes current velocities to decrease rapidly with depth, strong wind-driven currents are confined to the upper hundred meters or so of the ocean.

A well-developed ocean current takes a long time to respond to changes in the wind, and even waves produced by strong winds require many hours to reach their full development. Therefore, ocean currents reflect average wind conditions over periods of many months, and the circulation of the oceans is closely related to the general circulation of the atmosphere.

We have seen that if the earth were uniform, without any land and water contrasts, the winds would form well-defined belts. Assuming complete global cover by water, wind-driven ocean currents would flow around the earth in a similar pattern. But the actual distribution of land and water means that only the ocean that encircles Antarctica can circulate freely in a continuous belt. The Atlantic, Pacific, and Indian oceans are bounded by continents or island archipelagos on the east and west, blocking free flow of ocean water around the earth at most latitudes.

As a consequence, the surface currents of the oceans consist of closed circulation loops, or *gyres*, which correspond to the global wind patterns. The Atlantic and Pacific oceans ideally would include three gyres on either side of the equator, as represented in Figure 4.9. The trade winds drive the low-latitude east-to-west currents of the subtropical gyres, and the midlatitude westerly winds drive the higher latitude return flows from west to east. The actual oceanic circulation is shown in Figure 4.10. The strongest currents are on the perimeters of the gyres (near the coasts), with much less movement near the centers of the oceans. Gyres are clearly developed in the Atlantic and Pacific

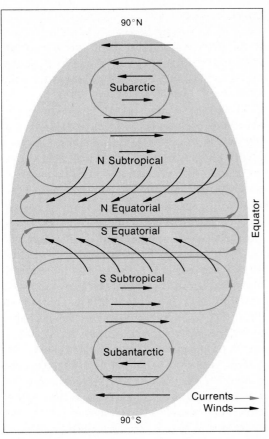

Figure 4.9 Idealized oceanic circulation in an ocean basin, consisting of loops, or gyres, which are driven by prevailing winds. (Doug Armstrong after P. Weyl, *Oceanography*, 1979, John Wiley & Sons)

Figure 4.10 This map shows the principal oceanic currents in the surface layer of the oceans. The currents form loops of circulation, with strong currents on the perimeters and relatively little movement internally (compare with Figure 4.9). Ocean currents move warm water poleward and cold water toward the tropics, which helps to equalize the distribution of energy over the earth.

The principal ocean currents along the east coast of North America include the Florida Current and Gulf Stream, which carry warm waters northward, and the Labrador Current, which carries cool waters southward past the coast of New England. (The Florida Current is the name given to the portion of the Gulf Stream from Florida to Cape Hatteras, North Carolina.) The principal ocean current along the west coast of North America is the southward flowing California Current. (Andy Lucas after L. Don Leet and Sheldon Judson, *Physical Geology*, 3rd ed., © 1965, by permission of Prentice-Hall, Inc.)

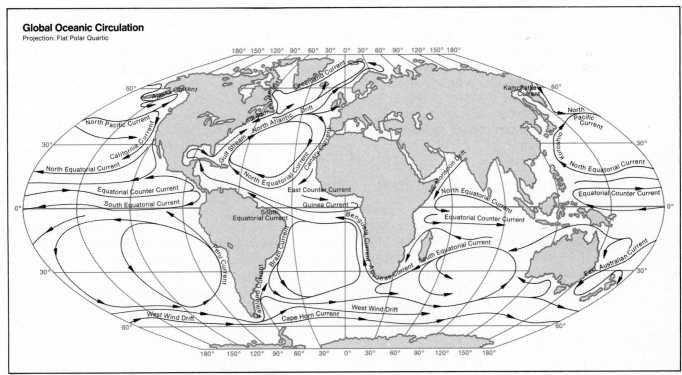

oceans, but near Antarctica an eastward moving current flows around the earth; the idealized southernmost gyre in Figure 4.9 is replaced by the Antarctic continent.

Ocean currents play key roles in redistributing heat around the earth. A current of warm tropical water flows poleward at the western edge of each ocean basin, and cool waters move toward the tropics along the eastern margins of the seas. Along the east coast of North America, the Gulf Stream is the warm, north-setting current. Its counterpart in the Pacific is the Kuroshio Current off Japan. Both are fast, narrow currents, moving several kilometers per hour. The volume of water transported annually by the Gulf Stream alone is more than 30 times greater than the total amount of streamflow from all the continents.

The Gulf Stream merges into the slower North Atlantic Drift near the Grand Banks off Newfoundland. The warm water of the North Atlantic Drift, which flows northeastward to Europe, causes the average winter temperatures of western Europe to be significantly higher than those of eastern North America, even though Europe lies closer to the pole. Thus Denmark and Sweden are well populated and developed although they are at the same latitude as Hudson Bay in subarctic Canada, and Murmansk is essentially an ice-free port in the Soviet Union on the Arctic Ocean. In addition, evaporation from such warm currents provides high-latitude maritime air masses with heat and moisture supplies for the cyclonic storms that bring rain to Europe, eastern Asia, and western North America.

The Deep Circulation and Upwelling

In the deep ocean, below the surface currents, there is a slow circulation of water, driven primarily by density differences. The density of seawater varies with both temperature and salt content. The densest water is cold and highly saline; such water is produced off Antarctica when surface water freezes, leaving its salt content in adjacent cold but unfrozen water. This "salted" water slowly subsides toward the ocean floor, mixing little with adjacent waters of different temperatures and densities. According to carbon 14 dating, water in the deep oceans may remain there for hundreds of years before returning to the surface.

The surface ocean currents shown in Figure 4.10 are driven by the atmosphere's general circulation. But the ocean currents do not exactly parallel the prevailing winds. The Coriolis effect tends to turn the currents a bit to the right in the northern hemisphere and to the left in the southern hemisphere. The surface water tends to drift off at an angle of about 45° to the wind. Along the west coasts of continents, in particular, where winds blow toward the equator around the eastern sides of the subtropical highs, the surface water moves away from the coasts toward the centers of the subtropical gyres. This water is replaced by deeper cold water that rises to the surface along these coasts, producing several important effects.

This coastal *upwelling* of cold water occurs along the west coasts of all continents, but it is pronounced off California, Peru, Morocco, Namibia, and Somalia. The resulting cold water chills the air above it, producing fog (see Chapter 5). The chilled air is denser than the warmer air above it. Such a surface temperature inversion prevents the vertical motion of air needed to build rain clouds. Therefore, coasts where upwelling occurs are foggy but receive very little rainfall. All have desert climates, either seasonally or the year around.

Upwelling is very important ecologically because it lifts nutrients up into sunlit surface waters, allowing the tiny marine plants and animals known as plankton to flourish and support a very rich marine food chain. This permits fishing industries to thrive along many desert coasts. The richest fishing grounds are off the Peruvian coast of South America. Here, however, disaster strikes every few years, when the waters become abnormally warm and upwelling seems to fail. Because they are not adapted to warm water, and lacking the nutrients supplied by upwelling, the plankton die off, and the entire food chain collapses. Fish die in great numbers, along with the birds that normally feed on them. Decomposing fish litter the beaches and the waters, giving off so much hydrogen sulfide that the white lead paint used on ships turns black. The phenomenon has been called the "Callao painter," after the port of Callao in Peru. Since this effect usually occurs around Christmas, the southward drifting warm current is called El Niño, the Christ Child.

The cause of the disaster seems to be periodic fluctuation in the strength of the easterly trade winds. The trade winds strengthen every few years, probably in response to sea surface temperature anomalies, and drive water into the western Pacific, actually raising the sea level there. When the trade winds slacken again, this warm water "sloshes back" eastward across the width of the Pacific, and a portion spreads southward along the Peruvian coast, overrunning the upwelling cold water and producing the disastrous El Niño phenomenon.

SUMMARY

The atmospheric circulation is a response to the energy surplus in low latitudes and the energy deficiency at high latitudes. Differential surface heating gives rise to horizontal differences of atmospheric pressure. Air flows toward lower pressure but is deflected by the Coriolis force, which is a result of the earth's rotation. In the upper atmosphere the Coriolis force is equal to and opposite the pressure gradient force, so the air flows parallel to pressure isobars. Near the surface, however, the effect of friction causes air to flow obliquely across isobars toward the lower pressure.

In the northern hemisphere a counterclockwise flow of surface air spirals into areas of low pressure, and a clockwise flow spirals out of high-pressure areas. These motions are reversed in the southern hemisphere, resulting in a mirror image of northern hemisphere patterns. On a uniform earth there would be several belts of high and low pressure related to the ascent and descent of air on a global scale. The effect of the continents and oceans is to break the belts of pressure and winds into a series of cells that change seasonally in size and strength. The circulation of the upper atmosphere is dominated by the effects of the polar and subtropical jet streams found in both the northern and southern hemispheres. Strong undulations in the jet streams cause latitudinal interchanges of energy and affect surface weather.

The oceanic circulations are driven by the circulation of the atmosphere above, and consist of gyres bounded by rapidly moving currents, with only slow currents at their centers. Warm currents moving poleward on the west sides of the oceans and cold currents moving from high to low latitudes on the east sides of the oceans help offset net radiation gains in equatorial latitudes and losses in subpolar and polar regions. Small changes in ocean temperatures appear to trigger major weather changes over land areas, accounting for unusual droughts, floods, and severe winters.

The cold currents of the eastern sides of the oceans are reinforced by the upwelling of cold water from greater ocean depths. Upwelling results from westward Coriolis deflection of currents moving toward the equator. Areas of upwelling are unusually rich in marine life due to the nutrients brought up into sunlit surface waters. Massive die-offs of fish and marine birds occur when easterly trade winds weaken, permitting warm equatorial water to invade the area of upwelling.

REVIEW QUESTIONS

1. What is the relationship of the geostrophic wind to high and low pressure in the northern and southern hemispheres?
2. What does Buys Ballot's Law tell us about the interaction of the pressure gradient force, the Coriolis force, and friction?
3. How does surface airflow in the vicinity of high- and low-pressure centers differ in the northern and southern hemispheres? Explain.
4. Compare and contrast the idealized patterns of atmospheric pressure systems and winds over the continents and ocean basins for winter and for summer.
5. How are the jet streams related to the energy balance of the earth-atmosphere system?
6. How do the continents and mountain barriers alter the idealized zonal patterns of pressure and wind systems over the earth?
7. Describe how ocean current patterns are related to atmospheric pressure and wind systems.

APPLICATIONS

1. Suppose your residence is in a North American urban area, with a bakery to your west, a chocolate factory to the east, an oil refinery to the north, and a stockyard to the south. Describe the sequence of odors you would sense if a strong low pressure center passed from west to east along a line 100 km north of your location. Would the odors be the same if the low passed to the south of you?
2. By means of radio and TV weather reports and your own observations of cloud movements, keep a record of wind direction each day for the rest of the term. Do the wind conditions help you predict weather changes? How far in advance of weather changes do wind shifts occur?
3. Contrast your own data from the preceding exercise with the prevailing winds shown in Figure 4.6. What significant differences appear? How can these differences be explained?
4. Look at the existing pattern of ocean currents in Figure 4.10. Would it be possible to send a message in a floating bottle to every coastal location in the world? Could this always be done from another continent, or would some messages have to be sent from ships at sea?
5. The continents have not always occupied the positions they do today. Global maps of the most likely configurations of the continents back to about 250 million years ago can be found in Chapter 12. Develop some ideas about how the general circulation patterns of the atmosphere and oceans (Figures 4.6 and 4.10) might have been different then than they are at the present time.

FURTHER READING

Gedzelman, Stanley David. *The Science and Wonders of the Atmosphere.* New York: John Wiley (1980), 535 pp. Relationships between pressure and wind are developed in Chapters 15 and 16 by drawing upon commonplace examples and simple explanatory equations.

Hare, F. Kenneth. *The Restless Atmosphere,* rev. ed. London: Hutchinson (1956), 192 pp. This brief introductory classic contains outstanding regional and continental chapters that focus on the dynamics of the general circulation.

Lutgens, Frederick K., and **Edward J. Tarbuck.** *The Atmosphere: An Introduction to Meteorology,* 2nd ed. Englewood Cliffs, N.J.: Prentice-Hall (1982), 478 pp. An introductory text with a strong geographical perspective. Chapters 6 and 7 are particularly helpful for pressure, winds, and the general circulation.

Neiburger, Morris, James Edinger, and **William Bonner.** *Understanding Our Atmospheric Environment,* 2nd ed. San Francisco: W. H. Freeman (1982), 453 pp. Basic fundamentals are presented in a nonmathematical framework.

Riehl, Herbert. *Introduction to the Atmosphere,* 3rd ed. New York: McGraw-Hill (1978), 410 pp. This nonmathematical text emphasizes the upper air circulation and its relationships to weather and climate.

Young, Louise B. *Earth's Aura.* New York: Avon Books (1979), 305 pp. A highly readable account of the various phenomena of the atmosphere, written for the general public. Full of anecdotes; informative and accurate.

CASE STUDY

The Winter of 1982–1983

In this chapter we discussed the discovery in the 1970s of the "pools" of warmer and cooler ocean water, and their apparent links with the unusual weather and climate of the 1970s. The identification of the so-called pools of oceanic water become possible when analyses of data from temperature sensors mounted on satellites were obtained. The sensors estimate longwave radiation upward from ocean surfaces adjusted for the water vapor content of the atmosphere between the satellite and the ocean surface. The technology has encouraged climatologists to enter the forecasting arena, but not enough is understood about the causes and effects of the geographical patterns of the warmer and cooler water to make a seasonal forecast of atmospheric circulation patterns and the resultant weather with a high degree of certainty.

The winter and early spring of 1982–1983 turned out to be seasons of unusual weather and climate, bringing great costs and some benefits to the inhabitants of the United States. Some climatologists had predicted another record cold winter for much of the nation, but the Climate Analysis Section of the National Weather Service, the federal agency responsible for official predictions, was not very far off when in late November it predicted a warmer than normal winter for the East and an unusually cold winter for the West. Americans east of the Rockies saved as much as 2.5 billion dollars in fuel costs beause the winter was so mild!

But the same atmospheric circulation pattern that produced the milder than normal winter for the East drove a long sequence of un-usually severe winter storms from the Pacific onto the southern and central California coasts along tracks much farther south than usual. The media were filled with graphic images of destructive landslides and coastal erosion and the potential impact of record snowpacks in the mountains was a great cause for concern at the beginning of the summer of 1983. The same storms that battered California crossed the southern Rockies, with runoff from snowmelts raising the level of Great Salt Lake in Utah to worrisome elevations. There is no outlet for this large lake, and its water level depends totally on the "balance" between incoming mountain runoff from streams and the evaporation from the lake itself. The storms from the Pacific were also rejuvenated over the northern Gulf of Mexico, producing record floods in southeastern Louisiana and troublesome

During the unusually long sequence of severe winter storms that struck the southern coast of California in the winter of 82-83, very large waves eroded the cliffed headlands rapidly. At a number of places piers were severely damaged, and expensive homes fell victim to the sea. This photo shows the evacuation of the Pacific Skies Estates trailer park at Pacifica, just south of San Francisco.

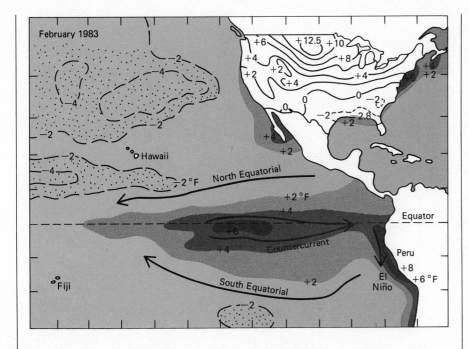

February 1983

+6 +12.5 +10
+4 +8
+2 +2
+4 +4
0
0 0 –2
0 –2 –2.8
–2
+2
+2

–2
–4
–4
–2
–2

Hawaii

–2
–4
–2 °F North Equatorial

+2 °F
+4

Equator

+6
+4
Countercurrent

Peru
+8

Fiji
South Equatorial +2
El
Niño
+6 °F

This analysis of sea-surface temperature anomalies, or departures from long term averages, is based on satellite data alone. The map shows the large pool of warm water associated with the El Niño and the southern oscillation, and the equally large pool of cool surface water northwest of Hawaii.

high water and more flooding over Florida and northeastward along the Atlantic Coast as far as New York City.

Meteorologists, climatologists, and oceanographers recognize the interconnectedness of ocean water temperatures, ocean currents, variations of the general atmospheric circulation, and climatic impacts in both the southern and northern hemispheres. There is much scientific interest now in the "southern oscillation," a sort of flip-flop reversal of small atmospheric pressure changes over the Pacific Ocean south of the equator. This is associated with weakened southeasterly trades that drive less warm water westward in the north and south equatorial currents, allowing the warm water to seep southward instead along the

west coast of Ecuador and Peru as the equatorial countercurrent and the El Niño. Related atmospheric abnormalities include the displaced storm track along the southern coasts of the United States and higher than normal atmospheric pressure over Australia, along with the most severe drought conditions in recorded history there.

At this writing the causes of the southern oscillation and the warmer and cooler pools of ocean water are unknown. The 1982–1983 El Niño appears to be the strongest and most persistent on record, but by May of 1983 there were some signs that the unusual oceanic and atmospheric circulation patterns were beginning to weaken and change. More about climatic variation can be found in the case study following Chapter 8.

Seascape Study with Rain Clouds by John Constable, c. 1824–1828. (Royal Academy of Arts, London)

Water enters the atmosphere through evaporation, is transported as water vapor, eventually condenses into clouds, and falls again to the earth as rain, snow, sleet, or hail. Condensation and precipitation are the results of a number of complex physical interactions that involve energy and moisture.

In the afternoon of June 9, 1972, clouds assembled over the Black Hills of South Dakota. Then, over large areas more than 30 cm (12 in.) of rain—equal to a full year's precipitation—poured down in a 6-hour period. Mountain streams became rampaging torrents, overflowing their banks and sweeping away cars and even houses. Automobiles and buildings blocked the spillway of a dam just upstream from Rapid City, causing water levels behind the dam to rise an additional 3 meters (10 ft). Late in the evening, the dam burst, and a flood wave swept through the city, causing enormous destruction and taking more than 200 lives. The total damage exceeded $120 million.

This type of disaster was repeated north of Denver, Colorado, on the night of July 31, 1976, when a similar deluge of rain fell in a 1- to 3-hour period, causing floodwaters to rush through the narrow canyons of the Big Thompson and Cache La Poudre rivers in the foothills of the Rocky Mountains. Within a few hours, the raging waters destroyed nearly all traces of human habitation in the canyons and drowned 144 persons. The amount of water discharged through some sections of the Big Thompson Canyon was nearly seven times greater than the maximum recorded in the previous 29 years.

How is it possible for so much rainfall to be poured over small areas in so short a time? The water that fell on both locations had evaporated from the Gulf of Mexico, thousands of kilometers away. Moving masses of warm, wet air carried the water vapor northward across the Great Plains. Along the way, some of the molecules of water vapor condensed to form droplets of water only micrometers in diameter—much too small to fall to the ground as rain. Finally, however, the moisture-laden air was forced to rise over the steep slopes of abruptly rising mountain ranges, where towering thunderclouds formed, and water droplets coalesced to form raindrops that fell in a deluge.

The movement of water from the earth's surface to the atmosphere and then back again to the surface constitutes the atmospheric part of the hydrologic cycle described in Chapter 2. The transfer of water between the surface and the

5
Moisture and Precipitation Processes

Figure 5.1 Because precipitation occurs only when certain conditions are present, the amount of moisture received by a given region may vary markedly from year to year. Rainfall, or the lack of it, is a central concern of farmers each year.

People in India rejoice in the coming of the summer monsoon rains, which supply the country with much of its annual precipitation. (Brian Brake/Rapho/Photo Researchers, Inc.)

This farm in Oklahoma was abandoned in the 1930s when a succession of drought years made farming impossible. The dry soil was transported by the wind, forming drifts of sand and turning the region into a dust bowl for many years. (The Bettmann Archive)

atmosphere occurs by means of three related processes. *Evaporation* is the transformation of liquid water at the surface into water vapor in the atmosphere. *Condensation* is the conversion of water vapor to water droplets or ice crystals in the form of fog or clouds in the air, or dew or frost at the surface. *Precipitation* involves the coalescence of water droplets, or transformation of ice crystals, into raindrops, snowflakes, or hailstones, heavy enough to fall to the ground. During the condensation process the latent heat absorbed during evaporation is released as sensible heat. Thus the hydrologic cycle can also be thought of as part of the energy balance of the earth-atmosphere system.

Under average conditions, nearly half of the earth's surface is covered by clouds. But only a fraction of the clouds produce significant rainfall. Precipitation is the result of complex atmospheric interactions, and the conditions that lead to it are so limiting that rain, snowfall, or hail should be considered an unusual rather than a commonplace event (Figure 5.1). This chapter describes how the atmosphere is continuously supplied with water vapor and how condensation, cloud formation, and precipitation occur.

TRANSFER OF WATER TO THE ATMOSPHERE

When water evaporates, water molecules become detached from the surface of the liquid and enter the air as water vapor, a gas, not a liquid. Molecules in liquids are in continuous motion but are kept from separating by attractive forces. The higher the temperature of the liquid, the greater the energy and speed of its molecules. Eventually, some molecules gain enough energy to break away from the liquid and enter the air. Because the escaping molecules carry kinetic energy with them, evaporation tends to cool the surface from which the molecules have left. The evaporation of water molecules on your skin is what makes you feel a chill when you step out of a swimming pool, even if the air is warmer than the water. The sun supplies most of the energy needed to vaporize water—as much as 590 calories for every gram of water transformed from the liquid to the vapor phase under average conditions.

Evaporation from Water Surfaces

If a jar filled with water is left open in a room, all of the water will eventually evaporate. Heat stored in the room's air provides all the energy required for vaporization of the water. Even if the jar is sealed, some water will evaporate into the air above the liquid; eventually, however, a condition of equilibrium is reached, and the water level in the jar remains constant. The air in the jar is then said to be *saturated*—it contains as much water vapor as it can hold at that particular temperature.

Water vapor exerts pressure, just like any other gas. The amount of pressure exerted by molecules of water vapor in the air is called the *vapor pressure*. When the air is saturated, the vapor pressure is at its maximum, but this maximum varies with the temperature. In other words, *saturation vapor pressure* depends on the temperature of the air. As the air temperature rises, the saturation vapor pressure increases rapidly so that more and more water vapor must be added before the air becomes saturated. Figure 5.2 shows that at the temperatures common near the earth's surface, the saturation vapor pressure nearly doubles for each 10°C increase in temperature. Hence, warm air has a much greater capacity to hold water vapor than does cold air.

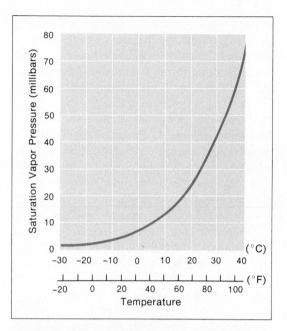

Figure 5.2 The curve shows that the saturation vapor pressure of water vapor in air increases rapidly with increased temperature. The data extend to temperatures lower than 0°C (32°F), the normal freezing point of water, because small droplets of water can remain liquid at temperatures as low as −40°C (−40°F); this is known as "supercooled" water. (Doug Armstrong after *Smithsonian Meteorological Tables*, edited by Robert J. List, 6th ed., 1971, by permission of The Smithsonian Institution)

Although the rate of evaporation from water surfaces depends especially on the energy supply, it is also affected by the degree of saturation of air above the water surface. Under normal atmospheric conditions, there is variation in the vapor pressure if it is measured at the water surface and at various heights in the air above. This is known as the *vapor pressure gradient*, which tells us the rate at which vapor pressure varies with distance. How much evaporation takes place depends on how "steep" the gradient is (how rapidly the vapor pressure decreases upward from the water surface).

The steepness of the vapor pressure gradient is determined mainly by turbulence in the lower atmosphere. When there is no wind, a layer of nearly saturated air forms immediately above water surfaces. The vertical change in vapor pressure within this layer is small, so the vapor pressure gradient is low, or "weak." Under such conditions the rate of evaporation is low. During windy conditions, however, water vapor is mixed through a much deeper layer of the atmosphere. This mixing maintains a steeper vapor pressure gradient near the surface, allowing evaporation rates to approach maximum values. To maintain high rates of evaporation, energy for vaporization must be supplied by warm water, warm air immediately above the surface, or absorbed solar radiation. Evaporation, then, depends on both the energy supply and the vapor pressure gradient.

There are few regular measurements of evaporation from the world's oceans or large lakes. The global distribution of evaporation has to be estimated indirectly by means of energy-budget calculations. In general, average annual evaporation is related to latitude and the corresponding solar radiation gains and long-wave radiation losses, as shown in Figure 3.10 (p. 62).

Evaporation rates also vary from place to place due to such factors as winds, cloudiness, water temperature, and water vapor content of the atmosphere. The highest annual evaporation rates occur over the subtropical oceans, where the water is warm, the air aloft is dry, and the weather is usually fair. Estimates of mean annual evaporation range from more than 100 cm (40 in.) of equivalent precipitation over equatorial and subtropical oceans to less than 10 cm (4 in.) over polar oceans. A very steep pressure gradient occurs in winter over the eastern seaboard of the United States, as warm subtropical waters of the Gulf Stream persistently move poleward under the colder and relatively dry air from the nearby land. The combination of warm water and dry air leads to very high evaporation rates throughout this area.

Evapotranspiration from Land Surfaces

Where plants cover the land, only a small part of the water vapor entering the atmosphere comes from direct evaporation of water in the soil. Instead, most is released through small openings in plant leaves called *stomata* (Figure 5.3). After a plant absorbs soil water through its roots, water and dissolved nutrients are

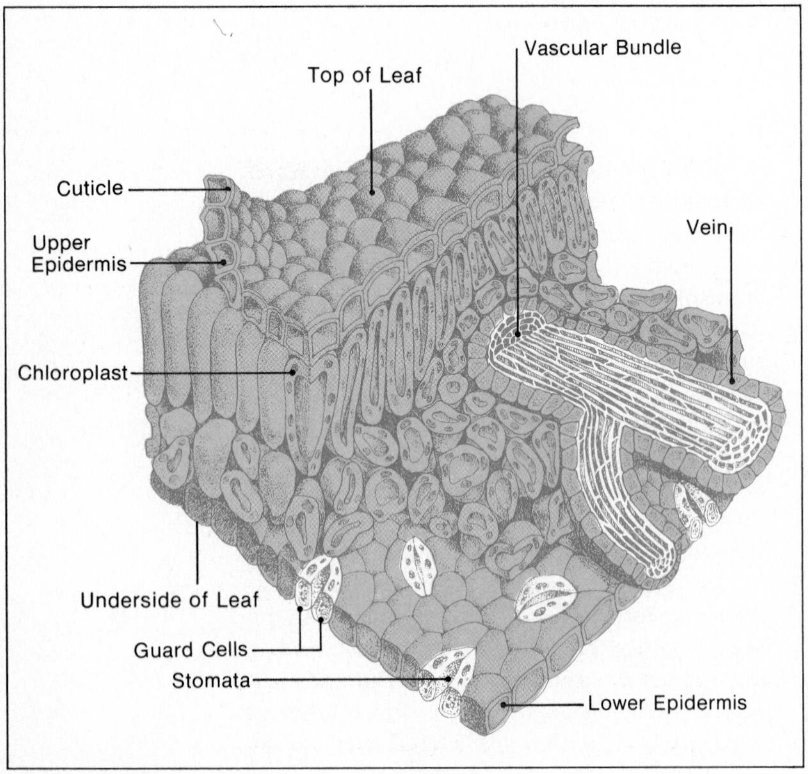

Figure 5.3 This vastly enlarged cross section shows the functional parts of a typical plant leaf. The leaves of green plants possess openings known as *stomata* in the bottom layer of protective epidermis. When the leaf is exposed to light, photosynthesis occurs in the chloroplasts, and the stomata are open to allow the entry of carbon dioxide and the exit of oxygen and water vapor. Soil moisture absorbed by the roots is transported to the leaves through the vascular bundles in the veins. The transpiration of water vapor is an essential process in green plants, and most of the water vapor entering the atmosphere in heavily vegetated regions is from plant transpiration rather than from evaporation.

The guard cells around the stomata close the openings if the plant lacks moisture and begins to wilt. In plants adapted to hot, dry climates, the guard cells keep the stomata closed during the hottest part of the day, when water loss by transpiration would be greatest. In some plant species adapted to dry climates, the covering, or cuticle, is particularly thick and waxy to prevent loss of moisture through cell walls. (John Dawson)

transported up the stem to the leaves, and excess water is "transpired" through the stomata as vapor. During the day, when a plant is photosynthesizing, the stomata are open to allow the entry of carbon dioxide; at the same time, water vapor escapes by *transpiration*. At night, the stomata are closed and there is little transpiration. The energy requirement for transpiration is the same as that for evaporation: up to 590 calories per gram of water vaporized at normal temperatures. Hence transpiration too is a cooling process.

Water loss from land surfaces is usually regarded as due to the combination of transpiration from vegetation and evaporation from soil. As we saw in Chapter 3, the two processes together are known as *evapotranspiration*. In areas covered by vegetation, water loss by transpiration is usually two to three times greater than water loss by evaporation directly from the soil. The clearing of forests noticeably decreases evapotranspiration and increases both streamflow and groundwater recharge, and forest regrowth clearly reduces both surface and subsurface water supplies. The ratio between transpiration and direct evaporation can vary considerably, depending on degree and type of vegetation cover, making it very difficult to evaluate annual patterns of evapotranspiration from the continents. Estimates of evaporation from the oceans are far less difficult. The effects of evapotranspiration on local water budgets are treated in detail in Chapter 7.

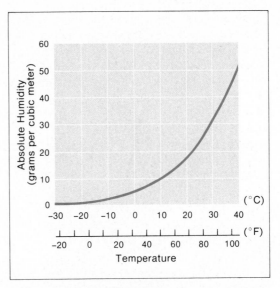

Figure 5.4 The graph shows the maximum amount of water vapor that can be contained in a cubic meter of air at various temperatures. If the air is not saturated, the absolute humidity lies below the curve. The absolute humidity of saturated air rises rapidly with increased temperature, so warm air is able to contain more water vapor per unit of volume than cool air. The shape of this curve is similar to that of the saturation vapor pressure curve in Figure 5.2. (Doug Armstrong after *Smithsonian Meteorological Tables*, edited by Robert J. List, 6th ed., 1971, by permission of The Smithsonian Institution)

MOISTURE IN THE ATMOSPHERE

There is always some water vapor in the air. At a given temperature, the air may contain any amount of water vapor up to the maximum value, at which saturation occurs. Figure 5.4 shows that air at 10°C (50°F) can contain up to 10 grams of water vapor per cubic meter. Ten grams, the weight of two U.S. nickels, may not seem like much water, but rain clouds that contain 10 grams of water vapor per cubic meter have the potential to produce significant rainfall on the ground below. The water vapor content of the atmosphere is referred to as *humidity*. Humidity can be described by several different measures, each one useful for specific purposes.

Measures of Humidity

The most direct measure of the air's moisture content is the *absolute humidity*, which is the weight of water vapor per volume of air. Absolute humidity is usually expressed in grams of water vapor per cubic meter of air. In the atmosphere, however, warming air expands and cooling air contracts, so even when there is no change in the water vapor content, the absolute humidity changes when the temperature changes. The use of absolute humidity, therefore, is usually restricted to controlled laboratory experiments.

Relative humidity is the ratio between the

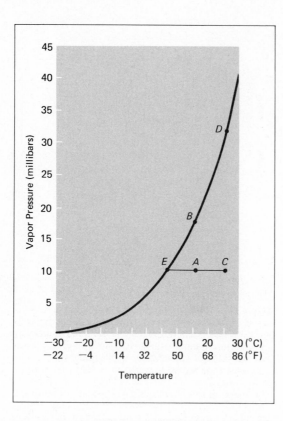

Figure 5.5 Changes in temperature can affect relative humidity even when the actual moisture content of the air remains unchanged. Consider a hypothetical sample of air taken at 10:00 A.M. with a temperature of 15°C and a vapor pressure of 10 millibars (mb). This sample is represented by point A. Saturation vapor pressure for 15°C is about 17 mb (point B); so the relative humidity is approximately ¹⁰/₁₇ × 100, or 59 percent. Point C represents a sample taken the same day at 3:00 P.M. The water vapor content of the air has remained unchanged, but the temperature has risen to 25°C. Saturation vapor pressure for 25°C is about 32 mb (point D); so the relative humidity is now ¹⁰/₃₂ × 100, or 31 percent. Relative humidity has decreased as the air has grown warmer. Point E represents a sample taken the following morning at 5:00 A.M. Water vapor content is still the same, but the temperature is now 7°C. Saturation vapor pressure for 7°C is about 10 mb; therefore the relative humidity is now ¹⁰/₁₀ × 100, or 100 percent. The sample has become saturated by cooling without any change in water vapor content. The rapid decrease of saturation vapor pressure with decreasing temperature has very important implications for condensation processes. (After Neiburger et al., 1973)

Figure 5.6 During fair weather, the vapor pressure of the air near the ground remains almost constant over a 24-hour period. However, the relative humidity changes markedly as the air temperature changes because the saturation vapor pressure of air is strongly dependent upon temperature. (Doug Armstrong adapted from *Hydrology Handbook*, 1949, published by the American Society of Civil Engineers)

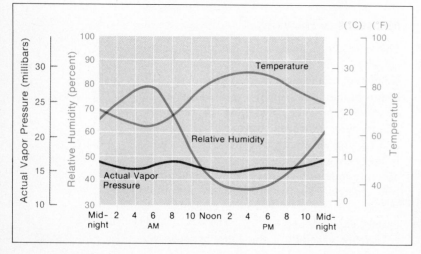

water vapor in the air and the amount the air could hold if it were saturated. The relative humidity is expressed only as a percentage, and does not tell us the actual amount of water vapor present in the air. For example, a certain sample of air at 40°C (104°F) has a relative humidity of 50 percent. This means it has 50 percent of the vapor it could hold if it were saturated. If we consult the graph in Figure 5.4, we find that the air contains about 25 grams of moisture per cubic meter. Similarly, cooler air at 20°C (68°F) with 80 percent relative humidity contains only about 14 grams of moisture per cubic meter. The cool air in these examples has a lower water vapor content but a higher relative humidity than the hot air because the cool air is closer to saturation.

Figure 5.5 shows that air can become saturated merely by cooling, without any change in water vapor content. As we shall see, cooling of air down to the temperature at which it is saturated is the most important factor in condensation and precipitation.

As a measure of atmospheric moisture, relative humidity has the same disadvantage as absolute humidity: whenever the temperature of a sample of air changes, the humidity measure also changes, even if there is no change in water vapor content. Figure 5.6 shows that the changing temperature over a 24-hour period during fair weather usually causes relative humidity to be highest overnight and in the early morning, and lowest in mid-afternoon.

Several other measures are used to describe the water vapor content of air. The *specific humidity* is the weight in grams of the water vapor present per kilogram of air. The *mixing ratio* is the weight in grams of the water vapor present per kilogram of dry air (air minus its water vapor). Both of these measures are useful because they change only when the amount of water vapor itself changes. They are used especially in studies of air masses, which are discussed in Chapter 6.

A final and increasingly important method of expressing the moisture content of air is by means of the *dew point*—the temperature at which saturation occurs and condensation begins. The higher the dew point, the greater the moisture content of the air. Dew point is discussed more fully in the section on condensation processes.

Distribution of Water Vapor in the Atmosphere

Only a tiny fraction of the earth's water supply is stored in the atmosphere at any one time. If all the atmospheric water vapor fell to the earth as rain, it would produce a layer of water only about 2.5 cm (1 in.) deep. Heavy rains are possible only because there is a constant recycling of liquid water and water vapor within the hydrologic cycle. In the atmosphere, water that is lost by precipitation at one place is at the same time being replaced by evapotranspiration at another place.

On land the transfer of moisture to the atmosphere by evapotranspiration proceeds most rapidly on a fair summer day when the ground is warm and moist and there is a large net gain of radiation. Convection carries water vapor aloft quickly, so a steep vapor pressure gradient (rapid vapor pressure change with height) is maintained near the ground (Figure 5.7). At night, however, the net radiation is negative and there is little or no energy available for evapotranspiration. At the same time, the air

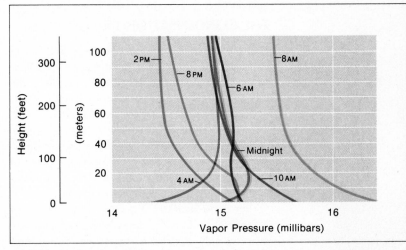

Figure 5.7 Each curve on this graph shows how the vapor pressure in the atmosphere near the ground varies with height throughout a clear summer day. At 8:00 A.M. the vapor pressure decreases rapidly with increased height, indicating a flow of water vapor from the surface into the atmosphere. The vapor pressure decreases from 8:00 A.M. to 2:00 P.M. because the air becomes heated during the morning and early afternoon, causing the onset of convection and consequently a more efficient removal of moist air from near the surface. Much of the water vapor remains in the lowest 50 meters of the atmosphere, forming a humid blanket. Water vapor continues to flow upward from the surface during the day while plants are transpiring. By 8:00 P.M. the vapor pressure is nearly constant through the lowest 15 meters of air, indicating that little water vapor is leaving the surface. Later at night, the vapor pressure immediately above the surface increases with increased height, indicating a flow of water vapor from the atmosphere toward the ground as dew or frost. By 6:00 A.M., after sunrise, the flow is once again from the surface to the atmosphere. (Doug Armstrong after R. Geiger, *The Climate Near the Ground*, © 1965, Harvard University Press)

immediately above the surface usually is cooled to its dew point. Then water vapor condenses on the ground, roof tops, and automobiles as dew or frost.

The water vapor content of the atmosphere varies horizontally as well as vertically. Average atmospheric storage of water over North America ranges from about 5 cm (2 in.) of equivalent precipitation near the Gulf of Mexico in the summer to less than 0.8 cm (0.3 in.) over central Canada during the winter. This pattern of water vapor distribution varies constantly, however. Daily rainfalls of as much as 60 cm (24 in.) fell near Houston, Texas, during tropical storm Claudette in July 1979, indicating that atmospheric circulations are capable of concentrating immense quantities of water vapor in a very short period of time.

THE CONDENSATION PROCESS

Condensation is the process by which water vapor in the atmosphere changes phase to become tiny liquid droplets or ice crystals; the phase change from vapor directly to ice crystals is often called *sublimation*. At the earth's surface, condensation on cool or cold objects produces dew or frost. When water vapor condenses in the atmosphere, the result is a mass of water droplets or ice crystals known as a cloud. Fog is simply a cloud at the earth's surface.

For condensation to occur, the water vapor content of at least one layer of the atmosphere must closely approach saturation. Saturation can be produced by adding water vapor to the atmosphere, by cooling the atmosphere, or by a combination of moisture addition and cooling. Most cloud formation and precipitation result from atmospheric cooling.

Condensation Nuclei

When moist air is cooled to saturation, condensation does not occur automatically. Fine particles, called *condensation nuclei*, must be present to act as centers for condensation and the growth of water droplets. In moist air that is completely free of particles of any size, condensation takes place only when the vapor pressure is three or four times the saturation value. Air containing more than its normal saturation amount of water vapor is said to be *supersaturated*. However, since air is usually well supplied with microscopic particles, condensation generally begins almost as soon as the air is cooled to saturation.

Figure 5.8 shows that condensation nuclei are extremely small. Few exceed a radius of 1 micrometer (0.001 mm), and most have radii smaller than 0.1 micrometer (0.0001 mm). Giant condensation nuclei, with radii greater than 1 micrometer, number only about one per cubic cm. Principal sources of condensation nuclei are industrial pollutants and smoke, forest fires, salt crystals from sea spray, pollen, and dust particles. Natural processes and industrial and agricultural activities renew the supply of atmospheric particles as fast as they fall to the

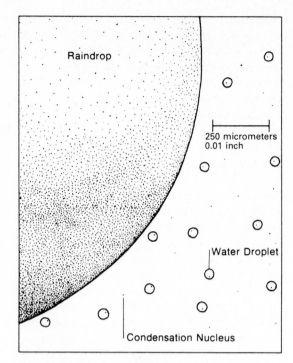

Figure 5.8 This drawing shows the relative sizes of condensation nuclei, water droplets, and a raindrop. Most condensation nuclei are less than 1 micrometer in radius; water droplets are usually less than 20 micrometers in radius; and raindrops have radii of 1,000 micrometers or greater. Each water droplet forms in saturated air by condensation of water vapor on a condensation nucleus, and other processes cause millions of water droplets to coalesce and form a raindrop. (John Dawson)

surface or are washed out by rain. Hence nuclei for condensation are almost always available when the air becomes saturated.

The water droplets that form clouds usually grow to a radius of only about 10 micrometers before the addition of water molecules ceases. In a typical cloud there may be a million water droplets per cubic meter, and the average water content of the cloud may be about 1 gram per cubic meter of air.

A cloud droplet with a radius of only 10 micrometers has a settling velocity (rate of fall) of less than 1 centimeter per second, and it would normally evaporate in the drier air below the cloud. Under average conditions, cloud droplets do not grow larger than about 10 micrometers in radius, nor can they easily combine into larger drops. Studies have shown that when such small droplets approach one another in a cloud, they usually are swept past each other in the airflow adjacent to the droplets. Thus most clouds never contain water drops large enough to fall as rain. Later in the chapter we shall see what processes are necessary to produce precipitation.

Dew and Frost

When a volume of air cools with no change in water vapor content, its relative humidity increases. The temperature at which the relative humidity becomes 100 percent is the *dew point*. When the air cools to the dew point, water vapor begins to condense into small drops of liquid water. In warm, humid weather the outside of a glass of ice water becomes covered with beads of moisture because the moist air near the glass has been chilled to the dew point temperature. The same process accounts for the

formation of dew itself. Longwave radiation cools plants and soil after sunset, and the air in contact with them is also cooled by conduction. If the air cools to the dew point, its water vapor condenses on the ground and on exposed plant surfaces, forming dew (Figure 5.9). In urban areas, dew formation is most apparent on automobiles parked outside overnight.

When the moisture content in the air is such that the dew point is at or below the freezing point of water, the water vapor sublimates as ice crystals, and frost appears on exposed surfaces. Because condensation releases latent heat, dew and frost formation keep plant and soil surfaces from cooling as much as they would if these phenomena did not occur.

Fog

Fog occurs when a thick layer of moist air near ground level is cooled to its dew point. This

Figure 5.9 Dew forms on plants and other natural and artificial surfaces when the surrounding air cools to the dew point. (David Cavagnaro)

can occur in several ways, so various types of fog are recognized.

Radiation fog occurs at night when the ground and the air immediately above lose heat by conduction and longwave radiation. If the ground and air are moist, the lowest 10 or more meters of air may be cooled to the dew point, resulting in condensation in the form of dense fog. Moist air and damp ground are particularly susceptible to radiation fog on clear nights, which promote longwave radiation losses to space. But even moderate breezes will prevent the formation of radiation fog by mixing the colder air near the ground with warmer air above.

Advection refers to the horizontal movement of air across the earth's surface. When warm, moist air passes over snow or cold ocean water, the air may be cooled to its dew point, resulting in *advection fog*. Warm air moving over cold ocean water off the coasts of California and New England frequently produces advection fog. San Francisco is famed for its summer fog, which forms over cold coastal waters and is drawn inland in the afternoon by thermal convection over the land (Figure 5.10). As Figure 5.11 shows, in the United States fog is most common along coastlines and over the Appalachian region.

When advection fog occurs over snow, there may be a marked increase in snowmelt. Normally, snow melts slowly because it has a high albedo and therefore reflects much of the incoming solar radiation back into space. However, the advection of warm, moist air over snow-covered areas causes condensation on the snow surface. Every gram of water vapor that condenses releases enough latent heat to melt more than 7 grams of snow. Thus when warm, moist air from the Gulf of Mexico moves northward over snow-covered landscapes in the American Midwest and Northeast, rapid melting of snow can occur even on a cool, overcast day. Quite reasonably, skiers hate fog!

Orographic fog is associated with mountain areas, such as the Appalachians. When moist air is forced up mountain slopes, it can be cooled to the dew point, resulting in the formation of orographic fog and clouds. Orographic cooling is a major cause of condensa-

Figure 5.10 This photograph shows San Francisco Bay and Oakland, California, shrouded in fog, with Mount Diablo in the distance to the east. Low-lying advection fog frequently occurs in the coastal regions around San Francisco Bay as moist air moving eastward over the Pacific Ocean cools to its dew point and streams through the Golden Gate. (David Cavagnaro)

tion and cloud formation and will be discussed in more detail later in this chapter.

Cooling of surface air by radiation or advection is restricted to a shallow layer of the atmosphere near the surface. The resulting fog and low clouds are simply not thick enough for the processes that produce rainfall, snow, or hail to be effective. The air may be filled with a mist of small droplets, sometimes falling as light drizzle, but significant precipitation can occur only when a much thicker mass of air is cooled throughout its depth by being lifted to higher elevations.

Fog of any type can be very hazardous to human activities. Throughout history the vast majority of ship collisions have occurred in fog. Many other accidents on the water, including frequent collisions of boats with bridge piers, are attributable to fog. On one day in September 1923, seven destroyers of the U.S. fleet ran

aground in dense fog, one on the heels of another, near Point Conception on the southern
California coast. Such accidents, which increasingly involve large tankers laden with enormous quantities of oil, gain even more importance because they can cause disastrous
pollution of coastal waters.

On the land massive chain accidents, involving dozens to more than a hundred vehicles, have occurred on fog-shrouded highways
in California and New Jersey. Airports are usually located to avoid fog, but many continue to
be affected by this hazard, causing frequent and
inconvenient diversions of flights to alternate
points. Airports serving Boston, New York, Los
Angeles, San Francisco, Seattle, London, Zurich, and many other important centers have
frequent fog problems.

Clouds

Clouds are condensation forms that develop
above the ground, usually by the lifting and
chilling of moist air. Although clouds seem to
appear in almost infinite variety, they can be
categorized fairly easily. Clouds occur in three
general types and are recognized as falling into
three elevation classes (Figure 5.12). At the
highest altitudes are *cirrus* clouds composed of

Figure 5.11   This map shows the average annual number of days of fog in the coterminous United States.
Fog occurs most frequently along the Pacific coast
and the coast of New England, where moist air is
cooled by cold ocean waters. Fog also occurs frequently in the Appalachian Mountains of the east-cen-
tral United States. It seldom occurs in the warm, dry
air of the western deserts. (Andy Lucas and Laurie
Curran adapted from Arnold Court and Richard Ger-
ston, *Geographical Review*, 56, © 1966, American Geographical Society of New York)

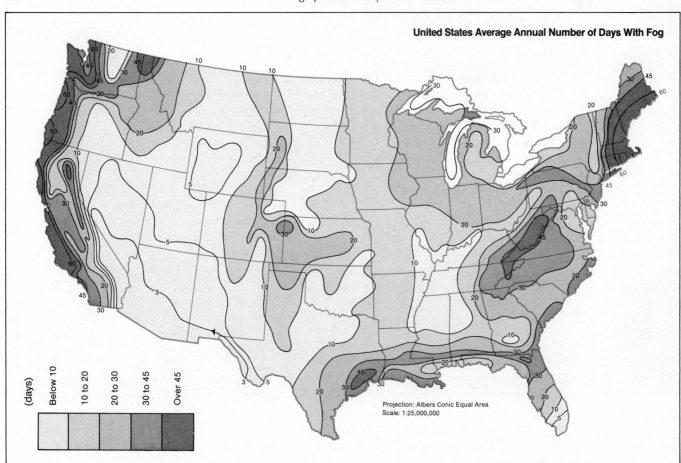

Figure 5.12 (a) Fair-weather cumulus clouds over Baton Rouge, Louisiana. Each cloud forms in a rising air bubble or thermal. Their flat, even bases mark the altitude at which the rising thermals are cooled to the dew point and saturation. These clouds tend to be short-lived, usually evaporating into the surrounding drier air within an hour or so. (Robert A. Muller)

(b) Cirrus clouds are feathery wisps of ice crystals that form at high altitudes. They are frequently the first indication in the sky of the approach of a midlatitude cyclone system from the west or south in the United States. (Robert A. Muller)

(c) An altocumulus cloud, or "mackerel sky," over Boulder, Colorado, takes the form of a layer of patchy cloud puffs. Altocumulus clouds develop at moderate altitudes and often herald the approach of a warm front and accompanying precipitation. (National Center for Atmospheric Research, Boulder, Colorado)

(d) Cumulonimbus clouds, or "thunderheads," rise to great heights and are sometimes associated with heavy rain, flash floods, hail, and tornadoes. This view shows massive thunderstorms organized in an orographic situation with warm, moist air being forced up and over the Rocky Mountains at Boulder, Colorado. (Robert A. Muller)

(e) The leading edge of a squall line passing over Baton Rouge. The squall line was just ahead of a vigorous cold front, and the dark clouds are indicative of severe turbulence and violent winds. (Robert A. Muller)

(f) Shallow radiation fog near sunrise during autumn in southeastern Wisconsin. (Robert A. Muller)

(g) Artificially produced cumulus clouds over petrochemical plants in the Los Angeles basin. These small clouds are associated with heat and water vapor added to the lower atmosphere as by-products of industrial processes. (Robert A. Muller)

e

f

g

ice crystals that form thin filaments, wispy puffs, or translucent veils. A second general type is the horizontal *stratus* cloud, which forms an extensive continuous sheet. The third general type is the *cumulus* cloud, which has vertical rather than horizontal development, appearing puffed up, billowy, and sometimes forming awesome white towers in the sky. Each cumulus cloud indicates a strong but localized current of rising air, whereas stratus types develop in air with widespread but gentle uplift. These simple cloud types can be combined: thus *cirrocumulus* clouds are high puffs; *cirrostratus* clouds are high thin veils; *stratocumulus* clouds are layers of coalescing puffy clouds.

Cloud types are also designated by altitude: as high clouds (cirrus types); middle-altitude clouds, including *altostratus* and *altocumulus* types; and low clouds (stratus and stratocumulus). The term *nimbo*, or *nimbus*, means that the clouds are producing precipitation. Low-level cloud blankets that result in rains are *nimbostratus* clouds. Towering cumulus clouds that bring torrential downpours, often including hail, are *cumulonimbus* clouds. The cumulonimbus type may reach to an elevation of 12,000 to 15,000 meters (40,000 to 50,000 ft) and frequently flares out at the top in the shape of a blacksmith's anvil. Meteorologists have more sophisticated names to distinguish the great variety of clouds that exist, but the names discussed here are the ones most widely recognized.

THE PRECIPITATION PROCESS

We mentioned earlier that most cloud droplets are only about 10 micrometers in diameter, and that they will tend to remain aloft in clouds because of updrafts associated with most developing clouds. An average-sized raindrop, however, has a radius of 0.1 cm (1,000 micrometers), and geometric calculations show that about 1 million 10-micrometer cloud droplets are needed to form a single raindrop. The processes that cause water droplets to coalesce to drops large enough to fall from clouds have been studied in detail since the 1940s. Scientists

have described two separate processes that are believed to be responsible for precipitation. These are known as the *coalescence model* and the *Bergeron ice crystal model*.

Formation of Rain and Snow

The generation of rain by the coalescence process depends on the occurrence of oversize water droplets that are larger than 20 micrometers in radius. An oversize droplet falls just a bit faster than the typical droplet, and it grows by colliding with and sweeping up smaller droplets in its path. Rising currents of air carry the swelling droplets upward faster than they can fall out of the cloud, allowing them more time to grow. A droplet requires about half an hour to grow to raindrop size by coalescence, and the rain cloud must be at least 2.5 km (1.6 miles) thick to contain the growing drops long enough for them to become raindrops. Thinner clouds limit the growth of drops by coalescence, resulting in *drizzle*, a form of precipitation that consists of very tiny drops that "float" rather than fall to the surface. Pavements made wet by drizzle can be very hazardous for motorists, but drizzle never produces significant quantities of precipitation.

Sea salt particles 1 micrometer or more in radius make particularly effective condensation nuclei for the formation of oversize droplets. This is because of the *hygroscopic* nature of salt—its natural tendency to absorb water vapor. Even far inland, salt particles blown off the ocean play an important role. Nor is it necessary for the salt particles, or other giant condensation nuclei, to be present in large quantities. Only about one large condensation nucleus is needed for each cubic meter of cloud.

The coalescence process is most effective in thick vertical clouds, which are common in the tropics. It is less effective in higher latitudes, where some process other than coalescence is needed to produce significant quantities of precipitation. In the middle latitudes, the tops of many rain clouds are at altitudes where the temperature is well below freezing.

In 1933 Tor Bergeron, a Swedish meteorologist, proposed a process of droplet growth to explain rainfall from cold clouds that were not necessarily thick. Although puddles of water on the ground freeze at 0°C (32°F), tiny water droplets in the atmosphere can remain liquid at temperatures down to −40°C (−40°F). Because of this *supercooling* effect, droplets often remain liquid even when the upper parts of a cloud are far below 0°C. Tiny clay particles in the air, known as *ice-forming nuclei*, promote the freezing of supercooled droplets into ice crystals, usually after the temperature has dropped below about −10°C (14°F).

In other words, at temperatures between 0° and −10°C (32° and 14°F), most clouds contain only supercooled water droplets. Between −10° and −40°C (14° and −40°F), clouds contain both supercooled water droplets and ice crystals; these clouds are usually called mixed clouds. At temperatures below −40°C (−40°F), only ice crystals are normally present.

Although ice-forming nuclei are less common than condensation nuclei, only about 100 per cubic meter of cloud are needed to trigger precipitation. In a mixed cloud, the vapor pressure of supercooled water droplets is greater than that of nearby ice crystals at the same temperature. Therefore, once ice crystals have formed, a small vapor pressure gradient is established between supercooled water droplets and nearby ice crystals. The ice crystals grow by sublimation at the same time that the droplets are losing water by evaporation. The ice crystals also grow by collision with other supercooled water droplets. Most summer rains in the middle latitudes therefore begin as ice and snow in the upper parts of towering clouds. In winter, the ice crystals fall as snow if the air is not warm enough to melt the snowflakes. The precipitation processes are illustrated in Figure 5.13.

Weather modification has evolved primarily for situations when supercooled clouds are present. Silver iodide crystals can be introduced into clouds as "artificial" ice nuclei, but sometimes powdered dry ice is used to cool clouds down to the temperature of spontaneous sublimation, −40°C (−40°F). Weather modification is discussed in the Case Study at the end of this chapter.

Formation of Hail

Hailstones are spherical balls of ice that fall from towering summer thunderheads (cumulonimbus clouds) that have strong updrafts. Hail ranges from tiny grains to spheres the size of a grapefruit. When cut open, hailstones show an onionlike layering of clear and opaque shells of ice. The layering suggests that hailstones grow by being cycled through a cloud many times (Figure 5.14).

It is generally believed that strong updrafts in a thundercloud carry small ice particles upward through the cloud's central column. If the particles rise through a section containing many supercooled water droplets, a wet layer will form around each ice particle. This water gradually freezes into a shell of clear ice. If the ice particles then rise into a region having relatively few small supercooled droplets, the collisions result in instant freezing. In this way another layer of ice forms on each ice particle, a layer that is cloudy due to pockets of trapped air. Near the top of the thunderhead, the ice particles emerge from the central updraft and fall through the outer areas of the cloud. As they near the base of the advancing cloud, however, they may be swept back into the central updraft.

With each cycle through the cloud, the particles gain new layers of ice. After a number of cycles, they grow too heavy to be supported by the updrafts, and they fall to the ground as hail. Large hailstones are associated with violent thunderstorms in cumulonimbus clouds containing updrafts with speeds of 20 to 30 meters per second (45 to 60 miles per hour). Hail is most common east of the Rocky Mountains on the high plains from Alberta to Kansas, where it can destroy late spring or early summer wheat crops in a few moments.

ATMOSPHERIC COOLING AND PRECIPITATION

To generate significant rainfall by either the coalescence or ice crystal process, large masses of

Coalescence Model

(1) Condensation of
vapor onto nuclei
and growth of droplets
from vapor

(2) Growth of raindrops
by coalescence with
droplets

Bergeron Ice Crystal Model

(1) Condensation
of cool vapor
onto nuclei and
growth of
supercooled
droplets

(2) Freezing of
supercooled
droplets onto
ice-forming
nuclei

(3) Growth of ice
crystals at
the expense of
water droplets

(4) Further growth
of ice crystals
by coalescence
with other ice
crystals and
supercooled droplets

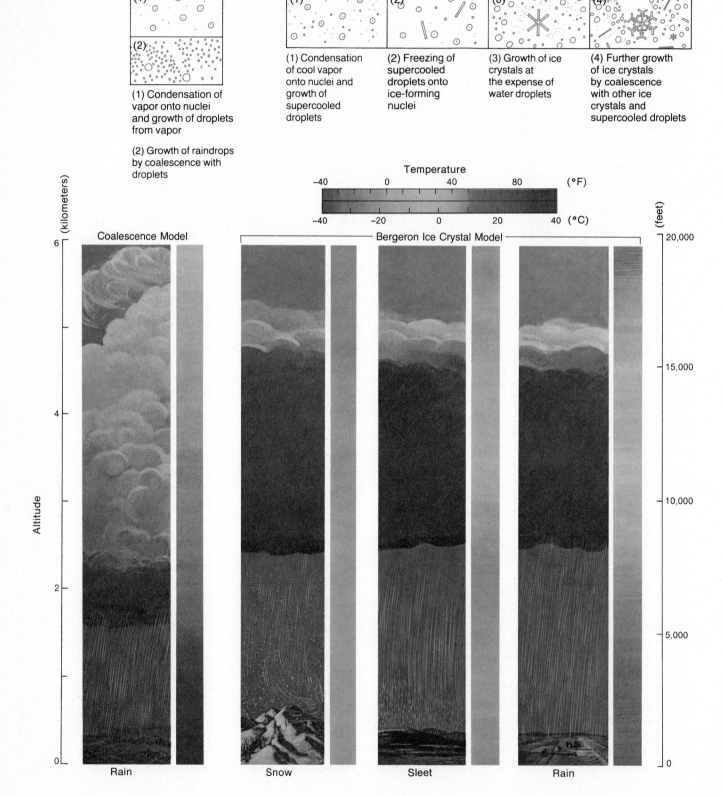

Temperature

−40   0   40   80   (°F)

−40   −20   0   20   40   (°C)

Coalescence Model

Bergeron Ice Crystal Model

Altitude

Rain   Snow   Sleet   Rain

Figure 5.13 (opposite) The principal models that successfully explain precipitation from clouds are the coalescence model and the Bergeron ice crystal model. The main steps of each process are outlined in this figure.

The coalescence process of precipitation is dominant in the tropics, where many clouds are too warm to contain supercooled water droplets and ice crystals. The Bergeron process dominates at higher latitudes, where the upper portions of many clouds are considerably colder than the normal freezing point of water, even in summer.

The type of precipitation that falls from a cloud depends on the mechanism involved and on the variations of temperature in the atmosphere. Precipitation is formed as snow in the Bergeron process, and if the atmosphere beneath the cloud is below freezing, the precipitation reaches the ground as snow. If the snow falls into warm air, however, the precipitation arrives at the ground in the form of rain. The coalescence process cannot produce snow, but either process can produce sleet or ice pellets if raindrops freeze while falling through a layer of cold air. (John Dawson)

dense. The air in the rising parcel expands in response to the decrease in the density and pressure of the atmosphere around it. Because the molecules of the expanding air become more widely separated, they collide less often, so the sensible temperature of the rising air decreases.

The rate of adiabatic cooling of rising air in which condensation is not occurring is known as the *dry adiabatic rate*. Its value is 10°C for every kilometer of increasing altitude (5.5°F per 1,000 ft), as illustrated in Figure 5.16.

Figure 5.14 This cross section of an intense thunderstorm illustrates the cyclic processes involved in the formation of hailstones. Powerful currents of air in the central updraft carry falling raindrops and ice crystals upward. Because of the forward motion of the storm, ice particles may be swept up in the updraft a number of times. On each passage through the cloud, the particles gain new layers of ice, thus becoming hailstones. Clouds with strong updrafts are able to carry hailstones through many cycles until they become too heavy for the air currents to support. Particles are also blown away from the central column by strong winds aloft, giving the cloud its characteristic anvil shape. (Doug Armstrong after Hermann Flohn, *Climate and Weather*, © 1969 by H. Flohn, used by permission of McGraw-Hill Book Company)

air must be cooled throughout their depth. This can be accomplished only by lifting the air to higher altitudes. Vertical displacement of large masses of air, which can produce heavy rainfall, snow, or hail from clouds, is brought about in several ways. The remainder of this chapter discusses the phenomena that trigger precipitation from clouds.

Adiabatic Cooling

Net radiation gains at the surface may either heat the surface or cause evapotranspiration. Largely because albedo and evapotranspiration vary from place to place, some locations heat up much more than others. Figure 5.15 shows how air above such local "hot spots" becomes organized into rising bubbles of warm air. Soaring birds and glider pilots make use of these rising bubbles, called *thermals*, to gain or maintain altitude. Circling within the updraft, they are carried upward until the air cools by expansion and loses its bouyancy.

The temperature decrease in rising air results from a process known as *adiabatic cooling*. The term "adiabatic" refers to the fact that the process occurs without energy gains or losses by the air. As a parcel of air rises to higher levels, the surrounding air is increasingly less

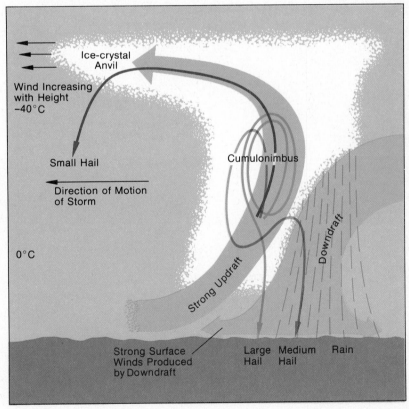

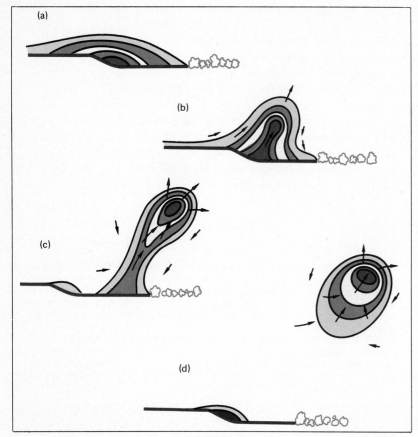

Figure 5.15 A thermal is a rising-bubble of air ema-
nating from a local "hot spot" on the earth's surface
where the temperature exceeds that of the surrounding area. This differential heating is associated with
variations in albedo, orientation of slopes to the sun,
and availability of water for evapotranspiration. The
rising thermal initially exchanges little heat with the
surrounding air. Each concentric circle, or isotherm,
represents an increase of 0.1°C; thus the outer circle
is 0.1°C warmer than the surrouding atmosphere, the
next circle is 0.2°C warmer, and so on. The highest
temperature occurs at the center of the thermal. (Vantage Art, Inc. after Herbert Riehl, *Introduction to the
Atmosphere*, © 1972 by McGraw-Hill Book Company)

The adiabatic cooling process can be reversed. When a parcel of cool air descends, it
becomes compressed. As a result, the molecules in the parcel collide more frequently and
the temperature increases. If no heat energy is
gained or lost, the process is called *adiabatic
heating*. A descending parcel of cool air, subject
to adiabatic heating, increases in temperature
at the dry adiabatic rate.

Rising air cools at the dry adiabatic rate only
until it reaches the temperature that is its dew
point and it becomes saturated. After saturation, the air may continue to rise, but it cools
at a reduced rate called the *moist adiabatic rate*.
The rate of cooling decreases at saturation because the water vapor in the parcel begins to
condense, releasing latent heat. The added heat
increases the buoyancy of the parcel of air and
enables it to rise to greater altitudes. This mechanism permits towering clouds to develop with
potential for high-intensity precipitation.

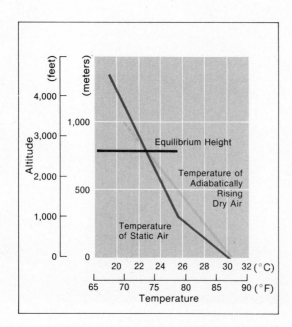

Figure 5.16 When a parcel of dry air rises, it expands
in volume to equal the pressure of the surrounding
atmosphere. The expansion causes the temperature of
the parcel to fall. The figure compares the temperature of a typical parcel of rising dry air with the temperature of the air in the surrounding static atmosphere. The temperature of the parcel and the
temperature of the static air both decrease with increased altitude, but in general the temperature of
the rising parcel falls more rapidly, as indicated in the
figure. The parcel experiences a net upward force,
causing it to continue rising, only as long as its temperature exceeds the temperature of the static air.
The equilibrium height of the parcel is near the
height where the temperatures are equal. There is no
upward convection past this level, but forced lifting
and cooling at the adiabatic rate can continue along
frontal surfaces and over mountain barriers. (Doug
Armstrong)

The water vapor in the air parcel does not all condense at once. Condensation continues as the parcel rises, and at any given altitude the parcel contains just enough water vapor to maintain saturation. When the rising air cools to the same temperature as the surrounding atmosphere, its upward movement by convection ceases.

The moist adiabatic rate varies with the moisture content of the rising air. In warm, moist air it may be 5°C per km (2.8°F per 1,000 ft), but at very low temperatures it approaches the dry adiabatic rate of 10°C per km.

Thermal Convection

On a fair day, puffy clouds form within thermals at the altitude where saturation is reached and condensation begins. Figure 5.17a shows that the thermal continues to rise until the temperature of the rising air equals that of the surrounding atmosphere. This *equilibrium point* marks the greatest height to which the cloud can grow. The vertical movement of thermals responsible for fair-weather cumulus cloud formation is known as *thermal convection*.

Figure 5.17 (a) When the land surface heats, it warms the lower atmosphere, causing the environmental lapse rate (red line) to bend to the right (higher temperature) at the surface. A rising thermal or parcel cools at the dry adiabatic rate (yellow line) until condensation begins. The base of the cloud at about 0.7 km marks the altitude at which the thermal cools to the dew point, initiating condensation. The parcel then cools at the moist adiabatic rate, which is less than the dry rate. The release of the latent heat of condensation enables the parcel to rise to much higher altitudes before it reaches equilibrium with the surrounding static air. Condensation continues as the parcel rises, and at any given altitude the parcel contains just enough water vapor to keep the air in it saturated.

(b) This shallow inversion, with warmer air above cooler air at the surface, is characteristic of radiational cooling at night. If a parcel were forced to rise (yellow line), it would come quickly into equilibrium with the surrounding static air. Inversions are associated with atmospheric stability.

(c) With daytime heating, the inversion is much higher aloft, permitting thermals to reach much higher altitudes. The equilibrium level is the limit of thermal convection. (Doug Armstrong)

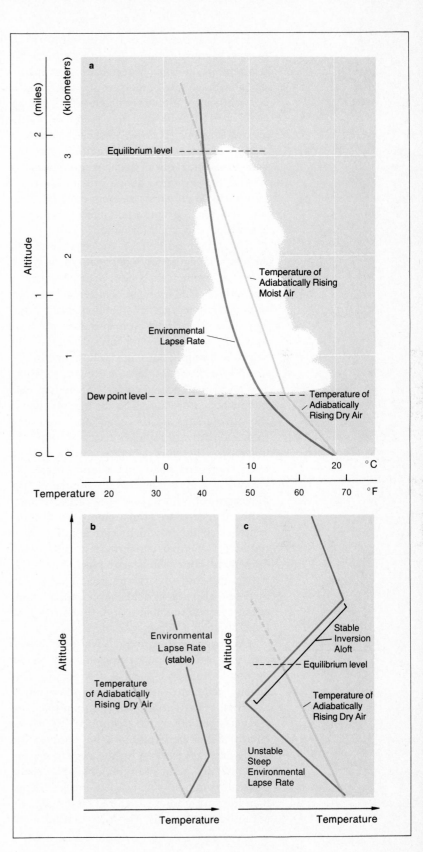

Since the atmosphere is heated principally by energy transfer from the earth's surface, air temperature normally decreases upward. A global average rate of temperature decrease for all weather conditions is about 6.5°C per km (3.6°F per 1,000 ft), and this is known as the *average lapse rate*. Locally, however, the rate of change of temperature with altitude may deviate considerably from that average. The actual vertical temperature change measured at any location, known as the *environmental lapse rate* (Figure 5.15a), has important effects on thermal convection and cloud formation.

The lower portion of the environmental lapse rate in Figure 5.17b represents a condition in which warmer air is aloft over cooler air at the surface. This *temperature inversion* is a common result of nighttime radiational cooling during fair weather. If dry air near the surface were forced to rise by flowing upward against the slopes of a mountain range (yellow line in Figure 5.17b), it would cool at the dry adiabatic rate and soon become cooler and denser than the air above the inversion. Being cooler than the surrounding air, the lifted air would tend to sink downward rather than rise through the inversion. Therefore a temperature inversion restricts upward motion in the atmosphere. Environmental lapse rates that discourage vertical movement of air are *stable lapse rates*. Temperature inversions are an extreme type of atmospheric stability, in which there is little opportunity for thermal convection or the development of thick clouds and precipitation. By preventing the upward movement of air, an inversion also acts as a lid that traps pollutants emitted by automobiles and industries and holds them in a dense layer of "smog."

On clear days the ground is warmed by solar radiation, and the lower atmosphere is heated by conduction and thermal convection. This usually eliminates any overnight temperature inversion at the surface. Air can then rise, and thermals become active (Figure 5.17a). If the environmental lapse rate is steep (temperature decreasing rapidly with height at rates greater than the dry adiabatic rate: 10°C/km), rising thermals continue to be warmer and less dense than the surrounding air an can continue moving upward buoyantly. The ascent of this air to great altitudes permits the development of towering cumulonimbus clouds that are associated with thunderstorms (Figures 5.17a and 5.12d). Steep environmental lapse rates leading to the development of thermal convection are *unstable lapse rates*.

During summer, when rapid heating of the ground is responsible for steep lapse rates and conditions of atmospheric instability, rain-producing cumulus-type clouds follow a regular pattern of development. Early in the day, the sky is usually clear. After a few hours, scattered wisps of cloud begin to appear at an altitude of about 1 km (3,000 ft). These are formed by condensation in convective updrafts. The wisps increase in size and eventually become large cumulus clouds whose bases are at the altitude where the rising air has cooled to its dew point. Cumulus clouds indicate the maximum heights of rising thermals. Between the clouds, air sinks to replace the air that rises in the thermals. For this reason, cumulus clouds never cover the sky completely, and precipitation from them takes the form of scattered showers and thunderstorms that typically break out in the afternoon and persist into early evening. An individual shower may produce more than 2.5 cm (1 in.) of rain locally, but thermal convection rarely results in widespread rains.

Orographic Lifting

When moisture-laden air is forced to rise over a mountain barrier, the air cools, and condensation and precipitation often result. This process, which may exert a strong local effect on weather, is known as *orographic lifting*.

Clouds, fog, and precipitation are characteristic of most mountainous regions, to the frequent disappointment of vacationers. In western North America, the air normally flows from west to east; thus the western, or windward, slopes of mountain ranges are often cloudy and wet. When the air descends the eastern slopes, however, it warms at the dry adiabatic rate. In the warmer descending air, water droplets evaporate and the clouds dissipate. The eastern slopes and adjacent lowlands are usually sunny and dry (Figure 5.18) and are said to lie in the "rainshadow" of the mountains. Other mountain areas are subject to different patterns of orographic lifting. In the eastern United States, moist air may advance against the Appalachian mountain system from either the east or the west, so those mountains have no dry side.

A spectacular example of the effects of orographic lifting is found on the island of Kauai in the Hawaiian chain. The average annual rainfall on the windward side of Mount Waialeale is 1,170 cm (460 in.), but it is only 51 cm (20 in.) on the lee side, just 25 km (15 miles) away. This extraordinary rainfall record is produced by orographic clouds maintained by persistent moisture-laden winds from the northeast.

It was orographic lifting of unusually moist unstable air, producing enormous cumulonimbus clouds, that caused the Rapid City and Big Thompson Canyon disasters noted at the beginning of the chapter. The truly unusual aspect of these storms was that the rain-producing clouds remained over the same areas for several hours, rather than drifting onward and spreading their effects, as normally happens.

Frontal Lifting and Convergence

The lifting of large masses of moist air, causing widespread clouds and precipitation, does not require mountains. In the middle and higher latitudes, the atmospheric circulation often brings large masses of warm and cold air side by side. Largely because of density differences, the warm and cold air masses do not mix but remain separated by a boundary zone known as a *front*. Fronts, which are discussed more fully in Chapter 6, are zones of rapid transitions in the temperature and humidity characteristics of adjacent air masses. Where the air masses are converging, the warmer, less dense air either rides up and over the cooler air or is wedged upward by invading colder air. As it rises, the warm air cools to its dew point, clouds form, and precipitation is likely to occur.

Idealized cross sections of fronts are shown in Figure 5.19. When warm air is advancing horizontally over a surface previously occupied by colder air, a *warm front* exists (see Figure 5.19a). The advancing warm air slides over the

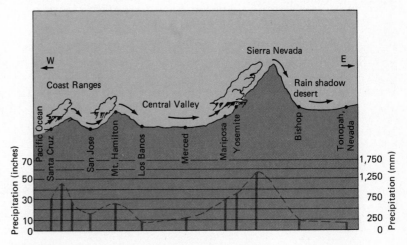

Figure 5.18 In mountainous regions average annual precipitation patterns are closely related to the terrain. This idealized transect across central California and western Nevada shows strong orographic effects on precipitation along the chains of mountains, and the rainshadow valleys between the mountain ranges east of the Sierra Nevada. The contrast between mountains and valleys is especially great in the Far West because just about all moisture-laden air moves from west to east. Around exposed summits of the Klamath Mountains in northwestern California, mean annual precipitation is estimated to be as high as 5,000 mm (200 inches). (From C. Donald Ahrens, *Meteorology Today.* St. Paul, Minn.: West, 1982.)

retreating cold air. The warm air rises only about 1 km in a horizontal distance of about 100 km; thus the ascent is gentle, unlike airflow over a mountain barrier. High cirrus clouds come first, followed by lower altostratus and nimbostratus clouds that produce widespread but gentle precipitation.

When cold air is advancing horizontally over a surface previously occupied by warmer air, a *cold front* is formed. Figure 5.19b illustrates typical cloud distribution along a cold front, where advancing cold air pushes warm air aloft. The

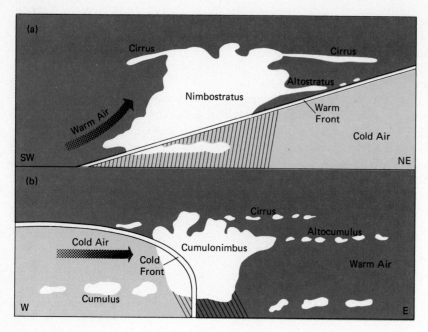

Figure 5.19 (a) When a large mass of warm air moves into a region occupied by colder air, the less dense warm air flows up and over the cold air. This atmospheric cross section shows that warm front slopes gently upward through the lower atmosphere. Characteristic clouds and precipitation from the warm air tend to be widespread, with precipitation usually at light to moderate intensities.

(b) When a cold air mass moves into an area of warmer air, the warm air is pushed upward more sharply. As a result, the slope of the cold front is steeper, and the precipitation associated with it tends to be more intense and localized. (Vantage Art, Inc. after Horace R. Byers, *General Meteorology,* 3rd ed., 1959, McGraw-Hill Book Company)

slope of a cold front is about twice as steep as that of a warm front, rising about 1 km over a horizontal distance of some 50 km. This more abrupt uplift causes cold-front precipitation to be more intense but shorter in duration than warm-front precipitation. In eastern North America, a cold front usually arrives in the form of a "squall line" of cumulonimbus thunderheads that produce violent rain. The thunderheads pass by rather quickly and are followed by clearing skies with scattered fair-weather cumulus clouds.

Surface air can also rise and cool by *convergence.* In areas of low atmospheric pressure, designated as "lows" on weather maps, surface air spirals toward the low-pressure center like the water moving toward a bathtub drain. As air converges into the low-pressure area, it is swept upward and escapes outward aloft. This convergence and ascent frequently results in cloud formation and precipitation, even though no air mass fronts are present.

The specific weather associated with atmospheric pressure systems, air masses, and fronts are the subject of the next chapter.

GLOBAL PRECIPITATION DISTRIBUTION

A knowledge of the atmospheric and oceanic circulations and their interactions allows us to predict the distribution of precipitation over the earth—a most important element of global climates. The processes producing thick clouds and precipitation are most often associated with air mass convergence in low-pressure systems, causing the uplift of moist, unstable maritime air. Where air is subsiding and diverging from surface high-pressure centers, cloud formation and precipitation are unlikely. It follows, therefore, that in equatorial and tropical latitudes the precipitation pattern should mirror the pattern of convergence of moist tropical air. In higher latitudes precipitation should be concentrated in zones of frontal interactions. Coasts washed by warm currents will be zones of heavy precipitation, as will the windward sides of mountain ranges. Cold-water coasts and the leeward sides of mountain ranges will be dry.

The map of average annual precipitation

over the continents, Figure 5.20a, bears out these expectations. Precipitation is especially high in three general locations: the equatorial region, where the intertropical convergence zone is present most of the year; the monsoon regions of India and southeast Asia; and the region of the polar front in the middle latitudes. Wet spots also appear wherever moist air is regularly forced up and over mountain ranges, as in the Pacific Northwest from northern California to southeastern Alaska, in the Andes of southern Chile, and on the south side of the Himalaya Mountains.

Conversely, precipitation is low over most subtropical regions, including Baja California and northern Mexico and the Old World desert region reaching from the Sahara Desert in North Africa through Arabia and Iran to Pakistan; the interiors of Asia and North America, which are far from oceanic moisture sources; and the polar regions, where the capacity of the atmosphere for water vapor is low.

SUMMARY

Moisture enters the atmosphere by evaporation from water surfaces and by evapotranspiration from the land. Condensation is the process by which water vapor in saturated air changes to water droplets or ice crystals, and precipitation follows the growth of water droplets and ice crystals to sizes large enough to fall to earth. Evapotranspiration is a cooling process in which heat is removed from the local environment and stored as latent heat in water vapor, to be released as sensible heat elsewhere in the atmosphere when condensation occurs.

The moisture content of air can be expressed in several ways, including absolute, relative, and specific humidity; the mixing ratio; and the dew point. The capacity of the atmosphere to hold water vapor declines as temperature decreases, and condensation normally occurs when air is cooled to its dew point, when it is saturated with water vapor. In the atmosphere, condensation, which forms clouds and fog, is not spontaneous at dew point temperatures, but requires microscopic particles as condensation nuclei.

Condensation at the land surface results in dew and frost. Condensation within a thick layer of air at the surface produces fog. The chilling required to cause this condensation occurs in various ways, which bring about radiation, advection, and orographic fog, all of which can be hazardous to humans. Condensation in the atmosphere above the land surface produces clouds of various types. Clouds are classified in terms of their form and the level at which they occur. High, middle, and low level clouds can be of the cirrus (high only), cumulus, or stratus types, or combination forms of these. Cumulus clouds arise from vertical currents of air, whereas stratus types indicate an absence of strong vertical motion.

Precipitation occurs only when water droplets become raindrops by coalescence in warm clouds, or when ice crystals in cold clouds grow by transfer of water from nearby supercooled droplets. Hailstones are formed in clouds with strong updrafts and are built up layer by layer, by being recycled through the cloud several times.

To generate significant rainfall, large masses of air must be cooled, which requires that the air be lifted to higher elevations. This lifting and cooling can be accomplished in several ways: by thermal convection and adiabatic cooling, resulting in cumulus clouds; by orographic lifting, in which air is forced to rise over mountain barriers; by frontal lifting, in which warm, moist air rises to flow over cooler surface air or is forced upward by invading cold air; and by convergence, in which air flowing into a low-pressure system ascends aloft. Warm ocean waters increase air mass instability, whereas cold ocean currents create the stable atmospheric conditions that allow deserts to border the sea. The global distribution of precipitation is largely controlled by the general atmospheric circulation, but is also strongly affected by the pattern of the continents and ocean basins and the form and elevation of the land surface.

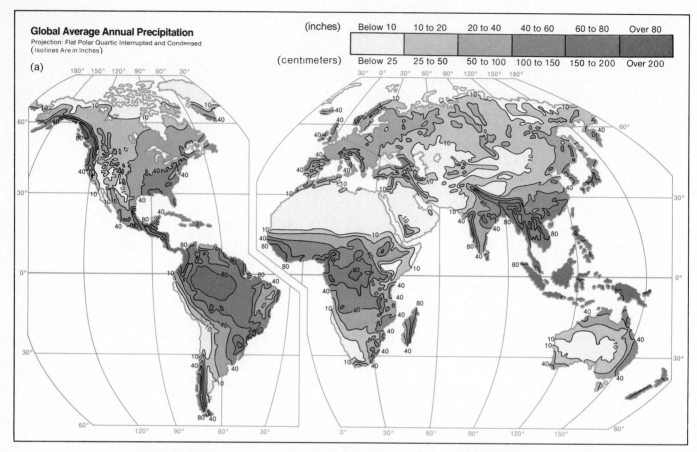

Figure 5.20 (a) This map of average annual precipitation illustrates the complex distributions associated with the positions of continents, ocean basins, and mountain barriers. In spite of the complexity, however, global patterns of high and low precipitation are evident, and these should be related to the global pressure and wind systems shown in Figure 4.6, pp. 88–89. Particularly evident are the subtropical deserts in both hemispheres, separated by rainfall zones associated with the intertropical convergence, and succeeded on their poleward sides by rainfall zones associated with the polar fronts.

(b) The pattern of average annual precipitation over North America displays some broad regularities. The East tends to be moist because of atmospheric moisture from the Gulf of Mexico and the adjacent Atlantic. Precipitation decreases westward from the Mississippi River Valley. Moist air from the Gulf of Mexico seldom flows westward, and moist air from the Pacific releases most of its moisture as orographic rainfall on coastal mountain slopes.

(c) The Great Plains are dry in winter because of the influx of cold dry air from the Canadian Arctic. The Far West receives most of its precipitation in winter, when midlatitude cyclones sweep the coast; a strong subtropical high over the eastern North Pacific inhibits precipitation during summer. The East receives precipitation during both winter and summer. (After *Goode's World Atlas*, 1970)

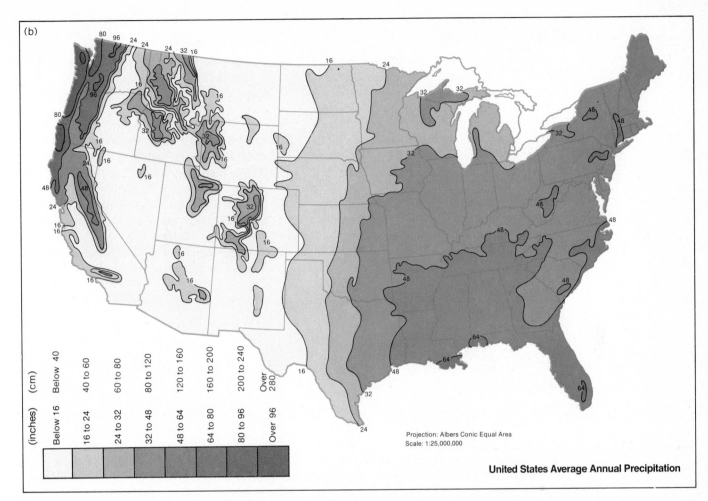

(b)

(inches) (cm)

Below 16	Below 40
16 to 24	40 to 60
24 to 32	60 to 80
32 to 48	80 to 120
48 to 64	120 to 160
64 to 80	160 to 200
80 to 96	200 to 240
Over 96	Over 280

Projection: Albers Conic Equal Area
Scale: 1:25,000,000

United States Average Annual Precipitation

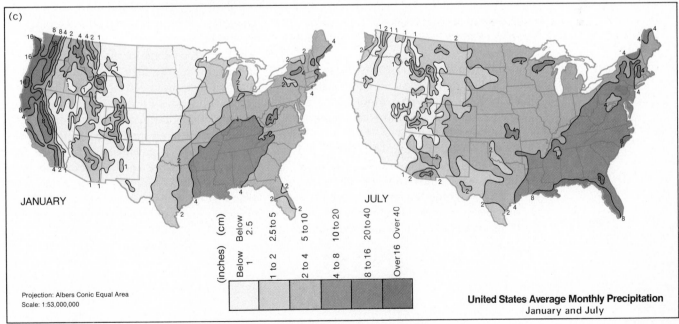

(c)

JANUARY

JULY

(inches) (cm)

Below 1	Below 2.5
1 to 2	2.5 to 5
2 to 4	5 to 10
4 to 8	10 to 20
8 to 16	20 to 40
Over 16	Over 40

Projection: Albers Conic Equal Area
Scale: 1:53,000,000

United States Average Monthly Precipitation
January and July

REVIEW QUESTIONS

1. Compare the factors affecting the transfer of water to the atmosphere over oceans and continents.
2. What is the significance of the relationship between temperature and the atmosphere's capacity for holding water vapor?
3. What are the mechanisms by which water is returned to the atmosphere from the earth's surface?
4. What variables determine the vapor pressure gradient? What is its significance in terms of evaporation rates?
5. Compare the various measures of atmospheric humidity.
6. Describe the diurnal regime of vapor pressure and relative humidity during fair weather.
7. Discuss the process by which water vapor is transformed into drops or ice crystals large enough to fall from clouds to the ground as precipitation.
8. Discuss some situations in which atmospheric cooling might eventually result in clouds and precipitation.
9. Why is there very little chance of significant precipitation during radiation fogs?
10. Discuss the diurnal regime of stability and instability near the earth's surface during fair weather.
11. What are the physical bases for atmospheric weather modification?

APPLICATIONS

1. From Figure 5.20, identify the regions of the earth where the annual precipitation averages less than 25 cm (10 in.). Which of these regions appear to be associated with the subtropical highs, rainshadows, continental interiors, and cold ocean currents? Explain the reason for each of the following deserts: The Gobi, Atacama, Thar, Namib, Taklamakan, Kalahari, Sahara, Simpson, Sonoran, and Great Basin.
2. Maintain a diary of cloud types for several weeks. Does there appear to be a relationship between cloud types, wind directions, and air temperature and humidity changes? What about precipitation?
3. Fill a small plastic container with water, and for the rest of the term measure the daily amount of evaporation from the container. Do this by weighing the container on a laboratory balance each day. Compute the depth of evaporation by remembering that 1 g of water is equal to 1 cm^3 of water. How do the rates of evaporation vary from day to day? What is the relationship of the rate of evaporation to the weather?
4. In your neighborhood, make a survey of the occurrence of dew after sunset. On what types of surfaces does dew appear first? Does dew play any role in the local ecological system in your region?
5. Are there fog-prone locations in your area? If so, how are the fogs there related to the types of fogs discussed in the text? Do you see any relationship between fog and the amount of dew formed on particular nights?
6. Why is it unlikely that there will ever be complete public agreement about plans for large-scale weather modification schemes? Are there any locations where it seems possible that a small-scale weather modification of some type would not provoke a controversy?

FURTHER READING

Barry, Roger G., and **Richard J. Chorley**. *Atmosphere, Weather, and Climate,* 3rd ed. London: Methuen (1976), 432 pp. This paperback is written largely from the perspective of the British Isles. Concepts are illustrated frequently by examples from the professional literature. Chapter 2 includes a particularly useful development of evaporation.

Battan, Louis J. *Fundamentals of Meteorology.* Englewood Cliffs, N.J.: Prentice-Hall (1979), 321 pp. Chapter 8 is especially helpful for its discussion of the growth of cloud droplets and the formation of rain and snow.

Breuer, Georg. *Weather Modification: Prospects and Problems.* Hans Mörth, trans. London: Cambridge University Press (1979), 178 pp. A very readable, up-to-date analysis of world-wide studies and projects.

Chagnon, Stanley A. "The La Porte Anomaly— Fact or Fiction?" *Bull. Amer. Meteor. Soc.,* Vol. 49 (1968): 4-11. This analysis of an unusual 40-year precipitation record southeast of Chicago has proved somewhat provocative.

Chagnon, Stanley A., Ed. "Metromex: A Review and Summary." *Meteorological Monographs,* Vol. 18, No. 40 (1982), 181 pp. This relatively technical volume is the summary of the most intensive study of the effects of metropolitan regions on weather and climate.

Holzman, B. G., and **H. C. S. Thom.** "The La Porte Precipitation Anomaly." *Bull. Amer. Meteor. Soc.,* Vol. 51 (1970): 335-337. The authors present a contrasting view of the original 1968 analysis of Chagnon.

Neiburger, Morris, James Edinger, and **William Bonner.** *Understanding Our Atmospheric Environment,* 2nd ed. San Francisco: W. H. Freeman (1982), 453 pp. This is an excellent source for basic meteorological concepts in a nonmathematical framework. Chapters 5, 6, and 15 are especially pertinent.

Schaefer, Vincent J., and **John A. Day.** *A Field Guide to the Atmosphere.* Boston: Houghton Mifflin Co., (1981), 359 pp. This outstanding book is a new addition to the well-known series of Peterson Field Guides. This guide contains about 400 black-and-white and color photographs, line drawings of clouds and atmospheric phenomena, and simple experiments.

CASE STUDY

Weather Modification— Cloud Seeding

Most regions of the earth experience occasional departures from their regular weather patterns: they become wetter, drier, warmer, or colder than usual for periods of weeks, months, and sometimes even years. The winter of 1976–1977, for example, was very dry from the northern Great Plains westward across the Rocky Mountains to the Pacific coast. Farmers in western Iowa, ski resort operators in Colorado, and water-resource managers in northern California wished that precipitation could be brought under effective control. At the same time, weary snow-removal crews in Buffalo, New York, wanted an end to the unusually persistent squalls that swept off Lake Erie, eventually bringing normal activities to a standstill during mid-February.

Unusually dry weather in the Soviet Union during 1972 forced the Russians to buy large quantities of American grain, and these purchases produced serious economic consequences within the United States. The effects of the severe winters of 1976–1977 and 1977–1978 across the northern United States persisted for several years.

Scientists have learned to exert some control over the weather through the process of cloud seeding to produce rain or snowfall. The key to modern rainmaking is the realization that many clouds possess all the conditions necessary for precipitation except a natural way to trigger the growth of cloud droplets into raindrops.

In 1946, Vincent Schaefer of the General Electric Research Labora-

This photograph shows the first conclusive field test of cloud seeding, carried out in 1946 by Irving Langmuir and Vincent Schaefer. The supercooled stratus clouds were seeded by dry ice pellets dropped from an airplane flying around an oval course. In less than an hour, the clouds in the seeded area cleared because of the induced precipitation. (General Electric Company)

ries discovered that he could change droplets to ice crystals by dropping dry ice (solid carbon dioxide below $-40°C$) into a chamber containing supercooled water droplets. Field tests by Schaefer and Irving Langmuir showed that a few pounds of crushed dry ice dropped from an airplane into a supercooled cloud could produce light precipitation. Precipitation is usually produced if the cloud is thick and cold enough,

and is long-lived—in other words, if the cloud already has the necessary conditions for producing rain.

Tiny crystals in the smoke from burning silver iodide also make excellent ice-forming nuclei. Supercooled droplets will form ice crystals on silver iodide at temperatures as high as $-4°C$ (24.8°F). An effective number of nuclei can be produced by burning only a few ounces of silver iodide in burners mounted on aircraft. By this method, commercial rainmakers are able to seed updrafts and the most promising cloud systems directly.

But the overall effectiveness of weather modification is difficult to judge. Experiments in seeding large cumulus clouds in southern Florida have produced dramatic changes in cloud structures and large but local increases in rainfall. Other experiments with cumulus cloud systems suggest that precipitation over large regions may not have been substantially increased. In the dry and mountainous West, where supercooled clouds are forced over mountain barriers, favorable results seem to depend on careful selection of weather situations. Analyses suggest that winter-spring snowfall in the mountains can be increased by between 10 and 30 percent. Whether increased snowfall in the mountains means less precipitation over downwind dry areas farther east is not known.

Rainmaking is not the only objective of weather modification. Cloud seeding has also been applied in efforts to modify conditions that lead to fog, hail, and hurricanes.

Fog is especially dangerous over airports and along highways. If fog droplets are supercooled, seeding with dry ice can dissipate the fog as snow crystals, and some airports are able to use this technique. Fogs in most areas are warm, however, and no inexpensive way of clearing a warm fog has been developed. Heating the air to a temperature above the dew point will clear a warm fog, but the energy required makes the method too costly for general application.

Hailstorms cause so much damage to crops that efforts to disrupt the formation of hail in turbulent clouds are believed to be worthwhile. Seeding with silver iodide seems to inhibit further growth of hailstones or to cause the cloud to produce only rain. Because hailstorms tend to be concentrated in certain regions it has been feasible to set up hail suppression facilities in particularly susceptible areas.

Some efforts have also been directed toward modifying the inten-

In this early attempt at rainmaking, electricity from a hand-cranked generator was pumped into clouds in the mistaken belief that the electricity of a lightning flash was somehow responsible for rain. (The Bettmann Archive, Inc.)

sity of hurricanes, but the erratic movements and growth of these storms make it difficult to determine whether a seeding experiment has reduced their winds. Hurricane Debbie was seeded by Project Stormfury planes on August 18 and 20, 1969. Maximum winds decreased 31 percent on the first day, increased on the second day, and decreased again (by 15 percent) after seeding on the third day. Much more research is needed, however, before seeding of hurricanes can become routine.

Wheatfields by Jacob Isaacksz Van Ruisdael (The Metropolitan Museum of Art, New York)

The panorama of the skies often changes from hour to hour as migrating systems of high and low atmospheric pressure generate flows of moist and dry air that bring us our daily weather. Air masses, fronts, local winds, and cyclonic storms are all aspects of the secondary circulations of the earth's atmosphere that directly affect our lives.

The day-to-day changes in the weather anticipated by people in much of the United States and Canada do not occur everywhere on our planet. In the tropics people do not experience the successions of fair and stormy or warmer and colder days that follow one another across much of North America. Yet even in the supposedly monotonous tropics, dramatic changes of weather do occur. Over large areas monsoon conditions cause extreme contrasts between wet and dry seasons. In late summer, hurricanes and typhoons batter many areas with destructive winds and torrential rains, and even under the prevailing calmer conditions, groups of wetter than normal days do come along now and then.

Whatever the geographical area—the midlatitudes, the tropics, or the polar regions—changes in weather are important in people's lives. These changes are not isolated events. They are linked to the broader phenomena of the general circulation described in Chapter 4. A summer shower that covers only a few square kilometers is a local weather event. But the local weather is controlled by larger scale weather systems, or *secondary circulations*, that move through the middle and low latitudes. These spiraling eddies are called cyclones and anticyclones, or the highs and lows of the daily weather maps seen in television newscasts. "Cyclones" and "anticyclones" are terms used by meteorologists to designate large systems of converging or diverging surface airflow, and they should not be confused with the popular use of the term "cyclone," employed in the North American Midwest as a synonym for tornado.

6
Secondary Atmospheric Circulations

Midlatitude Secondary Circulations

Air Masses
Fronts
Cyclones and Anticyclones
Local Circulations

Tropical Secondary Circulations

Easterly Waves
Tropical Cyclones
Tropical Monsoons

MIDLATITUDE SECONDARY CIRCULATIONS

Most midlatitude weather is dominated by the interaction of large air masses of unlike characteristics. In the northern hemisphere, especially during winter, warm moist air moving poleward out of the subtropical highs meets

131

cold dry air flowing equatorward out of the Siberian and Yukon highs. Where the air masses meet, a front is formed. The interacting airflows frequently become organized into vast, spiraling vortices that constitute the "storms" of the midlatitudes. These usually migrate from west to east and may be as much as 1,000 km (600 miles) in diameter. Such migrating secondary circulations and their associated air masses and fronts are responsible for much of the changeable weather of the middle latitudes.

Air Masses

An *air mass* is a large, nearly uniform segment of the atmosphere that moves as a unit, usually in association with secondary high or low pressure systems. Air masses often retain the temperature and moisture characteristics of their source region even after traveling thousands of kilometers over the continents. Air mass types

are therefore identified according to their source region: *polar* (*P*) or *tropical* (*T*), *continental* (*c*) or *maritime* (*m*). The capital letter indicates air mass temperature characteristics; the lower-case letter suggests relative moistness. Thus, four general air mass types are widely recognized: *cP*, *mP*, *cT*, and *mT*. Some classifications also include arctic (*A*) and equatorial (*E*) air masses. Air masses are further designated according to their stability at the surface and aloft: *K* and *W*, meaning colder and warmer, respectively, than the surface beneath, and *u* and *s*, meaning unstable or stable aloft. An *mTKu* air mass has maximum instability: it is colder than the surface, therefore subject to heating from below, which produces a steep lapse rate; and it is unstable aloft as well.

Air masses follow predictable paths as they move over the earth's surface. Figure 6.1 shows typical tracks of air masses in the United States in winter and summer. The pattern of movement of these air masses largely determines weather conditions throughout North America.

Extremely cold winter weather in the United States is associated with *continental polar* (*cP*) or *arctic* (*A*) air masses that form over the snow-covered plains of northern Canada. Arctic air masses actually originate closer to the polar region and are colder than those air masses called polar. In the following discussion of North American air masses, we shall regard all cold air masses as being the polar type.

Figure 6.1 The air masses that enter a region can strongly influence weather conditions there by replacing the existing air with air of different temperature and humidity. Maritime air is humid, and continental air tends to be dry. The maps in this figure depict the average movements of air masses over the coterminous United States during winter and summer. During winter much of the northern half of the United States is invaded by cold polar air, whereas during summer the northward flow of tropical air dominates the weather in most regions. (Calvin Woo after Dieter H. Brunnschweiler, *Geographical Helvetica*, vol. 12, 1957)

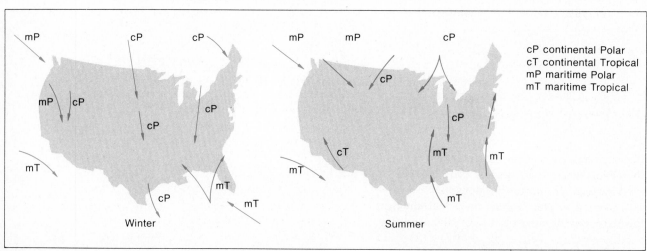

cP continental Polar
cT continental Tropical
mP maritime Polar
mT maritime Tropical

Winter

Summer

During the long subarctic winter, the sun remains below the horizon much of the time, and frequent clear skies promote intense radiational cooling of the snow-covered land surface. After a few days these conditions give rise to a nearly homogeneous cold, dry cP air mass that extends horizontally for more than 1,000 km (600 miles).

Eventually, a "polar outbreak" of cP air spreads out of the source region and moves in its characteristic path toward lower latitudes. The air is modified only slightly as it sweeps southward, traveling more than 4,000 km (2,500 miles) across the Great Plains and down the Mississippi Valley to the Gulf of Mexico. Figure 6.2 shows that during midwinter, even as far south as New Orleans, the mean temperature of the cP air is close to 0°C (32°F), with occasional temperatures lower than −5°C (23°F). However, when the cP air moves out over the nearby warm waters of the Gulf of Mexico, it gains moisture and is warmed rapidly. Modification is so quick that a "new" air mass results within 48 hours or so. This warm, moist air mass is designated *maritime tropical* (mT), just like air masses that originate over the tropical oceans. Most heavy rains and snows in the midlatitudes are produced by condensation of moisture within poleward-moving mT air masses.

The mountains of western North America prevent most cP air from spreading to the Pacific coast. Instead, the west coast is usually under the influence of *maritime polar* (mP) air that has moved eastward from the seas off Japan and Alaska. This air is never as cold as polar air that originates over the land, and it is much more moist than cP air. As a consequence, the Pacific coast receives heavy rain and wet snow in winter, when polar outbreaks are common, but seldom experiences freezing temperatures. Similar mP air from the North Atlantic occasionally invades the northeastern United States, bringing freezing rain in the fall and winter.

Occasionally, in summer, the Pacific coast is overrun by *continental tropical* (cT) air from the southeast. This air is hot and extremely dry, coming from the deserts of Mexico and the southwestern United States. Globally, the most

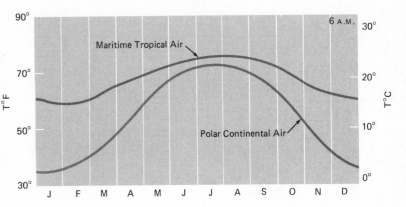

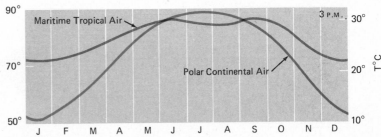

Figure 6.2 This graph illustrates the annual regime of air temperature at 6 A.M., near the coolest time of the day, and 3 P.M., about the warmest time of the day, at New Orleans, Louisiana, for continental polar and maritime tropical air masses. New Orleans is adjacent to one of the source regions of mT air, but far from the source region of cP air. Therefore, thermal characteristics of the mT air do not change greatly through the year. In contrast, the cP air remains cold, especially for a latitude only 30° from the equator, during the winter. In summer the cP air is slightly warmer, on the average, than the mT air, mainly because it is characterized by much less cloud cover and almost no local thundershowers. (Robert A. Muller and James E. Willis, 1983)

important source of cT air is the Sahara Desert of North Africa. Saharan cT air affects all the Mediterranean coastlands, occasionally crosses the Alps, and even penetrates as far as the British Isles and Scandinavia.

The weather associated with various air masses depends also on the nature of the surface over which the air masses move. For example, over the vast land areas of the Great Plains and Mississippi Valley, cP air is generally associated with weather that is cold and windy, but clear. However, over the Great Lakes, the

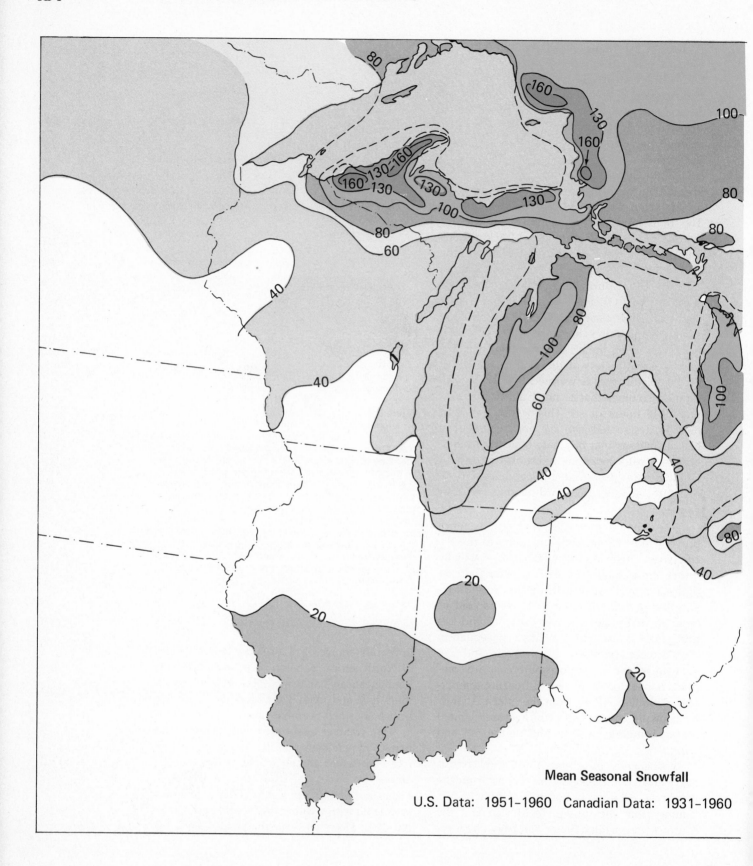

Mean Seasonal Snowfall

U.S. Data: 1951–1960   Canadian Data: 1931–1960

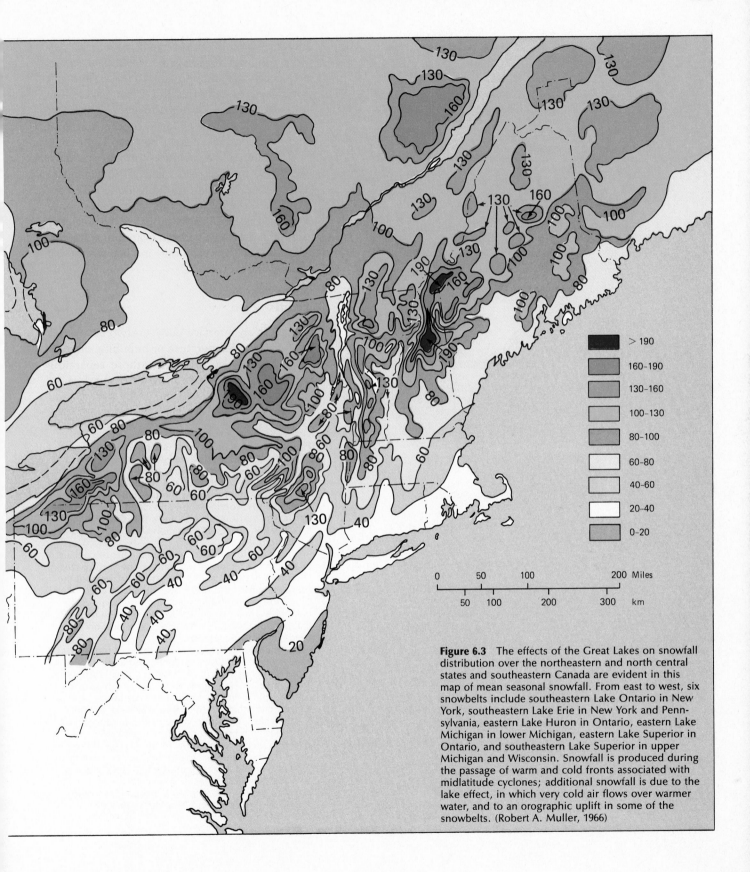

Figure 6.3 The effects of the Great Lakes on snowfall distribution over the northeastern and north central states and southeastern Canada are evident in this map of mean seasonal snowfall. From east to west, six snowbelts include southeastern Lake Ontario in New York, southeastern Lake Erie in New York and Pennsylvania, eastern Lake Huron in Ontario, eastern Lake Michigan in lower Michigan, eastern Lake Superior in Ontario, and southeastern Lake Superior in upper Michigan and Wisconsin. Snowfall is produced during the passage of warm and cold fronts associated with midlatitude cyclones; additional snowfall is due to the lake effect, in which very cold air flows over warmer water, and to an orographic uplift in some of the snowbelts. (Robert A. Muller, 1966)

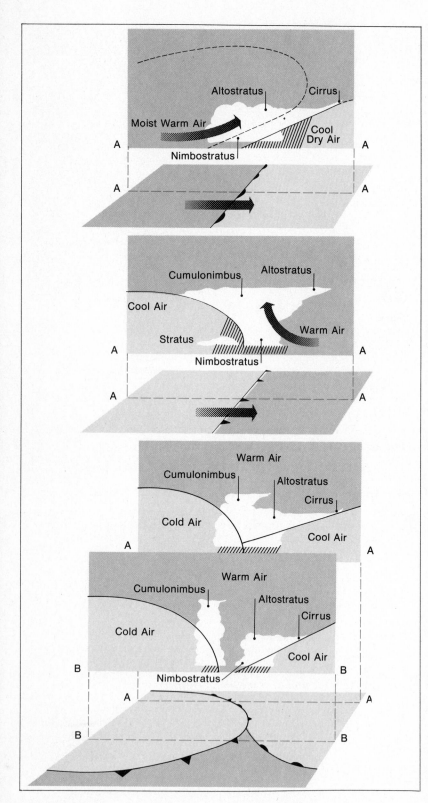

lower layers of *cP* air receive heat and moisture from the large expanses of water. The results are spectacular snow squalls on the eastern and southeastern shores of the Great Lakes. The impact of these snow squalls on the pattern of average seasonal snowfall over the northeast and northcentral states is shown in Figure 6.3.

Continental air masses normally produce very little precipitation. Instead, they tend to gain moisture from evapotranspiration. The opposite is true of maritime air masses; once formed, they lose more moisture by precipitation than they receive from evapotranspiration. They are the sources of rain and heavy snowfall around the world.

Fronts

A *front* is an interface between two air masses that differ in temperature or humidity, or both. The most distinct fronts are those separating air masses whose properties contrast most sharply—in particular, the cold, dry *cP* and

Figure 6.4   (a) A warm front is formed when a mass of warm air encounters a denser mass of cool air. The diagram shows a vertical cross section of a warm front; beneath it is the surface symbol as it would appear on a weather map. As the uplifted warm air becomes cooler, condensation and cloud formation may occur. The vertical section shows that the resulting precipitation arrives at a location ahead of the surface warm front itself.

(b) A cold front develops where a mass of cool air pushes into a region of warmer air. The vertical cross section of the front shows the warm air ahead of the front being forced upward over the incoming cool air. Cloud formation and precipitation may occur as the uplifted warm air cools. The vertical section shows that precipitation normally arrives at a location just ahead of the surface front.

(c) An occluded front involves three or more air masses of different temperatures, and it combines some of the features of both warm and cold fronts. The vertical cross sections show that the mass of incoming cold air forces warm air to rise above both regions of cool air. In the forward vertical section B–B' the occlusion is not complete, and the warm air is in contact with the ground in close proximity to a cold front and a warm front. In the rear vertical section A–A' the occlusion has formed, and the warm air is lifted completely above the ground. Precipitation may occur on both sides of an occluded front, as the diagrams indicate. (Calvin Woo adapted from Hermann Flohn, *Climate and Weather,* © 1969 by H. Flohn, used with permission of McGraw-Hill Book Company)

warm, moist *mT* types. Such fronts are common in eastern North America. On the Pacific coast most fronts merely separate successive *mP* air masses of varying temperature and humidity, although *mT* and *cT* air masses are occasionally involved, especially in southern California. Flooding caused by heavy rain in the coastal region and interior deserts of southern California often indicates the arrival of a rare *mT* air mass from the southwest; hot winds that fan forest fires mark the arrival of *cT* air from the east. On rare occasions cold *cP* air penetrates into central California and the Pacific Northwest, with disastrous consequences to ornamental vegetation and winter crops grown in these areas.

When warm air moves into a region previously occupied by a colder air mass, the forward edge of the warm air mass is designated a warm front. Figures 5.19 and 6.4a show how a warm front slopes forward from the surface as the lighter warm air slides over the denser cold air it is replacing. The cool air is shown retreating toward the right, corresponding on a map to the east or northeast, the usual direction in which a warm front advances in the middle latitudes. At an altitude of one kilometer the warm air front may be several hundred kilometers ahead of the front at the surface.

Weather conditions in the vicinity of a warm front depend on the properties of the air masses as well as the nature of the land surface. Nevertheless, there is a characteristic sequence of weather conditions associated with a warm front. Condensation and cloud formation begin where the warm, moist air rises over the cooler air and cools adiabatically to its dew point. Because the warm front slopes so gently, it is usually detected through the appearance of high cirrus clouds a day or more before the surface front arrives (Figures 5.19 and 6.4a). As the surface front approaches, sheet-like stratus clouds become thicker and lower. A broad band of precipitation is usually ahead of the surface front. Warm fronts move slowly and often bring steady rain or snow that may last a day or more.

A *cold front* develops when cold air advances into a region occupied by warm air. Figure 6.4b shows that the advancing cold air pushes under the warm air and forces the warm air to rise

Figure 6.5 Instability of the air in this squall line over western Texas is indicated by the development of cumulus clouds overtopped by giant cumulonimbus thunderheads. (Robert A. Muller)

abruptly, so the frontal slope is much steeper than that of a warm air front. Frequently there is rapid development of towering cumulonimbus clouds, with precipitation occurring just ahead of the surface front. The passage of a cold front in summer is usually associated with the sudden appearance of a line of thunderstorms and a rapid drop in temperature (Figure 6.5). The zone of precipitation along a cold front may pass by in only an hour or two.

In those cases where the surface boundary between tropical and polar air masses remains at about the same location for a day or more, the front is designated as *stationary*. More complex *occluded fronts* consist of three or more air masses, with one front overtaking another so that the intervening air mass loses contact with the ground. This creates a broad band of heavy precipitation falling from the occluded (upward displaced) air mass (Figure 6.4c). Occluded fronts will be treated more fully in the following discussion of cyclonic storms.

Cyclones and Anticyclones

Warm, cold, stationary, and occluded fronts usually are segments of the polar fronts in each hemisphere, which are components of the general circulation (Figure 4.5, p. 85). As such, they are associated with moving secondary circulations, including midlatitude cyclones and anticyclones.

Almost any daily weather map for a midlatitude region shows centers of high and low pressure. The spiraling flows of air into moving centers of low pressure (or merely "lows") are *cyclones*. The diverging flows around cells of high pressure (referred to simply as "highs") are *anticyclones*. The horizontal differences in pressure within large highs or lows usually amount to less than 10 millibars over a distance of 100 km (60 miles), only a small fraction of the normal sea-level atmospheric pressure of 1,013 millibars.

As shown in Figure 6.6, a traveling cyclone

Figure 6.6 In a region of high pressure, or an anticyclone (left), air descends from the upper atmosphere and diverges outward along the surface of the earth. The air streaming outward from highs therefore tends to be dry. In a region of low pressure, or a cyclone, (right), air converges inward along the surface of the earth and ascends to the upper atmosphere. If the air streaming into a low is warm and moist, condensation and cloud formation occur as the air rises and cools. The directions of circulation are shown for the northern hemisphere. (Calvin Woo)

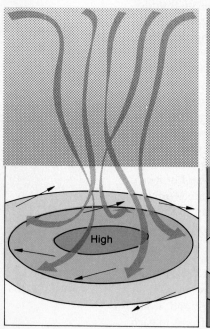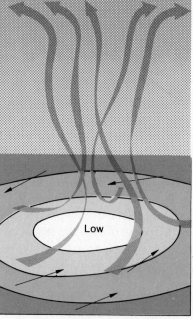

consists of converging surface air that ascends and then diverges in the upper atmosphere. As air rises in a cyclone, it cools adiabatically, resulting in cloud formation and perhaps rainfall. A traveling anticyclone, on the other hand, consists of air subsiding from aloft and diverging at the surface. The descending air in an anticyclone is warmed adiabatically, thus reducing its relative humidity and resulting in clear skies.

Midlatitude cyclones develop along the polar front where warm and cold air come into contact. In North America cyclones most often form just east of the Rocky Mountains and migrate toward the southeast, east, or northeast, eventually moving out over the Atlantic Ocean. Most precipitation in the central and eastern United States and Canada is associated with the ascent of warm, moist *mT* air over the cold *cP* air within the cyclonic circulation. Along the Pacific coast most cyclonic disturbances involve the interaction of *mP* air masses of differing character. Having originated far away off the coast of Japan or Alaska, these cyclonic storms commonly arrive at the west coast in the form of mature occluded fronts. On the other hand, most cold anticyclones in North America peel away from the Yukon high and migrate southeastward to the Atlantic Ocean off the Carolinas.

The evolution of midlatitude cyclones usually follows a characteristic pattern. Cyclones typically begin along stationary segments of the polar front. Figure 6.7a shows a stationary front with a northeast-southwest orientation. Such a front may remain over the same region for as much as 24 to 48 hours before a cyclone begins to develop; this may occur due to an upper air disturbance, such as a wave in the jet stream.

The isobars in Figure 6.7a show that pressure is lowest along the front, increasing away from the front on both sides. On the polar side of the front, cold air flows from the northeast; on the tropical side, warm air flows from the southwest. As is normal where masses of warm and cold air are in contact, the warm air is above the cold air at the surface. This produces a narrow band of clouds on the polar side of the front, with some light precipitation.

The cyclone begins as a small wave disturbance associated with a local fall in pressure, as shown in Figure 6.7b. As pressure falls even more, air begins to converge toward the center of the low in a counterclockwise circulation. Ahead of the center, the warm air advances into the former domain of the cold air, forming a warm front. The cold air begins to sweep southeastward behind the center, forming a cold front. A broad apron of warm-front drizzle, rain, or snow develops ahead of the center, and a narrow band of showers breaks out along the cold front.

Many small waves may move toward the northeast along the polar front, dissipating in 6 to 12 hours. A few waves, however, grow into well-developed cyclonic circulations. One of the difficult tasks of the weather forecaster is to predict the development of a significant midlatitude cyclone from a small wave disturbance.

Figure 6.7c shows the cyclone at a more advanced stage. The central pressure of the cyclone may fall to less than 1,000 millibars (29.5 in.), and the converging counterclockwise circulation may expand to a diameter exceeding 1,000 km (600 miles). In eastern North America the cyclonic system normally moves in a northeasterly direction with a speed between 40 and 80 km (25 and 50 miles) per hour. Steady and substantial precipitation usually occurs ahead of the surface position of the warm front, and occasionally severe thunderstorms and even tornadoes are associated with the cold front (see the Case Study following this chapter). The weather map in Figure 6.8 shows a small, intense midlatitude cyclone out of which several "killer" tornadoes were born.

Figure 6.7d shows the cyclone in its occluded phase. The cold front moves more rapidly toward the east than the cyclonic system does; therefore, the cold front eventually overtakes the warm front. This lifts the warm air away from the surface, forming an occlusion. Occluded fronts have properties of both warm and cold fronts. Heavy precipitation usually falls from the warm, moist air that is lifted during the occlusion. Thus precipitation occurs on both sides of the surface front. The intensity and size of cyclonic storms are often greatest at the beginning of the occluded phase. Occlusions are especially characteristic of cyclonic systems over northern California and the Pacific

Northwest, where the Coast Ranges, Canadian Rockies, Cascade Mountains, and Sierra Nevada get heavy rains, and snows at higher elevations, during winter and spring.

In its final stage, the cyclone's energy supply undergoes a significant alteration. This occurs because the developing occlusion eventually eliminates the surface air temperature differences that power the circulation. Then warm, moist air from the warm front stops feeding into the center of the system. It is not long before the low-pressure center begins to fill with cooler, drier air, and the cyclone loses its energy. The life cycle of a midlatitude cyclone ranges from 24 hours to as much as five days, so it is possible for a particularly long-lived cyclone to travel across most of the United States.

Favored regions for cyclone development in North America include the eastern slopes of the Rocky Mountains from Colorado to Alberta, as well as the Gulf of Mexico and the Atlantic coast. When a polar outbreak spills southward across the Great Plains, for example, a stationary front forms between the very cold polar air east of the Rocky Mountains and milder Pacific air over the western states. The cold, dense air does not normally flow up and over the mountains to the west; instead, frontal waves on the eastern slopes often develop into winter storms that then sweep eastward. On the southern and eastern margins of the polar outbreak, the cold front sometimes becomes stationary over the warm waters of the Gulf of Mexico or over the Gulf Stream off the Carolinas or Virginia. Storms that originate in these regions often become well developed as they sweep along the Gulf and Atlantic coasts toward New England, where they are known as "nor'easters" because of the blustery northeast winds that precede the passage of the storm center.

Cyclones that form along the eastern coast of Asia usually reach the Pacific coast of North America in an occluded state, but cyclones over central and eastern North America are in earlier stages of development. They follow paths that lead them toward the northeast and eventually to the region of the Icelandic low. Most of the late fall, winter, and spring precipitation over North America is due to the passage of midlatitude cyclones and associated fronts.

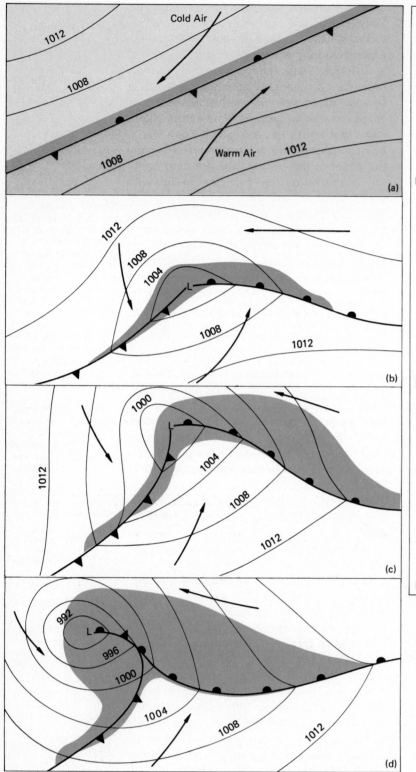

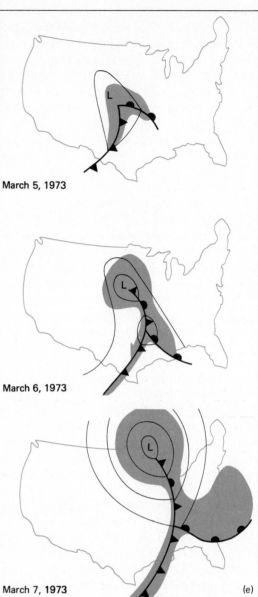

March 5, 1973

March 6, 1973

March 7, 1973   (e)

Figure 6.7 (opposite) Cyclones in the midlatitudes usually follow characteristic patterns of evolution along the polar front. This figure should be studied in conjunction with Figure 6.4, which shows other properties of fronts.

(a) The evolution begins along stationary segments of the polar front where cold and warm air stream in opposite directions.

(b) The stationary front tends to be unstable, and a bulge, or wave, usually develops within one or two days. The waves move along the polar front toward the northeast, and most dissipate within six to twelve hours.

(c) A few waves, however, grow into well-developed circulations with diameters of 1,000 km (600 miles) or more. In the northern hemisphere the counterclockwise converging circulation of the cyclone includes warm and cold fronts that separate polar and tropical air masses.

(d) The cold front eventually overtakes the warm front, lifting the warm air away from the surface and forming an occlusion. The occlusion eliminates the surface air temperature differences that provide energy for the system, and the circulation then weakens and finally dissipates.

(e) This sequence of development can be traced on the three map sketches for March 5, 6, and 7, 1973. As a cyclone occludes, a new wave disturbance may form farther back along the trailing stationary front. Over the oceans, the polar front often supports a cyclone family of three to five members in various stages of development. Most late fall, winter, and spring precipitation over North America is associated with midlatitude cyclones and associated fronts. (Vantage Art, Inc.)

Figure 6.8 The weather map (**left**) portrays weather conditions in the south-central United States on February 21, 1971, based on ground station reports at 9:00 A.M. Central Standard Time. The information on the map can be read with the help of the accompanying symbol table (**right**). The map shows isobars labeled with barometric pressure readings in inches of mercury, a cold front advancing eastward through Texas, a warm front advancing northward, and numerous station reports. The station symbols indicate temperature and dew point in °F, barometric pressure in inches of mercury, wind direction and speed in knots, cloud coverage, and precipitation events. Note that a station in Alabama, near the eastern edge of the map, reports a thunderstorm and heavy rain.

The station symbols on the weather maps prepared by the National Weather Service carry more detailed information than do the station reports on the simplified weather map shown here. Among other additional station data, the Weather Service maps present cloud type, height of the cloud base, change of barometric pressure in the previous three hours, and weather conditions during the previous six hours.

This weather map shows an intense low-pressure system over eastern Texas. Note that most of the surface winds cross the isobars at small angles and converge toward the low center. West of the cold front, winds are strong, with speeds of 25 knots. During the afternoon of February 21, the low-pressure area generated thunderstorms and tornadoes in Louisiana and Mississippi. Damage from the tornadoes was severe, and many people were injured or killed. (After Robert A. Muller, 1973)

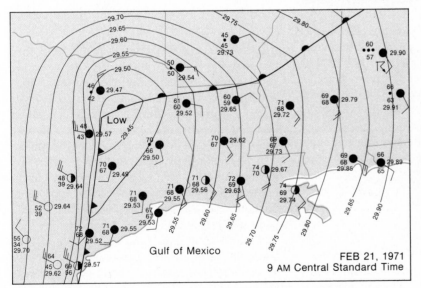

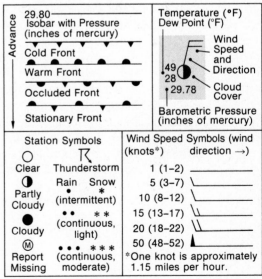

Jet streams are important in the formation of midlatitude cyclones. The waves in a jet stream aid in the ascent of surface air in the low-pressure center of a cyclone. Cyclones intensify when the surface flow of air coincides with favorable conditions aloft; when airflow aloft is unfavorable, cyclone development seems to be suppressed. That is why weather forecasters constantly monitor the upper air (Figure 6.9).

In North America during winter, cyclonic storms and anticyclones that bring clear skies often cross the continent in rapid succession, resulting in very changeable weather. In some winter weeks as many as three pairs of cyclones and anticyclones move down the St. Lawrence River valley toward the Atlantic Ocean or invade the Pacific coast from the west.

In the United States many of the anticyclones, or high-pressure systems, spill southeastward from the Yukon high with clear skies, dry air, strong winds, and cold temperatures. During fall, winter, and spring, meteorologists often refer to these fair-weather systems as polar outbreaks; the most severe "Arctic outbreaks" are capable of bringing near-zero temperatures (0°F or −18°C) almost to the Gulf Coast, from eastern Texas to the Florida panhandle. In a few hours these systems can wreak great damage on citrus orchards, winter vegetables, and suburban landscapes that feature exotic tropical plants.

During summer and fall, on the other hand, high pressure is often associated with the western extension of the "Bermuda" subtropical high, with warm and moist maritime tropical air settling over the eastern third of the United States. These anticyclones are notorious for their light winds and calms, for subsidence inversions aloft, and for the deteriorating air quality that results when a stagnant high stalls over the same region for days.

Local Circulations

The configuration of coastlines and the land surface significantly affect local circulations and weather. Under weak anticyclonic conditions, *land and sea breezes* dominate the weather along

Figure 6.9 This weather satellite image of the eastern Pacific shows the typical comma-shaped cloud pattern produced by an occluded front that has moved into British Columbia and is approaching Washington and Oregon. The low-pressure center is marked by the cloud spiral south of Alaska. The cloud band at the bottom of the view is the result of convection in the intertropical convergence zone. (National Oceanic and Atmospheric Administration)

coasts. Land and sea breezes result from the fact that the land warms during the day and cools at night, whereas the temperature of the adjacent ocean surface remains nearly constant. Heating of the land during the day generates localized low-pressure centers that draw air onshore in the form of sea breezes from the cooler ocean regions. Such breezes may be chilly in the midlatitudes but are very welcome along hot tropical coasts. Where cold water is present offshore, the sea breeze may bring fog inland daily in the late afternoon; this occurs in the San Francisco area in summer. Cooling of the land relative to the sea surface at night causes an offshore flow, or land breeze. Land and sea breezes involve closed circulation loops, as shown in Figure 6.10a and b, with currents of air aloft that are the reverse of those at the surface. The upper (reverse) flow occurs at a height of 1.5 to 2 km (5,000 to 7,500 ft).

A similar diurnal reversal of wind direction is experienced in mountain regions, where the upper mountain slopes heat more strongly than the valleys during the day, causing air to be drawn upward through the valleys. These weak flows are called *valley winds*. The ascent of air over mountain summits often produces clouds in the late afternoon. At night the summit areas cool by radiation, and cold dense air drains away from them, moving down and through the valleys as chilly *mountain winds* (Figure 6.10c and d). This cold air drainage often produces temperature inversions and frigid nights in mountains valleys.

TROPICAL SECONDARY CIRCULATIONS

Much less is known about tropical weather patterns than about the weather of the middle latitudes. There are far fewer observing stations in the tropics than in the middle latitudes, and weather data from tropical ocean areas are especially sparse. The Coriolis force is small near the equator, so winds aloft are not geostrophic. Rotating cyclones and anticyclones do not form near the equator, and air mass contrasts and frontal activity are absent.

In the continental tropics, the principal influence on the weather is the daily cycle of heating and cooling of comparatively homogeneous humid air. The temperature variation from day to night in the tropics often exceeds the variation in average monthly temperatures through the year. Study of the thermoisopleth diagram for Belém (Figure 10.7), near the mouth of the Amazon River in Brazil, shows that the average daily temperature range is about 10°C (13°F), but that the range in average monthly temperature is no more than about 2°C (4°F).

The tropics may lack variety in day-to-day weather, but many tropical regions have strong seasonal weather contrasts and exhibit weather phenomena found nowhere else in the world.

Easterly Waves

Poleward of the band of persistent clouds and precipitation that marks the intertropical convergence, weak troughs of low pressure occasionally form in the trade wind zone and drift slowly westward (Figure 6.11). These *easterly waves* extend roughly north and south for a distance of several hundred kilometers. They form most often over the seas during the high-sun season.

In the subtropics, a warm layer of dry subsiding air usually overlies the moist surface layer; this produces a weak temperature inversion and prevents the surface air from rising to higher altitudes. The subsidence inversion is temporarily destroyed by an easterly wave, resulting in weather disturbances that vary from mild to quite violent.

In the trade-wind belt of the northern hemisphere, the normal atmospheric flow is from the northeast towards the southwest. An easterly wave can be thought of as a trough line, with the *mT* air converging and turning a bit towards the northwest as it enters the trough from the east. As the *mT* air flows westward out of the trough it turns back again towards the southwest. The surface divergence to the west of the trough is also associated with subsidence from the upper air, and the weather is fair and nearly cloudless. In the surface convergence to the east of the wave, the inversion is broken, moist air ascends to great heights, and thunderstorms may be generated. Each year a few easterly waves increase in intensity and develop into severe tropical cyclonic storms.

Tropical Cyclones

A dangerous interruption of the monotony of tropical weather is the late summer appearance of intense cyclonic disturbances. If their wind speeds exceed 120 km (75 miles) per hour, such disturbances are classified as *tropical cyclones*. In the Caribbean area and on North American coasts, these are called *hurricanes*; in the western Pacific, they are known as *typhoons*; in the Indian Ocean and Australia, they are simply called *cyclones* (Figure 6.12).

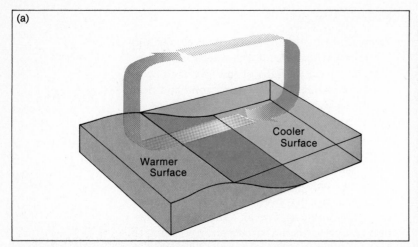

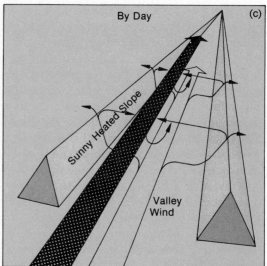

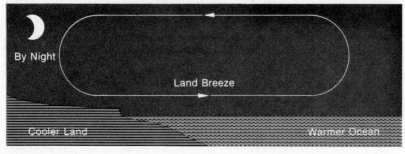

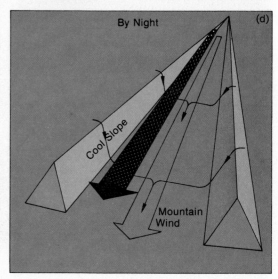

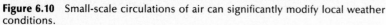

Figure 6.10 Small-scale circulations of air can significantly modify local weather conditions.

(a) This diagram shows a small convective cell generated near the boundary between warm and cool regions, such as land and water areas.

(b) Land and sea breezes develop because of the difference in temperature between the ocean and the land. During the day, the land heats more rapidly than the ocean, and a surface *sea breeze* develops from the ocean toward the land. At night, the land cools rapidly, and the surface flow of air is a weaker *land breeze* from the land to the ocean.

(c) Similar local winds develop because of the temperature differences between valleys and mountain slopes. During the day, as air rises up the warm slopes, a surface *valley wind* flows up the valley to replace the ascending air.

(d) At night, cool air descends the slopes as a *mountain wind* and flows down the valley. (Calvin Woo)

(e) This view shows cloud formation resulting from convective rise of air above heated mountain slopes adjacent to Lake Brienz, Switzerland. (T. M. O.)

(f) Orographic fog and cloud shrouding Neuschwanstein castle, near Füssen in Bavaria. (Robert A. Muller)

A tropical cyclone is an unusually compact and intense low-pressure center, much smaller in diameter than midlatitude cyclones, but having a pressure gradient that can exceed 30 millibars per 100 kilometers. The winds that whirl around the center of a tropical cyclone sometimes have velocities greater than 200 km (120 miles) per hour and are accompanied by driving rain and severe thunderstorms.

A hurricane begins as a rotating tropical storm generated from a strongly developed easterly wave. Why some storms die out and others continue to build to hurricane strength is not known, although airflows in the upper

(f)

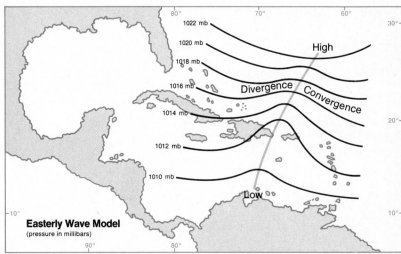

(e)

Figure 6.11 The kink in the isobars denotes the presence of an easterly wave. Easterly waves move toward the west with the trade winds. Ahead of a wave, to the west of the trough, warm, dry air descends from the upper atmosphere, and the weather is fair. Behind the wave, to the east of the trough, moist air from the surface ascends to great heights, and severe thunderstorms may be generated. Each year a few easterly waves increase in intensity and develop into the severe tropical cyclonic storms known as hurricanes. (Andy Lucas after Hermann Flohn, *Climate and Weather*, © 1969 by H. Flohn, used with permission of McGraw-Hill Book Company)

Figure 6.12 This figure shows the regions of occurrence, and the typical tracks, of tropical storms and hurricanes. These devastating tropical storms form only over warm ocean waters between the latitudes of about 5° to 25° in each hemisphere. Initially they are carried toward the west by the trade wind circulations, but they tend to curve poleward and toward the east if they move under the circumpolar flow of westerly winds aloft in subtropical latitudes. Note also that tropical storms and hurricanes do not occur in the South Atlantic Ocean.

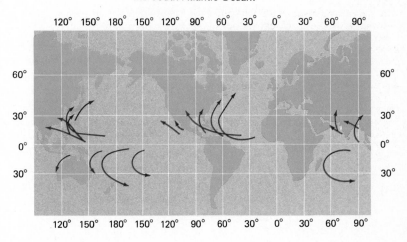

atmosphere may play a role. The characteristic structure of a fully formed tropical cyclone consists of a relatively cloudless central "eye" surrounded by a rapidly rotating wall of towering clouds, as illustrated in Figure 6.13. The eye is normally no more than 30 km (20 miles) in diameter. Centrifugal force associated with the rapid rotation prevents the converging surface air from "filling in" the eye. Drier air from the upper atmosphere descends into the eye and is heated adiabatically, keeping the eye relatively free of clouds. Moist air spirals upward around the eye, and massive condensation produces the wall clouds, releasing enormous amounts

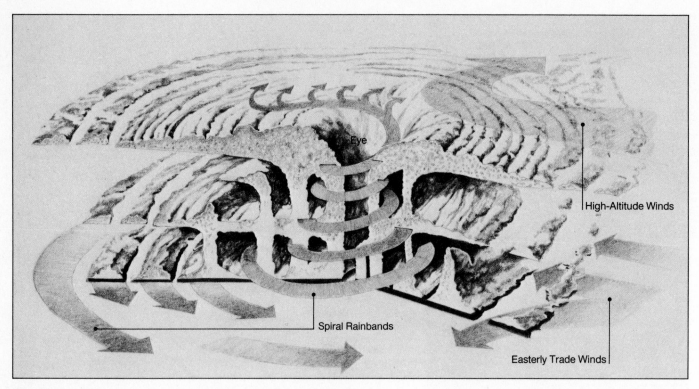

High-Altitude Winds

Eye

Spiral Rainbands

Easterly Trade Winds

Figure 6.13 This cross section of a hurricane (vertical scale much exaggerated) shows the central column, or eye, and the swirling cloud bands that give a hurricane its characteristic appearance as seen from above. A tropical cyclone in the northern hemisphere has an intense counterclockwise rotation. Wind speeds near the center may exceed 200 km (120 miles) per hour. Dry air descending into the eye from above makes weather there clear and calm. (Tom O'Mary after *The Atlas of the Earth*, p. 30, © 1971, Mitchell-Beazley, Ltd.)

of latent heat. The height of the wall clouds is typically 10 to 15 km (6 to 9 miles).

Because of the weakness of the Coriolis force at low latitudes, tropical cyclones rarely form within 10° of the equator; nor do they usually form at latitudes higher than about 25°. The warm ocean surfaces between these two latitudes provide the necessary conditions for the formation of tropical cyclones. Unlike midlatitude cyclones, which are powered primarily by the temperature differences between dissimilar air masses, the energy source for tropical cyclones is the latent heat released by massive condensation of water vapor. Water vapor is most abundant in warm tropical air over oceans with surface temperatures greater than 27°C (81°F). The oceans reach their maximum temperature in late summer or early fall, making that part of the year the hurricane season. Once

a tropical cyclone begins to travel over land or cold water, it becomes separated from the abundant supply of water vapor that is its source of energy, and its strength diminishes.

On a densely populated low-lying coastline, a direct hit by a strong hurricane is devastating (Figure 6.14). When gale-force or hurricane winds drive coastal waters up onto low shores, the sea may rise several meters above normal high tide, surging tens of kilometers inland. The resulting destruction and loss of life can be enormous. Some 250,000 people were drowned by such an event in Bangladesh in 1970. A rainfall of more than 25 cm (10 in.) is not unusual as a hurricane passes by, and if the hurricane moves over land, heavy rains soon bring rivers to the flood stage. The number of hurricanes generated in the Atlantic and the Caribbean varies from three or four to twelve per year, depending largely on the temperature of the sea surface and the behavior of the subtropical jet stream.

Although it is difficult to predict the course of a hurricane (Figure 6.15), its distinctive cloud formations are easy to detect on weather satellite images (Figure 6.16). Therefore, the moving storm can be tracked in time to give warning to threatened areas. Improved warning and communications systems have steadily reduced the loss of life resulting from hurricanes in the United States, even though those in peril often underestimate the danger and do not always take the warnings seriously. In less developed areas of the world, lack of adequate communication leaves great numbers of people unaware of the hazard, even after it is detected. When the cyclone in the Bay of Bengal struck Bangladesh in 1970, vast numbers received no warning, which accounts for the catastrophic loss of life.

Although every hurricane appears to be a potential disaster, these powerful storms are vital to the earth's energy balance. They are a means of transporting energy from areas of excess to areas of deficiency. If there were fewer hurricanes each year, those that occurred would have to be even larger and more violent to perform their vital function of carrying energy poleward.

Tropical Monsoons

As mentioned in Chapter 4, monsoons are secondary circulations that involve seasonal reversals in the direction of surface winds. This wind regime results from seasonal pressure changes on a continental scale. The most extreme case occurs in Asia. During winter, the Asian interior is cold, and an almost steady outflow of dry cP air from the Siberian high brings clear weather to most of the continent. The dry northwest monsoon effects China and Japan, and the somewhat warmer northeast monsoon prevails in India and Pakistan. But during the summer, the heating of southern Asia produces a thermal low, shown over the Indus Valley of Pakistan in Figure 4.6. This causes the regional winds to reverse as warm, humid mT air sweeps over the continent from the Indian Ocean and the southwestern Pacific. The monsoon circulation, therefore, can be thought of as a continental example of the differential heating of land and water by seasons.

The southwest winds of the Indian summer

Figure 6.14 This photograph illustrates the storm surge produced by Hurricane Beulah in 1967. Here the wind-driven waves are seen washing over the beach and inland at Corpus Christi, Texas. (Wide World Photos)

Figure 6.15 This map shows the tracks of some devastating North Atlantic hurricanes for the years 1954 to 1980. The path of an individual hurricane tends to be erratic, but as the map indicates, many of the hurricanes generated in the western Atlantic follow the same general course westward and northward. Some hurricanes turn northward early and skirt the east coast of the United States; others enter the Gulf of Mexico before heading north. Hurricanes lose strength over land because of friction and lack of water vapor, but some hurricanes, such as Hazel, 1954, and Agnes, 1972, traveled long distances across the United States, producing record floods along major river systems. (Andy Lucas after *The National Atlas of the United States of America*, 1970)

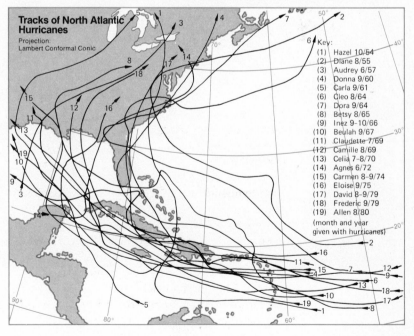

Tracks of North Atlantic Hurricanes

Projection: Lambert Conformal Conic

Key:
(1) Hazel 10/54
(2) Diane 8/55
(3) Audrey 6/57
(4) Donna 9/60
(5) Carla 9/61
(6) Cleo 8/64
(7) Dora 9/64
(8) Betsy 8/65
(9) Inez 9–10/66
(10) Beulah 9/67
(11) Claudette 7/69
(12) Camille 8/69
(13) Celia 7–8/70
(14) Agnes 6/72
(15) Carmen 8–9/74
(16) Eloise 9/75
(17) David 8–9/79
(18) Frederic 9/79
(19) Allen 8/80
(month and year given with hurricanes)

(a)

(b)

Figure 6.16 (a) This photograph of hurricane Gladys, 1968, was taken from the Apollo 7 spacecraft. The open central eye and the counterclockwise circulation of the hurricane are visible. (NASA)

(b) This satellite image of weather conditions on August 30, 1975, shows three hurricanes in various stages of development. All are moving westward in the zone of easterly winds. Hurricane Caroline, which originated in the western Atlantic and later caused destruction on the mainland, is seen moving onshore from the Gulf of Mexico. Hurricane Katrina appears near its point of origin off the west coast of Mexico, and Hurricane Jewell is dissipating farther to the northwest. An occluded midlatitude cyclonic storm can be seen decaying over British Columbia and Washington. (National Oceanic and Atmospheric Administration)

monsoon soak up moisture as they cross the warm Indian Ocean. Summer precipitation from this air accounts for 70 percent of India's annual rainfall. The mountains of India trigger very heavy orographic precipitation, producing annual totals of more than 1,000 cm (400 in.) in some locations—85 to 90 percent of which falls between May and September.

Weaker monsoon circulations also occur in China and Japan, Southeast Asia, northern Australia, and the Guinea coast of Africa. There is a slight monsoon effect over the Mississippi Valley, with frequent outbreaks of polar air (called "northers") in winter and invading waves of sultry air from the Gulf of Mexico during summer. The summer thunderstorms of the Arizona desert and semiarid New Mexico and Texas reveal that a weak monsoon also affects the dry southwestern regions of the United States.

SUMMARY

The general circulation of the atmosphere produces smaller scale secondary circulations that are the controls of day-to-day weather in the middle and low latitudes. This weather is associated with moving high and low pressure centers and accompanying air masses and zones of air mass interaction.

Air takes on the heat and moisture properties of the underlying surface when it occupies a homogeneous region for days or weeks at a time. Except in the tropics, most weather can be explained in terms of the interactions between cP, mT, mP, and cT air masses having differing temperatures and humidities. Continental and maritime air masses differ greatly in humidity, whereas polar and tropical types present strong temperature contrasts. Precipitation in the midlatitudes is produced mainly by frontal activity in cyclonic storms that move from west to east along the polar front. Cyclonic storms are generated by pressure drops along stationary fronts that separate air masses from dissimilar source regions. If a low-pressure center develops along a stationary front, cyclonic circulation is initiated, in which air behind a cold front overtakes warmer air advancing in a warm front. Warm air slides up over cooler air along the warm front, and cold air pushes under warm air along the cold front. Eventually the cold front overtakes the warm front, and the warm air loses contact with the ground, forming an occluded front. The development of cyclonic storms is assisted by the flow of air in the upper atmosphere.

Air mass contrasts and fronts rarely exist in tropical and equatorial regions. Convective showers and thunderstorms develop almost daily in the unstable air of the intertropical convergence zone. Weather disturbances in the tropics begin as easterly waves that produce clusters of rainy days. Heating of the oceans during the summer causes some easterly waves to intensify into tropical cyclones, also known as hurricanes and typhoons, that are enormously destructive. These powerful storms are energized by latent heat released by moisture condensation and cloud formation in extremely humid air. Tropical cyclones are part of the mechanism by which net energy gains in the low latitudes are transferred to higher latitudes to compensate for net energy losses.

Monsoon circulations are another tropical weather phenomenon of importance. The monsoon effect is a seasonal reversal of winds and associated weather conditions related to changes in atmospheric pressure over land and water areas. The most extreme example is seen in India, which is largely free of clouds in winter as dry air streams southward toward the ITC from the high-pressure center in the interior of Asia. Heavy rainfall follows in the summer as moist air from the Indian Ocean is drawn into a low-pressure center that develops over southern Asia. Minor monsoon effects may be observed along the Gulf coast and in the southwestern deserts of North America.

REVIEW QUESTIONS

1. Compare and contrast source regions and properties of cP, mP, cT, and mT air masses over North America.
2. Why is mT air frequently lifted over cP air?
3. Why might air-mass contrasts be greater over the southeastern United States than over eastern China?
4. Outline the stages of a midlatitude cyclone and associated weather over the eastern United States during winter.
5. Compare and contrast vertical and horizontal airflow in cyclones and anticyclones in the northern and southern hemispheres.
6. Contrast the form, processes, and properties of midlatitude and tropical cyclones.
7. Which oceanic areas and coastal regions are normally visited by tropical storms and hurricanes?
8. How are monsoon circulations related to the geographic distributions of land and water?

APPLICATIONS

1. Keep a daily log of the air mass types present in your area, using your own judgment and recollection of past extreme conditions to establish the actual air mass types. How abrupt are changes in air masses? Are some air masses transitional between the basic types discussed in the text? What is the nature of the weather during periods when air mass types change in your area?
2. On the basis of the midlatitude cyclone model described in Figure 6.7, determine the most likely progression of cloud types and weather events when a cyclonic storm approaches your area from the west in winter and passes to the north and northeast. What would change if the cyclonic disturbance passed to the south?
3. How far toward the equator do midlatitude cyclones penetrate? Does Hawaii experience such storms? Are there places that experience both midlatitude cyclones and tropical cyclones? If so, would both be possible at the same time of year? If the NOAA periodical publication *Environmental Satellite Imagery* is available on your campus, check the images to answer the above question.
4. What was the heaviest rainfall ever received in your area? What unusual meteorological conditions occurred or combined to produce it?
5. Study the hurricane history of a segment of the North American coastline. Annual summaries of hurricane tracks are published in the periodical, *Weatherwise* (February issue), as well as by NOAA, and newspaper files are excellent sources for day-by-day accounts of weather events. As an alternative do a similar study of tornadoes or of memorable midlatitude cyclones in some region—such as those that paralyze urban areas for days because they create snow removal problems.
6. What would be the long-range effect of a workable program to stop the growth of every storm that threatened to be destructive or costly?
7. What have been the greatest floods in your region's history? Were they produced by similar meteorological events?

FURTHER READING

Anthes, Richard A., et al. *The Atmosphere*, 2nd ed. Columbus, Ohio: Merrill (1978), 442 pp. This introductory text has an interesting historical perspective. Other unusual features include synoptic and seasonal analyses of midlatitude weather.

Eagleman, Joe R. *Meteorology: The Atmosphere in Action.* New York: D. Van Nostrand (1980), 384 pp. This very interesting recent text emphasizes aspects of the circulation, and interrelationships to the upper air flow. Part 3 is especially helpful.

Hidore, John J. *Workbook of Weather Maps*, 3rd ed. Dubuque, Iowa: Brown (1976), 81 pp. This paperbound compilation of official U.S. weather maps for eleven series of weather events includes the record polar outbreaks of January and February 1962 and Hurricane Agnes in June 1972.

Hughes, Patrick. *American Weather Stories.* Washington, D.C.: U.S. Dept. of Commerce, NOAA (1976), 114 pp. Brief accounts of weather and climate events that had major impacts on our lives.

Lehr, Paul E., R. Will Burnett, and **Herbert S. Zim.** *Weather.* New York: Golden Press (1975), 160 pp. This excellent paperback is one of the well-known Golden Nature Guide series. The color diagrams of fronts and midlatitude cyclones are especially informative.

Muller, Robert A. "Snowbelts of the Great Lakes." *Weatherwise,* vol. 19, no. 6 (1966): 248-255. Focus is on weather events that produce persistent and deep snowfalls over small areas of the Northeast. Anyone interested in weather and its consequences should enjoy each issue of this journal.

Neiburger, Morris, James Edinger, and **William Bonner.** *Understanding Our Atmospheric Environment*, 2nd ed. San Francisco: Freeman (1982), 453 pp. Basic fundamentals are presented in a nonmathematical framework. Includes a chapter on modern forecasting techniques.

Simpson, Robert H., and **Herbert Riehl.** *The Hurricane and Its Impact.* Baton Rouge: LSU Press (1981), 420 pp. An excellent new book by two of the outstanding American experts on tropical storms. In addition to meteorological processes, this book emphasizes what can happen to people and buildings along subtropical coastlines.

Stewart, George R. *Storm.* New York: Random House (1941), 349 pp. This fictional classic follows the evolution of a midlatitude cyclone, named Maria, across the Pacific and then over the United States. Much of the perspective is through the eyes of forecasters at San Francisco and managers and workers of transportation and communications networks. This novel is a must for students interested in interactions between weather and human endeavors. (Reprinted in paperback in 1974 by Ballantine Books, New York.)

Weems, John Edward. *A Weekend in September.* New York: Henry Holt (1957), 180 pp. This is a careful journalistic account of the 1900 hurricane that swept Galveston, Texas, with a great loss of life.

CASE STUDY

The Tornado Hazard

Of all the earth's winds, tornadoes are the most feared. Along the track of a severe tornado, destruction is nearly total, while just a few meters away there may be no damage at all. Tornadoes are especially frightening because they appear suddenly. At night or in densely forested country, it is difficult to see the funnel cloud of an approaching tornado; the sound, frequently compared to the roar of an express train, is sometimes the first warning. The National Weather Service has an effective program of tornado watches for alerting the public to areas where tornadoes may break out, but there is no way to predict the precise time and place one will occur.

A *tornado* is a narrow vortex of rapidly whirling air. It is almost always associated with severe thunderstorm activity. The rotating vortex extends downward from a cumulonimbus cloud, and it becomes visible when water vapor condenses, producing the familiar *funnel cloud*. Dust and debris swept up from the ground create a much darker and more ominous-looking funnel. Occasionally, several funnels may dangle from the same cloud, and many funnel clouds aloft never reach the ground. As the mother cloud moves on, the funnel is often retarded at the surface by friction, so that it becomes tilted or crooked. Tornado funnels over coastal waters and seas are called *waterspouts*.

Tornadoes usually advance at a speed of 32 to 48 km/hour (20 to 30 mph). Thus a tornado will normally pass a given point in a matter of seconds. The tornado and its track along the surface are usually only tens of meters wide, although occasionally extending up to several

A close view of the tornado which swept across sections of Dallas, Texas, on April 2, 1957. (ESSA Weather Bureau)

hundred meters. Some tornadoes skip across the landscape, leaving a broken track of destruction. Most tornado tracks are 5 to 10 km (3 to 6 miles) in length, but a few are much longer; on May 26, 1917, a single tornado tracked more than 400 km (250 miles) across Illinois and Indiana in a little less than eight hours.

Although tornadoes are short-lived, they are extremely violent. The winds of the tornado vortex have been estimated to reach speeds of up to 600 km (370 miles)/ hour. Within the funnel cloud, atmospheric pressure is as much as 50 millibars lower than that of the adjacent air, and pressure drops of up to 100 millibars have been estimated from damage patterns. As a tornado passes over a building, the strong winds of the vortex rip at the exterior, and the abrupt pressure drop causes the building literally to explode, with the roof and walls blown

out. The debris is then caught by the rotating winds and strewn along the tornado's path; each piece of debris becomes a flying missile of destruction. Cellars, interior halls, and bathrooms offer greater safety than rooms with outside walls. When severe tornadoes pass over modern slab homes, sometimes only the plumbing fixtures, such as bathtubs and toilets, remain. House trailers are especially vulnerable to tornado winds.

In the United States, weather situations conducive to tornado development occur most frequently over the Great Plains, the Mississippi Valley, and the Southeast. The region extending from the Texas Panhandle northeastward across Oklahoma and eastern Kansas is sometimes called "tornado alley." Tornadoes are very infrequent west of the Rocky Mountains and across northern New England and the upper Great Lakes region. The mean annual number of tornadoes reported has increased significantly in recent decades. Since 1970 the annual average has been about 900. The recent increase is attributed mainly to improved observations and detection.

Tornadoes break out most commonly along thunderstorm or squall lines ahead of the cold fronts, where temperature contrasts are large and instability great. In the United States instability is usually associated with a deep layer of warm, dry *cT* air from the Southwest above warm, moist *mT* air from the Gulf of Mexico. Tornadoes usually occur in the warm sector and track from the southwest toward the northeast.

The seasonal distribution of tornado outbreaks tends to follow the geographical distribution of fronts separating *mT*, *cP*, and *cT* air masses, midlatitude cyclones, and maximum instability through a deep layer of the troposphere. In the United States, therefore, tornadoes are most frequent in the South during late winter and early spring, with the hazard migrating northward into Kansas and Missouri, and finally to Iowa and Nebraska by June, when there are few tornadoes near the Gulf Coast. Many tornadoes occur during late afternoon, when instability tends to be greatest, but unfortunately occurrences remain relatively common at night, when visual detection is difficult.

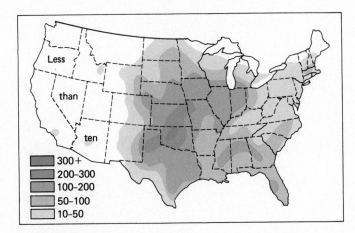

This map shows the number of tornadoes that were reported between 1955 and 1967 in the United States. The main features of the map include "tornado alley" over the Great Plains and the very small number of tornadoes westward from the Rocky Mountains.

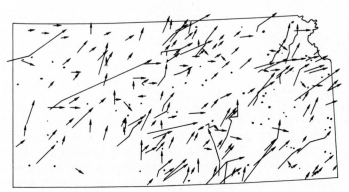

This map shows the direction and path lengths of tornadoes that occurred in Kansas between 1950 and 1970. Most tornadoes are dragged along within thunderstorms by the upper winds from the southwest that steer the thunderstorm cells. (Joe R. Eagleman, Vincent U. Muirhead, and Nicholas Willems, Thunderstorms, Tornadoes, and Building Damage, 1975, Lexington Books)

Rainy Season in the Tropics by Frederick E. Church, 1866. (Fine Arts Museum of San Francisco)

The sun's radiant energy provides the power for the atmospheric circulation to deliver energy and moisture to the earth's surface. On the land the water interacts with the rocks, soils, and vegetation. In time, most of it returns to the atmosphere in a never-ending cycle of renewal.

The processes by which water in its various states moves from the ocean to the atmosphere to the land and back to the ocean are components of the *hydrologic cycle*. The oceans are extremely vital not only to this cycle; water evaporated from them is the source of most of the precipitation that supports life on the land. The oceans are also enormous reservoirs of heat energy, and their temperature variations influence the general circulation of the atmosphere, which powers the hydrologic cycle.

The hydrologic cycle is much more complicated on the continents than over ocean areas, where we need think only of the processes of evaporation and precipitation. Consider, for example, the state of Washington. Inland from the gently rolling hills surrounding Puget Sound and the city of Seattle, the land rises abruptly to a spine of high mountains, the Cascade Range, which divides the state. To the west of the Cascades, rain falls much of the year, brought by storm clouds that sweep in from the North Pacific. As much as 2,500 cm (1,000 in.) of snow can accumulate in a year on major peaks like Mt. Rainier, and water is almost always present in the streams that rush westward toward the coast. But to the east of the Cascades is a region of dry grasslands, cut off from the coastal weather by the mountains. Here lie the sun-drenched fields and orchards of the Yakima and Wenatchee valleys and the Grand Coulee area. In this region irrigation is necessary to grow crops. The contrasts on the two sides of the Cascades, and of many similar mountain systems, illustrate how varied and complex hydrologic regimes can be on the land, and how human activities adjust to accommodate to these variations.

If we add up all the ways that water is used by humans, we find that the average daily consumption in the United States is equal to about 6 cubic meters (1,500 gal) per person—not including water used for the generation of hydroelectricity. The public uses only about 10 percent directly for such things as cooking, sanitation, and watering lawns and gardens. Industry uses another 10 percent. The remaining 80 percent is divided about equally between agriculture and thermal power plants that use water for cooling.

7
The Hydrologic Cycle and the Local Water Budget

The Global Water Budget

Water on the Land
Interception, Throughfall, and Stemflow
Infiltration and Soil Moisture Storage
Groundwater
Runoff and Streamflow

The Local Water Budget: An Accounting Scheme for Water
Potential and Actual Evapotranspiration
Calculating the Water Budget
A Local Water Budget: Baton Rouge
Problems in Application of the
    Water Budget Model

If all this water were truly consumed and permanently removed from the hydrosphere, disastrous water shortages would develop very quickly. But water is a renewable resource. Nearly all the water used on farms, in homes, and by industry takes some path back into the hydrologic cycle. Most of the water diverted from rivers or pumped from wells to be used in irrigation is returned to the atmosphere through evapotranspiration in the fields. The water used in cities and in suburban homes is usually channeled into sewage systems to return to surface streams and finally to the ocean.

Throughout history people have labored to ensure reliable water supplies. As early as five thousand years ago, the Egyptians channeled the floodwaters of the Nile River through long canals to vast basins surrounded by clay dikes. Aqueducts were used to provide ancient Rome with a water supply that was ample even by modern standards. Today, similar aqueducts carry water hundreds of kilometers to the Los Angeles basin and to agricultural areas in the California desert (see the Case Study following this chapter). Water projects of amazing scale, involving gigantic reservoirs and transfers over thousands of kilometers, have been proposed, but costs and environmental issues have so far prevented their development.

The need for fresh water grows ever more pressing as populations increase and expand into regions where natural supplies are scarce. Because of the growing need to manage available water, there is increasing emphasis on the study and analysis of water on and below the earth's surface—a field known as *hydrology*.

In this chapter we shall first review the global water budget before examining in detail the pathways of water on the continents. A key concept in this chapter is use of the local water budget as a tool for evaluating the availability of water at a particular place.

THE GLOBAL WATER BUDGET

As we saw in Chapter 2, the water of the earth's hydrosphere is stored in several different con-

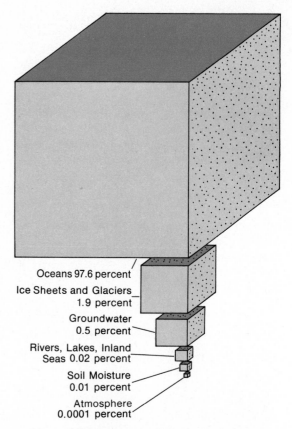

Oceans 97.6 percent
Ice Sheets and Glaciers 1.9 percent
Groundwater 0.5 percent
Rivers, Lakes, Inland Seas 0.02 percent
Soil Moisture 0.01 percent
Atmosphere 0.0001 percent

Figure 7.1 The volumes of the cubes show the relative amounts of free water in storage on the earth. Nearly 98 percent of the water is stored in the oceans, which contain an estimated volume of 1.3 billion cu km (0.3 billion cu miles) of water. Glaciers contain the largest store of fresh water, but the turnover is slow. Most of the readily available fresh water is stored in porous rock beds as groundwater. The amount of water stored in the atmosphere is relatively small, but because it is actively transported and released, it plays a key role in the hydrologic cycle. (Tom Lewis after R. L. Nace, *Water, Earth and Man*, edited by R. J. Chorley, 1969, Methuen & Co., Ltd., Publishers)

ditions. Figure 7.1 shows that the oceans, which cover about 70 percent of the earth's surface, contain nearly 98 percent of the total water supply. Because ocean water contains dissolved minerals, primarily salt, it is unfit for consumption by humans, land animals, and even land plants. But when water evaporates from the oceans, the dissolved salts are left behind. Hence, water vapor evaporated from the oceans is a source of liquid water that is fresh and essentially "pure." Life on the land relies almost entirely on the fresh water produced by this natural desalination process.

About three-fourths of the earth's nonsaline fresh water is stored as glacial ice, mainly in the polar ice sheets covering Antarctica and Greenland. These ice sheets receive an annual snowfall equivalent to only about 10 cm (4 in.) of liquid water. Water thus stored in polar ice sheets returns to the oceans centuries or even thousands of years later, when icebergs break away from coastal glaciers and melt in the sea. The ice sheets, then, represent long-term storage of fresh water. Water storage in smaller glaciers in mountain regions is trivial by comparison, but nevertheless it supplies much of the flow for the rivers issuing from high mountains.

The next largest reservoir of nonsaline water, accounting for only one half of 1 percent of the total water supply, is groundwater. Much of it is unusable because of mineral contamination or problems of extraction. Surface water supplies (rivers, lakes, and so on) are an even smaller part of the total, as is moisture stored in the soil.

The amount of water stored in the atmosphere at any one time is even less. As noted previously, all the water in the atmosphere at any moment would form a layer only 2.5 cm (1 in.) deep over the earth. Over an entire year, however, the atmosphere transports and recycles enough water to cover the earth with a layer about 95 cm (37 in.) deep.

For the global water budget to remain in balance, the amount of water that leaves the atmosphere as precipitation must return to the atmosphere through evapotranspiration from the continents and evaporation from the oceans. Figure 7.2 shows how the global water budget is kept in balance. If we consider the oceans alone, we find that precipitation into them is less than evaporation from them. Over the continents precipitation exceeds evapotranspiration. But ocean levels are not falling, and the continents are not becoming flooded. The apparent imbalance is offset by water that is continually moving from the continents to the oceans. The land sheds its excess precipitation by the flow of rivers to the sea and by

contributing moisture to continental air masses that move out over the oceans.

WATER ON THE LAND

What happens to precipitation that falls on the continents? Depending on the characteristics of the land surface, this water can take several different pathways back to the atmosphere or the seas to complete the hydrologic cycle.

Interception, Throughfall, and Stemflow

Not all the precipitation that falls reaches the soil. In urban environments, rain strikes roofs, building walls, and areas paved with concrete and asphalt. If the rain is heavy, sidewalks, parking lots, and streets become flooded very quickly, a twentieth-century phenomenon known as urban flash flooding. Most of this floodwater runs off within a few hours into gutters and subsurface storm drains that empty into streams, lakes, or the ocean. The little that remains in puddles eventually evaporates, returning directly to the atmosphere.

Beyond urban areas, plant growth covers the surface (except where it is arid) much of the year. This too prevents some rain from reaching the soil. When rainfall commences, leaves *intercept* and store much of the water, as shown in Figure 7.3. If the rain is heavy or lasts very long, the capacity of the leaves to retain water is exceeded, and water begins to drip down to the soil as *throughfall*. Thus trees become less effective shelters as rainfall continues. Water also reaches the soil as *stemflow*, that is, by trickling along branches and down the trunks of trees. This deviation from the ordinary route of rainfall diverts an above-average amount to the soil around the bases of plants.

After the rain ends and leaves stop dripping, some water remains on the leaves and evaporates back into the atmosphere. Because of interception, the amount of water reaching the soil is less than the total precipitation, and the

Transport of Water
Vapor from the Oceans
(94)

(12)

Storage as Ice and Snow

Precipitation over the
Continents (106)

Evapotranspiration
from the Continents
(69)

Interception by Plants

Temporary Surface Storage

Surface Runoff

Infiltration

Soil Moisture Storage

Percolation

Storage in Rivers and Lakes

Groundwater Storage

Groundwater Runoff to Streams

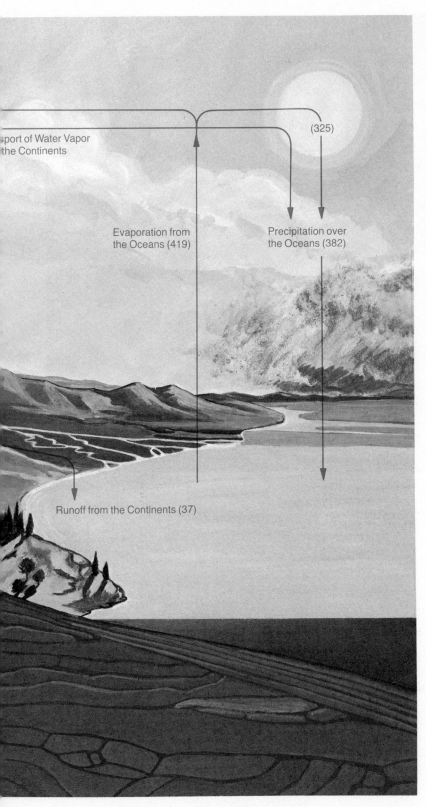

Transport of Water Vapor the Continents

(325)

Evaporation from the Oceans (419)

Precipitation over the Oceans (382)

Runoff from the Continents (37)

Figure 7.2 The movement of water through the hydrologic cycle involves numerous interactions and storage processes. Values in parentheses in the figure are water volumes in 1,000 cu km. Each year about 419,000 cu km of water evaporate from the oceans into maritime air masses. Evapotranspiration from the continents into continental air masses amounts to an additional 69,000 cu km. This total volume of 488,000 cu km is equivalent to a mean annual precipitation over the globe of about 95 cm (37 in.).

Precipitation back to the oceans amounts to only 382,000 cu km, with 325,000 cu km originating from maritime air masses and 57,000 cu km supplied by continental air masses that move over ocean areas. Over the continents, precipitation (106,000 cu km) is greater than evapotranspiration (69,000 cu km); the difference (37,000 cu km) represents stream runoff that eventually returns to the oceans. Note that precipitation over the continents is supplied largely by water vapor from the oceans.

When precipitation falls on the land, a portion of the moisture is intercepted by vegetation and evaporates from temporary storage on leaves. The moisture that reaches the ground either infiltrates the soil (inset), runs off across the surface, or evaporates from temporary storage in depressions. Some of the water that infiltrates the soil is stored as soil moisture, while a portion percolates deeper and enters groundwater storage. The flow of streams is maintained by both direct surface runoff and groundwater contributions. (John Dawson)

Figure 7.3 Some of the precipitation that is intercepted by vegetation and held in temporary storage by leaves returns to the atmosphere by evaporation and does not reach the soil. (David Cavagnaro)

timing and distribution of water arrival at the ground surface are varied. Compare the effects of interception on a continual, steady rain totaling 1 cm with the effects on ten brief showers falling at the same intensity and totaling 0.1 cm each. During the steady rain, the leaves' capacity to retain water is soon exceeded, and throughfall carries water to the soil. But between the brief showers the leaves have a chance to dry so that a significant portion of the rain from each shower is intercepted and subsequently returns to the atmosphere; very little water reaches the soil.

Infiltration and Soil Moisture Storage

A few days after a rainfall the surface soil has usually dried out, but the soil a few centimeters below the surface remains moist. Water enters soil, or *infiltrates* it, from the surface downward. The maximum rate at which a soil can absorb water is its *infiltration capacity*, which depends on the soil's porosity, permeability, surface condition, and moisture content.

Soil consists of particles of mineral and organic matter of various sizes (Chapter 11). Pulled by the force of gravity, water moves downward through the spaces, or *pores*, between these particles. Clay soils consist of particles of microscopic size and have much less pore space than sandy soils. If the pores are large and interconnected, as in sandy soils, the soil is *permeable* (Figure 7.4); clay soils tend to be nearly impermeable.

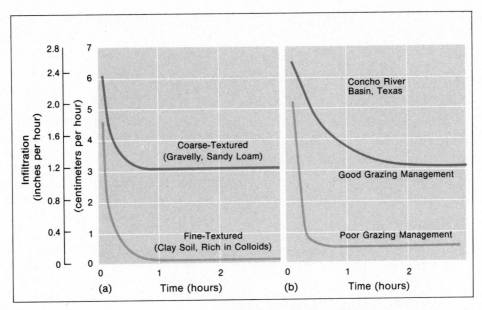

Figure 7.4 The rate of infiltration of water into a soil depends on such factors as the texture, porosity, permeability, and moisture content of the soil and the condition of the surface layer. The graphs show the infiltration rates of various soils measured from the time at which water is added to their surfaces. The infiltration rate falls sharply at first in all cases as the top layer of soil becomes well moistened; then a constant rate of infiltration is attained.

(a) Water infiltrates a coarse, permeable soil more easily than it does a dense clay soil, in which the water passages are small.

(b) The surface of well-managed grazing land retains an open texture and has a high infiltration rate. The infiltration rate is lower on poorly managed land because overgrazing exposes the soil, which allows the bare surface to become compacted by raindrops and animal hooves. (Doug Armstrong after *Yearbook of Agriculture*, U.S. Department of Agriculture, 1955, and E. E. Foster, *Rainfall and Runoff*, © 1949, Macmillan Co.)

The infiltration rate depends, among other factors, on the condition of the surface layer. If it has been compacted by vehicles, animals, or human traffic along a path, the soil may not be able to absorb moisture. Bare soil can also be compacted by rain itself, so a cover of vegetation and associated plant litter absorbs the impact of raindrops and helps to maintain higher infiltration rates. Heavy grazing by sheep or cattle reduces the vegetation cover and significantly lowers infiltration rates (see Figure 7.4b).

If water arrives at the surface at a rate less than the soil's infiltration capacity, all the water will be absorbed. Once the upper layers of soil are saturated, additional water cannot enter until water already in the soil begins to drain to lower levels. As Figure 7.4a illustrates, dry porous loam (a mixture of coarse and fine soil particles) can absorb water at an initial rate of over 6 cm (2.4 in.) per hour. As the soil becomes wet, infiltration rapidly decreases to a slow but constant rate.

As the water in a saturated soil drains, or *percolates*, downward, molecular attraction causes some water in the pores to cling to soil particles. Such remaining water is held most effectively in small pores, as in clay soils. In the large pores of sandy soils, molecular attraction is not strong enough to hold the greater masses of water against the pull of gravity; this water drains away over several days. The maximum amount of moisture that can remain stored in a soil after percolation is known as the *field capacity* of the soil (Figure 7.5). Soil moisture eventually returns to the atmosphere, either by direct evaporation from the soil or by transpiration through the leaves of plants.

Vegetation withdraws soil moisture for cell growth and transpiration throughout its rooting zone. After a time, however, the remaining water cannot be taken up by plant roots because

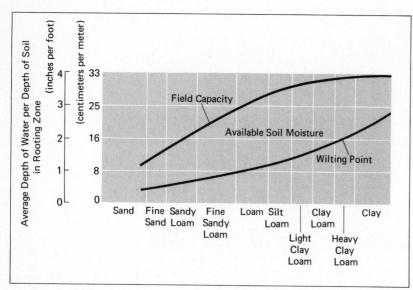

Figure 7.5 The *field capacity* is the amount of moisture stored in the pores between soil particles after drainage of excess water by gravity. The *wilting point* represents the amount of soil moisture still retained by molecular attraction after plant roots have taken up all the water available to them for transpiration. The available soil moisture is the difference between field capacity and the wilting point; it represents the maximum storage in the soil of water available for evaporation and transpiration.

The field capacity, the wilting point, and the available soil moisture all vary according to the texture, or size, of the soil particles. The curves show that more water is held by fine, clay-rich soils at both field capacity and the wilting point, than in coarse, sandy soils. But the shape of the curves indicates that the medium textured loams, soils, have the greatest available soil moisture capacities per unit of depth. Both the field capacity and the wilting point are related to the soil texture, or size of the soil particles. For example, nearly 33 cm of water can be stored in a layer of clay 1 meter deep (or 4 in. in a layer 1 ft deep), but only a little more than 8 cm in sand 1 meter deep. The soil moisture available in the rooting zone averages about 16 cm per meter (2 in. per ft) in silt loam, a little more than 8 cm per meter (1 in. per ft) in clay, and less than 8 cm per meter in sand. (After Smith and Ruhe, 1955)

it is held strongly to the soil particles by molecular attraction. Most plants begin to wilt when the soil moisture is reduced to this level, which is called the *wilting point*; if soil moisture is not replenished by precipitation or irrigation, the plants will eventually die. Too much soil moisture is also harmful to plants, except for specially adapted swamp and marsh vegetation. Soil pores must contain air as well as water because roots require oxygen and space in which to eliminate waste carbon dioxide.

Groundwater

Below the surfaces of the continents lies a vast and accessible reservoir of fresh water. A well-driller can strike water within 100 meters (330 ft) of the surface just about anywhere—even in desert areas where there is little surface water. The top of these underground water zones, in which all voids are saturated with water, is called the *water table*, and the water in the fully saturated *phreatic zone* below the water table is *groundwater*. This water is held in underground beds of sand and gravel and porous types of rock.

The water table rises and falls in response to precipitation, evapotranspiration, and lateral groundwater flow. It represents the minimum depth to which a well must be drilled for a reliable supply of water. Subsurface materials from which groundwater may be obtained either naturally (from springs) or artificially (from wells) are *aquifers*. Subsurface aquifers contain about 30 times the amount of fresh water that is in the streams, lakes, and swamps of the earth.

Under natural conditions the quality of groundwater is usually good. The major exceptions are in arid and coastal regions, where aquifers may be contaminated by dissolved minerals. Generally, the porous rock of an aquifer filters the water and removes suspended particles and harmful bacteria. But it is possible for urban and industrial contaminants and agricultural pesticides to seep into aquifers. This is a particular problem where geological materials are very permeable, as in areas underlain by sand or gravel deposited by streams or past glaciers, or by limestone that contains interconnecting cavities due to the dissolving action of groundwater. From New York's Long Island to California's coastal valleys, wells have been declared unsafe as a result of recent contamination by pesticides and industrial wastes.

The problem of maintaining groundwater quality is one that is growing in importance, because it is more practical for cities to utilize comparatively pure groundwater than to purify badly polluted river water. In the midlatitudes the temperature of groundwater from depths of

10 to 20 meters (30 to 60 ft) is usually only 1° to 2°C higher than the average annual temperature. The relatively constant cool temperature makes groundwater very desirable for urban and industrial users.

A variety of subsurface materials are porous and permeable enough to serve as useful aquifers. Most of the aquifers tapped in North America are beds of sand and gravel. Some were deposited as outwash from glaciers during the ice ages, and others are the result of much earlier deposition. Individual sand and gravel aquifers are commonly about 50 meters (160 ft) thick and often cover thousands of square kilometers. Among solid rocks the best aquifers are sandstones, limestones, and lava beds, because they have interconnected openings that collect and transmit water.

All the water stored in subsurface aquifers comes originally from precipitation. After field capacity is reached, additional rainfall that does not flow directly over the surface percolates through the ground to the water table. In arid regions water may also seep downward from streambeds and lake bottoms. In humid areas, by contrast, groundwater seeps *out* into streams and lakes.

The surface region from which water drains down into an aquifer is called the *recharge area*. Groundwater usually moves laterally in an aquifer at rates varying from meters per year to kilometers per day. When an extensive aquifer slopes gently for a long distance, the principal

recharge area may be hundreds of kilometers from the wells that extract the water (Figure 7.6).

Aquifers may be either *unconfined* or *confined*. In an unconfined aquifer the water is not under pressure so it will not rise above the level of the water table unless it is pumped to the surface. In confined aquifers the water-bearing layer is covered by a layer of impermeable material. A confined aquifer may lie far below the level of the water table. If the impermeable *confining layer* slopes downward, the difference in elevation between the upper and lower portions of an aquifer can result in a considerable difference in water pressure. In *artesian wells*, which are drilled into the lower section of a confined aquifer, where the pressure is high, the water

Figure 7.6 This diagram shows the principal features of aquifers in schematic form. If the rock above an aquifer is permeable enough to allow the vertical movement of water, the aquifer is said to be unconfined. The water table, or the water level in an unconfined aquifer, is the level to which water will rise in a well sunk into the aquifer. If the rock above an aquifer is impermeable, the aquifer is confined and must be replenished from a recharge area that is permeable to water from above. When a well is sunk into a confined aquifer, the level to which the water rises in the well is called the *piezometric surface*. The piezometric surface can be a considerable height above a confined aquifer, particularly above the lower portion of a sloping aquifer, where the water pressure is high. The piezometric surface shown here slopes to the right, toward a region where the aquifers drain slowly into surface streams. (John Dawson after Raphael G. Kazmann, *Modern Hydrology*, 2nd ed., © 1972 by R. G. Kazmann, used by permission of Harper & Row)

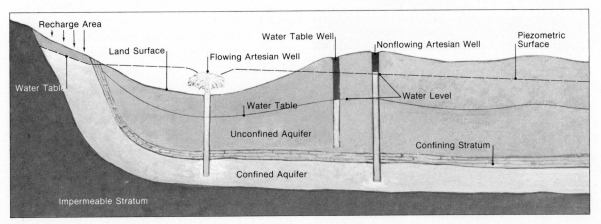

is forced upward considerably above the level of the aquifer itself, as shown in Figure 7.6. The elevation to which water will rise in such wells is known as the *piezometric surface*. In some places the piezometric surface lies above the land surface. This is the cause of *flowing artesian wells*, from which water gushes out at the surface with no pumping necessary. Natural *artesian springs* occur where fractures in rocks permit water to escape to the surface from a confined subsurface aquifer.

Water can also collect on top of impermeable layers that lie above the normal water table. This results in *perched water tables*. These sometimes feed springs along canyon walls. Perched water tables are common in areas of layered sedimentary rocks (Chapter 12) and in thick masses of sediment deposited by streams and glaciers.

Figure 7.7 This diagram illustrates two methods that have been used to recharge aquifers and raise the level of the water table in regions where supplies of groundwater have been depleted by excessive pumping.

(a) Water pumped into shallow surface depressions in the natural recharge area of an aquifer seeps through the permeable rock into the aquifer.

(b) Water pumped into boreholes sunk into an aquifer seeps into the permeable rock to recharge the aquifer. (John Dawson)

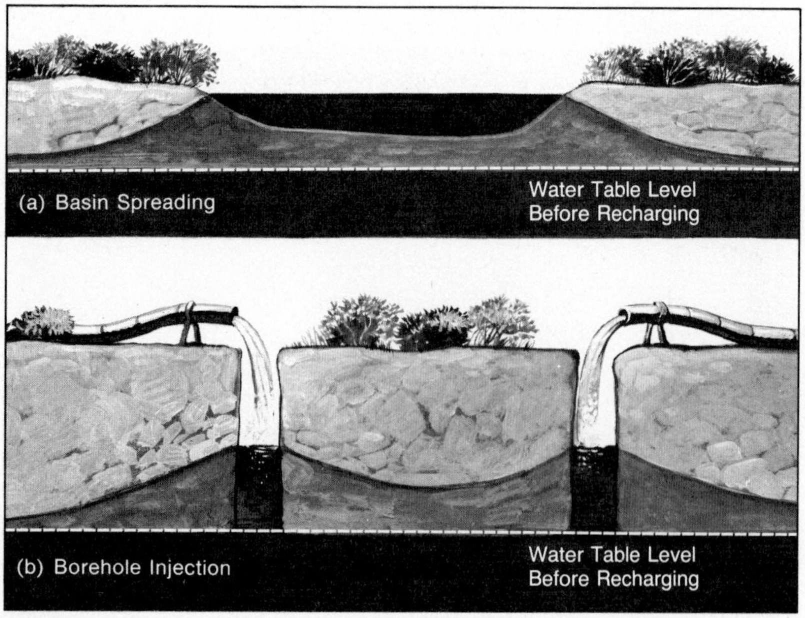

(a) Basin Spreading   Water Table Level Before Recharging

(b) Borehole Injection   Water Table Level Before Recharging

Approximately 50 percent of all the groundwater extracted in the United States is used for agriculture in Texas, Arizona, and California. Groundwater also supports extensive agricultural development and livestock industries on the Great Plains east of the Rocky Mountains and in eastern Australia, North Africa, Arabia, Iran, and other arid regions.

Groundwater must be carefully managed so that the amount of water pumped out does not exceed the net flow into the aquifer. Some aquifers were recharged naturally under climatic conditions that no longer prevail, and are receiving little or no input of water today. This is especially true in desert regions. Unfortunately, in many areas the groundwater is being "mined"—the rate of pumping greatly exceeds natural recharge, so water tables and piezometric surfaces are falling. In such places pumped wells must be deepened, and artesian wells must be pumped. Groundwater withdrawal in some areas has caused the land surface to subside by as much as 8 meters (25 ft). This disrupts irrigation canals and has caused damage to streets and buildings in scattered locations such as Venice (Italy), Mexico City (Mexico), Shanghai (China), Tokyo (Japan), and Phoenix (Arizona).

In some places groundwater is being recharged artificially (Figure 7.7). This is common in southern California, where water may be brought some 1,000 km (600 miles) by aqueduct. Artificial recharge is also employed in coastal regions where excessive pumping of groundwater has lowered the water table, allowing saline seawater to seep under the land, contaminating aquifers. This problem has appeared on Long Island (New York), in California, in Israel, and in many other coastal areas.

Runoff and Streamflow

If rain falls at a rate greater than the infiltration capacity of the soil, water begins to collect on the surface. Surface irregularities and vegetation store some of the water, which is then subject to evaporation or later infiltration. But water also begins to trickle across sloping ground or move in sheets under the influence of gravity. This *surface runoff* increases from small rivulets at the beginning of a heavy rainstorm to a steady torrent when no more opportunities for storage are available. Toward the end of a prolonged rain, nearly all the rainfall may become surface runoff.

Virtually all land surfaces not covered by ice or loose sand are laced with networks of small erosional channels that were made and are maintained by surface runoff. The initial streams carry water to a smaller number of larger streams that, in turn, feed major rivers. The characteristics of such drainage systems are discussed in more detail in Chapter 14. The runoff carried to the sea by rivers is that part of precipitation that did not go into subsurface storage or return to the atmosphere by evapotranspiration.

Hydrologists are concerned with the variable flow of rivers as time passes, and especially with the effect of precipitation on runoff. *Runoff*, as a technical term, is a measure of the average depth of water that flows from a drainage basin during a specified amount of time. It is normally expressed in units equivalent to precipitation—mm or cm or inches per day, month, or year.

Imagine that an intense rainstorm lasting one hour drops a 2.5 cm (1 in.) depth of rain on an impervious parking lot equipped with storm drains. The storm drains are efficient, and 15 minutes after the storm an average depth of only 0.05 cm of water, concentrated in a few depressions and cracks, is left on the parking lot. Eventually this water will return to the atmosphere by evaporation. The runoff for the day from the parking lot amounts to 2.45 cm, or 98 percent of the rainfall. At the same time, we know that the runoff from an equal area of nearby parkland would be far smaller, if any runoff occurred at all, for much of the rainfall there would percolate into the soil, eventually returning to the atmosphere by evapotranspiration.

Streamflow, in contrast with runoff, is the measured volume of water passing any point in a stream channel during a specified unit of time. It is usually expressed as *stream discharge* in cubic feet per second (cfs) or cubic meters per second (Chapter 14). A plot of stream dis-

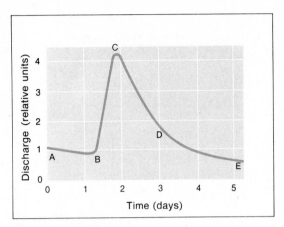

Figure 7.8 This hydrograph shows the effect of an upstream rainstorm on the amount of water carried by a stream, measured from the time of the storm. From *A* to *B*, water from the rain has not reached the stream and the discharge, or rate of flow, measures the base flow supplied by groundwater. From *B* to *C*, the discharge rises rapidly as direct surface runoff from the upstream drainage basin reaches the stream. From *C* to *D*, the discharge falls slowly as the last of the surface runoff, including the runoff retarded by the vegetation cover, makes its contribution. From *D* to *E*, the discharge again primarily measures groundwater supplies, which have been newly recharged by the rain. The discharge of the stream gradually decreases as groundwater inflow to the stream diminishes. (Doug Armstrong after R. C. Ward, *Principles of Hydrology*, © 1967, McGraw-Hill Book Co. (U.K.) Ltd., used with permission)

charge fluctuations over a period of time (hours, days, or months) is called a *stream hydrograph*; Figure 7.8 presents a hypothetical example for a stream of intermediate size in a humid region. The stream continues to flow between rains due to an almost steady input of groundwater that seeps into the channel. This minimum flow between storms that produce surface runoff is known as the *base flow*. During or shortly after each rainstorm, the water table rises and surface runoff reaches the stream. Figure 7.8 shows the characteristic rapid rise and slower recession associated with storm runoff, groundwater outflow, and recharged groundwater supplies.

Daily fluctuations in discharge for three small streams in different climatic regions are shown and explained in Figure 7.9. The peaks rising above the base flow are caused by surface runoff during storms.

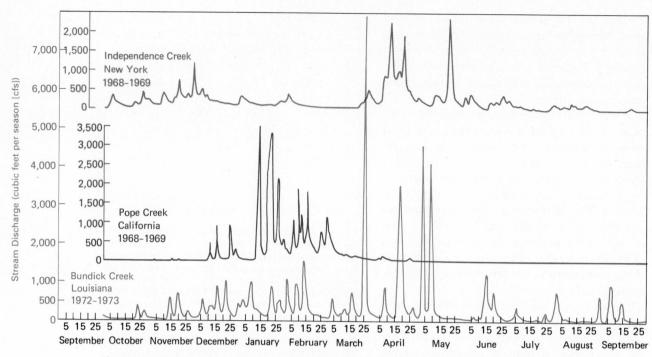

Figure 7.9 Hydrographs for three small streams in different climatic regions, each draining an area between 194 and 310 sq km (75 and 120 sq miles). The hydrographs are organized by water years, which begin October 1 and end September 30. The water year is a useful calendar for water-resource management because streamflow tends to be lowest in late September, with minimum groundwater outflow to support base flow. Typically, storm runoff is superimposed as spikes, or peaks, on the base flow contributed by groundwater. Storm runoff tends to rise quickly and to recede more slowly.

Bundick Creek is located in the warm and humid climate of southwestern Louisiana. Groundwater provides for some base flow all year, but streamflow is largest on the average during winter and spring. A considerable proportion of the annual flow is produced in just a few days by the very large spring flood flows, and the maximum daily discharge of 7,980 cfs on March 25, 1973, was the highest daily flow in 17 years of measurements.

Independence Creek is located on the western flanks of the Adirondack Mountains in northern New York. This drainage basin, or watershed, is representative of climates where persistent low temperatures during winter allow most of the precipitation to accumulate as snowpack. Much of the annual flow occurs during spring as the snow melts, but there is usually a secondary discharge peak in autumn before the winter snows begin to accumulate.

Pope Creek drains a low mountainous region of the coastal ranges of northern California west of Sacramento. The climate is hot and dry during summer with no precipitation. Groundwater contributes to streamflow only during winter and spring, and normally there is no water in the creek from June through October. (Vantage Art, Inc.)

THE LOCAL WATER BUDGET: AN ACCOUNTING SCHEME FOR WATER

Water is not always available in the desired amounts just when and where we need it. The supply of moisture is so variable that a special technique is needed to estimate moisture availability. That is why climatologists and hydrologists make use of the *water budget* concept. The water budget is the local version of the hydrologic cycle (Figure 7.2). It takes into account four principal components of water distribution: precipitation, soil moisture storage, evapotranspiration, and runoff. When appropriate, snow accumulation and snowmelt may also be included. Of all these components, evapotranspiration is the most difficult to estimate accurately because of its dependence on complex meteorological and biological factors.

Potential and Actual Evapotranspiration

Measurements of actual evapotranspiration are limited to those involved in detailed studies of small field plots, using expensive instrumentation. To overcome the difficulties of actual measurement, and to allow water budget components to be studied over large regions, the American climatologist C. Warren Thornthwaite introduced the concept of *potential evapotranspiration* (*PE*) and developed formulas for estimating *PE* under varying conditions. Potential evapotranspiration, the key to the water budget, is worth examining in detail.

Potential evapotranspiration is the rate at which water would be lost to the atmosphere from a land surface completely covered by growing vegetation that has been supplied with all of the soil moisture it can use. *PE* is normally expressed as the depth of liquid water that is converted to vapor in a given time. A typical value for *PE* in the eastern United States during summer is about 15 cm (6 in.) per month. During winter in the same region, *PE* is usually less than 2.5 cm (1 in.) per month.

Evapotranspiration normally proceeds at the potential rate as long as moisture is readily available in the soil. After a number of days without rain, however, soil moisture becomes partly depleted, and it is increasingly difficult for plants to extract the remaining moisture. The actual rate of evapotranspiration will then fall below the potential rate. For this reason Thornthwaite distinguished between *PE* and *actual evapotranspiration* (*AE*). During wet seasons, *AE* is the same as *PE*, but in prolonged dry periods, *AE* is less than *PE*. The term "actual evapotranspiration" is somewhat misleading; in Thornthwaite's water budget analysis, *AE* is an estimation, rather than an "actual" measurement. Many environmental scientists simply use the term evapotranspiration (*ET*), rather than Thornthwaite's term, *AE*.

Solar radiation is the principal factor that determines *PE*. In fact, one way to think about potential evapotranspiration is in terms of solar energy input and utilization. Both evaporation and transpiration require about 590 calories per gram of water. At average temperatures near the earth's surface, one gram of water has a volume of one cubic centimeter. Therefore, we can calculate the energy needed to evaporate a pan of water 1 cm deep if we know the surface area of the pan in square centimeters. For example, if the pan absorbs 1,500 calories of solar energy for each square centimeter of water surface, we calculate that the "potential" evaporation amounts to about 2.5 cm (1 in.) of water.

PE, therefore, can be thought of as both an index of energy supply to an area and, at the same time, an index of the climatic demand for water in the landscape. Rigorous estimates of *PE* from atmospheric data require the measurement and analysis of the terms of the radiation budget and energy balance, especially solar and net radiation, as well as temperature, humidity, and wind at two or more levels above the surface.

Potential evapotranspiration from a field, or even from an entire landscape, is almost independent of the type of plant cover. Imagine looking down from an airplane on a forest or field during the middle of the growing season. The plants usually present a nearly uniform cover of overlapping green leaves. The albedo of almost all green fields or forests is between 10 and 25 percent. Therefore, an acre of forest and an acre of soybeans are not very different as regards solar radiation absorption on a clear day.

Thornthwaite's estimation of monthly *PE* depends mostly on two factors: (1) average monthly air temperature, and (2) latitude, which controls the length of the daylight period from month to month. Data on temperature and duration of daylight are available for most places on earth. Together, these two factors give an indication of solar energy input, which is actually measured in only a few places. Estimated average annual *PE* variations over the United States, calculated from Thornthwaite's formula, are shown in Figure 7.10. Thornthwaite's formula is widely used for regional climatic analyses, but much more complex formulas that require more data have been devised for detailed local studies.

Potential evapotranspiration at a particular location can be measured in a device called an *evapotranspirometer*, which is an open tank about

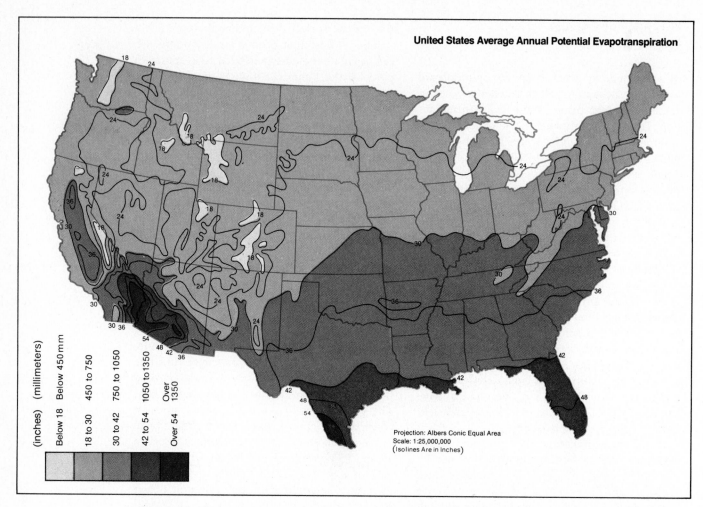

Figure 7.10   This map shows the average annual potential evapotranspiration, or *PE,* for the coterminous United States, calculated from Thornthwaite's formula. The Southwest, Texas, and Florida have high values of *PE* because of high solar radiation income and very warm weather. *PE* is much lower farther north because of decreased solar radiation income in winter and much cool and cloudy weather. Across the high mountain areas of the West, where temperatures are low, the *PE* is also low. There is also a strong seasonal regime of *PE* throughout the United States: *PE* is low in winter and high in summer, although this seasonality is least marked along the West Coast and across the South. (Andy Lucas and Laurie Curran adapted from *Geographical Review,* vol. 38, © 1948, American Geographical Society of New York)

60 cm (2 ft) in diameter and 90 cm (3 ft) deep. The tank is sunk into the ground so its top is flush with the surface (Figure 7.11). The tank is filled with soil and usually planted with a cover of grass. A weight increase represents water added to the tank by precipitation or irrigation; a weight loss, on the other hand, represents evapotranspiration, and percolation, which is measured as it collects in a false bottom. The various measurements constitute a water budget of the tank. As long as the grass in the tank does not experience water shortage, *AE* is equal to *PE;* the measured *PE,* averaged by months, is assumed to represent the climatic energy available in the surrounding area.

Regular operation of an evapotranspirometer is expensive and requires skilled personnel. Therefore such measurements have been limited to experimental studies, mostly in dry regions where it is important to know how much water is needed for particular types of vegetation or crops.

plus represents surface water runoff and groundwater recharge, and eventually finds its way to streams and rivers.

A Local Water Budget: Baton Rouge

To illustrate a local water budget, let us consider the example of Baton Rouge, Louisiana, in 1962. In an average year, Baton Rouge receives more than 125 cm (50 in.) of rainfall. Although Baton Rouge experiences one of the highest average

Figure 7.11 This is a weighing lysimeter, or evapotranspirometer, as it was being installed at Lompoc, California. When ready for measurement, plants growing in the lysimeter should be exactly the same as those around the lysimeter, so that the sun will not "see" the lysimeter. (Robert A. Muller)

Calculating the Water Budget

The local water budget simply represents a systematic accounting of the input, output, and storage of water at a location (Figure 7.12). The computation is essentially a comparison between PE and AE. If PE exceeds AE, there is a water deficit—there is no soil moisture recharge and no runoff; there is not enough water available to satisfy the climatic demand for it. If AE equals PE, there may be additional water that can result in moisture storage in the soil or surface runoff. In the computation, the incoming precipitation (P) during a given time period is allocated to evapotranspiration (AE), soil moisture recharge (ΔST), and surplus (S). Sur-

Figure 7.12 This schematic diagram illustrates the principal components of the local water budget. The input to the system of soil and vegetation is the amount of moisture supplied by precipitation. A large portion of the input returns to the atmosphere by evaporation and plant transpiration. Some of the moisture input is stored in the soil. However, the storage capacity of the soil is limited; when the storage is full, a moisture surplus becomes available to supply surface runoff and groundwater recharge. (Tom Lewis)

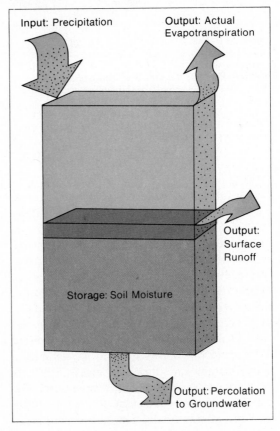

Input: Precipitation

Output: Actual Evapotranspiration

Output: Surface Runoff

Storage: Soil Moisture

Output: Percolation to Groundwater

	JAN	FEB	MAR	APR	MAY	JUN	JUL	AUG	SEP	OCT	NOV	DEC	Total
1. Precipitation (P)	6.4	0.7	3.3	9.7	1.6	11.4	2.0	4.5	4.3	5.2	0.9	2.9	52.9
2. Potential Evapotranspiration (PE)	0.4	1.7	1.3	2.6	5.3	6.1	7.4	6.8	5.2	3.4	1.1	0.6	41.9
3. Precipitation minus Potential Evaporation (P-PE)	6.0	−1.0	2.0	7.1	−3.7	5.3	−5.4	−2.3	−0.9	1.8	−0.2	2.3	11.0
4. Change in Stored Soil Moisture (ΔST)	0	−1.0	1.0	0	−3.7	3.7	−5.4	−0.6	0	1.8	−0.2	2.3	−2.1
5. Total Available Soil Moisture (ST)	6.0	5.0	6.0	6.0	2.3	6.0	0.6	0	0	1.8	1.6	3.9	—
6. Actual Evapotranspiration (AE)	0.4	1.7	1.3	2.6	5.3	6.1	7.4	5.1	4.3	3.4	1.1	0.6	39.3
7. Deficit (D)	0	0	0	0	0	0	0	1.7	0.9	0	0	0	2.6
8. Surplus (S)	6.0	0	1.0	7.1	0	1.6	0	0	0	0	0	0	15.7

Figure 7.13 This is the local water budget for Baton Rouge, Louisiana, during 1962, in inches of equivalent precipitation. The water budget equation is $P = AE + S \pm \Delta ST$. For 1962 the equation works out: 52.9 = 39.3 + 15.7 − 2.1. Similarly, the energy budget equation is $PE = AE + D$, and for 1962 this is 41.9 = 39.3 + 2.6. Each equation balances, so we can be quite confident that we have not made any calculation errors. (Doug Armstrong)

annual rainfall totals for cities in the 50 states, a water budget analysis shows that water deficits did exist and that crops in the Baton Rouge area needed irrigation water during several months of 1962.

The 1962 water budget for Baton Rouge in Figure 7.13 considers water income, output, and storage, month by month. Weekly or daily data could also be shown. The first row in Figure 7.13 shows monthly precipitation in inches. The second row shows monthly PE estimated from Thornthwaite's formula. Since PE tends to follow the seasonal regime of temperature, it is much smaller in winter than in summer. The third row shows the difference between precipitation and PE. When $P - PE$ is positive, the locality has a "wet" month, and when $P - PE$ is negative, it has a "dry" month. In Baton Rouge during 1962, six months were wet, and six were dry.

Available soil moisture is the storage component between field capacity and the wilting point (Figure 7.5). Because 1961 had been a wet year, 1962 began with the soil moisture storage in Baton Rouge at its full capacity of 6.0 in. This storage capacity figure is representative of the Baton Rouge area; on a global basis, soil moisture storage capacity ranges from about 2 in. to 12 in. Our simple water budget assumes that plants take all the water they need from the soil until there is no more moisture in storage. Row 4 shows the change in soil moisture storage during each month, and row 5 the amount of soil moisture storage at the end of each month.

During wet months, AE is equal to PE. During dry months, plants draw on soil moisture as needed (note the storage change in July). AE represents water passing through the plant sys-

Figure 7.14 Average water budget for Baton Rouge, Louisiana, based on standard climatological data for 1941–1970. The calculations are shown in the table (a), with the same data shown graphically in (b).

(a) This average budget is based on 30 years of temperature and precipitation data. Average monthly precipitation is much less variable than monthly precipitation in individual years, such as 1962 in Figure 7.13. In this particular average water budget, *AE* is always equal to *PE*, and there is no deficit (*D*). The distinction between the components of an average water budget and the budget of an individual year (Figure 7.13) should be kept clear.

(b) This graph of the average water budget for Baton Rouge is an example of standardized graphs that have appeared in research publications. It shows the seasonal regimes of *P*, *PE*, *AE*, and *S*, as well as soil moisture withdrawal and recharge. (Vantage Art, Inc.)

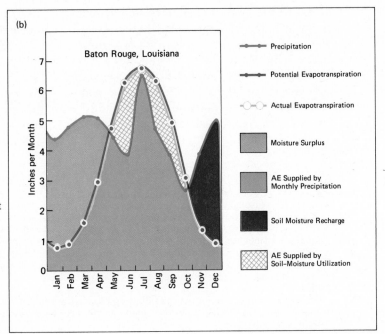

(a)	Jan	Feb	Mar	Apr	May	Jun	Jul	Aug	Sep	Oct	Nov	Dec	Year
1. Precipitation (*P*)	4.4	4.8	5.1	5.1	4.4	3.8	6.5	4.7	3.8	2.6	3.8	5.0	54.0
2. Potential Evapotranspiration (*PE*)	0.7	0.9	1.6	3.0	4.7	6.2	6.7	6.3	4.9	2.8	1.2	0.8	39.8
3. Precipitation minus Potential Evapotranspiration	3.7	3.9	3.5	2.1	−0.3	−2.4	−0.2	−1.6	−1.1	−0.2	2.6	4.2	14.2
4. Change in Stored Soil Moisture (Δ*ST*)	0	0	0	0	−0.3	−2.4	−0.2	−1.6	−1.1	−0.2	2.6	3.2	0
5. Total Available Soil Moisture (*ST*)	6.0	6.0	6.0	6.0	5.7	3.3	3.1	1.5	0.4	0.2	2.8	6.0	——
6. Actual Evapotranspiration (*AE*)	0.7	0.9	1.6	3.0	4.7	6.2	6.7	6.3	4.9	2.8	1.2	0.8	39.8
7. Deficit (*D*)	0	0	0	0	0	0	0	0	0	0	0	0	0
8. Surplus (*S*)	3.7	3.9	3.5	2.1	0	0	0	0	0	0	0	1.0	14.2

tem, and the amount of *AE* is related in a general way to the building of plant tissue. According to the table, *AE* was less than *PE* during August and September. This indicates a *deficit* (*D*), shown in row 7. The deficit represents the amount of additional water that the plants could have used. Therefore, it is an index of irrigation needed to maintain crops at their full growth potential.

Any excess water that remains after soil moisture is brought back to capacity is *surplus* (*S*). During 1962, surpluses occurred during only four months: January, March, April, and June. Surplus water is available for surface run-off and groundwater recharge. It is this water that enters streams and changes the land surface by erosion and sediment deposition.

Figure 7.14 shows both tabular and graphic

Figure 7.15 Calculated monthly water budget components for Baton Rouge between 1960 and 1967. This graph emphasizes seasonal consequences of the variability of precipitation, which itself is not shown.

A more realistic, but at the same time more complex, model of soil moisture storage and depletion was used for these calculations. In the water budgets for Figures 7.13 and 7.14, it was assumed that soil moisture was "equally available" to meet the climatic demand of *PE* during months when *P − PE* was negative. In the model used for this figure, it was assumed that soil moisture would become "decreasingly available" as it was depleted; in other words, the vegetation would not be able to withdraw all needed soil moisture even though the soil still contained some available moisture. The decreasing availability model of soil moisture depletion requires special tables or equations.

The water budget graphics in the following chapters are based on the decreasing availability model. Note especially that this model shows that small deficits recur each year at Baton Rouge. (Vantage Art, Inc., after Robert A. Muller, 1976)

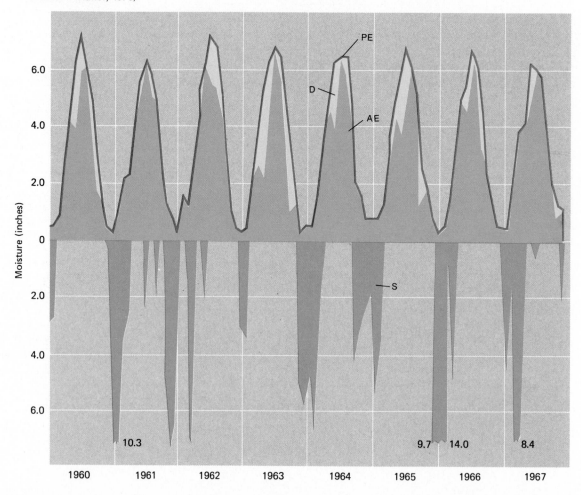

representations of the *average* annual water budget for Baton Rouge over a 30-year period. Long-term average monthly precipitation is much less variable than actual monthly precipitation in individual years. This is because a wet July in one year compensates for a dry July in another year, and so on. Long-term average monthly precipitation in Baton Rouge, therefore, is never low, and the city's average water budget shows no deficits.

The seasonal and annual variations in energy and moisture regimes revealed by water budgets are extremely important to plants, wildlife, and humans. To illustrate this variability, a more complex water budget model was used in Figure 7.15, which shows monthly *PE*, *AE*, *D*, and *S* for Baton Rouge between 1960 and 1967. Deficits occurred each summer, but they ranged from small deficits with little environmental consequences to large ones during 1962, 1963, and 1965. Surpluses, on the other hand, were extremely variable from year to year

(compare the large surplus during 1961–1962 with the small one for 1962–1963). This type of variability, which is due to changes in the upper atmospheric circulation, has far-reaching impact on the environment, as well as on economic activities.

Problems in Application of the Water Budget Model

The water budget model that has been presented is a simple representation of a very com-

Figure 7.16 This graph compares the stored soil moisture estimated by the daily water budget method to the measured amount. The excellent agreement between them indicates the suitability of the local water budget approach to agricultural problems.

The amount of stored soil moisture attains its maximum in late winter and spring, when precipitation is heavy and transpiration is small due to lack of plant cover. Soil moisture decreases in April and May and reaches a minimum during summer, when high temperatures and dense plant cover cause evapotranspiration to be high. The local peaks in the soil moisture are caused by rapid recharge during rainstorms. Can you explain why the average level of soil moisture increases in late autumn? (Doug Armstrong after C. W. Thornthwaite and J. W. Mather, *Publications in Climatology*, vol. 8, 1955)

Measured Soil Moisture

Soil Moisture Estimated by the Water Budget Method

plex natural system. Better estimates of *PE* are possible, but they require detailed measurements of each of the radiation and energy components discussed in Chapter 3. The model assumes that all rainfall infiltrates unsaturated soil; allowances are not made for intense rainfalls that exceed infiltration capacities or for variations in specific soil types. The model used to develop the water budget tables in Figures 7.13 and 7.14 assumes that soil moisture, until it is depleted, is "equally available" to meet the demands of *PE*. But most plant scientists agree that plants are less able to utilize soil moisture for *AE* as it is depleted from the rooting zone. Therefore a more sophisticated "decreasing availability" model of soil moisture depletion was used to generate the data for Figure 7.15, and the remaining water budget graphs in this text are based on this more complex model.

For irrigation and flood forecasting, data on monthly variations are inadequate—a daily water budget is necessary. Modern computer technology permits analyses of daily water budgets for research and environmental monitoring. Most of the simplifications discussed in the preceding paragraph can be taken into account by the computer models. Figure 7.16 is a comparison of a water balance estimate of the daily regime of soil moisture during an entire year and the actual measurement of soil moisture changes. It can be seen that the daily water balance estimate closely approximates reality.

Another problem with the water budget model is that precipitation data are not always representative of conditions in the region. Precipitation is normally measured by noting the accumulation of water in a standard rain gauge, a cylinder 8 in. (20 cm) in diameter, exposed under standardized conditions. A gauge, however, does not catch a fully representative sample on windy days or when the precipitation falls as snow. With fewer than 15,000 gauges in the United States, there is an average of only one official rain gauge for every 500 sq km (200 sq miles); all the official rain gauges in use in this country could fit between the 40-yard lines of one football field! Because storm rainfall can vary by more than 5 cm (2 in.) in just a few kilometers, the rain gauge network often fails to provide accurate precipitation data during life-threatening storms.

Despite these problems, the water budget model is a valuable tool for a wide range of objectives. In its more complex form, it can be used for research in fields ranging from agriculture to flood hydrology. Various water budget components can be related to particular objectives. For example, evapotranspiration (*AE*) is related to plant growth and the yields of many crops; the deficit (*D*) is an index of irrigation needs; and the surplus (*S*) represents water potentially available for water-resource projects, sewage disposal, and industrial uses. Calculations of surplus have revealed that streamflow, or water yield, is affected by changes in land use and vegetation cover. Urban areas and some croplands cause the largest water yields per unit of precipitation. Forested regions, on the other hand, contribute the smallest proportions of precipitation to streamflow but sustain higher base flows (Figure 7.4) between rainfalls.

SUMMARY

The movements of water from place to place and its transformations into liquid, solid, and vapor phases are all part of the hydrologic cycle. The oceans are the largest reservoir of water on the earth, and the evaporation of water from tropical oceans and warm currents is the beginning of the hydrologic cycle. The cycle comes full circle when streamflow returns surface runoff and groundwater outflow to the sea.

The amount of water gained by the atmosphere as a whole over a year must equal the amount that falls from the atmosphere as precipitation. When the oceans or continents are considered separately, however, imbalances appear. The oceans lose more water by evaporation than they gain by precipitation. The continents gain more water by precipitation than they lose by evapotranspiration. Runoff to the sea from streams and rivers and the horizontal transport of moisture in the atmosphere compensate these imbalances.

In unpaved land areas a small proportion of the precipitation is intercepted by vegetation and returns to the atmosphere by evaporation. The remainder reaches the ground by through-fall and stemflow. Water infiltration into the soil is determined by the soil's porosity, permeability, and moisture content. When the soil is saturated, surplus water either runs off over the surface or percolates downward to the water table, entering storage as groundwater. Groundwater outflow contributes to the flow of surface streams between rainfalls. Groundwater occurs below the water table in unconfined and confined aquifers. Chemical contamination of groundwater by human activities and the removal of groundwater faster than it is recharged are increasingly important problems.

Surface runoff and stream discharge are two different measures. Runoff refers to the average depth of water removed from an area of the earth's surface, while stream discharge is the amount of water passing through a portion of a stream channel per unit of time.

The availability of water as soil moisture or as surplus, producing runoff, can be estimated by using the local water budget model. *PE* represents the maximum rate of evaporation and transpiration from a vegetation-covered landscape with no water shortage. The water budget method compares *PE* with precipitation and soil moisture storage for varying time periods, and permits the estimation of water surpluses or deficits on a daily, monthly, or annual basis. The water budget model is a useful tool in a variety of situations ranging from agricultural planning to flood forecasting and water-resource development.

REVIEW QUESTIONS

1. Discuss some of the environmental problems which the water budget model can be used to study.
2. How does the pattern and timing of rainfall, as distinct from quantity, affect the amount of moisture absorbed by the ground?
3. How does land use affect infiltration and surface runoff?
4. Compare field capacity, wilting point, and available soil moisture for soils of different textures.
5. What is groundwater, and how is it recharged? What is the water table?
6. Explain the terms "runoff" and "streamflow."
7. Describe the operation of an evapotranspirometer.
8. Compare potential evapotranspiration (*PE*) and actual evapotranspiration (*AE*)
9. What is meant by the term "deficit" in regard to the local water budget, and what is its significance for irrigation procedures?
10. How is surplus (*S*) of the local water budget related to a hydrograph of a river in a humid region?

APPLICATIONS

1. Where are there stream gauging stations in your area? What do they look like? How do they work?

2. Using graph paper, plot daily or monthly stream discharge for local gauged streams in your area for the wettest and driest years on record. Stream discharge data are available from U.S. Geological Survey Water Resource offices in each state and from state departments of water resources, and are published in annual bulletins covering both states and major drainage basins. Remember that stream discharges represent mostly the surplus, or "leftover," water within local water budgets.

3. Observe the ground during a heavy rainfall. Do you see runoff occurring on "natural" surfaces? On surfaces modified by human activity? Go to an area of bare ground with no vegetation cover during a heavy rain—a large roadcut or a spoil heap will do. What do you see there? How does the timing, as distinct from the quantity, of rainfall affect the amount of moisture absorbed by the ground?

4. How do the water supply agencies in your region obtain their water? Are there times when supplies are critically short? Has there been a water supply emergency in your area? What measures were taken to overcome it?

5. Produce an "artificial rainfall device" by perforating the bottom of a large juice can with holes approximately 1 mm in diameter and 1 in. apart. Pouring a measured amount of water into the perforated can, observe how rapidly this is absorbed by different types of ground—wet, dry, sandy, clayey, grass-covered, forest-covered. Over vegetation note how much water is intercepted and retained by plant parts. Where is it easiest and hardest to produce runoff?

6. Compute a continuous monthly water budget using data from the nearest weather station, or some other location of interest. The computation procedure can be found in John R. Mather, *The Climatic Water Budget in Environmental Analysis* (see Further Reading). Then prepare a graph illustrating the variability of water budget components such as *AE, D,* and *S,* as in Figure 7.15. As an alternative, maintain a daily water budget, using temperature and precipitation data given each day by radio, TV, or newspapers.

FURTHER READING

Chorley, Richard J., ed. *Water, Earth, and Man.* London: Methuen (1969), 588 pp. This unusual book is a well-organized collection of essays on various aspects of the hydrologic cycle and their applications in hydrology, geomorphology, and socio-economic geography. Many of the contributors are British, giving the book a perspective from the British Isles.

Dunne, Thomas, and **Luna B. Leopold.** *Water in Environmental Planning.* San Francisco: W. H. Freeman (1978), 818 pp. This is the most thorough treatment of all aspects of water in the environment, by two of the leading authorities in the fields of hydrology and water management. Very well illustrated with maps, graphs, and photographs.

Goudie, Andrew. *The Human Impact: Man's Role in Environmental Change.* Cambridge, Mass.: MIT Press (1982), 316 pp. The focus in this very readable softcover book is on how human culture has modified the physical geography of the earth. Chapter 5 is devoted to water, but all of the chapters complement the various topics of this text.

Kazmann, Raphael G. *Modern Hydrology,* 2nd ed. New York: Harper & Row (1972), 365 pp. Emphasis in this work is on critical evaluation of the hydrologic variables for water-resource management.

Leopold, Luna B. *Water: A Primer.* San Francisco: W. H. Freemen (1974), 172 pp. This excellent little book stresses the interrelationships between hydrologic principles and environmental responses.

Mather, John R. *The Climatic Water Budget in Environmental Analysis.* Lexington, Mass.: Lexington Books (1978), 239 pp. A most useful summary of ideas and applications of the water budget, including appendices for calculations.

Thornthwaite, C. W., and **John R. Mather.** "The Water Balance." *Publ. in Climatology*, Vol. 8 (1955): 1-86. This is a summary of the early water budget research and applications by Thornthwaite and his colleagues at the Laboratory of Climatology.

————. "Instructions and Tables for Computing Potential Evapotranspiration and the Water Balance." *Publ. in Climatology*, Vol. 10, 3 (1957): 185-311. This instruction manual provides explanations and tables for calculation of daily and monthly water budgets using the decreasing availability model of soil moisture depletion.

Ward, R. C. *Principles of Hydrology*, 2nd ed. London: McGraw-Hill (1975), 367 pp. Various components of the hydrologic cycle and the water budget are treated systematically. This text incorporates a geographic perspective, and it also includes an extensive bibliography organized by topic.

CASE STUDY

California Water Transfers

The management of fresh water is often a controversial subject. According to one view, there is plenty of water on the earth—it simply needs to be redistributed. Others argue that major economic development should not be extended into areas having a shortage of water, but should be concentrated where water is naturally available.

A case to consider is southern California, which has become more populous by the importation of water from remote areas. In the early 1900s it became clear that the growth of Los Angeles—then a small coastal town—would be stimulated by irrigating larger areas and developing a more abundant supply of urban water. For a source of water, Los Angeles turned to the Owens Valley, 400 km (250 miles) to the north, at the eastern foot of the Sierra Nevada. The farmers of the Owens Valley had hoped for their own irrigation system, to be developed by the U.S. Bureau of Reclamation. But the city of Los Angeles convinced the Bureau to step aside, and covertly began to purchase all land with water rights in the valley, at the same time constructing an aqueduct to convey the water to the San Fernando Valley, near Angeles.

The Los Angeles Aqueduct from the Owens Valley was completed in 1913. Soon more water was needed, however, and aqueducts were built to carry water across the Mojave Desert from the Colorado River to Los Angeles and San Diego. Today more than a hundred cities in southern California rely on water from the Colorado. This supply has recently been supplemented by water from northern California in the world's

longest artificial water transfer. Rather than continuing to rely on local and regional initiative, the government of California has taken over water resources development for the entire state.

The California State Water Project is founded on the recognition that most of the state's population, voting power, and irrigated land are in the south, whereas most of the water is in the north. The State Water Project includes aqueducts, canals, dams, reservoirs, and power stations needed to transport the water from the north to the south. The project begins at the massive Oroville Dam on the Feather River north of Sacramento. The current

This simplified map shows the conduits of the California State Water Project in bold blue lines. Other water transfers are shown in lighter blue. The project extends nearly the length of California, bringing water 1,000 km (600 miles) from the north to the more densely populated south. (Steve Harrison and Louis Neiheisel)

source of water for the present stage of the plan is the Sacramento River delta, formed by a confluence of streams from the west slope of the Sierra Nevada that carry over 40 percent of the state's runoff to San Francisco Bay through the only opening in the coastal range. A future project, bitterly opposed by

178

Farmlands in the Sacramento River delta. Smoke is from intentional burning of crop residues. (T. M. O.)

residents of the delta area, is to route Sacramento River water around the delta, rather than through it. This could allow salt water intrusion into the delta, which would destroy delta farmlands and wildlife habitats. The southern end of the water transfer system is Perris Lake, an artificial reservoir located southeast of Los Angeles. Never before has a human project sent so much water flowing so far. At one point in its 1,000-km (600-mile) journey southward, the water is lifted nearly 600 meters (2,000 ft) over the Tehachapi Mountains—a record lift for so much water.

In addition to supplying water for southern regions, the aqueducts provide irrigation for the dusty, windblown San Joaquin Valley, which is currently the major beneficiary of the project. Eventually, a million acres of formerly unproductive land will be made available for agriculture.

The water development program has instigated an angry sectional feud between the moist "north" and the dry "south," but the more numerous votes in the south have carried the issue. Environmentalists claim that the project has upset the natural balances of streams, estuaries, vegetation, and wildlife, and that it will destroy the ecology and economy of the Sacramento delta. It is also argued that providing more water to Los Angeles will promote population growth, more congestion, and associated problems.

Many of the questions and controversies over the water plan center on whether the water is really needed. Ninety percent of the water used in southern California is for irrigation; there is at present abundant water for domestic and industrial use. There is currently no attempt to conserve water in either urban or agricultural areas. However, under the 1982 Boulder Canyon Act much of the water now taken from the Colorado River by Californa will soon have to be relinquished to Arizona. If the water used for irrigation in California were cut to 80 percent, the amount available for other uses would double from 10 to 20 percent. However, what effect would reducing irrigation have on food supplies for the entire country? California is now the leading supplier of dozens of agricultural commodities, almost all of them grown by irrigation.

The allocation of water resources is a matter of economics as well as of technology. Water demand is, in reality, often a demand for water at a sufficiently low cost to make irrigation profitable. The emphasis on higher-priced specialty crops, such as fruit, nuts, vegetables, and cotton, in California agriculture reflects the need for profitable crops to offset the cost of irrigation. However, vast amounts of water are also used to grow alfalfa, a low-value crop that is a heavy consumer of water. The California water project has been criticized for using public funds to subsidize and increase the value of privately held farmland.

Although more aqueducts, canals, and pumping plants are planned, it is uncertain when and if they will be completed. The pace of the water plan is influenced by projected water needs, the development of new technology, ecological impacts, and political considerations.

Sky Above Clouds II by Georgia O'Keefe. (Collection of Mrs. Potter Palmer, Lake Forest, Illinois; with permission of Georgia O'Keefe)

Dynamic processes in the atmosphere cause the weather patterns that give each region of the earth a distinctive climate. Climate can be viewed in two ways: as the delivery of energy and moisture by the atmosphere, or as the interaction of energy and moisture at the earth's surface with systems of soils, vegetation, and landforms.

A region's climate may be defined as the average condition of its weather over a period of years. Regional climates are usually described in terms of average monthly temperatures and precipitation, with additional information about extremes. For example, the average temperature for Tallahassee, Florida, during February is 13°C (56°F), but the temperature there fell to −19°C (−2°F) during February, 1899. Another approach to regional climates is to focus on the dynamic circulation patterns of the atmosphere—the semi-permanent high and low pressure systems and winds depicted in Figure 4.6 (p. 88), the general circulation model of the atmosphere, and the secondary circulation systems (Chapter 6)—which collectively make up the day-to-day weather.

The climate of a region is responsible for such surface conditions as evapotranspiration, surface runoff, and soil moisture availability, and is fundamental to the development of the region's soils, vegetation, landforms, and agricultural possibilities. Knowledge of the global distribution of the earth's varying climates is a key to understanding the environments of various places and the human activities in them.

From time to time we hear of proposals to change the climate artificially in some part of the world. One such scheme illustrates the *teleconnections* (long-distance connections) that influence climates. This scheme was intended to increase precipitation in the arid southwestern part of the United States, where potential evapotranspiration far exceeds precipitation. The annual moisture deficit at Phoenix, Arizona, the largest of the desert cities, is greater than 102 cm (40 in.) in an average year, and extensive irrigation is required if crops are to be grown. The idea in this case was to change the climate by building a gigantic dam across the far-off Bering Strait between Siberia and Alaska. The dam would have to be 100 kilometers (60 miles) long, far surpassing in size any dam existing on the earth. How could a dam at the Arctic Circle affect the climate in a desert 5,000 kilometers (3,000 miles) away?

If such a dam could be built, it would disrupt the oceanic circulation by preventing the flow of cold arctic waters into the Pacific Ocean. This

8
Global Systems of Climate

would cause the Pacific Ocean to become warmer. We have seen (Chapter 4) that unusual cooling in the Pacific Ocean seems to trigger drought conditions in the western United States. Accordingly, advocates of the Bering Strait dam anticipated that *warming* the Pacific Ocean would *increase* rainfall in the American Southwest. Russian proponents of the plan hoped that it would indirectly improve the cold, dry climates over much of Siberia as well. But warming the Pacific would almost certainly induce a longer hurricane season in the tropics, and perhaps would bring hurricanes into higher latitudes. Furthermore, the subsequent cooling of the polar region could produce stronger outbursts of polar air and more violent cyclonic storms along the polar front. Changing the temperature of the Pacific Ocean would change the temperature of the air over the ocean and would affect the general circulation of the atmosphere. Because the general circulation controls the climates of the earth, the effects of the dam would be far-reaching. Beneficial effects in the American Southwest would quite likely be outweighed by harmful effects elsewhere. It is impossible to predict all the consequences of such an enormous scheme to modify climate.

The point of this rather far-fetched example is that the earth's climates are not independent phenomena—they are determined by the general atmospheric circulation. To change climates significantly, the circulation system as a whole must be changed, and this would cause climates everywhere to be affected. It is much more sensible to learn how existing climates are produced and how to take advantage of what they have to offer.

This chapter focuses on an understanding of global climatic systems. Two different but related methods for classifying climates are developed. The first focuses on atmospheric dynamics—the climates associated with the general atmospheric circulation and the "delivery" of energy and moisture *to* the surface of the earth. The second focuses on the interactions of energy and moisture *at* the surface of the earth that make possible plant and animal communites and the earth's agricultural systems.

MEASURING CLIMATE

The comparative stability of the general circulation gives rise to characteristic regional weather patterns that make each region distinctive. Winter vacationers in Florida, for example, have every reason to expect a succession of warm, sunny days during their stay. Because the distribution of energy and moisture has implications for other systems in addition to weather, it is useful to classify the types of climate experienced in the various regions of the earth. Climate classification helps reveal the general patterns in other environmental systems and serves to organize a wealth of information about the earth's surface.

To be most useful, a classification scheme should have enough categories and distinctions to account for the system's principal features, but not so many that a general understanding is lost in a welter of detail. The classification schemes discussed later in this chapter arrange the earth's climates according to no more than five main categories and ten to twenty subcategories. No two places on earth have exactly the same climate, however, so any classification scheme inevitably submerges individual details. Because climatic characteristics do not change abruptly, the boundaries of climatic regions should usually be interpreted as broad zones of transition rather than as sharp divisions. On a global scale, however, the transition zones are frequently narrow compared to the size of the climatic regions. It is meaningful to speak of boundaries only if those concepts are kept in mind.

Climatic Indexes

Climatic indexes are numerical measures used to distinguish one climatic type from another. The choice of an appropriate index of climate depends on its intended use. Someone interested in constructing tall buildings might want to consider wind speed and direction to assess the possibilities of wind damage. If the concern is transportation, interest would center on the frequency of fog and icing conditions. An irrigation engineer would be concerned with the amount of evapotranspiration from cropland. To assess a region's potential for air pollution, wind regimes and the timing and duration of temperature inversions must be known.

Classifications by physical geographers of regional climates on a global basis generally employ temperature and precipitation as indexes of climate. These are easily measured, data about them are generally available for most countries, and they clearly influence the distribution of vegetation and water supplies and the potential for various types of agriculture.

Time and Space Scales of Climate

The general factors that control climates are those we have considered in previous chapters. They include *latitude,* which determines intensity of solar radiation and day length and influences temperature; proximity to warm and cold *ocean currents,* which affect temperature and air mass stability; proximity to *moisture sources* and *rain-producing mechanisms,* including the polar front and the intertropical convergence zone; *topography,* which causes ascent of air and high rainfall or descent of air and rainshadows; and the *general circulation,* which determines the sources of air masses and the general directions in which they move.

The movement of air masses can change local weather conditions significantly within a few hours. Weather conditions can also vary markedly over short distances. Even if the indexes of climate are accepted as being temperature and precipitation, geographers must still decide when and how often to analyze these factors. The following example shows how important the time scale—the "when"—of measurement is.

The average annual temperatures at Aberdeen, Scotland, and at Chicago, Illinois, differ by less than 2°C (3°F). The average annual precipitation at both locations is the same (84 cm or 33 in.). Despite similarities in such annual averages, the climates at the two locations are quite different. The monthly averages give a much more accurate indication of the climate in each city. During January, Chicago averages 8°C

(14°F) colder than Aberdeen. During July, Chicago averages 9°C (16°F) warmer than Aberdeen. The climate at Chicago is therefore considerably more extreme than the climate at Aberdeen. It should be evident that the seasonal distributions of temperature and precipitation, as well as the average annual values, are important in characterizing the climate of a place. The Case Study at the close of this chapter focuses on some of the causes and effects of climatic variability, a topic frequently in the news during recent years.

The "where" of measurement—the location and size of the climatic region—is also important. Again, the best areal scale depends on the intended use of the information. For example, on a world map most of the west coast of the United States is considered to be in a single climatic region. But someone concerned with water resources on the Pacific coast would need to see a much more detailed picture, showing the gradation from the abundant rainfall in the Pacific Northwest to the general dryness of southern California. The complex patterns of orographic precipitation and rainshadows, illustrated for California in Figure 5.18 (p. 121), need to be known and understood in detail.

Even a single residential lot has several microclimates, with a wide range of temperatures over a small area. Figure 8.1 shows temperatures that were read at 11 P.M. at various points in a backyard located in Baton Rouge, Louisiana. The microclimates of the yard are too small to appear on even a city-wide climate survey, but climatic conditions in different parts of the yard are important to someone trying to protect subtropical plants from freeze damage.

This chapter focuses on global climatic types. Excessive detail is not desirable at this scale. The Köppen system of global climatic classification (discussed later in this chapter) assigns New York City and Nashville, Tennessee, to the same climatic region. On a nationwide scale, the climates of the two cities are obviously different, but on a global scale their climates are more like each other than they are like climates of tropical rainforests or Asian steppes, for example, which is the scale of distinction appropriate for world maps.

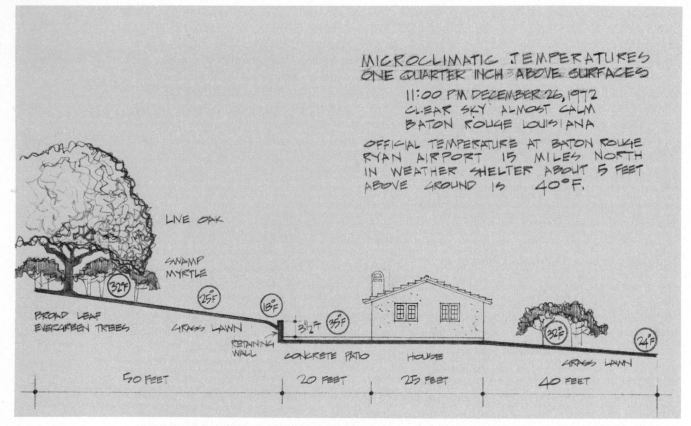

Figure 8.1 The nighttime temperatures in this Louisiana backyard show a large degree of variation from place to place because of such factors as local movements of air, differences in the amount of heat stored in the ground and buildings, and differences in cooling rates. Climate measured on such a small scale is called the *micro-climate* of the area. Such details are lost when the climates of large regions are classified into broad categories. (Ron Wiseman after Robert A. Muller, 1973)

ONE VIEW OF GLOBAL CLIMATES: ATMOSPHERIC DELIVERY OF ENERGY AND MOISTURE

The availability of energy and moisture in different regions of the earth is a result of the global distribution of solar radiant energy and the general circulation of the atmosphere. To understand how solar energy and the general circulation control the gross features of the earth's climates, we need to analyze the climatic regions of an idealized hypothetical continent, on which no surface features impose distorting effects.

Distribution of Climatic Regions on a Hypothetical Continent

The hypothetical continent shown in Figure 8.2 is flat and featureless with no mountains, seas, or gulfs. Nevertheless, it embodies some of the

Figure 8.2 The climatic regions represented on this hypothetical continent are determined by the input of solar radiant energy and the delivery of moist air masses circulated by the atmosphere. In this "computer-style" map the regions are depicted with interlocking and overlapping boundaries to suggest that climate changes gradually from one location to another across the surface of the earth. The distribution of the climates should be studied in conjunction with the text and with the general circulation maps in Figure 4.6, pp. 88–89. (Doug Armstrong)

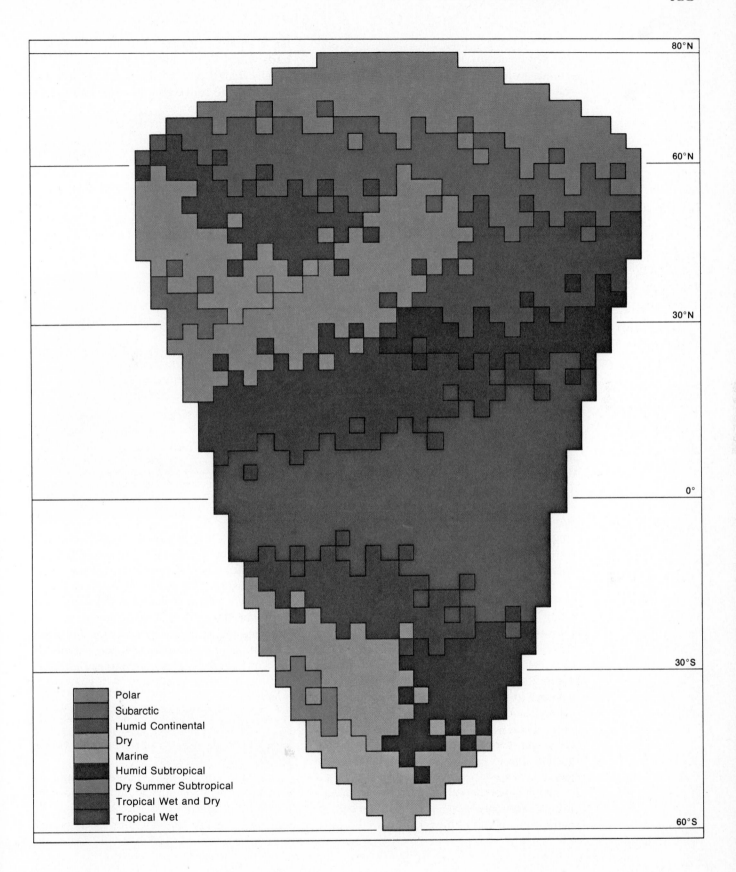

features of the actual continents: it is surrounded by oceans; it is broad at the north and extends to high latitudes, like North America and Eurasia; and it tapers toward the south and ends near latitude 60°S, like South America.

The distribution of average temperature on the hypothetical continent is directly related to the patterns of incoming solar radiation. Tropical and subtropical regions receive the greatest annual amounts of solar energy and are warmest year-around (see Figures 3.9 and 3.17, pp. 60 and 70). Regions at higher latitudes receive less solar energy and have lower average temperatures.

The annual temperature range varies according to latitude and distance from the oceans. Because the tropical regions receive a nearly uniform seasonal input of solar energy, they maintain comparatively constant temperatures all year. But because of the tilt of the earth's axis, the higher latitudes receive a large amount of solar energy in the summer and a small amount in the winter. As a result, the difference between average summer and winter temperatures on the hypothetical continent is greater the farther a region is from the equator. The annual temperature range also increases with distance from the moderating effect of the sea. This phenomenon is known as *continentality*. The continental interiors heat strongly in the summer because of the low heat capacity of land areas and become cold in the winter as a consequence of rapid heat loss by longwave radiation.

The distributions of atmospheric moisture and precipitation are controlled by air masses, wind patterns, and pressure systems. Air masses reaching the hypothetical continent after crossing the ocean would be moisture-laden as a result of evaporation of water from the ocean. This humid, maritime air would become a source of precipitation if it were cooled sufficiently to cause condensation. Most cooling takes place when the air is forced to rise by convergence in a low-pressure center or by convection due to heating at the continental surface.

On the hypothetical continent the equatorial trough of low pressure (or intertropical convergence zone) occupies the region near the equator, as shown in Figure 4.5 (p. 85). The easterly trade winds that converge into this trough are usually moist due to evaporation of ocean water. As the converging air rises into the circulation of the Hadley cells, cloud formation and rainfall occur. Rainfall is frequent year-around near the equator; thus the climate of the equatorial region is classified as *tropical wet*. Because the trade winds sweep onshore and across the eastern side of the hypothetical continent between about 15°N and 15°S, the eastern side receives more moisture than the western side in equatorial latitudes. This is the climate of the equatorial portion of the Amazon Basin in Brazil.

The tropical wet zone astride the equator is succeeded both north and south by *tropical wet and dry* climatic regions. These zones receive rain in the summer high-sun season as a result of the presence of the intertropical convergence zone with its rising moist air. Drought occurs in the winter low-sun season when the ITC moves into the opposite hemisphere and is replaced by the subsiding air and temperature inversions of the subtropical high-pressure cells, which are below the polar margins of the Hadley cells. Here winter is a time of drought rather than cold. Tropical wet and dry is the climate of the African Sahel, the broad region south of the Sahara plagued by drought in recent years.

Near 30°N and 30°S on the hypothetical continent are *dry* climates that are under the permanent influence of the subtropical zones of high pressure, where dry air from the upper atmosphere subsides almost to the surface. As it subsides the air warms adiabatically, resulting in a persistent temperature inversion and stable environmental lapse rates at middle altitudes of the atmosphere. Despite intensive surface heating, the dry air and stability aloft combine to inhibit precipitation processes. Thus subtropical deserts occur on the poleward sides of the tropical wet and dry climates. The Sahara Desert of North Africa and the Kalahari Desert in Southern Africa are examples.

In the interior of the hypothetical continent in both hemispheres, dry regions extend poleward of the subtropical deserts. These midlatitude *dry* climates are not directly associated with the subtropical zones of high pressure, but occur because the regions are far from oceanic moisture sources. Weak midlatitude cyclones are unable to produce much rainfall from the relatively dry air masses in the continental interiors. The midlatitude dry regions are much cooler during winter than their subtropical counterparts. Such cold deserts are most strongly developed in Inner Asia, the outstanding example being the Gobi Desert of Mongolia.

Along the western margins of the hypothetical continent in the subtropical latitudes, the delivery of moisture is governed mostly by the subtropical high-pressure cells centered over the oceans to the west. In the winter the high pressure shifts to lower latitudes, allowing midlatitude cyclones to bring moist maritime air and frequent precipitation. In the summer the high-pressure cells move poleward and stabilize the atmosphere over the subtropical region so that it remains quite dry. As a consequence, this climate is known as *dry summer subtropical*. Cold ocean currents offshore further strengthen the inversion and stability. This is the climate of central and southern California.

While the dry summer subtropical climatic region has a wet winter and a very dry summer, the *marine* climatic region farther poleward is beyond the influence of the subtropical high-pressure cells, and receives moisture from midlatitude cyclones throughout the year, as in the case of western Washington State and coastal British Columbia.

On the eastern side of the hypothetical continent at subtropical latitudes, precipitation occurs year-around, producing a *humid subtropical* climate. Atmospheric subsidence is less well-developed over the west sides of the subtropical highs than over the east sides. The oceanic circulation brings warm water to the east coasts at these latitudes, in contrast to the cold water along west coasts. Maritime tropical air is swept around the western margins of the subtropical highs and over the coastlines toward the interior. In the summer this air has a steep lapse rate due to heating by the land surface. Summer showers and thunderstorms recur frequently in the unstable, warm, humid air. Precipitation is also heavy during winter, when midlatitude cyclones occur along the polar front. The south-eastern United States has such a humid subtropical climate.

Like the actual northern hemisphere continents, the hypothetical continent is broad in the middle latitudes of the northern hemisphere. Here we find two regions of *humid continental* climate, separated by the interior dry climate, as shown in Figure 8.2. Because of the minimal moderating effect of the far-off oceans, these interior climatic regions are characterized by very wide annual temperature ranges, or continentality. In the southern hemisphere the continent is narrow and the moderating effect of the oceans is greater, so there is no development of continental climate. In the northern hemisphere, radiational heating of the land causes hot summers, and radiational cooling leads to cold winters. The prevailing westerly winds carry the continentality effect from the interior to the east coast. During winter and summer, precipitation is associated mainly with midlatitude cyclones, but precipitation over most areas is greater during summer, when the warmer air masses are supplied with abundant moisture. Midlatitude cyclones tend to become more intense during winter and spring, but the stormy periods occur between stretches of cold, fair weather. This humid continental climate, is found in the North American Plains east of the Rocky Mountains.

Poleward of the humid continental climates in the northern hemisphere is the *subarctic* climatic region, which extends from coast to coast. In winter the subarctic climatic region is dominated by the polar high-pressure cell displaced southward over the colder continent. The winter air is always cold and snowfall is light except near the coasts. Continentality again produces a large annual temperature range. Precipitation falls mainly during the mild summer in association with midlatitude cyclones. Most of northern Canada and the major part of the Soviet Union have such a subarctic climate.

Beyond the subarctic climate region in the northern hemisphere is the region of *polar* climate. The region poleward of latitude 60°N is dominated by the polar high-pressure cell much of the year and is also influenced by the nearby cold and often ice-covered ocean. Winters are very cold with meager snowfall. Most precipitation occurs during the short summer when temperatures are above freezing over all but extensive ice-covered areas. Only the fringes of the Arctic Ocean experience a true polar climate.

Distribution of Climatic Regions on the Earth

The more complex patterns of climatic regions of the real continents shown in Figure 8.3 differ from those of the hypothetical continent mainly as a result of variations in the sizes, shapes, and topographic features of the land areas. For example, the mountain ranges in western North and South America prevent the west coast marine climates from extending as far inland as they do in Europe. The regions to the east of the Rocky Mountains are dry because moist air from the Pacific is blocked off or dried out in its passage across the mountains. The Gulf of Mexico provides a fortunate source of warm moist air that keeps these dry regions from reaching farther eastward. Similarly, the Andes Mountains of South America block the easterly trades, preventing moisture from reaching the Pacific coast of South America between the equator and latitude 30°S. Thus the coasts of Peru and northern Chile are almost completely rainless.

The tropical wet and dry climate reaches unusually far northward in Asia because of the monsoon effect (Chapter 6). As the vast continent of Eurasia heats up in the summer, the resulting low pressure pulls the ITC far north of its average summer latitude, causing heavy rain over India and Southeast Asia. Drought follows when the ITC moves back into the tropics. But the causes of the monsoons are complex, and include the high-pressure cell over Siberia in winter, low pressure over the Indus Valley in summer, the presence of the Himalaya Mountains, and sudden changes in the location of the jet stream over Asia.

Nevertheless, many climatic regions on the continents (Figure 8.3) follow the simple patterns of climates on the hypothetical continent (Figure 8.2). Near the equator are tropical wet regions, with tropical wet and dry immediately

to the north and south. Subtropical deserts sprawl across North Africa, the Arabian peninsula, Iran, and Pakistan. Smaller dry regions occur in the southwestern United States and western Mexico. Africa and Australia show the expected dry regions near latitude 30°S. The dry summer subtropical climates are found on the western sides of the continents in five locations that would be expected to have them: California and Oregon; central Chile; Portugal, Spain, Morocco, and the northern Mediterranean coastlands; the Cape Town region of South Africa; and small areas of western and southern Australia. The higher-latitude west coasts of North America, South America, and Europe show the expected marine climates. The interiors of the United States and Europe are humid continental climatic regions. Subarctic and polar climates occupy northern Asia and North America.

Figure 8.3 correctly suggests that a resident of Seattle would find familiar weather in such far-off places as the British Isles and New Zealand, and an Alabaman or Georgian would find the weather in South China or eastern Australia more like that at home than the weather only a long day's drive to the north or west. The occurrence of similar climates and associated environments in widely separated locations on the earth is one of the most important facts to be understood in the study of physical geography.

ANOTHER VIEW OF GLOBAL CLIMATES: ENERGY AND MOISTURE INTERACTIONS AT THE SURFACE

The preceding view of climate emphasizes the atmosphere as a delivery system for energy and moisture but does not consider the interaction of energy and moisture with environmental systems on the surface of the earth. However, systems on the earth's surface do share the energy and moisture that the atmosphere delivers; plant growth, for example, depends on the amount of soil moisture available to vegetation and not directly on delivered precipitation. So to emphasize the relationships between climate and other systems, the view of climate can be extended to include the interactions of delivered energy and moisture at the earth's surface.

Penck's Three Functional Realms of Climate

In the late nineteenth century, the German geographer Albrecht Penck recognized that in terms of the interaction of climate with environmental systems, three distinct *realms of climate* can be designated. The criteria that distinguished Penck's three natural realms of climate can now be stated in terms of potential evapotranspiration.

The *frozen* realm embraces regions where potential evapotranspiration equals zero. Plants cannot transpire moisture at freezing temperatures; thus the realm of no potential evapotranspiration is characterized by low temperatures. The *dry* realm includes all regions where precipitation is less than potential evapotranspiration. Characteristic of the dry realm are dry grasslands and deserts, where deficiencies of soil moisture are common. The *moist* realm, in which forests are located, includes all regions where precipitation exceeds potential evapotranspiration.

Different systems of climate classification use different methods to distinguish between dry and moist realms. Two widely accepted classification systems based on energy-moisture interactions are discussed next.

The Köppen System of Climate Classification

The most widely used system of climatic classification was developed and refined in the early decades of this century by Wladimir Köppen, a German botanist and climatologist. Köppen's system was influenced by the work of nineteenth-century plant geographers who had mapped the world's vegetation on the basis of extensive field studies, and by the implications of Penck's three functional realms of climate.

Köppen's classification is related to vegetation types that he thought to be responses to climate; it recognizes five general climatic types, designated *A*, *B*, *C*, *D*, and *E*. The types *A*, *C*, and *D* represent the moist climatic realms in which precipitation exceeds evapotranspiration, so that there is an annual water surplus. Areas having these climates support forests under natural conditions. The dry realm is represented by the *B* climates where precipitation is exceeded by potential evapotranspiration, so that there is an annual water deficit. The *E* climates are polar types of the frozen realm, in which low temperatures may reduce potential evapotranspiration to near zero. The *B* climates are subdivided into the *BW*, or arid (desert) climatic type (*W* from German *Wüste*, or desert), and the *BS*, or semiarid (steppe) climatic type. The *E* climatic type includes the *ET* or tundra climate, and the *EF* frost (perpetual ice) climate.

Köppen's definition of the *ET* climatic type illustrates his use of vegetation as an indicator of climate. In the *ET* climate the average temperature of the warmest month falls between 0°C and 10°C (32°F and 50°F). The 10°C (50°F) limit for the warmest month corresponds approximately to the poleward boundary of tree growth. By choosing this limit, Köppen ensured that regions with the *ET* climate would be essentially treeless. Still, the 0°C (32°F) lower limit ensures that temperatures above freezing do occur, so that some plant growth is possible, which is not true in the *EF* climate.

Köppen established specific temperatures and precipitation amounts and ratios to distinguish between his principal climatic types and their subdivisions. Figure 8.4 is a schematic diagram of his classification; the specific criteria are given in Appendix III. The *B* climatic types include a wide span of temperature and moisture values; the sloping boundary between the humid and a dry climatic realm in the figure is not an absolute precipitation value, but is a ratio between temperature and precipitation. The boundary moves to greater values of annual precipitation as the temperature increases. The reason for this shift is that the boundary is intended to fall where annual precipitation and potential evapotranspiration are equal, and potential evapotranspiration increases with temperature.

Figure 8.3 This global map shows a climate-classification system based on solar radiation input and atmospheric delivery of energy and moisture to the earth's surface. Compare it with the distribution of climates on the hypothetical continent shown in Figure 8.2. The landmasses of the high northern latitudes have polar and subarctic climates because of the low input of solar radiant energy. The tropical wet climates of equatorial Africa, South America, and Southeast Asia are flanked by tropical wet and dry climate regions. North America, Africa, and Australia have dry climates near latitude 30°. The effect of mountains on climate distributions is apparent in the way the tropical wet region of South America does not cross the Andes to the west coast, which is a desert in tropical latitudes.

The distribution of continents and oceans on the earth affects the movements of air masses. The southeastern United States, for example, has a humid subtropical climate instead of a humid continental climate because of the northward movement of moist tropical air from the Gulf of Mexico and the Caribbean. Parts of western Europe and the west coast of North America have marine climates. In Europe the marine climate gradually merges into a humid continental climate region, but in North America the west winds from the Pacific Ocean are blocked by mountains. The region immediately inland from the high mountains is desert. Central and eastern North America have a humid continental climate because of cold winds from the north in winter and warm winds from the south in summer. (Andy Lucas and Laurie Curran after Vernon C. Finch and Glenn Trewartha, *Physical Elements of Geography*, 1949, McGraw-Hill Book Co.)

Major Global Climatic Regions

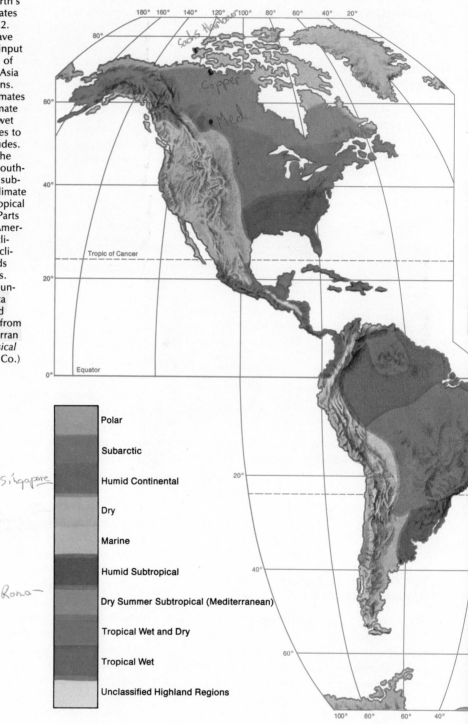

Polar

Subarctic

Humid Continental

Dry

Marine

Humid Subtropical

Dry Summer Subtropical (Mediterranean)

Tropical Wet and Dry

Tropical Wet

Unclassified Highland Regions

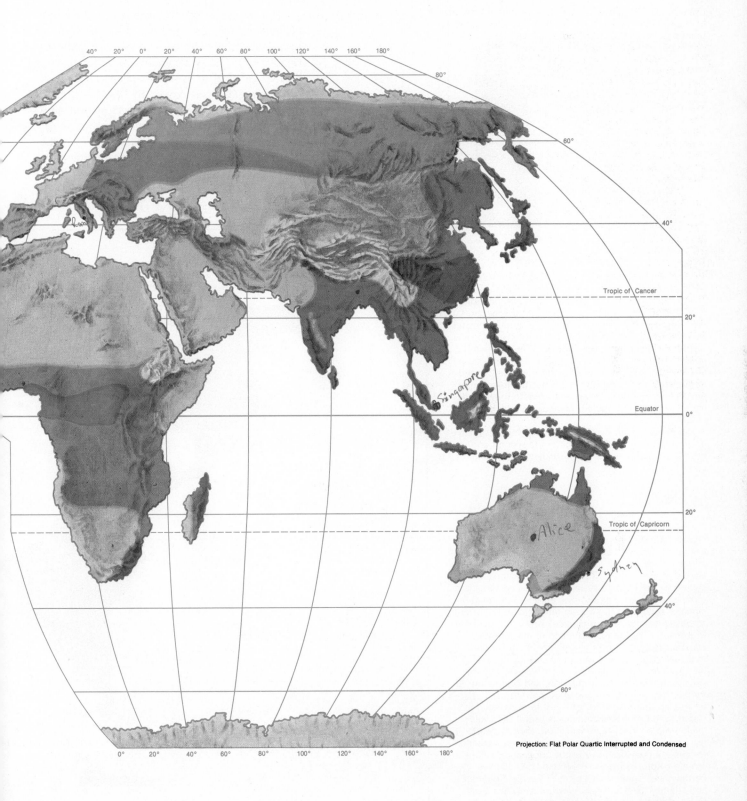

Projection: Flat Polar Quartic Interrupted and Condensed

MISSING
Verkhoysansk, USSR
Vostok, Antarctica

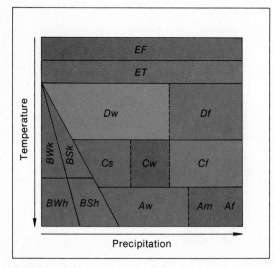

Figure 8.4 This schematic diagram represents the principal climatic types of the Köppen classification in terms of average annual temperatures and precipitation. The *E* climates across the top of the diagram represent the frozen realm. The *D*, *C*, and *A* climates make up the humid realm, with the *A* climates, towards the lower right, the warmest. The *B* climates make up the dry realm. Precipitation increases from left to right, and the boundary between the dry and humid climatic realms angles toward the lower right because more precipitation is needed to meet *PE* demands for water in warmer climatic regions than in colder ones.

The names of all the climatic types, their respective properties, and the definitions of the second and third letter symbols are given in Figure 8.5, Table 8.1, and Appendix III.

Figure 8.5 This map depicts the broad global distribution of climates according to the Köppen system of climatic classification. The definite boundary lines shown between the principal climatic regions are assigned by the Köppen system; in actuality, climatic regions shade gradually into one another. The map of climates according to Köppen differs in an important respect from the map of climates presented earlier in this chapter (Figure 8.3). The earlier map shows a classification based on the delivery of energy and moisture. The Köppen system of classification takes into account very general relationships among seasonal temperatures, precipitation, and vegetation. Although Köppen's estimates of evapotranspiration are far less accurate than more recent formulations, his classification of global climates is useful for a relatively simple global regionalization. (See Appendix II for a more detailed explanation of Köppen's criteria for climate classification.) (Andy Lucas and Laurie Curran after Köppen-Geiger-Pohl map [1953], Justes Perthes, and Köppen-Geiger in *Erdkunde*, volume 8; and Glenn T. Trewartha, *An Introduction to Climate*, fourth edition. New York: McGraw-Hill, 1968)

The Köppen system was an outstanding achievement for its time, and it still provides a useful introduction to world climatic patterns (Figure 8.5). The system allows additional symbols to be used to describe special features of climatic regions (see Appendix III). The symbol *f*, for example, means that precipitation adequate to offset potential evapotranspiration occurs in every month of the year. Therefore, the symbol *Af* specifies a tropical wet climate, as in the equatorial rainforests of South America. The symbol *Cf* describes the warm, moist climate of the eastern United States. The symbols *w* and

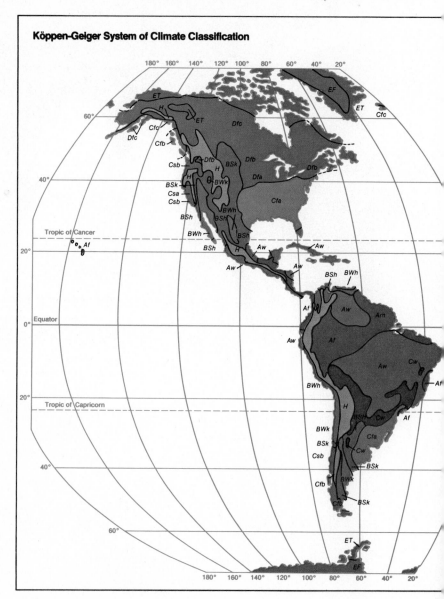

Köppen-Geiger System of Climate Classification

s signify short winter and summer drier seasons within the *C* and *D* climates of the humid realm, respectively. Thus the tropical wet and dry climate is an *Aw* type, and the dry summer subtropical climate is the *Cs* type. Additional symbols are added to characterize special conditions, such as a monsoon climate or frequent fog.

The Köppen climatic system represents an intermediate historical step in the evolution of climatic classification from the point of view of the availability of energy and moisture at the surface of the earth. The concept of potential evapotranspiration had not yet been developed, and Köppen vastly overestimated the areas of the continents included within the humid climatic realm. Later in this chapter we shall see that Thornthwaite's concepts of *PE* and the water budget provide more appropriate information about the climatic relationships between energy and moisture, and their interactions with plants, soils, and other environmental systems.

Nevertheless, the Köppen climatic system is comprehensive and flexible. The global patterns of climatic types in Figure 8.5 can also be used

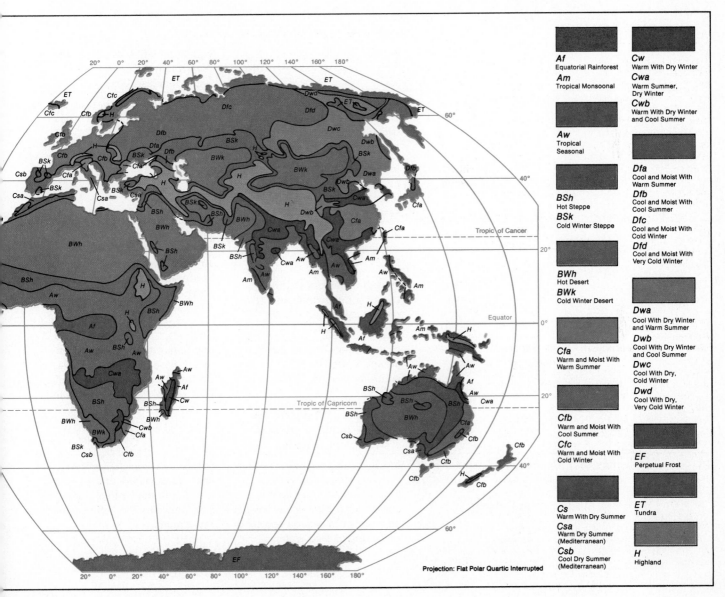

Projection: Flat Polar Quartic Interrupted

Af Equatorial Rainforest
Am Tropical Monsoonal
Aw Tropical Seasonal
BSh Hot Steppe
BSk Cold Winter Steppe
BWh Hot Desert
BWk Cold Winter Desert
Cfa Warm and Moist With Warm Summer
Cfb Warm and Moist With Cool Summer
Cfc Warm and Moist With Cold Winter
Cs Warm With Dry Summer
Csa Warm Dry Summer (Mediterranean)
Csb Cool Dry Summer (Mediterranean)
Cw Warm With Dry Winter
Cwa Warm Summer, Dry Winter
Cwb Warm With Dry Winter and Cool Summer
Dfa Cool and Moist With Warm Summer
Dfb Cool and Moist With Cool Summer
Dfc Cool and Moist With Cold Winter
Dfd Cool and Moist With Very Cold Winter
Dwa Cool With Dry Winter and Warm Summer
Dwb Cool With Dry Winter and Cool Summer
Dwc Cool With Dry, Cold Winter
Dwd Cool With Dry, Very Cold Winter
EF Perpetual Frost
ET Tundra
H Highland

as a teaching model of climatic types associated with the atmospheric delivery of energy and moisture *to* the surface of the earth. The Köppen climatic types were developed originally in terms of interactions of energy (represented by temperature) and moisture with global vegetation patterns. Indeed, many of the climatic types are designated in terms of vegetation. However, the sytem is now most useful as a global scheme of dynamic atmospheric climates. These dynamic characteristics of the individual Köppen climatic types are summarized in Table 8.1, which integrates in capsule form by climatic regions many of the fundamental concepts developed in Chapters 2 through 7. The photographs in Figure 8.6 illustrate some examples of associated landscapes.

Table 8.1   The Köppen Climatic Types: Dynamic Atmosphere Properties

Köppen Symbol	Köppen Name	Climatic Type	Season	Dynamic Properties (from Chapters 4 and 6)	Primary Locations
HUMID CLIMATIC REALM					
Af	Tropical Rainforest	Tropical Wet	year-round	*mT* or *E* air masses and convergence associated with ITC	equatorial latitudes, especially Amazon and Congo Basins, and Southeast Asia
Am	Tropical Monsoon	Tropical Wet	high-sun season	*mT* or *E* air masses and ITC with strong onshore flow and occasional tropical storms, typhoons, and cyclones	Vietnam, western coast of India, and West African coast
			low-sun season	*mT* or *E* air masses, but offshore flow of somewhat drier air produces short dry season	
Aw	Tropical Savanna	Tropical Wet and Dry	high-sun season	*mT* or *E* air masses and convergence associated with ITC	poleward of the tropical rainforest climates in South America and Africa; India, Southeast Asia, and northern Australia
			low-sun season	subsiding air, upper-level dynamic inversions, and *cT* air masses associated with subtropical highs	
Csa	Mediterranean	Dry-Summer Subtropical	summer	*cT* air masses due to subsiding air and upper air inversions of subtropical highs	margins of the Mediterranean Sea in Europe, Asia, and northwestern Africa; California, central Chile, South Africa, and southwestern Australia
			winter	*mT, cT,* and *mP* air masses asssociated with midlatitude cyclones and high-pressure systems along the polar front	
Csb	Mediterranean	Dry-Summer Subtropical	summer	much cooler temperatures than *Csa*, low stratus clouds and fog, due to onshore flow of marine air chilled by cold water offshore (cold ocean currents or upwelling of cold water)	coastal locations adjacent to cold ocean waters in *Csa* regions; in the United States, San Francisco is celebrated for this climate
			winter	same as *Csa*	

Table 8.1 The Köppen Climatic Types: Dynamic Atmosphere Properties (continued)

Köppen Symbol	Köppen Name	Climatic Type	Season	Dynamic Properties (from Chapters 4 and 6)	Primary Locations
Cfa	Humid Subtropical	Humid Subtropical	summer	mT air masses carried onshore around western margins of subtropical highs over the oceans, some midlatitude cyclones and weak frontal activity, and occasional tropical storms over coastal areas except in South America	southeastern United States, eastern China and Japan, southern Brazil, Uruguay, the Pampas of Argentina, and small areas of South Africa and southeastern Australia
			winter	mT and cP air masses alternate in association with the passage of midlatitude cyclones and polar outbreaks along the polar front	
Cwa	Winter-Dry Subtropical and Upland Savanna	Tropical Wet and Dry	summer	onshore flow of mT air masses related to rainy summer monsoon	margins of humid subtropical regions of Asia associated with drier monsoon winters, and cooler upland areas of tropical savanna climate regions of Asia, Africa, and South America
			winter	dominated by cP or cT air and dry-winter monsoon	
Cfb	Marine West Coast	Marine	year-round	mostly mP air masses swept onshore in association with midlatitude cyclones along polar front and westerly flow aloft	northwest Europe, Pacific northwest of the United States, British Columbia, southern Chile, southeastern Australia and New Zealand
Dfa Dfb Dwa Dwb	Humid Continental	Humid Continental	summer	varying frequencies of cP, mP, mT, and even cT air masses, in association with milder continental high-pressure systems and midlatitude cyclones	only in the northern hemisphere; northern United States and southern Canada, central Europe, northern Japan, Korea and the Soviet Far East, with Dwa and Dwb, winter dry, due to winter monsoon over eastern Asia in the Soviet Union and China
			winter	cP air masses associated with polar outbreaks, and some mP and mT air masses associated with midlatitude cyclones along the polar front	
Dfc Dfd Dwc Dwd	Boreal or Taiga	Subarctic	summer	cP and mP air masses in relation to weaker continental high-pressure systems and midlatitude cyclones	only in the northern hemisphere; interior Alaska, Canada, and the Soviet Union, with the Dw types in eastern Asia
			winter	cP and A air masses with the polar high displaced southward in the northern hemisphere as the Siberian and Yukon highs	

(Continued)

Table 8.1 The Köppen Climatic Types: Dynamic Atmosphere Properties (*continued*)

Köppen Symbol	Köppen Name	Climatic Type	Season	Dynamic Properties (from Chapters 4 and 6)	Primary Locations
DRY CLIMATIC REALM					
BSh BWh	Subtropical Steppes and Deserts	Dry	high-sun season	cT air masses associated with subsiding air of the subtropical highs with some light rainfall over steppe regions on equatorial sides of the dry realm marginal to ITC	North African and Arabian deserts, southwestern United States and northern Mexico, Great Australian Desert, Kalahari Desert of South Africa, and steppe regions of northern Argentina
			low-sun season	cT air masses predominate with some rainfall over steppe regions on poleward sides of the dry realm marginal to midlatitude cyclone passages	
BSn BWn	Low-Latitude Coastal Steppes and Deserts	Dry	year-round	restricted low-latitude coastal locations with very cold ocean waters immediately offshore, usually equatorward of Mediterranean climatic regions; cool, damp, fog and low-stratus cloud, but little precipitation because of temperature inversions	Baja California, coastal Peru and northern Chile, coastal Morocco and Mauritania, coastal Somaliland, and coastal Southwest Africa
BSk BWk	Interior Steppes and Deserts	Dry	year-round	cT and cP air masses is summer and winter respectively, with little atmospheric moisture because of remoteness of oceans and rainshadow effects	steppes and deserts of Mongolia, China, and Soviet Central Asia, the Great Basin and the Great Plains of the United States and Canada, Patagonia, and portions of the Kalahari of South Africa
FROZEN CLIMATIC REALM					
ET	Tundra	Polar	summer	mP and some cP air masses and frontal activity associated with midlatitude cyclones	limited mostly to margins of Arctic and Antarctic oceans
			winter	mP, cP, and A air masses associated with the displacement of the polar high in the northern hemisphere over the continents, and occluded midlatitude cyclones along coastal regions adjacent to the persistent Icelandic and Aleutian subpolar lows	
EF	Ice Cap	Polar	year-round	mostly Arctic and Antarctic air masses associated with snow- and ice-covered surfaces and subsiding air of the polar highs	Antarctica and Greenland

(Continued)

Table 8.1 The Köppen Climatic Types: Dynamic Atmosphere Properties (*continued*)

Köppen Symbol	Köppen Name	Climatic Type	Season	Dynamic Properties (from Chapters 4 and 6)	Primary Locations
UNDIFFERENTIATED HIGHLANDS					
H	Highlands	Highland		complex localized patterns of orographic precipitation and rainshadows with very high precipitation totals where warm, moist, onshore winds are forced over coastal mountain barriers; lower temperatures with increasing elevations so that tundra and ice-cap climates are present locally even in equatorial regions	

The Thornthwaite System of Climate Classification

In an attempt to overcome the limitations of the Köppen system, C. Warren Thornthwaite in 1948 devised a method for classifying climates according to water budget evaluations of energy and moisture. In Thornthwaite's classification system, energy is specified by potential evapotranspiration (*PE*), and moisture is expressed by a *moisture index*, which is defined in Figure 8.7.

Thornthwaite's moisture index depends on the difference between precipitation and calculated values of potential evapotranspiration. The moisture index has the value of − 100 when there is no precipitation, and may exceed + 100 where rainfall far exceeds potential evapotranspiration. At the boundary between the dry and moist realms, the moisture index is zero.

Figure 8.8 shows schematically how basic climatic types are defined in Thornthwaite's classification system. Each climatic region corresponds to a range of energy and moisture values that are calculated according to *PE* and moisture index classes.

One of the principal differences between the Köppen and Thornthwaite classification systems is in the way the dry and moist realms are divided. Like Köppen, Thornthwaite divided the nonpolar climates into five general types: two in the moist realm, two in the dry realm, and one, the *subhumid*, that occupies the transitional zone where precipitation and potential evapotranspiration are about equal. But when the Köppen and Thornthwaite climatic classifications are applied to the United States, the difference between them emerges clearly. Consider the North American grasslands east of the Rocky Mountains, which are so distinctive that they appear to occupy a separate climatic region between the moister areas to the east and the drier regions to the west. When Köppen's criteria are used, the dry grasslands do not emerge as a distinct climatic region. However, the dry grasslands correspond closely to the subhumid climatic type of Thornthwaite's system, as shown in Figure 8.9.

Perhaps because its computations are time-consuming, the Thornthwaite system is not often used to define climatic regions on a global or continental scale. It does not give a graphic portrayal of atmospheric dynamic climates as does the Köppen system. Still, Thornthwaite's concept of the moisture index based on month-by-month water budget calculations (Chapter 7) represents an improvement over Köppen's temperature and precipitation relationships. World maps have been constructed that show individ-

Figure 8.6 Some "climatic" landscapes from west to east across the United States.

(a) Orographic lifting of persistent easterly trade winds creates a nearly continuous cloud canopy over the Hawaiian island of Kauai. Precipitation is so frequent and intense that the summit area in the background is one of the rainiest places on earth.

(b) The dry summer subtropical landscapes of coastal California are striking mosaics of dark evergreen oaks, golden brown grasses, bright green plants in irrigated areas, and—in this view from Mt. Tamalpais just north of San Francisco—a very blue lake. (Robert A. Muller)

(c) During the rainless summer and summer seasons of the dry summer subtropical climates, the vegetation dries out and brush fires are common. The fires in this view burned for several days in August 1977, in the hills east of Oakland, California. (T. M. O.)

(d) Lush tropical landscaping on this island in San Diego Bay is possible only because of the total absence of freezing temperatures and because of frequent irrigation during the dry summers. North of Mexico, similar exotic tropical landscaping is limited to southern California, lowland places in Arizona such as Phoenix and Yuma, the Rio Grande valley in southern Texas, and southern Florida. (Robert A. Muller)

(e) The steppe climates of the Great Plains are much too dry for tree growth, but dryland wheat production is very successful during years with normal rainfall. This nearly featureless plain in the Oklahoma Panhandle experiences runs of dry years, such as the Dust Bowl years of the 1930s, which can be disastrous for the regional economy. (Robert A. Muller)

(f) Roofs need to be cleared of heavy snow loads in the snowbelt regions of the Great Lakes. This scene was repeated over and over again after snow squalls produced 182 cm (72 in.) of snow in 6 days in December 1958, in a narrow band between Oswego and Syracuse, New York. Ladders to the roofs and caved-in old buildings are characteristic features of the landscapes where big snow squalls are common. (Robert A. Muller)

(g) A massive thunderstorm towers to more than 12.2 km (40,000 ft) inland from the coast near Pensacola, Florida. Some vacationeers do not appreciate that these thunderstorms tend to form inland along the leading edge of the cooler sea breeze, leaving the beaches along the Atlantic Coast sunny and breezy much of the time. (Robert A. Muller)

(a)

(b)

(c)

(d)

(e)

(f)

(g)

Moisture Index = $100 \left[\dfrac{P - PE}{PE} \right]$

Moisture Index = $100 \left[\dfrac{S - D}{PE} \right]$

P = Precipitation

PE = Potential Evapotranspiration

S = Surplus

D = Deficit

Figure 8.7 The average value of Thornthwaite's moisture index can be calculated from either of these two equations. The value of the moisture index is positive at places where mean annual P is greater than PE, and it is negative where P is less than PE. The moisture index is also positive at places where S is greater than D. Even though mean annual P is greater than PE, monthly moisture deficits can occur where there is a strong seasonality of precipitation, or because of dry months associated with climatic variation, such as in dry summer subtropical regions.

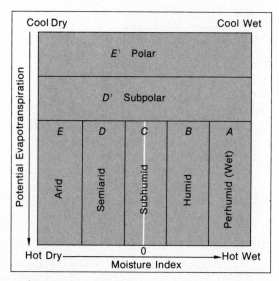

Figure 8.8 This schematic diagram represents the principal climate regions in the Thornthwaite classification system. This idealized diagram is for the northern hemisphere with polar latitudes at the top and the equator towards the bottom. Potential evapotranspiration is used as a measure of moisture utilization by environmental systems on the earth's surface. The climates run from cool and dry at the upper left corner of the diagram to hot and wet at the lower right. The climates are divided according to orderly ranges of values for potential evapotranspiration and the moisture index. In the higher latitudes where energy is in short supply, the emphasis is on energy availability and the climates are not divided into moisture classes. In lower and middle latitudes where large amounts of energy in terms of radiation and heat are normally available, the emphasis is on moisture availability. Five principal moisture regions, A through E, are recognized for middle and lower latitudes. The subhumid climatic region, C, is centered where the moisture index equals 0. (Doug Armstrong after D. Carter and J. Mather, *Publications in Climatology*, vol. 19, no. 4, 1966)

ual parameters of water budgets: potential evapotranspiration, the moisture index, annual water surplus, and annual water deficit. These maps are useful for study of energy and moisture interactions in terms of natural vegetation, agricultural capability, and water resources, on the continents.

WATER BUDGET CLIMATES: INTERACTIONS WITH ENVIRONMENTAL SYSTEMS

The local water budget discussed in Chapter 7 is a useful tool for understanding climate in terms of energy and moisture. The individual components of the water budget that can be used as indexes of climate include potential evapotranspiration, precipitation, soil moisture, actual evapotranspiration, water surplus, and water deficit. By analyzing these components, it is possible to understand how climate affects various systems in the environment, such as vegetation, soils, streams and rivers, and landforms. Even more sophisticated versions of water budget models are used by planners, engineers, and scientists for the design of water projects and the implementation of agricultural strategies.

To illustrate, we consider the average local water budgets at two locations: Cloverdale, California, and Manhattan, Kansas. Both locations have a total annual potential evapotranspiration of nearly 80 cm (31.5 in.), so their energy supplies are essentially equal.

On the basis of annual precipitation, Cloverdale (100 cm, or 39.4 in.) appears to be more

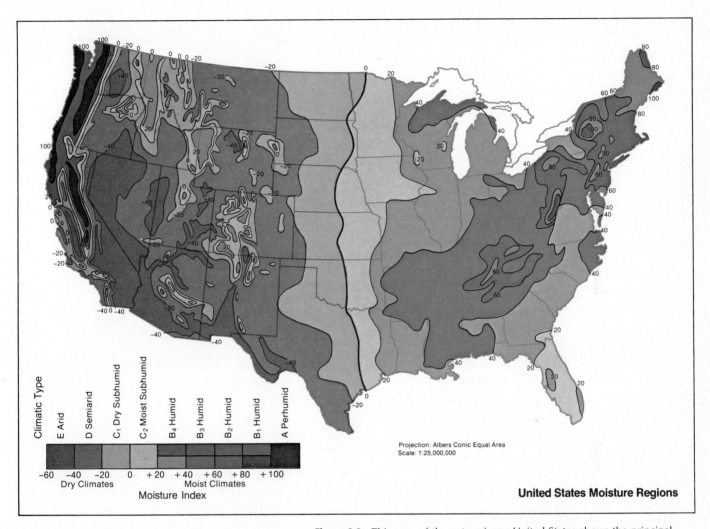

Climatic Type

| E Arid | D Semiarid | C₁ Dry Subhumid | C₂ Moist Subhumid | B₄ Humid | B₃ Humid | B₂ Humid | B₁ Humid | A Perhumid |

| −60 | −40 | −20 | 0 | +20 | +40 | +60 | +80 | +100 |

Dry Climates   Moist Climates

Moisture Index

Projection: Albers Conic Equal Area
Scale: 1:25,000,000

United States Moisture Regions

humid than Manhattan (80 cm, or 31.5 in.). But as noted in Chapter 7, the utilization of moisture at a place depends on the timing of precipitation. Figure 8.10 shows the water budgets for the two locations. At Cloverdale, large deficits of soil moisture are generated in the summer months when potential evapotranspiration is high and precipitation is low. Cloverdale's vegetation has evolved adaptations to survive a very dry growing season. In the winter and spring, however, heavy precipitation results in a large surplus of water. Therefore, Cloverdale's climate supports strong seasonal streamflow.

At Manhattan, the water budget shows that precipitation amount and timing tend to mirror the variation of potential evapotranspiration

Figure 8.9 This map of the coterminous United States shows the principal moisture realms as classified according to the Thornthwaite moisture index. Precipitation exceeds potential evapotranspiration where the index is positive, and it is less than potential evapotranspiration where the index is negative. The line where the moisture index is 0 extends from western Minnesota south to the Gulf of Mexico. To the east of this line there is a surplus of moisture, and to the west there is a deficiency, under average conditions. The moisture index of a region bears a close relation to the form of its vegetation. Forests, for example, can maintain themselves only where the moisture index is positive and has a value of at least 20 or so. Only grasslands and desert vegetation can survive where the moisture index is negative. (Andy Lucas and Laurie Curran after C. W. Thornthwaite and J. Mather, "The Water Balance," *Publications in Climatology*, vol. 8, 1955)

throughout the year. Therefore, only small deficits and surpluses are generated at Manhattan. There is moisture for vegetation throughout the growing season, on the average, but little to support streamflow at any time.

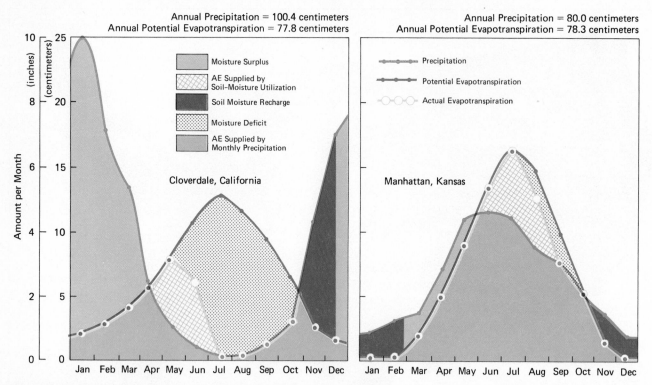

Figure 8.10 Cloverdale, California, and Manhattan, Kansas, receive similar amounts of precipitation annually and have nearly the same total annual potential evapotranspiration. The climates at these locations appear to be similar on the basis of the annual inputs of moisture and energy. But their local water budgets indicate that the timing of precipitation compared to the demand for moisture at each location makes the climates quite different. At Cloverdale, heavy precipitation and a low moisture demand by vegetation during the winter generate large moisture surpluses. Deficits occur during the dry summer. Thus Cloverdale is moist for runoff and dry for vegetation. At Manhattan, the timing of precipitation through the year is in close accord with the demand by vegetation, making surpluses and deficits small. Manhattan is therefore dry for runoff and moist for vegetation. (Vantage Art, Inc. after Douglas B. Carter, Theodore H. Schmudde, and David M. Sharpe, "The Interface: As a Working Environment: A Purpose for Physical Geography," *Tech. Paper No. 1*, Comm. on College Geog., Association of American Geographers, 1972)

The examples of Cloverdale and Manhattan show that when one considers the effects of climate on other environmental systems, simply knowing the total amounts of energy and moisture delivered is not enough. How the distribution of energy and moisture varies with time must also be taken into account. The examples indicate also that a given location offers particular problems and possibilities that are revealed by various components of the water budget.

The surplus in the water budget is a good

index of the variability in space and time that can exist within a given climate. Figure 8.11a shows the average winter-spring surplus across Louisiana for a 30-year period, 1941-1970. Although Louisiana has the highest average precipitation of the 50 states, the average seasonal surplus shown on the map ranges from 56 cm (22 in.) near the central part of the state to 36 cm (14 in.) along the Gulf Coast in the south. Even in this wet state, the spatial variability of the water supply is considerable.

Variability from one year to the next, which depends on small changes in the upper atmospheric circulation, can be greater than variation from place to place. The surpluses across Louisiana for three winter-spring seasons are shown in Figure 8.11b–d. These diagrams represent first a season that is "wet" everywhere, then a season with extreme spatial variability, and, finally, a "dry" season everywhere. Clearly the range between extreme years is very large. This variability exerts stress on plant and animal life; the populations of some species may be significantly reduced or even eliminated

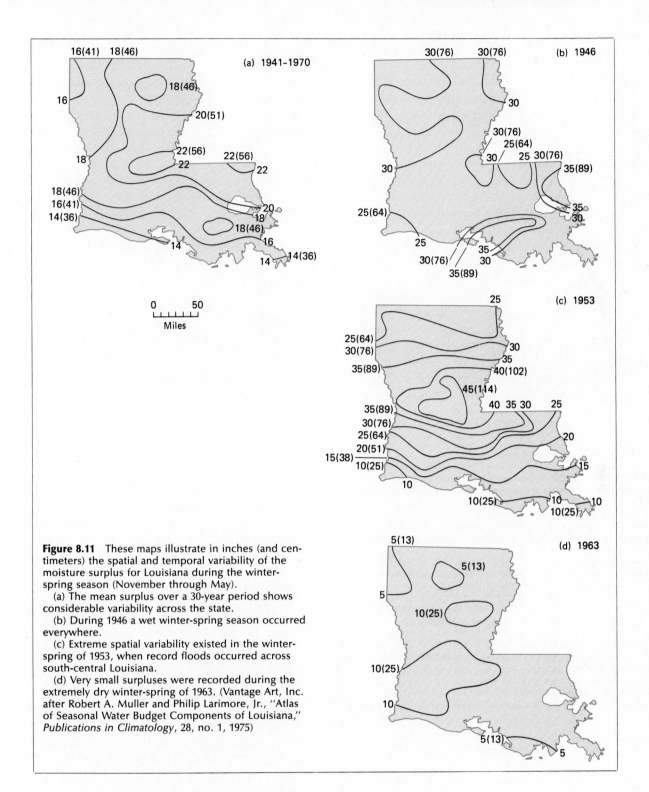

Figure 8.11 These maps illustrate in inches (and centimeters) the spatial and temporal variability of the moisture surplus for Louisiana during the winter-spring season (November through May).

(a) The mean surplus over a 30-year period shows considerable variability across the state.

(b) During 1946 a wet winter-spring season occurred everywhere.

(c) Extreme spatial variability existed in the winter-spring of 1953, when record floods occurred across south-central Louisiana.

(d) Very small surpluses were recorded during the extremely dry winter-spring of 1963. (Vantage Art, Inc. after Robert A. Muller and Philip Larimore, Jr., "Atlas of Seasonal Water Budget Components of Louisiana," *Publications in Climatology*, 28, no. 1, 1975)

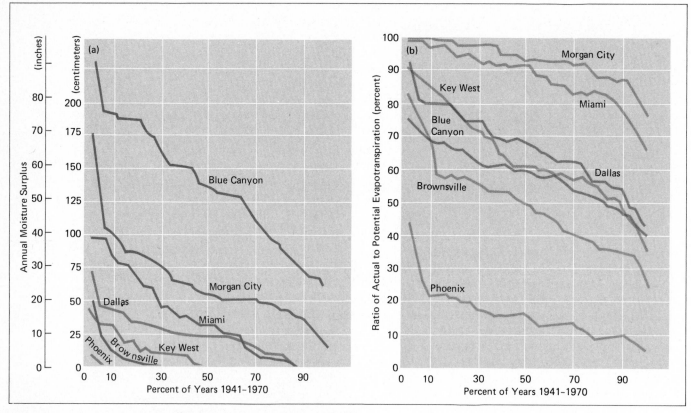

Figure 8.12 Components of the local water budget can be used as climatic indexes. These graphs emphasize variability of the climate at seven locations across the United States for the 30-year period between 1941 and 1970. The annual values of the components are plotted with the largest value to the left and the smallest value to the right.

(a) Annual moisture surpluses are shown here. The largest annual surplus over the 30 years at Blue Canyon, California, amounted to more than 229 cm (90 in.); the smallest surplus was only about 64 cm (25 in.). Over the 30-year period, the average surplus was about 140 cm (55 in.). On the basis of the cumulative frequency curve, we can see that the surplus amounted to 157 cm (62 in.) or more during 30 percent of the years. Similarly, during 90 percent of the years, the surplus was equal to or greater than 73 cm (29 in.); we can transpose the perspective and also say that during 10 percent of the years the surplus amounted to less than 73 cm (29 in.). The largest surpluses were at Blue Canyon; Morgan City, Louisiana, ranked second of the seven.

(b) The ratio of actual to potential evapotranspiration is a measure of the degree to which precipitation and soil moisture meet the climatic demand for water. Most crops and forest vegetation do poorly when the ratio is low over the growing season. For this index, Morgan City ranked first and Blue Canyon only fifth! (Doug Armstrong after R. Muller, "Frequency Analyses of the Ratio of Actual to Potential Evapotranspiration for the Study of Climate and Vegetation Relationships," *Proceedings*, Association of American Geographers, vol. 3, 1971)

by a run of very wet or dry seasons, or unusually warm or cold periods (the Case Study at the end of this chapter focuses on climatic variation).

In 1972, the year of "the great freeze" in central California, it was apparent that native evergreen vegetation was unaffected by several successive days of very unusual sub-freezing temperatures. Imported plants, including whole forests of non-native eucalyptus trees, were either temporarily or permanently damaged, because their evolution had occurred in other regions where such weather never occurred. Past "freezes" in central California, despite their infrequent occurrence, have eliminated such plants from the local flora.

Figure 8.12a shows the annual water surplus generated at seven locations in the United States. The data are presented in terms of frequency distributions over a 30-year period. The plots show the degree of variability that can occur at each place. At Blue Canyon, California, a moisture surplus greater than 135 cm (53 in.) was generated in 50 percent of the years. But in individual years the surplus ranged from over 229 cm (90 in.) to under 64 cm (25 in.). From the standpoint of moisture surplus, Blue Canyon should be classified as exceptionally humid; Morgan City, Louisiana, as humid; and Phoenix, Arizona, as arid.

In Figure 8.12b the same seven locations are arranged according to the ratio of actual to potential evapotranspiration—meaning water supply relative to water demand. Vegetation that is native to places with high water supplies relative to demands will not do well in dry climates without irrigation, as in Brownsville, Texas, where actual evapotranspiration can rarely supply more than 60 percent of the water demand. On this climatic scale, Morgan City and Miami are best supplied with moisture, Blue Canyon is only intermediate, and Phoenix is again lowest. If the stations were plotted in terms of annual water deficit, Phoenix would rank first. Depending upon which water budget components are selected, locations change their relative positions along the climatic scale.

SUMMARY

Climate is more than just the average condition of the atmosphere near the earth's surface over a period of years. Its most common measures are temperature and precipitation. Climates show significant spatial and temporal variations whether they are viewed on the scale of a continent, a region, or even within the confines of a suburban backyard. The range of possible climates can be divided into a small number of specific types in order to classify geographical areas according to their dominant climatic characteristics. Different schemes of climatic analysis and classification are useful for different purposes.

The geographic distribution of climatic types resulting from the delivery of energy and moisture by the sun and the general atmospheric circulation can be demonstrated on a hypothetical continent with a smooth surface and a regular coastline. The influences of mountain ranges and continental shapes are clear on a global map of actual climatic patterns resulting from the general circulation. The occurrence of similar climatic types in widely separated parts of the earth is one of the most important facts of physical geography, and affects the form and distribution of vegetation, soils, landforms, and human activities.

The climatic classifications of Köppen and Thornthwaite represent progressive attempts to evaluate energy–moisture interactions at the earth's surface. The Köppen system has the advantage of specifying the nature of annual temperature and moisture regimes, but has little application beyond showing the global distributions of climatic types. The Thornthwaite system, based on the local water budget, analyzes interactions between energy and moisture in terms of precipitation, evapotranspiration, soil moisture, runoff, and water deficit. This system is a useful tool for understanding the ways in which climate affects various environmental systems. It is also helpful in the analysis of spatial and temporal variability within climates.

By evaluating water budget components, we can see that areas having similar annual inputs of moisture and energy may, in fact, have quite different climates. This may be reflected in such phenomena as streamflow, vegetation cover, soil characteristics, and erosional processes and resulting landforms.

REVIEW QUESTIONS

1. What climatic indexes are generally used in physical geography?
2. What factors determine variations within a microclimate? How do they differ from those that determine global climates?
3. Why are different types of climate developed on the east and west coasts of the hypothetical continent at subtropical latitudes?
4. What are the main differences in the climatic pattern over the northern and southern hemispheres of the hypothetical continent?
5. Why does the marine climate of Western Europe extend further inland than the marine climates on the west coasts of North and South America?
6. According to Penck's system of climate classification, what conditions characterize the boundary between the moist and dry realms?
7. Why was dependence on vegetation as an indicator of climate a drawback to Köppen's system?
8. What are the principal ways in which Thornthwaite's system of climate classification differs from Köppen's?
9. Where would you expect farmers to depend more on irrigation—Manhattan, Kansas, or Cloverdale, California? How would a comparison of the water budgets of the two areas reflect this difference?

APPLICATIONS

1. Along portions of the coasts of North and South America, Southern Africa, Portugal, and Australia there are climates that are generally regarded as "Mediterranean" in type. However, the climates of these coasts do not exactly duplicate the climates of the Mediterranean coastlands themselves. What is the difference?
2. The largest high latitude land area that was not covered by an ice sheet during the Ice Age was Siberia, which now has the coldest winters of any area outside of Antarctica. How do you explain this contradiction?
3. Suppose the Gulf of Mexico and the Caribbean Sea did not exist and the North American coastline extended continuously from Florida through the Bahamas, Puerto Rico, and the Lesser Antilles and Trinidad to Venezuela. How would this affect the climates of North America? What would happen to the geographic location of Thornthwaite's zero moisture index that separates dry and moist climates? Draw a map showing hypothetical climatic regions on such a continent, using the Köppen system of classification.
4. The Mississippi River valley interrupts an otherwise continuous belt of hilly terrain and low mountains extending from Oklahoma to New England. In some places these mountains were much higher in the past. If they were as lofty today as the Rockies are, how would the climates of North America be affected?
5. Climatic variability from year to year can have significant impact on environmental systems and economic prosperity. Using Appendix II and National Weather Service data on monthly temperature and precipitation for some place of interest, determine the Köppen climatic type for each of five successive years. What portions of the earth would be most likely to fall into different Köppen climates from year to year?
6. Imagine the effect on the earth's climates if the axis of the earth's rotation were perpendicular to the plane of the earth's orbit around the sun, rather than being inclined at the present angle of 23½ degrees. What differences in climatic types would result? Would the earth be more or less habitable? You might be assisted by Figure 3.5.

FURTHER READING

Carter, Douglas B., and **John R. Mather.** "Climatic Classification for Environmental Biology." *Publ. in Climatology,* Vol. 19, No. 4 (1966):305–395. The evolutionary sequence of the several Thornthwaite climatic classifications and some of their relationships to vegetation distribution are presented in this technical paper. Average water-budget tables from thousands of places around the world are included in other issues of this specialized journal.

————, **Theodore H. Schmudde,** and **David M. Sharpe.** "The Interface as a Working Environment: A Purpose for Physical Geography." *Tech. Paper No. 7,* Comm. on College Geog., Assoc. of American Geog. (1972), 52 pp. This monograph stresses the relationship between water budget components and environment responses.

Critchfield, Howard J. *General Climatology,* 4th ed. Englewood Cliffs, N.J.: Prentice-Hall (1983), 453 pp. An introductory text that emphasizes the interrelationships of climate to global patterns of vegetation and soils, as well as to a wide range of economic activities.

Griffiths, John F., and **Dennis M. Driscoll.** *Survey of Climatology.* Columbus, Ohio: Charles E. Merrill (1982), 358 pp. This introductory text has a meteorological perspective and focuses on applications.

Hare, F. Kenneth. *The Restless Atmosphere,* rev. ed. London: Hutchinson (1956), 192 pp. This brief introductory classic contains outstanding regional and continental chapters that focus on the dynamics of the general circulation.

Lamk, H. H. *Climate, History, and the Modern World.* London: Methuen (1982), 387 pp. An up-to-date overview of the impacts of climatic variation and change by the pioneer scholar of these interesting and important topics.

Roberts, Walter O., and **Henry Lansford.** *The Climate Mandate.* San Francisco: W. H. Freeman (1979), 197 pp. This little paperback provides an outstanding current perspective on climatic variation and on some of its causes and consquences.

Thornthwaite, C. Warren. "An Approach Toward a Rational Classification of Climate." *Geographical Review,* Vol. 38, No. 1 (1948):55–94. Thornthwaite first set out the potential evapotranspiration and water budget systems in this classic paper.

Trewartha, Glenn T., and **Lyle H. Horn.** *An Introduction to Climate,* 5th ed. New York: McGraw-Hill (1980), 416 pp. Temperature and precipitation properties of climatic regions over the globe are stressed in this introductory text.

Wilcock, Arthur A. "Köppen After Fifty Years." *Annals Assoc. of American Geog.,* Vol. 58, No. 1 (1968):12–28. This article presents a concise review of Köppen's development of his climatic classification and the subsequent modifications by other climatologists.

CASE STUDY

How Much Do Climates Change?

Although the weather seems to be relatively the same from year to year, one summer may be hot and dry, and another cool and wet. Climatic fluctuation on this scale is very costly to technological civilizations, especially in terms of energy consumption and agricultural production. An early season freeze on Labor Day weekend in 1974 devastated crops across much of the upper Midwest. The record-breaking heat in Texas, the Plains states, and the South during the early summer of 1980 increased energy use, seared pastures, and withered crops. Fluctuations in water budgets are especially significant. Above normal deficits can greatly reduce crop yields (the "dust bowl" years of the 1930s are a dramatic example), and along the Mississippi, alternations in the surplus have resulted in both floods and low-water problems (see Case Study following Chapter 13).

Only in recent decades have scientists begun to unravel the detailed outline of the earth's climatic history for the past million years, using studies of tree rings, fossil plant pollen, ancient soils, lake and deep-sea sediments, mountain glaciers, and cores from ice caps on Greenland and Antarctica. The figure on this page illustrates the major trends of global climate based on these analyses. Perhaps the most astounding features are the glacial-interglacial cycles that last about 100,000 years, with shorter interglacial separated by long glacial climates, as shown in graph a. Glacial conditions appear to develop slowly and irregularly, with interglacial climates evolving suddenly. Recent studies indicate that climatic variability on time scales of about 100,000 and 40,000 years may

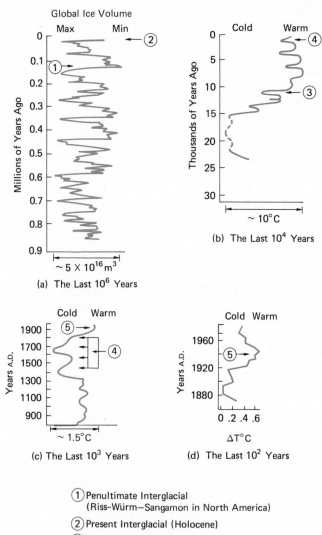

① Penultimate Interglacial
(Riss–Würm–Sangamon in North America)

② Present Interglacial (Holocene)

③ Younger Dryas Cold Interval

④ Little Ice Age

⑤ Thermal Maximum of 1940's

The main trends of global climate during the past million years. Graph a is based on isotope analyses of deep sea floor sediments; b on alpine tree lines, fluctuations of alpine glacier snouts, and fossil pollen; c on historical records of alpine glacier snouts and wine and grain production; and d on measured temperature data. (Vantage Art, Inc., modified from Understanding Climatic Change: A Program for Action, *National Academy of Sciences, 1975, Washington, D.C.)*

208

be related to systematic variations of the earth's orbit about the sun, which produce seasonal and latitudinal changes in the distribution of incoming solar radiation. Graph *b* shows the rapid warming trend that began about 15,000 years ago. During the very sudden and brief Younger Dryas cold interval about 10,500 years ago, much forest in Europe was destroyed. Graph *b* also shows that the mildest post-Pleistocene temperatures occurred about 5,000 to 7,000 years ago.

Temperature variations in eastern Europe over the past 1,000 years are shown in graph *c*; the temperature reconstruction is based on historical records of the positions of the snouts of alpine glaciers and of the wine production at various monasteries. The lower temperatures associated with the "Little Ice Age" between 1400 and 1850 resulted in social and economic dislocations. The thermal maximum of the 1940s represents the highest average temperatures of the past 1,000 years. Graph *d* shows the cooling that began about 1945, but recent data suggest that the global cooling may have leveled off since the mid-1970s.

Much shorter-term climatic variability has direct impact on environmental processes and economic activities. The next figure shows monthly temperature and precipitation variability for southeastern Louisiana from 1911 through June 1982. Each month's deviation from its respective average for the 30-year period 1931–1960 is shown in graphs *a* and *d*. Graphs *b* and *c* show short-term variations in terms of six-month "running averages."

There are two outstanding features of these graphs for southeastern Louisiana. One is the downward trend of temperature beginning in the fall of 1957 and continuing, with the exception of several winters, to the present. The other is the tendency for clusters of months or of years to be warmer or colder, or wetter or drier, than normal. Especially obvious are the dry periods of 1915–1918, 1933–1938, and 1951–1955, all associated with droughts on the Great Plains; other intense dry periods include 1924–1925, 1962–1963, and 1967–1971. Climatic variability on these time scales significantly affects agricultural yields, flooding and water-resource management.

Explaining why climates change is extremely difficult because many interactions of the general circulation are not understood in detail. Furthermore, slight climatic changes can produce disproportionate effects in the environment. If the average global temperature were to drop a few degrees, ice and snow would cover a greater proportion of the earth's surface. Because of its high albedo, the ice would reflect more radiant energy back to space, which would cause further cooling and further spread of ice and snow.

A number of different explanations have been proposed to account for climatic variability over geologic time, but none is entirely satisfactory. Glacial climates appear to have been associated with continental drift and plate tectonic processes (see Chapter 12), which caused continents to move into polar regions, allowing ice sheets to develop. In addition to Pleistocene glaciation, there is evidence of at least three widely separated glacial periods over the last billion years. Within these glacial periods, it is now believed that systematic variations in the earth's orbit account for the 100,000-year glacial-interglacial cycle. Furthermore, mountain uplift is believed to be connected with colder temperatures and increased snowfall over uplands. Stratospheric dust from intense volcanic activity has been linked to periods of somewhat diminished solar radiation income and lower temperatures at the earth's surface. The early summer freezes and snows in northern New England in 1816 followed the massive eruption of the volcano Tambora in the Dutch East Indies in 1815, and several years of spectacular red sunsets followed the eruption of Mt. Agung in Indonesia in the early 1960s. Some scientists believe that gases and particles ejected into the stratosphere by the Mexican volcano El Chichón in 1982 had an effect on the unusual atmospheric and oceanic circulation patterns during the winter and spring of 1982–1983.

Numerous attempts have been made to relate sunspot activity to climatic variation, especially drought periods in the Great Plains. Although there is some correlation between sunspots and dry periods on a time scale of several years, no cause-and-effect relationship has been demonstrated. Climatic variability on shorter time scales, such as those shown for southeastern Louisiana, are obviously related to changing patterns of the circumpolar vortex of westerlies in the upper atmosphere. Meteorologists and geophysicists continue to debate the causes of these circulation changes and the eventual return to more "normal" patterns. Precise forecasting of the circulation patterns a season or more in advance remains an elusive objective at present.

In predicting the climates of the near future, most concern centers on the effect of human activities, particularly the use of the atmosphere as a dumping ground. The atmosphere is a sensitive controller

CASE
STUDY

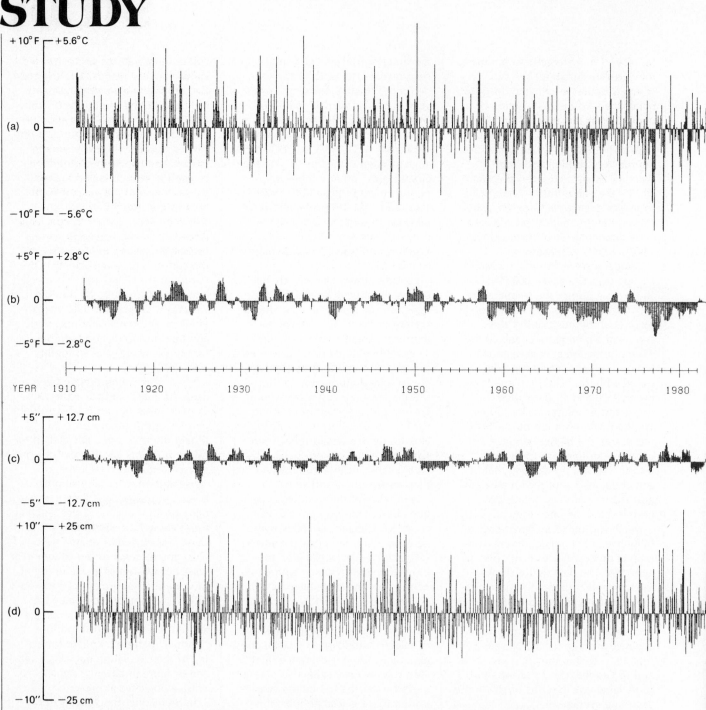

Temperature and precipitation variability in southeastern Louisiana. The temperature deviation of each month from the averages of the 30-year period between 1931 and 1960 is shown in graph a, and precipitation deviations are shown in graph d. For example, January 1940 was 7°C (12°F) below normal, and the rainfall during October 1937 was 28 cm (11 in.) above normal. Twelve-month running averages of temperature deviations are used in graph b to illustrate a "smoothed" interpretation of short-term temperature variation, and graph c is a similar smoothed interpretation of precipitation variation. Note especially the colder temperatures beginning in late 1957 and the fluctuations between warmer and colder, and wetter and drier conditions. (From Robert A. Muller and J. E. Willis, "Climate Variability in the Lower Mississippi River Valley," Geoscience and Man, 1978)

of climate; cloudiness and dust in the air directly affect the amount of energy reaching the earth, and carbon dioxide in the air affects the amount of energy that is radiated away. Industry and agriculture add smoke and dust particles to the air, increasing the albedo of the atmosphere and decreasing the amount of solar radiation received at the surface, and possibly lowering surface temperatures as well.

The interconnections of the earth's systems make it difficult to predict the net effect from a particular change. The amount of carbon dioxide in the atmosphere is expected to increase by at least 20 to 30 percent in the last half of this century, primarily because of industrial activities. The direct effect of the increase in carbon dioxide will be a decrease in longwave radiation losses from the earth's surface, which will raise the average temperature. The higher temperatures will increase evaporation, perhaps leading to increased cloudiness. Because it is difficult to determine the degree to which each of these modifications will affect other environmental properties, the final temperature change caused by the increase of carbon dioxide cannot be predicted.

Any small, long-term change in the atmosphere is critical. The present state of the environment is a function of its past history; there is no fresh start each year because the environment accumulates the effects of minor changes. Orbit wiggles, volcanic dust—these and other small and complicated interactions influence climatic trends. Now human activities can be added to the list.

La Charmeuse des Serpents (The Snake Charmer) by Henri Rousseau. (Art Resource)

All interactions among the parts of an ecosystem are united by the energy flow through the system. This intimate interdependence of living things implies a certain stability, a certain dynamic reciprocity, and a certain harmony of nature.

The miracle of our planet is its life—mysterious in origin and wondrously diverse—inhabiting every portion of the surface of the lands and the soils below, the surface waters of the continents, the oceans, and the lower atmosphere. All of these are components of the *biosphere*: the realm of living organisms on our planet.

The organisms inhabiting the biosphere are incredibly diversified, consisting of more than 50 primary subdivisions known as *phyla*. The system for classifying organisms descends from phyla through *classes, orders, families*, and *genera*, to the *species* level. A species is composed of related individuals that closely resemble one another and are able to breed only among themselves. About one and a half million species have been named and described in some manner. Two-thirds of these are animals—most of them insects. It is estimated that as many as 8 or 9 million unnamed and undescribed additional species exist—as many as a million species of mites alone!

Although we often speak of the plant and animal kingdoms, this is an oversimplification; some authorities recognize four or five kingdoms at the broadest level of classification (Figure 9.1). Vast numbers of unicellular bacteria, algae, slime molds, fungi, and related life forms are neither plant nor animal according to usual definitions, and must be assigned to separate categories. The criteria on which such broad classifications are based include cell characteristics, mode of nutrition (absorption, photosynthesis, or ingestion), degree and mechanism of motility (movement), mode of reproduction, level of organization and degree of tissue differentiation, and development of sensory-neuro-motor systems and muscle fibers.

All forms of life require energy and liquid water. The ultimate energy source for all but certain chemolithotrophic bacteria is solar radiation. On planets beyond the earth's orbit the low level of solar energy received causes any water present to be frozen and therefore unable to react chemically, thus precluding biological activity. The two planets within the earth's orbit and nearer to the sun (Mercury and Venus) receive so much energy that their water has been vaporized and is present only in the upper

9

The Biosphere and Ecological Energetics

Plant and Animal Interactions

Ecosystems and Biomes
Ecological Niches
Plant and Animal Dispersal

Energy Flow Through Ecosystems

Food Chains and Food Webs
Energy Flow Pyramids

Energy Capture and Plant Production

Photosynthesis and Plant Growth
Plant Respiration
Measures of Plant Production
Global Patterns of Plant Production

Natural and Agricultural Ecosystems

Natural Ecosystems
Agricultural Ecosystems and Yields
Climate and Agricultural Yields

Problems of World Agriculture

213

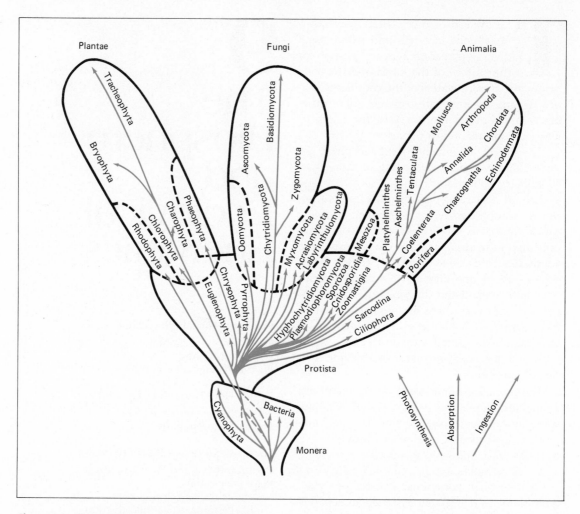

Figure 9.1 This five-kingdom classification of the earth's life forms was devised in 1969 by R. H. Whittaker, and is in common use. The major differences among the groupings are levels of organization and modes of nutrition. Keyed at the lower right, the different modes of nutrition include photosynthesis (by plants), absorption (by fungi), and ingestion (by animals). The two lower kingdoms are composed of very primitive and simple organisms with varying nutritional modes. (R. H. Whittaker, *Science*, 1969, p. 157)

atmosphere of Venus (Mercury has too little gravity to have retained an atmosphere). The earth's nearest planetary neighbors, Venus (nearer to the sun) and Mars (farther from the sun) differ enormously. The surface temperature of Venus is high enough to melt lead. This is a result of the planet's proximity to the sun and of a "runaway" greenhouse effect produced by an atmosphere rich in CO_2. By con-

trast, Mars is a cold and nearly airless desert, swept by dust storms, with evidence of a permanently frozen layer below the land surface (Figure 9.2). Neither planet has any trace of liquid water. Thus it is not surprising that instrument packages landed on their surfaces have failed to sense any indication of life.

On the fortunately situated earth, where liquid water is uniquely abundant, solar energy is converted to chemical energy by the earth's green plants through the process of photosynthesis (Chapter 1 and p. 223). This chemical energy is stored in plant tissue; it causes plant growth and, because plants are ingested as food by animals, it also sustains all animal life, including humans. Even the fossil fuels—coal, petroleum, and natural gas—have chemical en-

Figure 9.2 (a) The surface of the planet Mars is gashed by erosional features apparently produced by enormous floods of water at some time in the remote past. Since Mars is almost airless, it is thought that the water was released by sudden melting of a layer of ice below the land surface, producing conspicuous "collapse topography" as well as erosional valleys. Both features are evident here, along with crustal fractures and a few craters produced by meteorite impacts. This image was made by a Viking Orbiter that reached Mars in 1976.

(b) This view from the Viking Lander shows the hostile nature of the lifeless, rock-strewn Martian landscape, which resembles the driest and most barren of the earth's deserts. The dominant red coloration results from the slow oxidation of iron-bearing minerals. This may signify a great span of time, or some unknown surface process. (NASA and Jet Propulsion Laboratory)

(a)

(b)

ergy derived from solar energy and concentrated in former plants and microscopic marine animals.

Physical geographers are concerned with the processes of the biosphere, including the ways in which solar energy and precipitation control biotic productivity, the resultant interactions among various life forms, how these interactions vary geographically, and how life forms have spread over the earth's surface and adapted to various habitats. These are all extensive topics, and we can give only the briefest introduction to them here. They are the subject matter of the subfield of *biogeography*.

PLANT AND ANIMAL INTERACTIONS

The preceding chapters have shown that the atmospheric circulation patterns can be thought of as global delivery systems for energy and moisture. In the next chapter we shall relate the patterns of natural vegetation types on the earth to associated climatic characteristics. We shall also see that plants and animals live in associations, or communities, that offer mutual benefits to the individual members. Animals rely directly or indirectly on plants for food (chemical energy), but they also help to propagate and spread plants. Squirrels not only eat nuts and acorns; they also bury them for future use, thereby planting future trees. Animals that eat fruit distribute the indigestible seeds in their fecal droppings. In the wet tropics even fish eat fallen fruit, convey the seeds, and thereby help to propagate forest trees. Flowers, with their perfumes and bright colors, can have evolved only to attract animals to the plants. Insects that seek flower nectar inadvertently transport the pollen that is essential for plant reproduction. Without such insects many kinds of plants could not survive. Clearly, plants have evolved mechanisms to take advantage of animal activity, and animals have evolved processes to utilize plant resources.

Ecosystems and Biomes

The interactions of plants and animals, and their relationships with their physical environment, give the communities definite patterns of organization. However, the task of tracing and understanding all of these interactions is extraordinarily difficult. For example, a typical midlatitude forest may contain fifty or more species of plants and thousands of species of animals, primarily insects. Each seems to play a role in the functioning of the community as a whole.

The study of the interactions among organisms with their particular habitat or environment is known as *ecology* (from the Greek *oikos*: house). One way to organize the character of these interactions is to focus on the basic ecological unit, which is known as an *ecosystem*. An ecosystem is a community of plants and animals generally in equilibrium with the inputs of energy and materials in a particular environment. Thus an ecosystem has both biotic (living) and abiotic (non-living) components. An ecosystem can be as small as a tidal pool or a single tree, or as large as a lake or a forest. Its boundaries may be either sharply defined or gradational, and it may include several different biotic communities. The concept is a flexible one, defined by function rather than scale.

An ecological system on the largest scale is known as a *biome*. Biomes are highly generalized ecosystem types recognized on a global scale, such as deserts, temperate grasslands, or tropical rainforests, with their faunas and associated physical characteristics. The concept of biomes includes broad climatically influenced associations of plants and animals, that can be portrayed on a world map. Every biome is a synthesis of many separate ecosystems, unified by some common characteristic such as the general morphology of the vegetation (e.g., forest, grass, or shrubs) and associated animal types. Biomes are designated in various ways and may be divided into subtypes. Chapter 10, which concerns global vegetation, is organized in terms of biomes.

Ecological Niches

Associated with the concept of the ecosystem is the idea of the *ecological niche*, which focuses on the specific ways an organism actually functions in its particular habitat. The ecological niche of a species is the combination of environmental factors under which the species can exploit a source of energy sufficiently to be able to survive, reproduce, and successfully colonize other similar habitats. The scale of an ecological niche varies with the organism under consideration. For bacteria it may be a microscopic pit on a rock surface. A single grizzly bear's niche might be the full extent of a sprawling mountain range and the surrounding lowlands.

It is interesting to note that organisms do not occupy all of the ecological niches to which they are well adapted. For example, there are no polar bears in the Antarctic region, despite the availability of appropriate climatic conditions and abundant food supplies in Antarctic waters. Conversely, penguins are not found in the northern hemisphere, though they should do well in the Arctic. Likewise, the large apes (chimpanzees and gorillas) of the equatorial forests of Africa are absent from similar environments in Central and South America, which have plentiful resources that these animals normally exploit. African giraffes would be at home in the Australian bush and the tree-studded llanos of Venezuela, but they are not there, and never have been. These are rather glaring large-scale examples of ecological niches that seem underutilized.

Plant and Animal Dispersal

Why some ecological niches are fully exploited and others are not is a matter of interest to biogeographers. The explanations involve the geographical locations of the centers of evolution of the organisms adapted to the niche, their mode of dispersal, competition from other organisms, and geographical barriers to their diffusion. These barriers include impassable topography, large water bodies, and unsuitable climates.

Many ecological niches are more effectively occupied by plants than by animals. Animals can colonize new areas rapidly, but must do so by their own locomotion, whereas plant dispersal takes place by means of winds, running water, ocean currents, and animals. Clearly, cold-adapted polar bears and penguins could

not cross the low latitude deserts and humid equatorial region to go from one hemisphere to the other, though water barriers would present no difficulty to them. In order for gorillas to migrate outward from the African forests, they would have to swim oceans or cross deserts lacking water and suitable food resources. But had they evolved some 200 million years ago in Triassic times rather than within the past 10 million years, gorillas could have wandered overland into the Americas, which for millions of years were connected to the African continent (Chapter 12). Indeed, the dinosaurs of the Triassic period were common to both the Old and New Worlds.

Some of the mysteries of plant and animal distributions are explained by the horizontal movements of the continents mentioned in Chapter 1 and discussed at more length in Chapter 12. The continents appear to have been in motion on the earth's surface since their initial formation: colliding and separating, opening up migration routes and then sundering them—permitting multitudes of organisms to spill outward from centers of evolution at certain times, and locking them in isolation at other times.

Australia, long isolated from other land masses, has the most primitive and unusual of all existing faunas, including the pouched marsupial animals epitomized by the kangaroo and duck-billed platypus (Figure 9.3). The marsupials evolved more than 100 million years ago in South America and somehow spread to Australia, perhaps by way of an ice-free Antarctica. But the highly diversified marsupials remaining in South America were unable to compete with the large numbers of more modern placental mammals entering from North America after the link at Panama was established only 2 or 3 million years ago. In isolated Australia the marsupials experienced no such competition and have lingered on to the present as a biological curiosity.

Changes of sea level also affect animal distributions. A lowered sea level can cause chains of islands to become connected and to link with a nearby continent, thus opening migration routes, whereas a heightened sea level can isolate land areas and islands, blocking animal dis-

Figure 9.3 The semi-aquatic duck-billed platypus is an extremely primitive mammal found in the isolation of Australia and Tasmania. A furred animal hatched from soft-shelled eggs, it has webbed feet and a leathery toothless muzzle like a duck's bill. (Douglass Baglin Photography).

Like the platypus, the kangaroo represents an ancient evolutionary line surviving only in locations remote from the centers of plant and animal evolution. The kangaroo, a marsupial, is born the size of a honeybee and must mature in a pouch on its mother's belly. (San Diego Zoo).

persal. The classic example of faunal discontinuities resulting from sea level fluctuations is the area of Malaysia, Indonesia, and Australia, illustrated in Figure 9.4.

Some scientific observers recognize various Biogeographical Realms, groupings which reflect the evolutionary centers and patterns of dispersal of specific plant or animal associations. These realms reveal the close proximity of the northern hemisphere land masses to each other over a long period of time, the barrier imposed by the subtropical deserts, and the isolation of certain landmasses and oceanic islands.

Biological realms contain related species. They are different from the biomes resulting from the *convergent evolution* of different species toward similar forms suited to particular habitats. Thus the very similar appearing plants and animals of the desert biome in various parts of the world are actually genetically unrelated species that have evolved virtually identical strategies to minimize heat stress and water loss. Such adaptations give the flora and fauna within the various geographic components of a global biome very similar appearances, though entirely different evolutionary tracks and Biogeographical Realms are represented.

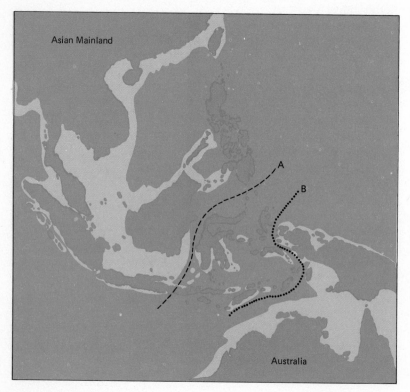

Figure 9.4 More than a century ago, Alfred Russel Wallace chose the islands of Southeast Asia as a key area in his attempt to explain the origin and distribution of species. Wallace was the first scientist to note that one association of plant and animal species was common to the Asian mainland and nearby islands west of line *A*, while another association of quite different species occurred south and east of line *B*. The light blue areas were all above sea level during the Pleistocene Ice Ages when sea level was more than 100 meters (330 feet) lower than at present. This permitted free movement of plants and animals between areas that are now large and small islands in the seas of Southeast Asia. However, the much deeper water present between lines *A* and *B* has remained a permanent barrier separating the two dissimilar biotic realms representing completely separate evolutionary streams. (After Time-Life Nature Library, *Tropical Asia*, 1969)

ENERGY FLOW THROUGH ECOSYSTEMS

The interactions within ecosystems can be analyzed in terms of energy flows and materials cycles. Solar energy enters and cascades through the ecosystem; it is partitioned among physical processes such as longwave radiation exchanges and evapotranspiration, and among biological processes such as photosynthesis and the production of chemical energy. Nutrients and other materials are continually recycled within the system.

Ecologists view the flow of energy in an ecosystem as a pyramid composed of several *trophic levels,* in which energy is stored in an organism that serves as food for an organism at a higher trophic level. Each successively higher level supports a smaller number and mass of organisms. There are various types of trophic levels. At the base of the pyramid are the *primary producers* consisting principally of green plants that can convert solar energy to organic energy in the form of living tissue. This in turn is usable as an energy source by other organisms, none of which can manufacture their own food. Certain bacteria are also primary producers, having the ability to use the chemical bonds of rock and soil minerals as energy sources. All other organisms are dependent directly or indirectly upon the primary producers (Figure 9.5). These other organisms are either *consumers,* which ingest live plant material or the prey that they or others have killed, or *decomposers,* such as bacteria, molds, and fungi, which make use of the energy stored in already dead plant and animal tissues. The animals that ingest plants are called *herbivores,* and these animals may in turn be ingested by other animals, called *carnivores.* Some animals, such as bears, blue jays, and humans, are both herbivorous and carnivorous.

Food Chains and Food Webs

A *food chain* is a sequence of consumption and represents energy transfer through the environment. There are two basic types of food chains. The *grazing food chain* begins with green plants and goes on to herbivores and then to carnivores. A grazing food chain can be symbolized as a linear relationship: Plant → herbivore → carnivore → second carnivore → top carnivore. The *decay food chain* begins with dead organic

Figure 9.5 (a) This schematic diagram illustrates the flow of energy in a natural ecosystem. Following the flow from left to right, solar radiant energy input is converted to food energy by green plants. The plants use some of the energy for respiration, herbivores consume some of it, and organisms of decay feed on some of it. Part of the energy in the herbivores is transferred to successive carnivores. However, the amount of transferred energy decreases at each step because energy is used for respiration and other purposes.

(b) This diagram indicates schematically how materials circulate between the principal units of an ecosystem and between the ecosystem and the environment. (After Whittaker, 1970)

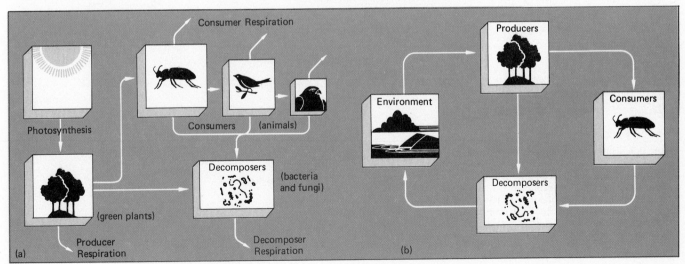

Figure 9.6 This diagram shows the principal members of a food web in a salt marsh near San Francisco, California, during winter. The energy producers are various salt marsh and marine plants, which grasshoppers, snails, fish, and marine invertebrates feed upon. Birds and rodents are intermediate carnivores, and the top carnivores of the web are hawks and owls. (After Smith, 1966, and Johnston, 1956)

matter and goes on to microorganisms and then to their predators, bacteria and fungi. These decomposers live within and on organic materials, especially dead tissues, breaking them down and returning minerals to the soil. Ordinarily, much less than half of the plant material on the continents is consumed directly by animals; the rest is recycled by decomposers.

Some food chains involve only a few links. In an agricultural ecosystem, for example, cattle eat grass and grain, and are in turn eaten by human carnivores. However, most feeding relationships in nature do not take the form of simple, isolated chains. Many food chains are interconnected, forming complex *food webs* (Figure 9.6). For this reason, the tracing of feeding relationships is not a simple task. In a midlatitude forest, there are numerous species of herbivores, each of which may feed on several plant species. Carnivores in the forest may also have a varied diet, feeding on herbivores and other carnivores.

One way to analyze a food chain is to take samples of all relevant animal species in the ecosystem and examine the contents of their digestive tracts. An owl's recent diet, for instance, is revealed by the bits of bone and hair in the pellet that the bird spits up every few days. The droppings of both herbivores and carnivores contain clear indications of their recent meals. The complexity of the food web is apparent when we consider that a few acres of grassland may harbor several hundred different species of insects and a dozen or more different kinds of vertebrate animals.

absorbed at one level is available to be transferred to the next feeding level. This is because no organism can convert the energy in the food it eats into an equal amount of stored energy. An adult person uses most of the energy obtained from food for body heat, motion, and work. Only a small amount of energy is stored for growth, and only the stored energy is available to a predator. Another reason for the low efficiency of the energy transfers along food chains is that the chemical reactions required for life are always accompanied by the transformation of energy into forms of heat that cannot be utilized. In addition, at each step in the grazing food chain, energy in the form of waste products is lost to decay food chains.

Consider a simple plant–herbivore–carnivore food chain consisting of grass plants, mice, and snakes. The mice obtain approximately ten percent of the energy absorbed earlier by the grass plants, and the snakes obtain approximately ten percent of the energy absorbed by the mice. Thus the carnivore receives only about one percent of the energy originally absorbed by the plants. The fraction of the original energy available to a succeeding carnivore stage—a hawk, for instance—is still less. The rapid decrease of available energy along a food chain limits such chains to four or five links. Large carnivores, such as lions, which are the last natural link in a food chain, obtain only a small fraction of the energy absorbed by the primary producers in their habitat. Lions must roam over large areas to obtain their food, and one region cannot support many of them.

Energy Flow Pyramids

Both field and laboratory studies have indicated that at each trophic level there is a significant loss of useful energy. Only a fraction of the organic production of one level becomes available as food at the next level, as indicated in Figure 9.7. Food chains in natural ecosystems exhibit approximately the same fractional transfer of useful energy as the laboratory ecosystems show. In a grazing food chain, for example, only about ten percent of the energy

ENERGY CAPTURE AND PLANT PRODUCTION

We can begin a closer study of the energy flow in an ecosystem by looking at the factors that determine how a primary producer captures and utilizes energy. We shall consider green plants, which support most terrestrial ecosystems. Of the solar radiation energy falling on a plant leaf, a small amount is immediately re-

Figure 9.7   As this energy flow diagram for the Silver Springs ecosystem shows, the energy input to the ecosystem is principally solar radiant energy, with a small input of organic matter that has entered the section of stream under study. Energy is expressed in kilocalories per sq meter per year. Most of the solar energy incident on the system is utilized for respiration and energy budget components, such as reflection, evapotranspiration, and heating the air and water. Only a small fraction is converted to food energy by various water plants. Some of the net production is absorbed by herbivores, some is exported in the form of plant matter that is carried downstream out of the system, and some is transferred to the chain of decay. The small energy loop in the decay chain represents the decomposition of decay organisms. Note the rapid decrease in energy transferred along the main food chain. The top carnivores receive only 0.6 percent of the energy absorbed by the herbivores. The *biomass pyramid* for Silver Springs shows that at any given time, most of the biomass of the ecosystem is in the form of primary producers, with only a small fraction in the form of herbivores and carnivores. The *energy flow pyramid* summarizes the energy flow data from the main diagram. (After H. Odum, 1957, and E. P. Odum, 1971)

flected and approximately 80 percent is absorbed, as Figure 9.8 illustrates. Some of the absorbed energy functions to warm the leaf and is then given off as longwave radiation, while much of the absorbed energy is used in the evaporation and transpiration of water stored in the plant. Only about one percent of the solar radiation is used in the process of photosynthesis to produce the chemical energy the plant requires for growth and maintenance.

The chemical processes involved in photosynthesis are complex. In terms of energy, however, photosynthesis may be thought of as a process in which simple molecules, water (H_2O) and carbon dioxide (CO_2), are joined with the aid of solar radiant energy to form more complex carbohydrate (sugar or starch) molecules (CH_2O):

$$H_2O + CO_2 + \text{solar energy} \rightarrow CH_2O + O_2$$

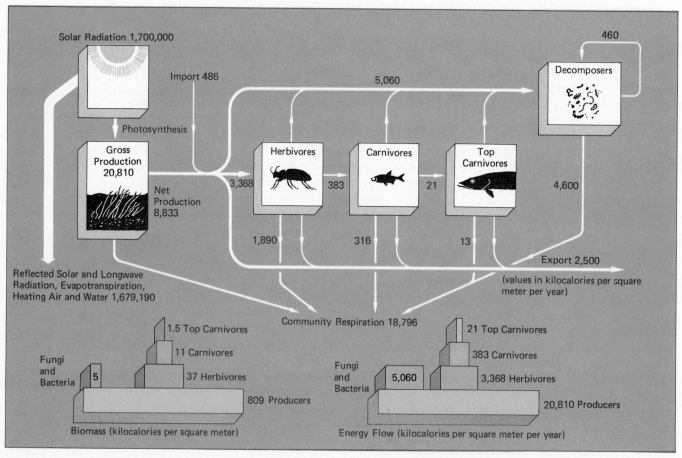

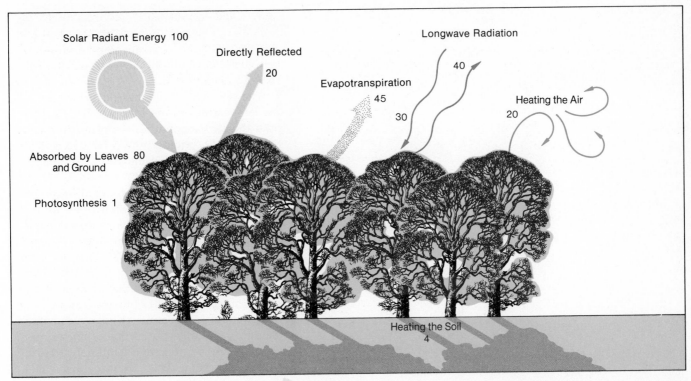

Figure 9.8 For every 100 units of solar radiation incident on a forest in a humid climatic region, only about 1 unit becomes stored chemical energy through photosynthesis. For this hypothetical midday situation, about 80 percent of the incoming solar radiation is absorbed by the vegetation and the ground. However, the forest crowns also receive longwave radiation redirected downward from the atmosphere. This diagram shows how the 130 units of energy received (100 solar radiant, 30 longwave) are utilized for various components of the local energy budget: reflection, 20 units; evapotranspiration, 45; longwave radiation, 40; heating the air, 20; heating the soil, 4; photosynthesis, 1.

The solar radiant energy is stored as chemical energy in carbohydrates. Further chemical reactions use the carbohydrates and nutrients from the soil to produce complex protein molecules that the plant's cells require for growth. In addition to synthesizing carbohydrates, photosynthesis produces free oxygen gas. The formation of oxygen is incidental to photosynthesis, but it is important to the ecosystem as a whole, as noted in Chapter 1. Without photosynthesis, there would be very little free oxygen in the atmosphere, and animal life as we know it could not exist (Figure 9.9).

Photosynthesis and Plant Growth

A significant question for agriculture is: How much chemical energy, or carbohydrate, can a plant produce photosynthetically for a given amount of incident solar energy? In 1926 Edgar Transeau measured the carbohydrate-producing efficiency of a cornfield. He recognized that all the carbon in a plant comes from the carbohydrate produced during photosynthesis,

and by measuring the amount of carbon in corn plants, he was able to estimate the amount of carbohydrate produced during the growing season. The amount of energy required to produce one kilogram of carbohydrate by photosynthesis had already been determined in laboratory studies, so Transeau was able to conclude that only 1.6 percent of the total solar energy incident on a field of corn during the 100-day growing season became stored chemical energy through photosynthesis. This percentage is the energy conversion efficiency of corn. No plants have energy conversion efficiencies of more than a few percent.

Photosynthesis is dependent on only the visible light portions of the total solar spectrum (see Figure 3.1, p. 46), and the rate of photo-

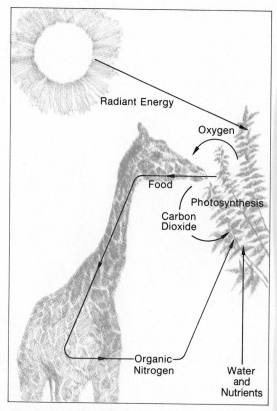

Figure 9.9 Photosynthesizing plants combine carbon dioxide from the air with water and nutrients from the soil, using radiant energy to produce stored food energy. Animals rely on plants for food, and they breathe the oxygen released as a by-product of photosynthesis. Carbon dioxide is returned to the atmosphere by the respiration and decay of plants and animals; nutrients such as organic nitrogen compounds also cycle between organisms and the environment. Photosynthesis thus plays a significant role in the cycles of water, oxygen, carbon, nitrogen, and nutrient minerals between the soil, plants, animals, and the atmosphere.

synthesis in a leaf varies with the intensity of the incoming light. If the intensity of light is increased, the rate of carbohydrate production will also increase, up to the maximum value for each plant species. Further increases of light intensity beyond this point will not result in increased photosynthesis. Leaves that are partially shaded, or that receive only indirect light, are able to carry on photosynthesis near the maximum rate. Because plants that have adapted to the tropics receive more solar energy, they generally have higher maximum rates of photosynthesis than plants native to midlatitude regions, where less solar energy is available. Graphs depicting the relationships between photosynthesis and solar radiation are shown in Figure 9.10.

The rate of photosynthesis depends also on the temperature of the leaf, which may differ from the air temperature. For plants in midlatitude regions, photosynthesis for a given light intensity reaches a maximum at a leaf temperature of about 25°C (77°F). For arctic plants, maximum photosynthesis occurs at lower leaf temperatures. The rate of carbohydrate production decreases above and below a plant's optimum leaf temperature, and production stops if leaf temperatures rise above 40°C (104°F) or so. If there is no wind to cool the leaves, leaf temperatures may rise so high that photosynthesis stops during the middle of the day, when solar energy input is maximum. Under such conditions, carbohydrate production is limited to a period in the morning and to a brief period in the late afternoon. It is partly for this reason that the warmest regions of the tropics tend to have lower agricultural yields than midlatitude regions.

A number of other factors also influence the rate of photosynthesis. Adequate supplies of water and carbon dioxide are necessary for efficient photosynthesis. The availability of nutrients from the soil, particularly nitrogen and phosphorus, also affects the rate of carbohydrate production. Nitrogen is required for the synthesis of the plant proteins necessary for cell growth. Phosphorus, a comparatively rare element in the earth's crust, is required in plants as they make chemical compounds important in the photosynthetic process. Phosphorus deficiency is often the limiting factor for plant growth in moist climatic regions, lakes, and coastal areas.

Plant Respiration

Not all of the carbohydrate a plant produces in photosynthesis is available to animals as food. The plant's own cells require a continuous supply of energy, day and night, for maintenance of its life systems. In the process of respiration, carbohydrate combines with oxygen and is reduced to carbon dioxide and water, releasing stored chemical energy for use by the living cells. Hence, respiration is a necessary process that provides energy to the plant to maintain its vital functions.

Photosynthesis stops in the absence of light or when leaves become too warm. Respiration, however, continues both night and day, although it is greater during the day when the temperature is higher and the leaves are exposed to light (Figure 9.11). Higher temperatures increase respiration, and as respiration rises, the amount of net energy stored by photosynthesis declines. Cool nights help plants conserve the chemical energy obtained during the day. Therefore, seasonal rates of respiration and the utilization of stored chemical energy in plants tend to be greater in climatic regions with high temperatures than in climatic regions with moderate or even cool temperatures.

One way to find the amount of energy a plant uses in respiration is to seal the plant in a dark box and measure the amount of carbon dioxide it produces by respiratory activity. Studies show that a wide variety of plants, including phytoplankton (floating microscopic green plants in water), use approximately one-fourth of their stored energy for respiration.

Measures of Plant Production

The rate at which a plant converts light to chemical energy is called its *gross productivity*. Since the plant uses some of the chemical energy for respiration, only a portion of the total chemical energy remains stored in plant tissues. The net rate at which a plant stores energy, exclusive of the energy it uses for respiration, is called the *net productivity*. Stated another way, net productivity equals gross productivity minus respiration over a period of time. Net productivity is the quantity important to the consumer of a plant because it represents the amount of organic material produced by the plant.

The *standing crop*, or *biomass*, represents the amount of energy stored in plants at any given time, expressed in grams of dry matter per

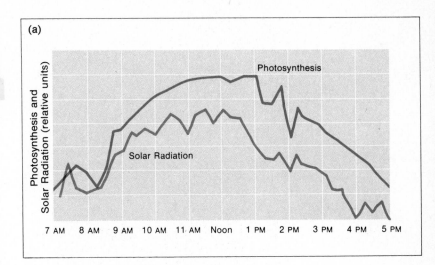

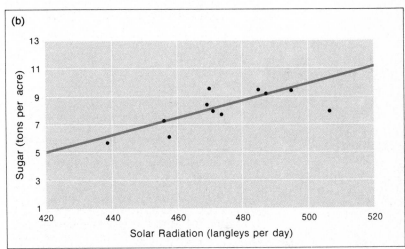

Figure 9.10 (a) The course of photosynthesis through the day rises and falls with the variations of incident solar energy.

(b) This graph shows the yield of sugar from a Hawaiian sugarcane plantation during different growing seasons. Each dot shows average daily solar radiation input and associated yield of sugar during a season. The solid line represents the general trend; it shows that yield is high during seasons when fair weather causes the average radiant energy input to be high and is low during cloudy seasons. (After Jen-hu Chang, 1968)

square meter of ground surface. Biomass should not be confused with net productivity, which is the rate at which a plant produces food. A mature, unlogged redwood forest has a large biomass but a comparatively low productivity. A large amount of energy is stored in the trees, but not much additional energy is added to storage each day. Phytoplankton in the ocean have a high productivity, but their

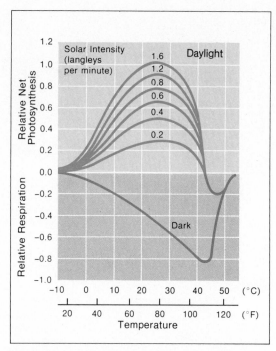

Figure 9.11 The upper portion of this graph shows the trend of net photosynthesis (that is, photosynthesis minus respiration) in a leaf for various leaf temperatures and various rates of solar energy input during daylight hours. Note that photosynthetic activity is small at both low and high temperatures and that it reaches a maximum at an intermediate temperature. For a leaf at a given temperature, photosynthesis increases with increased solar energy input, but not proportionately. Doubling the solar energy from 0.8 to 1.6 langleys per minute increases photosynthesis by only 30 percent or so.

The lower part of the graph shows the energy consumed by respiration at various temperatures for a leaf that is kept in the dark. (Note that in order to emphasize the relation of respiration to net photosynthesis, the lower graph shows respiration in negative quantities that increase *down* the scale.) Respiration increases with increased temperature until the leaf is too warm to function. High temperatures at night therefore promote the use of stored energy for respiration and tend to decrease the net productivity of the plant. (After Gates, 1968)

biomass is small because they are continuously being consumed by predators. *Yield* represents the amount of energy stored during the growing season in the desired portion of a crop, such as the fruit. It is usually determined by weighing the harvested portion of the crop.

The productivity of a particular plant is ex-

pressed in terms of the amount of energy stored during a given time interval, such as a day or a year. When the plant is exposed to light, photosynthesis and respiration both occur. Photosynthesis causes carbon dioxide to be absorbed at the same time that respiration causes carbon dioxide to be given off. The net rate at which the plant absorbs carbon dioxide is therefore a measure of its net productivity.

Because it is very difficult to carry out direct measurements of energy flow, productivity is usually measured indirectly. The net productivity of agricultural crops is often estimated by harvesting the complete plants, roots and all, at the end of the growing season. The plants are then dried and weighed, and the productivity is expressed in terms of the dry matter that a given area produces during a specified period of time. New productivity measured by the harvest method can be expressed as grams of dry matter per square meter of field per year, as in Table 9.1. When the harvest method is applied to natural plant communities, measurements of dry matter must be corrected for losses caused by the shedding of leaves, the depredations of insects, and the death of some plants during the growing season.

The energy content of a gram of dry matter differs among plant species, depending on the relative amounts of proteins and carbohydrates in the plant tissues. To correct for these differences, productivities expressed in terms of dry matter can be converted to energy units, as in Figure 9.12, by burning the dry matter in a device that measures the heat energy released by complete combustion.

Global Patterns of Plant Production

The average global net productivities of all important food crops, such as wheat, rice, corn, and potatoes, are of the order of 500 to 700 grams of dry matter per square meter per year. Surprisingly, mean agricultural productivity is no greater than that found in many "natural" ecosystems (Table 9.1). Productivities two or three times greater are attained in areas where intensive mechanized agriculture is practiced. The world average net productivity of sugar-

Table 9.1   Net Productivity of Major Ecosystems

Type of Ecosystem	Area (millions of sq km)*	Net Productivity per Unit Area (dry grams per square meter per year)† NORMAL RANGE	MEAN	World Net Productivity (billions of dry tons per year)
Tropical Forest	24.5	1,000–3,500	2,000	49.4
Temperate Forest	12.0	600–2,500	1,250	14.9
Boreal Forest	12.0	400–2,000	800	9.6
Woodland and Shrubland	8.5	250–1,200	700	6.0
Savanna	15.0	200–2,000	900	13.5
Temperate Grassland	9.0	200–1,500	600	5.4
Tundra and Alpine	8.0	10–400	140	1.1
Desert and Semidesert	42.0	0–250	40	1.7
Cultivated Land	14.0	100–3,500	650	9.1
Swamp and Marsh	2.0	800–3,500	2,000	4.0
Lake and Stream	2.0	100–1,500	250	0.5
Total Continent	149.0		773	115.0
Open Ocean	332.0	2–400	125	41.5
Continental Shelf, Upwelling	27.0	200–1,000	360	9.8
Algal Beds, Reefs, Estuaries	2.0	500–4,000	1,800	3.7
Total Marine	361.0		152	55.0
World Total	510.0		333	170.0

* One square kilometer is equal to about 0.39 square mile.

† One gram per square meter is equal to about 0.0033 ounce per square foot.

Source: Whittaker, Robert H., 1975. *Communities and Ecosystems*, 2nd ed. New York: Macmillan.

Figure 9.12   This diagram compares the productivity of various environments in terms of energy units (kilocalories per sq meter; 1 kilocalorie equals 1,000 calories) rather than as dry matter produced. The deep oceans and deserts, which cover approximately 80 percent of the earth's surface, have low productivities; the deep oceans lack nutrients, and the deserts lack moisture. The most productive areas include tidal estuaries. If all agricultural areas had a productivity as high as that of estuaries, the annual agricultural productivity of the earth would approach 10^{18} kilocalories. The present world productivity is estimated to be 10^{14} kilocalories per year. (After E. P. Odum, 1971)

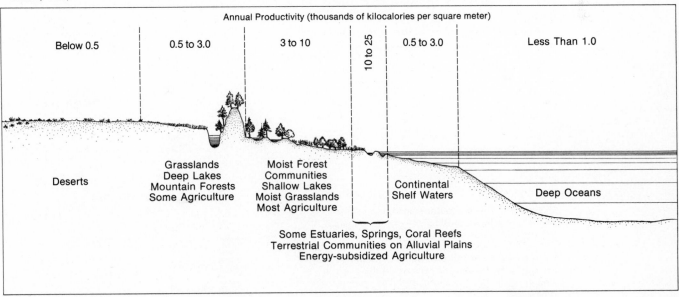

cane is high—about 1,500 grams per square meter per year. Productivities of more than 9,000 grams have been measured for sugarcane in some tropical regions, where the growing season lasts the entire year and the soil is moist and rich in nutrients.

Estimates of the mean annual net productivity of the land areas are shown in map form in Figure 9.13. The map data are estimated by means of a climatic model of the relationships among climate elements and gross and net productivity; therefore, the map patterns of net productivity are similar to the global maps of climatic regions in Chapter 8. The map patterns are also relatively similar to the net productivity data in Table 9.1.

The map and table show that productivity is greatest in equatorial and tropical regions where moisture availability meets the demands of *PE*. As both the *PE* and moisture availability decrease, so does net productivity. Because of the large areal extent and high productivity rates of tropical and midlatitude ecosystems, these regions account for more than half of the total productivity of the continents. In turn, the average annual productivity of the continents is four to five times that of the oceans, even though the oceans comprise more than two-thirds of the surface area of the globe. The most productive regions of the oceans are the shallow, sunlit waters of the continental shelves, and especially those river estuaries where nutrient-rich streamflow mixes with ocean waters across broad expanses of brackish and saline marshes. Such areas are too small to show on a world map.

NATURAL AND AGRICULTURAL ECOSYSTEMS

Despite the difficulties, the energy flows in several ecosystems have been studied in detail. The results of these studies are often summarized conveniently in energy flow diagrams, which show the main energy pathways through the systems.

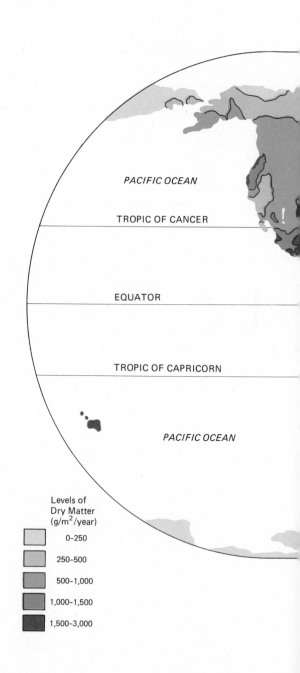

Figure 9.13 This computer map shows estimates of annual net primary productivity of the land areas of the earth in grams of dry matter per meter² per year. The estimates are based on relationships of productivity to temperature and precipitation, so the patterns are similar to the climatic regions on the maps in Chapter 8. (E. Box, *Radiation and Environmental Biophysics*, vol. 15, 1978)

Levels of
Dry Matter
(g/m²/year)

	0–250
	250–500
	500–1,000
	1,000–1,500
	1,500–3,000

Natural Ecosystems

The energy flow diagram in Figure 9.5a is that of a typical ecosystem. The diagram illustrates how the energy input to each stage in a food chain is divided among various energy flows and outputs. The principal outputs include the energy used in life processes, or respiration; the energy transferred to the next stage of the food chain; the energy transferred to the decay food chain; and any energy utilized for other purposes. In a pond, for example, the energy in plant and animal materials that sink into the bottom sediment is exported from the system. One of the classic energy flow studies of a natural ecosystem was carried out at Silver Springs, Florida, by the American ecologist Howard Odum and his colleagues. Their results are summarized in the energy flow diagram of

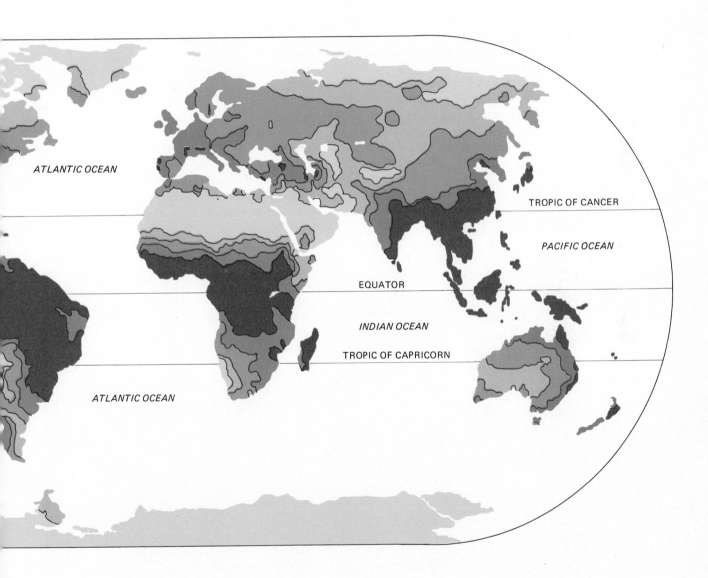

ATLANTIC OCEAN

TROPIC OF CANCER

PACIFIC OCEAN

EQUATOR

INDIAN OCEAN

TROPIC OF CAPRICORN

ATLANTIC OCEAN

Figure 9.7, where energy is expressed in kilocalories per square meter per year.

In the Silver Springs ecosystem, radiant energy from the sun provides almost all the energy input to the producer forms of life in the spring. A small additional contribution of energy is imported through plant and animal debris that enters the system. Only a small fraction of the absorbed radiant energy is employed in photosynthesis.

There are four links in the Silver Springs grazing chain: plant, herbivore, carnivore, and top carnivore. The significant reduction of energy from stage to stage is evident in the flow diagram. Note that the decay chain is somewhat more important than the grazing chain as regards amounts of energy transferred: the energy input to the herbivores is 3,368 kilocalories, but the energy passed to the decay chain is 5,060 kilocalories. Plant and animal materials

moving downstream out of the region of study account for the exported energy flow.

When plant communities have achieved a stable, comparatively unchanging system called *climax vegetation*, very little energy is used for the production of new plant material. The energy released by respiration and by the decay of dead plant and animal tissues balances the energy stored in new growth, so that the total energy stored in a climax forest remains effectively constant. Although an individual tree in a climax forest uses energy for growth, the forest ecosystem as a whole uses essentially all the input energy for respiration. Researchers who studied the energy flow for an oak and pine forest near the Brookhaven National Laboratory on Long Island, New York, found that of the annual gross production of dry matter per square meter, only about one-fifth was stored as new growth, litter, and humus. The major part of the gross production was used in respiration (Figure 9.14). Respiration accounted for more than 80 percent of the energy used, which indicates that the Brookhaven forest is in a late stage of succession toward an idealized stable climax condition.

Figure 9.14 This diagram traces the energy flow in an oak and pine forest on Long Island, New York, in units of grams per sq meter per year. Less than 20 percent of the gross productivity is used for net forest production, indicating that the forest is nearly at the climax stage, the point at which most of the new growth in the forest is balanced by the death and decay of old growth. The gross productivity in this forest is used primarily for the respiration of plants, animals, and decomposers. (After Woodwell and Whittaker, 1968)

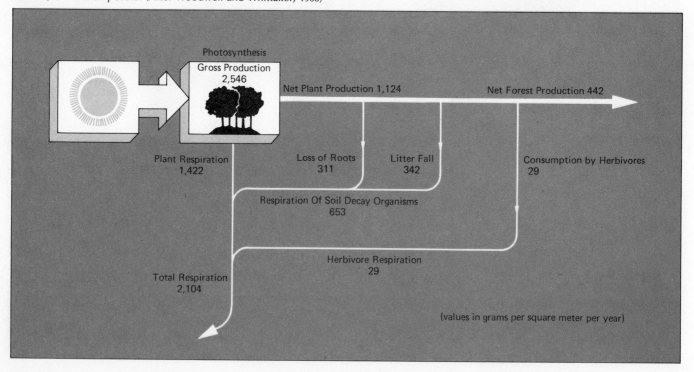

Agricultural Ecosystems and Yields

Early human societies followed a hunting and gathering mode of existence, but the eventual domestication of animals and plants represented more effective ways of assuring an available food supply. Both patterns persist today, and in both it is the humans who are the top carnivores. However, neither a hunting nor a grazing economy can support a dense population of top carnivores, because so much useful energy is lost at each stage of the long food chains involved. Societies that depend on hunting or grazing are therefore small in populations that are widely dispersed.

The cultivation of selected plant species in agriculture is more efficient than hunting or grazing because the food chain is much shorter. In a carefully managed agricultural ecosystem, humans often play the part of herbivore, although some agricultural output becomes food for domestic animals. Although people need proteins in their diets, only members of affluent societies can afford the cost in energy of eating meat every day—meat is at the end of a long food chain. In the United States, much of the energy available from cereal grains is consumed indirectly as meat because grain is the food of livestock. The average American consumes directly or indirectly more than four times the amount of grain available to the average person in developing countries.

The efficiency of agricultural ecosystems has been improved remarkably in the past hundred years or so. In the Middle Ages in Europe, a return from a cereal crop of eight or ten grains for every grain planted was considered good. Today the average return from most cornfields in the United States is greater than 500 grains for each grain planted. There are several reasons for the marked improvement in food production. The principal factors are highly specialized and controlled ecosystems including the development of improved and specialized plant varieties; the use of irrigation; the extensive use of fertilizers, weed-killers (herbicides), and insecticides; and the increased mechanization of agriculture by utilization of a large energy input from fossil fuels.

Figure 9.15b shows that from 1800 to 1940, corn yields in the United States averaged 25

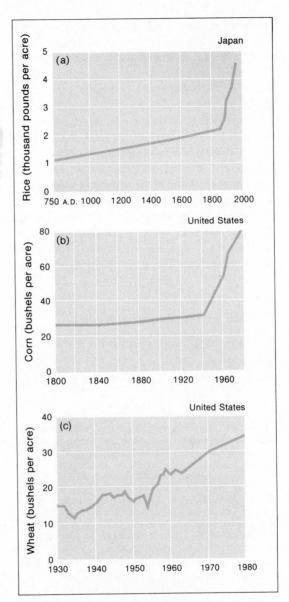

Figure 9.15 Estimated (a) rice yields in Japan and (b) corn yields in the United States have shown rapid increases in the past few decades because more productive species have been developed and because efficient, energy-intensive agricultural practices have been used. Such a rate of increase cannot be sustained indefinitely for a given crop.

(c) Since 1950, wheat yields in the United States have not increased as rapidly as corn yields; nevertheless, wheat yields have increased between 2 and 4 percent per year.

bushels per acre. Average present-day yields exceed 80 bushels per acre. A primary reason for the improvement was the introduction of high-yield varieties, principally hybrid corn. The recent development of highly productive strains of wheat and rice for use in tropical and subtropical regions has contributed greatly to the food supplies available to countries such as India. The new strains mature early and respond well to fertilizer applications. They are also relatively insensitive to seasonal variations in the duration of daylight, so a new crop can be planted at almost any time during the year. This characteristic allows three crops a year to be harvested regularly in tropical countries, which increases the annual productivity of the land.

Climate and Agricultural Yields

A plant's basic requirements include solar radiant energy, carbon dioxide from the atmosphere, water, and nutrients extracted in solution from the rooting zone of the soil. Insufficiency of any of these factors inhibits a plant's growth. Normally, each plant is adapted to the regional solar energy regime, but unusual spells of cold weather or freezes can retard plant growth or "kill back" tender vegetation that cannot tolerate freezing temperatures. Of course, factors other than climate, such as disease or insects, sometimes reduce plant growth drastically.

Because evapotranspiration, photosynthesis, and absorption of nutrients from the soil all depend on soil moisture availability, components of the local water budget are climatic indexes of crop growth and yield. An example of the general relationship between crop yields and evapotranspiration is shown in Figure 9.16a; once evapotranspiration exceeds an initial base level of 46 to 51 cm (18 to 20 in.) per year, crop yields tend to increase with greater evapotranspiration. The relationship is not simple, however. Figure 9.16b shows how moisture deficits affect yields of sugarcane. When soil moisture is low during the growing season, the yield is low, and when soil moisture is high, the yield is high. However, the dashed lines show that if a drought occurs early in the growing season during the period of active growth, the yield declines, but if the same degree of drought occurs late in the season, the yield increases. This seasonal drought relationship is characteristic of many but not all crop plants.

Figure 9.16 (a) This graph shows the relationship between annual evapotranspiration and the yield of several crops grown in evapotranspirometers near Columbia, South Carolina. Significant yields begin only after evapotranspiration is greater than 46 cm (18 in.).

(b) This graph illustrates the relationship between soil moisture availability and sugarcane yields in Hawaii. The curve shows that moisture deficits early in the growing season tend to decrease yields, but that deficits late in the growing season tend to increase yields. (After Jen-hu Chang, 1968)

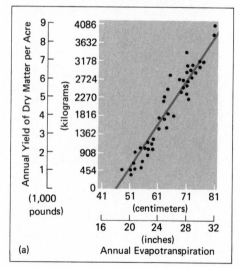

(a)

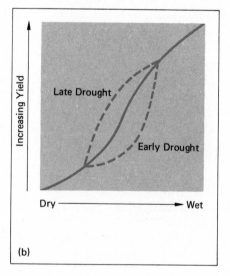

(b)

PROBLEMS OF WORLD AGRICULTURE

An agricultural ecosystem is not stable in the way that a climax forest is, a fact well-known to anyone who tries to maintain a beautiful green lawn. Humans have artificially stabilized agricultural ecosystems by supplying them with large amounts of external energy. In industrialized nations, the stored energy of fossil fuels drives the machinery required to increase agricultural productivity. So the true energy input for food production includes not only radiant energy but also the cost of building tractors, producing fertilizer and pesticides, training agricultural field agents, and so on. In highly mechanized agriculture, humans may supply more than one kilocalorie of energy for each kilocalorie of food energy produced.

The rapid increases in agricultural yields that have occurred in the past few decades cannot be expected to continue indefinitely because there are natural limits to agricultural performance. A plant may respond well to moderate applications of fertilizer, but applying ten times as much fertilizer will not increase the yield tenfold. Although further improvements in productivity will undoubtedly be made, the yield of a given plant species cannot be increased indefinitely.

The climatic limits set by photosynthesis also affect yields. In recent years, the total amount of land devoted to agriculture in the world has declined because some of the land is unsuitable for raising crops. The 100 million acres of virgin land in Central Asia put to agricultural use by the Soviet Union are often too dry to produce good harvests, so much of this land has been released from agricultural activity.

Industrialized nations, such as Japan and the United States, now grow more food on less agricultural land than they did 30 years ago. It appears that as advanced agricultural methods are introduced into less developed countries, agricultural production can be increased. Many social and economic problems are involved, however. There are social and political upheavals when a society changes from subsistence farming, in which one farm produces barely enough to support its own workers, to modern productive agriculture, in which one farm produces surpluses for other sectors of society. Fewer farm workers are needed, and those forced to leave the land to seek new employment usually migrate to cities. Marketing and transportation systems spring up to distribute farm products. Modern agriculture requires costly equipment and supplies, and the new methods necessitate intensive educational programs for farmers.

It is expected that the population of the world will continue to grow beyond the end of this century. Agricultural production will not be able to keep pace with such growth indefinitely. The food shortage in the Soviet Union and a number of other countries after the poor harvests of the early 1970s shows that global food production is not adequate during years of severe climatic stress. Figure 9.17 shows that world grain reserves have recovered somewhat from the dangerously low levels of the early to middle 1970s. But the complex interrelationships of climatic variation and national economies can threaten global world grain reserves at almost any time. The problem at hand is to feed people adequately until stable population sizes are attained, and to find an economic means to redistribute food among the many rich and poor nations.

Humans have always interacted with natural ecosystems and have created artificial ecosystems to fit their needs. Throughout most of history, and in many countries today, our goal in dealing with nature has been to obtain the materials and foods that are immediately necessary for survival. Sometimes, however, these activities have resulted in making survival more difficult.

A vivid example is the human catastrophe centered in the semiarid transition zone between the Sahara Desert and the tropical forests of Africa, known as the Sahel. In the 1960s an unusual period of moist years there had encouraged agriculturalists, who had already shifted from subsistence food crops to commercial crops, to expand their cropland into formerly dry areas. At the same time, wells were drilled that encouraged owners of cattle to double their herds. When the atypical wet period

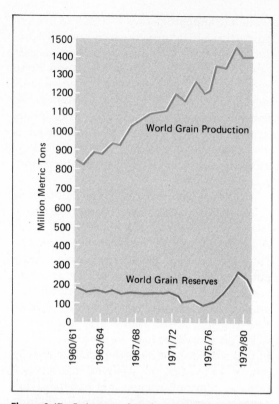

Figure 9.17 Estimates of total world grain production beginning with the marketing year 1960–1961 are shown by the upper curve. The upward trend is due largely to technology rather than increasing acreage, but the variability in the 1970s has been associated mostly with climatic stress, such as drought in the Soviet Union in 1972. The lower curve shows world grain reserves, which decreased substantially during the early 1970s because of population growth. Reserves during the middle 1970s fell to only about 10 percent of annual world production, but reserves increased rapidly in the late 1970s because of favorable climatic conditions. (Modified from *World Agricultural Situation and Agricultural Outlook*, U.S. Dept. of Agriculture, 1976–1982)

ended in 1968, to be succeeded by sustained drought, local economies collapsed. Agriculture failed, and pastures that had been overstocked during the wet period could no longer support the livestock population. Across the breadth of sub-Saharan Africa some 5 million cattle died of starvation, under a pall of dust. As many as 100,000 people perished. The surviving human population was reduced to a pathetic condition, and became dependent upon international relief organizations (Figure 9.18). Many trekked

out of the region, which had become a "dust bowl," and became refugees in already overcrowded urban centers to the south.

The Sahel crisis, which even now has not truly ended, provoked much scientific interest in the phenomenon of "desertification"—the expansion of deserts due to human disturbance of delicate natural balances in the semiarid zone. All subsequent studies concur in the verdict that every historic example of desertification has been more a consequence of human errors than of natural causes.

Despite such examples, our interaction with nature has been comparatively successful. The proportion of people who are well clothed and well fed today is large in comparison with earlier ages—and even past decades. It is becoming increasingly clear, however, that human activities in the drive for survival have important repercussions in nature. As we attempt to turn a greater part of the earth's productivity to our own uses, we must avoid acting in ways that will make that productivity less available to us in the future.

Another case in point pertains to the exploitation of tropical forests. The traditional method of agriculture in the rainforests of Central and South America and Southeast Asia has been to clear a section of forest by cutting and burning, to farm the land for a few years, and then to move on to a new section. The people do this because in a rainforest the nutrients are stored in the natural vegetation rather than in the soil, so after a few years cleared land becomes less productive.

This *slash-and-burn, swidden,* or *milpa,* cultivation was carried on for thousands of years with no apparent long-term ill effects, because population densities were low and the forest was allowed time to reestablish itself in a region before clearing was repeated. Today, however, increased population pressures and the need for food have led people to shorten the period between successive clearings. Furthermore, the cleared land is often used to graze cattle after it has been farmed for a few years. Grazing destroys secondary growth and prevents the reestablishment of the forest. The cattle pack the earth to a nearly impervious layer that seriously

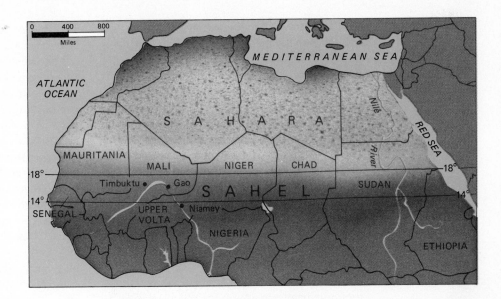

Figure 9.18 The drought in the African Sahel region, shown on the map, so reduced forage for livestock that several million cattle died of starvation. These cattle supported many populous nomadic tribes who maintained the animals for their yield of milk and blood. During the drought, which became a crisis in 1973, animals near death from hunger were slaughtered and their meat stripped and sold, as in this view near Niamey, Niger. Here meat is being stripped by the buyers, while a group of the nomads looks on. (UPI)

diminishes infiltration of rainfall and increases runoff and erosion. Eventually the soil becomes so depleted that even edible grasses disappear, to be replaced by coarse unpalatable grasses.

The accelerated pace of swidden cultivation and cattle pasturing may make large tracts of land in the rainforests economically useless and difficult to restore to productivity.

The search for food has caused many countries to turn to the sea. The sea as a whole is not a highly productive region, as indicated in Figure 9.12. Photosynthesis is confined to the sunlit upper 50 to 100 meters (160 to 330 ft) of the ocean, and life in this layer is usually limited by lack of nutrients in the water. An acre of midlatitude grassland is several times more productive than an acre of ocean. The net production of the ocean is less than half the net production of the land, and much of this production is concentrated in areas of upwelling (Chapter 4), which comprise less than one percent of the ocean surface.

The strategy of many nations is to maximize the investment made in modern fishing fleets and methods. Often the pressure to intensify fishing increases as the catch declines. For this reason, the oceans are being overfished, and many species, including some types of whales, have almost disappeared. In 1935, commercial fishermen in California landed 60,000 tons of Pacific sardines. The catches decreased with time, and by 1961 the sardine industry of California no longer existed. Similar overfishing is evident as regards Pacific perch, mackerel, and tuna.

The best long-term strategy for an intelligent predator is to leave enough animals so that good use can continue to be made of the available food. However, the desire of many nations for immediate increases in their living standards becomes more important than fear of the inevitable decline in ocean productivity.

The misuse of the Sahel and rainforests and the overfishing of the oceans are only three examples that indicate how immediate benefits can be accompanied by large hidden losses and costs. The farmers and the fishing fleets gain access to an inexpensive source of energy, but restoring the ecosystems that produced the energy will eventually be costly to society as a whole. Because the true costs of human actions are seldom considered, we may eventually suffer the dire consequences of bankrupt ecosystems. With the rapid increase in world population, humans are far from being in a stable long-term relationship with natural and agricultural ecosystems.

SUMMARY

The earth is the only known planet having the appropriate temperature and moisture conditions to support life. In the earth's biosphere various organisms interact with one another and with their physical habitats to form complex ecosystems. The largest-scale ecosystems are the global biomes that can be portrayed on a world map. Each organism has an appropriate ecological niche from which it can derive energy enabling it to survive and reproduce. Not all ecological niches are fully utilized. This is a result of the history of geographic diffusion of plants and animal species from their centers of evolution; the diffusion was affected by past climates, biotic competition, and geographic barriers, some of which relate to the changing positions of the continents.

Life is organized in food chains that can be separated into different trophic levels. The primary producers, dominantly plants, form the base of the terrestrial ecological pyramid, with consumers and decomposers at higher levels. Animal consumers consist of plant-eating herbivores and animal-eating carnivores. Energy flows through ecosystems can be evaluated in terms of transfers of energy along food chains and webs. Because of complex interactions, only about ten percent of the energy absorbed at one stage of a food chain is available to the succeeding stage, which severely restricts the population density of carnivores at the highest level of the chain.

Primary production on the land begins with the photosynthesis process, by which green plants make use of carbon dioxide from the air, water, nutrients in solution from the soil, and solar radiation, to produce complex molecules of plant carbohydrates (sugars and starches). Only a very small proportion of incident solar radiation is utilized directly; usually much less than five percent is converted to stored chemical energy.

Photosynthesis is restricted to daylight hours, but plants use energy day and night in respiration to maintain life processes. The net rate at which a plant stores energy is measured as photosynthesis minus respiration and is called its net productivity. Productivity tends to be greatest in tropical rainforests and to decrease toward drier and colder climatic regions. In water budget terms, productivity is related directly to evapotranspiration. Oceanic productivity is generally low except for regions of cold water upwelling and the continental shelves, especially estuaries, where nutrient-laden freshwater runoff mixes with ocean waters. The productivity of agricultural ecosystems is about equal to that of the more productive natural systems, but more of the agricultural productivity is available, directly or indirectly, for human consumption.

Population growth has exerted great pressure on the agricultural lands of the earth. Until early in this century, agricultural yields had remained almost constant for hundreds of years. Recently, yields of many grains and other crops have increased spectacularly, but the energy input in terms of fossil fuels has been very costly. Agricultural lands will have to be managed carefully and wisely if food production is to keep pace with population growth and the rising expectations of people everywhere. Examples of poor management are seen in the Sahel famine of 1973 and increasing pressure on the resources of tropical forests and oceans.

REVIEW QUESTIONS

1. Compare and contrast the diurnal cycles of photosynthesis and respiration.
2. Describe the global patterns of net productivity.
3. Compare the productivities of oceanic and continental ecosystems.
4. How is plant growth related to water budget components?
5. Discuss the various ways plants and animals interact within an ecosystem.
6. How is photosynthesis related to the energy budget of a forest stand?
7. Describe energy flows through food chains.
8. Discuss factors that affect global food production and consumption.

APPLICATIONS

1. What are the species names for your official state plant and animal? Classify each in terms of genus, family, order, class, and phylum. At each level indicate some related organisms that illustrate the degree of diversity within each category.
2. Are there important unresolved biogeographical questions pertaining to your area? Examples might concern the time or mode of arrival of particular species of plants or animals, or the effects of changing climates on species distributions.
3. Referring to Shelford's *Ecology of North America* or a similar source, determine what natural biome existed in the area of your home or campus. How important are the diagnostic plants and animals of the natural biome currently? How might you label the human-dominated biome existing in the area at present?
4. Describe several ecosystems on your campus. How are these present ecosystems different from "natural" ecosystems at the same sites before the time of European settlement?
5. How would you try to document ecosystem changes associated with climatic variation since 1970?
6. Some cultural geographers have specialized in the study of the domestication of wild plants and animals and their diffusion to other regions. Where are the primary regions of domestications, and how is the diffusion of plants and animals related to global patterns of climates?
7. Describe three very different food chains found in ecosystems characteristic of your state.
8. What can you learn about the productivity of land in your state? Which areas are generally believed to have the highest and lowest productivities respectively?
9. Agricultural yield data for significant crops are normally published for each county. Map the variation of the yield by counties of a significant crop in your state, and speculate about the reasons for the variation from one region to another.

FURTHER READING

Bennett, Charles F., Jr. *Man and Earth's Ecosystems: An Introduction to the Geography of Human Modification of the Earth.* New York: Wiley (1975), 331 pp. This unique text is organized by world regions in order to focus on the ecological impacts of human use of the earth. The book specifically analyzes the geographical and historical background of the environmental crisis.

Eckardt, F. E., ed. *Functioning of Terrestrial Ecosystems at the Primary Production Level.* Proceedings of the Copenhagen Symposium, July 1965. UNESCO (1968), 516 pp. This collection of technical papers by some of the world's leading scholars includes several studies that are especially useful at the introductory level.

Golley, Frank B., and **Ernesto Medina, eds.** *Tropical Ecological Systems: Trends in Terrestrial and Aquatic Research.* New York: Springer-Verlag (1975), 398 pp. This is Volume II of a series which focuses on analysis and synthesis of ecological research. It includes a wide range of papers on tropical environments, which probably present the most difficult resource-management dilemmas.

National Academy of Sciences. *Productivity of World Ecosystems.* Washington, D.C. (1975), 166 pp. This short book is a collection of papers presented at a symposium held in Seattle during 1972. The papers, written at beginning and intermediate levels, analyze the world's major ecosystems.

Odum, Eugene P. *Fundamentals of Ecology,* 3rd ed., Philadelphia: W. B. Saunders Co. (1971), 574 pp. This is a comprehensive work by an outstanding American ecologist.

————. *Ecology: The Link Between the Natural and Social Sciences,* 2nd ed. New York: Holt, Rinehart and Winston (1975), 244 pp. This paperback introduces ecological principles. More emphasis is placed on humans interactions with ecosystems than in the preceding work.

Phillipson, John. *Ecological Energetics.* New York: St. Martin's Press (1966), 57 pp. This very small book is often considered a classic; most of the basic themes of ecology are introduced succinctly.

Whittaker, Robert H. *Communities and Ecosystems,* 2nd ed. New York: Macmillan (1975), 385 pp. This is another outstanding paperback on ecology, especially pertinent to this chapter.

Asteroids, Overkill, and Species Extinctions

Two of the most debated questions in the life sciences concern the pace of evolution of new species and the extinction of older species that once were important or even dominant on the earth's surface or in its waters. In both instances the controversies are between scientists who believe the processes consist of gradual changes and those who favor more dramatic explanations—sudden "spurts" in the evolution of species or catastrophic events leading to the simultaneous extinctions of large groups of organisms. Interestingly, the seemingly related questions of evolution and extinction grow out of two entirely independent issues debated by different groups of scientists.

While both controversies are very lively, the focus recently has been on the causes of animal extinctions on the lands and in the seas. Almost everyone is familiar with the question of the causes of the seemingly sudden disappearance of the dinosaurs, which had been the dominant animals on the earth for 150 million years. These great reptiles suddenly vanished from the fossil record about 65 million years ago, to be replaced by the ancestors of modern mammals. In fact, some 50 to 75 percent of the plant and animal life of the Cretaceous period, from giant reptiles on the land and in the seas to land plants and microscopic marine organisms, seem to have disappeared at about this time, ushering in the *Cenozoic* era (from the Greek for "new life")—also called the *Tertiary*, meaning the third era of life.

What rejuvenated the old argu-ment about the Cretaceous/Tertiary mass extinction was the 1973 discovery of a peculiar fossil-free clay layer overlying fossiliferous Cretaceous limestones. The find was made in Italy by a research team led by geologist Walter Alvarez of the University of California at Berkeley. Analysis of the clay revealed a surprisingly high content of the rare element iridium—about 30 times the amount normally expected. Luis Alvarez, Walter's father and a Nobel Laureate in physics, concluded that the iridium, which is uncommon in the earth's crust but fairly abundant in outer space and in meteorites, could have been emplaced by an extraterrestrial event such as a nearby supernova (star explosion) or the impact of a meteorite or a comet. The subsequent discovery of the iridium anomaly at the Cretaceous/Tertiary boundary in dozens of locations in all parts of the world seemed to indicate a major event with global effects.

The most likely possibility appeared to the two Alvarezes to be the impact of an asteroid 5 to 15 km (3 to 8 miles) in diameter. The impact of an asteroid of this size would be expected to throw up a massive amount of pulverized rock, including fine dust that would reach the stratosphere, where it would form a veil blocking solar radiation for several years. This would halt photosynthesis all over the earth's surface, causing plant life to die off and terrestrial and marine food chains to collapse, resulting in the extinction of many species.

Numerical modeling of the earth's radiation budget indicates that the postulated dust veil could have caused the continents to cool as much as 40°C (70°F) within two to five months after the impact of an asteroid, although heat transfer from the oceans might have mitigated this effect somewhat, especially in coastal zones. Nevertheless, large animals that were not cold-adapted could not survive reduced food and light and severely depressed temperatures. Only the smaller burrowing ancestors of the Cenozoic mammals were equipped for survival under the drastically altered surface conditions.

Kenneth Hsu of the Geological Institute in Zurich, Switzerland, has suggested a different possibility—that a gaseous comet struck the earth. Such a collision could heat the earth's atmosphere to a temperature greater than large terrestrial animals could tolerate. It would also release cometary CO_2 and cyanide into the oceans. The CO_2 would make ocean waters more acidic and decrease the depth at which the calcium carbonate of marine microorganisms would dissolve, while the cyanide would be poisonous to marine life.

These frightening scenarios are vigorously attacked by scientists who do not believe that Cretaceous life vanished quite as abruptly as is often suggested. These investigators propose that the anomalous iridium may have been concentrated by some organic process, since iridium in nonmarine sediments seems concentrated in coal or rocks rich in organic debris. Additionally, they claim

CASE STUDY

Two giant animals whose extinction remains a puzzle: **(top)** a pleistocene mastodon and **(bottom)** a mesozoic carniverous dinosaur.

that the iridium anomaly may have occurred either well before or well after changes in life forms at a location. Opponents of the extraterrestrial theory also tend to scoff at the possibility of finding the effects of a single instantaneous event represented in rocks all around a 4.6-billion-year-old planet. They stress that the late Cretaceous era was a time of change, with oceans receding, accentuated volcanic activity, and unusual climatic instability. According to this group, the pattern of extinctions was complex, ecologically controlled, variable from place to place, and not synchronous in time. Proponents of the astronomical theory have responses to all of these points. And so the argument continues, with members of the same university departments often on opposite sides.

A typical Clovis spear point of the type used by Paleo-Indian hunters at the end of the Pleistocene. Stone spear points are found among the bones of many of the large mammals that disappeared from North America at the end of the Ice Ages.

Not all problems concerning the extinction of species are as remote in time as the age of the dinosaurs. Only some 10,000 to 12,000 years ago several of the most characteristic of the Ice Age mammals of North America suddenly disappeared. What happened to them?

The fossil record in North America shows an abundant Pleistocene fauna of large grazing animals, including primitive horses and camels, giant bison, elephants (mammoths), and mastodons. All of these vanished abruptly at the end of the Ice Age. Archaeologists investigating early human artifacts in the New World have found stone spear points associated with the bones of many of these animals. Were the Pleistocene fauna slaughtered to extinction by accomplished big-game hunters who entered North America near the end of the Ice Age?

The question of the cause of the extinction of a large portion of the North American fauna at the end of the Pleistocene Epoch is bound up with arguments concerning the time of arrival of humans in the New World. Paul Martin of the University of Arizona believes that the first humans moved into Alaska from Asia about 13,000 years ago and only then began peopling the New World. Martin believes that the Paleo-Indian immigrants from Asia slaughtered the Pleistocene megafauna as they advanced and multiplied, using in particular the deadly 3- to 5-inch-long, fluted "Clovis" spear point, found nowhere outside of North America, and no longer in use among Indians at the time of European contact.

According to Martin, the Paleo-Indian hunters, entering an entirely unexploited ecosystem of 26 million sq km (10 million square miles), swept over it and extinguished 31 genera of large mammals between Alaska and southern Argentina in a mere 1,000 years. Martin's computations make this appear quite possible. In Eurasia, however, Pleistocene hunters with a history extending back hundreds of thousands of years were able to eliminate only four genera of large mammals: the mammoth, musk ox, woolly rhinoceros, and giant deer. The discrepancy, according to Martin, proceeds from the fact that the Paleo-Indian hunters of the Americas found their prey pathetically vulnerable owing to the animals' lack of wariness of humans, whom they had never previously encountered! Only the more solitary and cryptic (nocturnal and burrowing) species escaped destruction.

However, a growing number of scholars believe that humans have been on the scene in the Americas for 25,000 to 50,000 years, or even longer, evolving and perfecting their deadly hunting tools and techniques over a much lengthier span of time than that proposed by Martin. This would have permitted slower expansion, perhaps less easy slaughter, but the same result—animal extinction on a scale not imaginable under natural conditions, unless we invoke a cosmic disaster as has been done in the case of the dinosaurs.

Studio di cielo e albieri (Study of Sky and Trees) by John Constable. (Art Resource)

Where both moisture and solar energy are abundant the earth's land surfaces are cloaked with a living blanket of green vegetation. This phenomenon, unique in our solar system, varies remarkably in form and behavior in response to our planet's diverse climates and surface characteristics. The appearance of plant life on the continents made possible the emergence of all other terrestrial life forms, whose food chains are ultimately based on vegetal matter.

When land plants evolved on earth, some 400 million years ago, they soon spread to all parts of the planet except those regions covered by glacial ice. To do so they had to evolve adaptations to every combination of energy, moisture, and soil conditions found on the earth. This had to be accomplished through genetic variation and natural selection. Although these adaptations have caused plant life to vary widely in form, from algae to giant forest trees, most of the higher plants (excluding algae, fungi, and lichens) have certain similarities. Most have a root system to gather chemical nutrients dissolved in soil moisture, green leaves to convert solar energy into chemical energy for plant growth, and stems or branches to support the leaves and to channel the nutrients and chemical energy throughout the plant.

These basic features must vary greatly to be effective over a wide range of environmental conditions. The depth and spread of roots reflect the supply of moisture and nutrients. Leaves may be large or small depending upon light levels and moisture stress. Leaf form varies in response to environmental constraints. Leaves are retained all year by *evergreen* plants and are dropped seasonally by *deciduous* plants to prevent frost damage or moisture loss when resupply is impossible. Under conditions of severe moisture stress plants may do without leaves altogether, with stems and branches taking over the function of photosynthesis, as in the case of the cactus family. Stems too are adapted to local conditions and take many forms. In later pages we shall see how these adaptations assist plants in specific environments.

In terms of their relationship to moisture supplies, plants may be classified as *hydrophytes*, which grow in water, *xerophytes*, which are structurally adapted to survive in extremely dry soils, and *mesophytes*, which grow where the water supply is neither scant nor excessive. Likewise, plants may be classified as *perennial*, persisting from year to year and enduring seasonal climatic fluctuations, and *annual*, dying off during periods of temperature or moisture stress but leaving behind a crop of seeds to

10
Vegetation and Climate

Ecology of Vegetation
Plant Communities
Plant Succession and Climax

Principal Vegetation Formations and Their Climates
Tropical Rainforests
Mixed Tropical Forest and Shrubland
Tropical Savannas
Deserts
Chaparral
Midlatitude Forests
Midlatitude Grasslands
Northern Coniferous Forests
Tundra
Highland Vegetation

germinate during the next favorable period for growth.

By their many adaptations, plants have developed the variety allowing them to grow and reproduce from the edge of wind-whipped snowfields to black-water swamps and sun-baked deserts. This chapter first summarizes some of the factors that are responsible for the development of natural vegetation types found in varying environments. Following this, the general characteristics of the principal global vegetation formations and their associated climates are described. It must be kept in mind that the natural vegetation that has evolved in equilibrium with undisturbed environments has been vastly altered by human activities. These activities have caused the virtual disappearance of some natural vegetation formations and threaten the extinction of others in the not-too-distant future.

ECOLOGY OF VEGETATION

In Chapter 9 the emphasis was on ecological systems, with attention to the energetics of photosynthesis, respiration, growth, and ecosystem productivity. In the initial portion of the present chapter we shall consider the sequential development, or succession, of plant communities with the passage of time. The communities themselves are the subject of the majority of the chapter, which is principally a description of the major natural vegetation realms and their climates.

Plant Communities

The associated plant species that form the natural vegetation of any one place are known as a *plant community*. In any midlatitude forest, for example, many kinds of trees, shrubs, ferns, grasses, and flowering herbs all live together in one plant community. Numerous genetically unrelated species of plants not only live in association with one another, but also with bacteria, fungi, insects, and burrowing, seed-collecting, grass- and herb-eating (grazing), and twig- and leaf-nibbling (browsing) animals. Besides providing food and shelter for animals, a plant community affects its local environment by modifying soils and moisture storage conditions, by shading the ground, and by cooling the air through the process of transpiration and latent heat transfer. In such ways, individual plants also affect one another, so that the plant community behaves as an ecological system composed of many interacting parts.

Each species in a plant community has its own particular way of utilizing energy and moisture. In a broadleaf forest, ferns are able to use the subdued light that filters through the high leafy canopy. They do not compete with trees for direct sunlight. Mosses, lichens, and fungi growing on rocks utilize a moisture supply that is different from that of ferns growing on the forest floor. Because of this principle, farmers in the tropics often successfully grow several dissimilar crops—such as bananas, a tall open plant, and cassava, a low root crop—in the same field. As long as the crops do not compete for the same moisture and sunlight, the crop yield per acre can be greater than if the fields were planted in a single crop.

Plant Succession and Climax

Hikers walking through the New England woods sometimes come upon the foundation stones of farm buildings that were abandoned more than a hundred years ago. Although the land was once cleared for agriculture, it is covered by forest again. But the forest is not immediately reestablished when cleared land is abandoned. Instead, a *succession* of plant communities occurs. Each succeeding community alters the local microclimate and surface condition, making possible the appearance of a community that is more demanding in terms of nutrients and specific soil and moisture conditions. The usual trend in plant succession is toward taller, more diverse, more permanent vegetation. Eventually a stable community is attained, and there is no further change in its composition.

The particular order in which plant com-

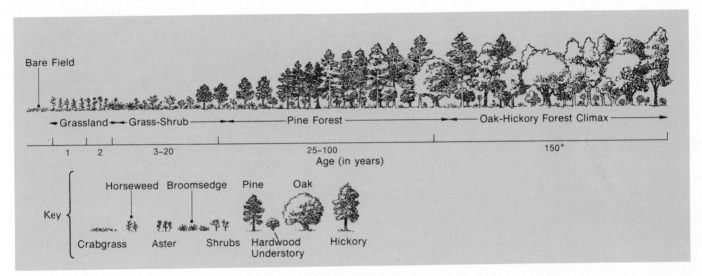

Bare Field

←Grassland→←Grass-Shrub————→←————Pine Forest————→←——Oak-Hickory Forest Climax——→

| 1 | 2 | 3–20 | 25–100 | 150⁺ |

Age (in years)

Key {

Horseweed   Broomsedge   Pine   Oak

Crabgrass   Aster   Shrubs   Hardwood   Hickory
                                Understory

Figure 10.1 This diagram illustrates a typical plant succession from open land to forest in the middle latitudes. In the initial stages of the succession, low, fast-growing grasses and shrubs dominate. Each stage alters the soil and micro-climate of the land, enabling other species to establish themselves and become dominant. The final stage consists of high trees that overshade and force out some of the low shrubs of earlier stages. (John Dawson after E. P. Odum, *Fundamentals of Ecology*, 3rd ed., © 1971, W. B. Saunders Co.)

munities succeed one another depends on whether the succession begins on cleared land, filled-in marsh, burned-over forest, or some other condition. It also depends on the climate, and on the particular stable plant community characteristic of the region as a whole. In a succession that eventually converts cleared land to forest, the early communities are dominated by grasses and low shrubs, as shown in Figure 10.1. In time, the shrubs increase in number, and small-statured trees appear. These, in turn, are replaced by larger trees. As more advanced communities become established, the pace of succession slows because the larger individual plants mature later and have longer lives. The larger trees of deciduous forests may live several hundred years and may not produce seeds during their first two decades.

The stable plant community that results at the end of a long undisturbed succession is called the *climax vegetation*. It may take hundreds or even thousands of years to establish a climax vegetation. Environmental change may disturb the climax once it has developed, initiating a renewed successional sequence, or a true climax may never become established. Such disturbances include fires, floods, severe droughts, hard freezes, insect or parasite infestations, and freezing rain that breaks down ice-laden trees. Pleistocene glaciation and associated climatic changes produced world-wide disturbance of plant communities. Stable communities have not yet been reestablished over large areas in the middle latitudes. More recently, human activities, especially logging, agriculture, and livestock grazing, have resulted in interruption of the plant succession or in complete replacement of the climax vegetation.

PRINCIPAL VEGETATION FORMATIONS AND THEIR CLIMATES

Vegetation that has evolved in response to the average climatic and soil conditions of a region, as well as its occasional temperature and moisture extremes, is called *natural vegetation*. As Figure 10.2 shows, energy availability is crucial in determining natural vegetation at higher latitudes, whereas moisture supply is more important at lower latitudes. In subhumid climates where the Thornthwaite moisture index

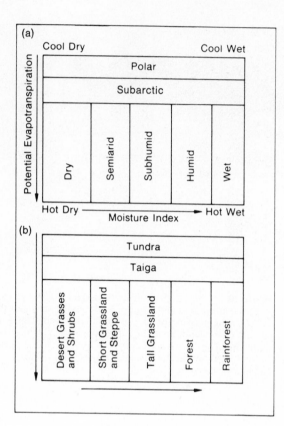

Figure 10.2 (a) The schematic diagram shows the principal climatic regions according to the Thornthwaite system of classification. (After Blumenstock and Thornthwaite, 1941)

(b) The principal vegetation regions are diagrammed according to the same *PE* and moisture index scales used to show Thornthwaite's climates. At higher latitudes, energy availability determines the presence of tundra or forest (taiga); at middle and lower latitudes, the vegetation is determined primarily by the availability of moisture. (Doug Armstrong adapted from Blumenstock and Thornthwaite, "Climate and the World Pattern," *Yearbook of Agriculture*, 1941, Government Printing Office)

Figure 10.3 This figure shows relationships between *PE*, the moisture index, and selected natural vegetation in North America and the tropics. Not all of the vegetation types discussed later in the chapter are included. The figure shows that the distribution of natural vegetation is quite similar to the simple theoretical scheme in Figure 10.2b and that there is some overlap among the types. (Vantage Art, Inc. after John R. Mather and Gary A. Yoshioka, "Role of Climate in the Distribution of Vegetation," *Annals of the Association of American Geographers*, vol. 58, 1968)

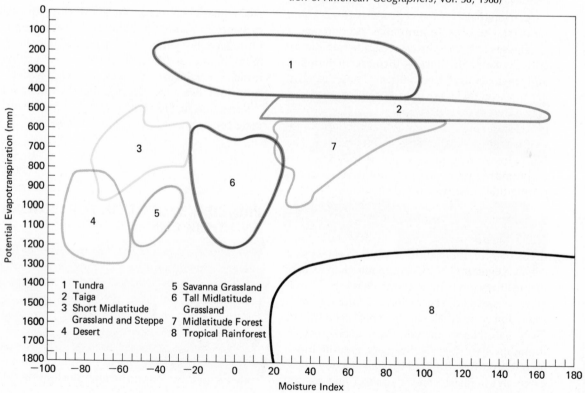

1 Tundra
2 Taiga
3 Short Midlatitude Grassland and Steppe
4 Desert
5 Savanna Grassland
6 Tall Midlatitude Grassland
7 Midlatitude Forest
8 Tropical Rainforest

is near zero, grasslands are predominant (Figure 10.3). Where the moisture index is somewhat greater than zero, the natural vegetation is usually forest. In polar and subarctic climates, where solar energy input is more critical than moisture, the natural vegetation is tundra or subarctic needleleaf evergreen forests.

As noted in Chapter 9, vegetation in widely separated regions with similar climates has evolved very similar characteristics. Many species of the New World cactus family and the Old World euphorbia family are so alike in appearance that only an expert can distinguish between them, yet these species result from completely separate streams of evolution. Their common characteristics allow species of both families to thrive in desert climates that exclude most other types of plants. This type of convergent evolution of different plant families toward similar forms is especially noticeable in climates having temperature or moisture extremes that require special plant adaptations for survival.

A global map of natural vegetation types is shown in Figure 10.4. This simplified map does not include natural and artificial disturbances and variations in terrain and soil conditions. Indeed, there is very little "undisturbed" natural vegetation on the continents, even in sparsely populated regions. In Figure 10.4 the very complex distribution of plant communities has been simplified into nine general types, which are commonly recognized as biomes (Chapter 9) that include not only vegetation associations but related animals as well. Regional boundaries should be regarded as zones

Figure 10.4 This very generalized global map of the distribution of major natural vegetation types should be compared with the global map of climates (Figure 8.3, pp. 190–191). Climate, soils, and vegetation are strongly interacting systems, and their global distributions bear marked similarities. (John Odam and Andy Lucas after Vernon C. Finch and Glenn T. Trewertha, *Physical Elements of Geography*, © 1949, McGraw-Hill Book Co.)

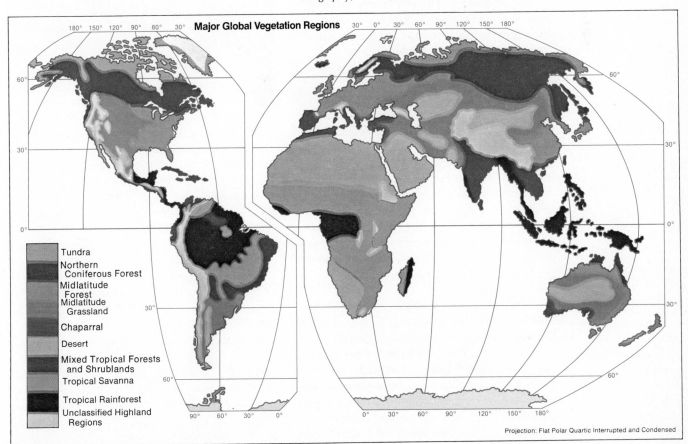

Major Global Vegetation Regions

Tundra
Northern Coniferous Forest
Midlatitude Forest
Midlatitude Grassland
Chaparral
Desert
Mixed Tropical Forests and Shrublands
Tropical Savanna
Tropical Rainforest
Unclassified Highland Regions

Projection: Flat Polar Quartic Interrupted and Condensed

of intermixture between neighboring vegetation formations.

Many of the major vegetation regions shown in Figure 10.4 and described in the following pages correspond to climatic regions discussed in Chapter 8. Here we shall outline the nature of the relationships between climate and the vegetation formations on the earth, beginning at the equator and progressing poleward.

Tropical Rainforests

The tropical rainforests of the Amazon Basin, equatorial Africa, and the peninsulas and islands of Southeast Asia are the vegetative response to a climate that exerts virtually no constraints on plant growth—a climate that provides abundant energy and moisture in every month of the year. The resulting vegetation is dominated by tall trees that manufacture and shed leaves throughout the year. Below the uppermost canopy of leaves, smaller trees have large leaves that permit photosynthesis at reduced light levels. The large leaf mass produces high regional rates of evapotranspiration.

The absence of climatic constraints is one reason that the number of plant and animal species per unit of area reaches its maximum in the tropical rainforest. Another reason is that plant and animal reproduction occurs all year long, increasing the opportunity for genetic variation. Here competition among species and predation of one species upon another are the chief problems for all organisms. With no winter freezes, insects are present in incredible numbers and multiply the year around, consuming enormous amounts of plant material and providing vast amounts of food for insect predators, such as birds, bats, and ground-dwelling animals.

Plant species are adapted to differing light intensities and reach varying heights. The tall tree species form a comparatively dense canopy more than 30 meters (100 ft) above the ground. A few even taller light-seeking trees protrude above the canopy here and there. A dense jungle of smaller trees and vines commonly makes a wall along sunlit riverbanks and around forest clearings, but the true rainforest is fairly open

at ground level, with little underbrush because of the reduced light at the forest floor. Exceptions occur where large and small ferns utilize the dim light and abundant moisture near ground level.

Tropical rainforests usually have an abundance of vine-like *lianas* that twine upward around the trunks of large and small trees in their quest for sunlight. Arboreal plants, known as *epiphytes,* form clumps and rosettes along tree branches (Figure 10.5). They contain their own fauna of insects, frogs, and crustaceans, and derive their nutrients from the decay of organic material before it reaches the ground. Some epiphytes are *stranglers* that send aerial roots downward to extract water and nutrients from the forest floor. In time they may completely envelop and replace the host tree.

The major limitation in the environment of the wet tropics is the poverty of the soils, which usually are deficient in mineral nutrients. In Chapter 11 we shall see that the heavy rainfall of the wet tropics dissolves soluble mineral matter in the soil and flushes (leaches) it away, leaving behind a very infertile residue in which plants must grow. The most fertile part of the soil is the surface layer, which receives nutrients from the plant litter (leaves, branches, fruit, etc.). The roots of rainforest trees are generally shallow because of this nutrient distribution and because roots will not penetrate soils that are perpetually waterlogged. Many roots do not enter the soil at all, but creep under the decaying litter on the forest floor, extracting nutrients directly from the organic debris. In all rainforests certain tall tree species have buttressed trunks that flare widely at the base to support a shallow-rooted, massive structure reaching high toward the sunlight.

It may be hard to believe that the survival of the once vast rainforests of the wet tropics is not at all certain, but such is the case. Human activities such as logging and clearing for agriculture and cattle pasturing are destroying the rainforests at an alarming pace. Reestablishment of a cleared rainforest may be next to impossible because the soil deteriorates rapidly as soon as the forest cover is removed (Chapter 11).

In tropical rainforest areas, most of the pre-

Figure 10.5 These three views illustrate the coastal rainforest of Brazil. At the top left, epiphytes with aerial roots cling to the trunk of a large tree. At the top right, a host tree (rough bark) supports a large climbing vine that sends out aerial roots. When these dangling roots strike the soil they develop into stout woody stems that gradually enclose the host tree. Experiments show that climbing vines first grow laterally along the ground toward trees that *block* the light. Meeting a vertical surface, the vines then climb upward *toward* the light. The bottom photo shows a portion of the flared base of a tree with a buttressed trunk, a mechanism that compensates for roots that are shallow because of waterlogged soil or lack of nutrients below the surface layer of the soil. (Robert Voeks)

cipitation falls as heavy showers and thunderstorms, between late morning and early evening, when solar heating at the surface makes the humid tropical air most unstable. Clusters of wet days are sometimes followed by days with only scattered showers. The wetter periods are associated with the westward passage of weak low-pressure disturbances within the trade-wind systems and intertropical convergence zone.

Air temperatures vary little through the seasons because midday solar altitudes are high throughout the year and polar air masses and fronts are absent. For example, the average monthly temperature at Singapore, located at latitude 1°N, shows little variation from season to season, and the rainfall exceeds 15 centimeters (6 in.) in every month (Figure 10.6). Such a climate can be described by a *thermoisopleth* diagram (Figure 10.7), which shows average air temperatures throughout the day for each day of the year. The pattern of the temperature contour lines, or *isotherms*, on a thermoisopleth diagram reflects the climatic characteristics. When the isotherms on the graph are primarily horizontal, temperature variations during a day are greater than variations from season to season. The thermoisopleth diagram for Belèm, Brazil (Figure 10.7), located in a tropical rainforest near the mouth of the Amazon River, shows that the temperature changes by 5°C (9°F) or more during each day but varies by no more than 2°C (3.6°F) at any given hour through the year.

The local water budget can be used to provide another perspective on climate. The water budgets presented in this chapter are based on a model of soil-moisture availability in which the rate of removal of soil moisture by evapotranspiration diminishes as the amount of soil moisture decreases. Thus, both a utilization of soil moisture and a moisture deficit may occur in a given month. Figure 10.8 gives the water budget for Kribi, Cameroon, located in the western section of the tropical rainforest region of Africa. This water budget shows no deficits, on the average, and large surpluses nearly every month. The large moisture surpluses are clearly reflected in the richness of plant and animal life in the tropical rainforest regions.

Mixed Tropical Forest and Shrubland

Despite the overall wetness of the climate, large portions of the tropics experience a short dry season. This is largely the result of seasonal migration of the intertropical convergence zone (ITC). The vegetation of such regions must adapt to moisture deficits during these brief dry seasons. Many of the trees are deciduous, shedding their leaves for a month or two when soil moisture is unavailable for evapotranspiration (Figure 10.9). The canopy of this forest is lower and more open than that of the rainforest, so that more light penetrates to the ground. As a result there may be dense thickets of low-growing shrubby vegetation. Where the dry season is accentuated, the forest gives way to low thorny trees and shrubs with small hard-surfaced leaves that resist heat stress and water loss. Figure 10.4 and Figure 8.5 (p. 192) show that this vegetation type corresponds to Köppen's tropical monsoon (*Am*) and portions of the tropical wet and dry (*Aw* and low-latitude *BS*) climates. Mixed tropical forest and shrubland vegetation is widespread in the monsoon regions of Southeast Asia and India, with smaller areas in Central and South America.

Tropical Savannas

A *savanna* is a tropical grassland, generally with scattered trees (Figure 10.10). Although not all savannas correspond to a single climatic region, most are located within the tropical wet and dry climatic type, approximately equivalent to Köppen's *Aw* climate and low-latitude areas of *BS* climate. Thus, the savannas lie between the tropical rainforests and the deserts centered near 30°N and S latitudes. Extensive savannas are located north and south of the rainforests of both the Zaire (Congo) Basin in Africa and the Amazon Basin in South America. Summer in the savannas is equivalent to the wet season and winter is the dry season. The rainy summer season is associated with the ITC, which invades the region during the high-sun period. The dry season results from subsidence of air in the subtropical highs that cover the region during the period when the sun's rays are most oblique.

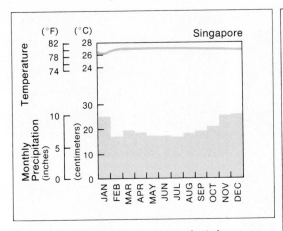

Figure 10.6 Singapore is in a tropical rainforest region (Köppen *Af* climate). As the graph shows, the average monthly temperature and precipitation are high and nearly constant through the year. (Doug Armstrong after H. Nelson, *Climatic Data for Representative Stations of the World*, © 1968, by permission of University of Nebraska Press)

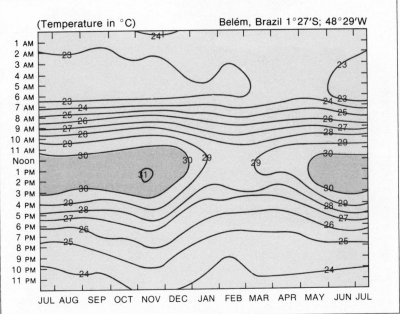

Figure 10.7 Belém, near the mouth of the Amazon River, is located in a tropical rainforest (Köppen *Af* climate). This thermoisopleth for Belém gives the average hourly temperature through a year. The scale of months begins with July because Belém is in the southern hemisphere, so January and February are summer months. The noon temperature through the year, which can be read by tracing across the diagram from left to right, is between 30° and 31°C (86° and 88°F) until December, when it falls 1° or 2°. The diagram can also be used to trace the temperature through a given day. On January 1, for example, the temperature after midnight remains at about 23°C (73°F) until the early morning, then rises to nearly 30°C (86°F) at midday. The temperature falls again in the evening. In a tropical rain forest, the temperature range through a day is greater than the range of average monthly temperature through the year. (Doug Armstrong after C. Troll, *Oriental Geographer*, vol. 2, 1958, by permission)

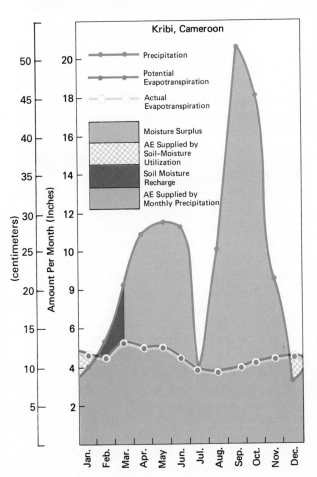

Figure 10.8 Kribi, Cameroon, is located at 3°N latitude in the tropical rainforest of Africa (Köppen *Af* climate). Monthly *PE* is high throughout the year, but during most months precipitation is even greater and large surpluses are generated. The ITC and associated showery precipitation tend to follow the seasonal migration of the sun; thus the two very rainy seasons at Kribi occur when the sun is overhead at noon near the equator. (Doug Armstrong after D. Carter and J. Mather, *Publications in Climatology*, vol. 19, 1966)

Figure 10.9 These photographs of an ancient Maya site in Honduras show a mixed tropical forest during the wet season and the dry season. (James S. Packer)

Figure 10.10 The grass savanna in Amboseli Park in Kenya, Africa, is an open grassland dotted with acacia trees. (John Lewis Stage/Photo Researchers, Inc.)

Savannas vary from open woodland with a ground cover of grass to open grassland with isolated trees. Most of the trees shed their leaves during the dry season. Savanna trees commonly have thick fire-resistant bark and small drought-resistant leaves. During the long winter drought the grass dies off above the surface, but the root systems survive and send up new shoots when moisture conditions improve.

Most savanna vegetation occupies flat plains. Many of these plains are ancient land surfaces whose soils are relatively infertile because the heavy rains of the high-sun season deplete them of soluble nutrients, often causing them to be capped by an iron-rich residue or crust several meters in thickness (Chapter 11). Some savannas, then, may be *edaphic* features—related to soil conditions that exclude forests rather than to climate alone. The absence of forest also results from widespread grass fires that sweep over the plains in the dry season, destroying seedling trees. Human use of savanna land for livestock pasturing also tends to prevent forests from becoming established.

The African savannas are famed for their enormous herds of hoofed animals, including zebras, giraffes, buffalo, and many varieties of gazelle—always followed by predators such as lions, cheetahs, and, of course, humans. This host of short and tall creatures is collectively equipped to chew at everything from the lowest herbs to the leaves and twigs of the spreading tree tops. Strangely, no such collection of herbivores has survived in the extensive savanna regions of South America, though it is present in the fossil record for the area. The great African herds of game animals seem doomed, due to hunting and rapid agricultural encroachment, and may shortly be restricted to a few protected reservations.

Savanna regions may have an annual precipitation as high as 150 centimeters (60 in.) and as low as 50 centimeters (20 in.). Figure 10.11 shows monthly temperature and precipitation data at Cuiabá, Brazil, located in a tropical savanna region at latitude 16°S. As in tropical rainforests, the temperature at Cuiabá exhibits little seasonal variation. However, the rainfall greatly diminishes during the low-sun season from May through September. Figure 10.12 shows the water budget for Caracas, Venezuela, located at latitude 10°N in the savanna (or *llanos*) of Venezuela. The annual precipitation at Caracas is 80 centimeters (32 in). The water budget shows some deficiency of moisture during the winter months, but soil moisture is never completely exhausted.

Deserts

In Chapter 8 we saw that dry climates occur systematically in several geographic settings. In the subtropical latitudes, deserts are produced

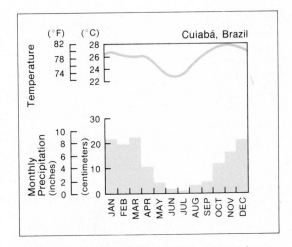

Figure 10.11 Cuiabá, in the tropical savanna of western Brazil, has a tropical wet and dry climate (Köppen *Aw* type). The graphs show a dip in average monthly temperature and precipitation during the lower sun months of May through September. Precipitation is nearly zero during June and July. (Doug Armstrong after H. Nelson, *Climatic Data for Representative Stations of the World*, © 1968, by permission of University of Nebraska Press)

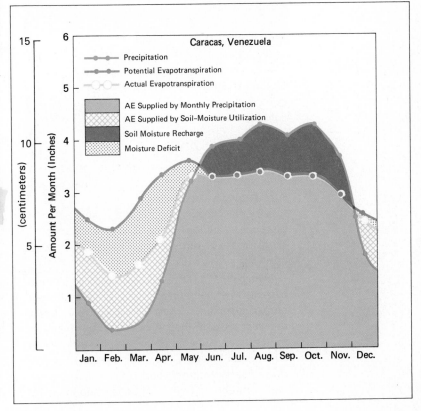

Figure 10.12 Caracas, Venezuela, has a tropical wet and dry climate (Köppen *Aw* type). The average annual precipitation is somewhat less than the annual potential evapotranspiration, but as the water budget shows, the seasonality of precipitation causes moisture deficits to occur during the months of December through May. Soil moisture stores are recharged during the remaining months. This budget indicates that no rainfall is left over for runoff or streamflow. (Doug Armstrong after D. Carter and J. Mather, *Publications in Climatology*, vol. 19, 1966)

by subsidence of air on the eastern sides of the subtropical highs. In the middle latitudes, deserts occur in continental interior locations that are distant from oceanic moisture sources. Deserts also occur along coasts bathed by cold upwelling waters and in rainshadow areas on the lee sides of mountain systems where air descends and is warmed adiabatically.

The vegetation cover developed in areas of dry climates consists of some plants that are drought-resistant and some that are drought-evading. Drought-resistant species have specialized adaptations that allow them to conserve moisture during long periods without rain. Drought-evading plants either germinate and complete their life cycles quickly during those brief periods when rain dampens desert soils, or they drop their leaves and become dormant until water is again available.

The drought-resisting plants are mainly shrubs. These are widely-spaced (Figure 10.13), with extensive root systems to gather moisture, small and often waxy-surfaced leaves to conserve moisture, and, in some cases, tissues that can store moisture. In a few instances there are no true leaves at all, but only enlarged green stems that take over the function of leaves in the photosynthesis process. The giant saguaro cactus of southern Arizona (Figure 10.13) is the prime example. Since the moist, fleshy tissue of cacti is attractive to animals, most cacti are defended with sharp spines. Some desert shrubs, such as the creosote bush and sagebrush, send their roots many meters downward in search of moisture. Others, such as the saguaro cactus and Joshua tree, have shallow wide-spreading roots that take advantage of the surface moisture resulting from light rain. Some remarkable shrubs can undergo almost complete dehydration without injury, becoming "resurrected" from a dormant state only when rains fall.

Much of the drought-evading vegetation consists of ephemeral annual grasses and other herbaceous plants. These vanish entirely during dry periods, and their seeds resist germination until guaranteed a moisture supply sufficient to take them through their life cycle. In many cases this is accomplished by chemical germination inhibitors that must be leached away by

Figure 10.13 (**above**) Little or no plant life can survive in this portion of Death Valley, California, because it receives so little moisture. The shrubs here are bursage (*Franseria dumosa*).

(**below**) These 5- to 8-meter-tall (15 to 25 ft) saguaro cactus (*cereus giganteus*) near Phoenix, Arizona, have no leaves, a waxy skin, and are able to store water within their tissues, enabling them to survive in the desert. (T. M. O.)

water before germination can proceed. Following a damp period, the desert surface may be covered by flowers as the annuals spring into activity. Once their seeds are produced and distributed, the annuals wither away, leaving the ground bare until the next wet period causes the new crop of seeds to germinate. While the perennial shrubs remain, the cover of ephemeral annuals varies enormously from season to season and from year to year. This makes desert life difficult for animals (rodents, rabbits, antelope, bighorn sheep) that depend upon vegetation or seeds for food, as well as for predators dependent upon the population of smaller animals, and for desert nomads whose sheep, goats, and camels are at the mercy of the unreliable rains.

The most common characteristic of regions of the dry realm is that precipitation is so much less than potential evapotranspiration. Most precipitation goes briefly into soil moisture storage, but then is quickly returned to the atmosphere by evapotranspiration. During heavy rainshowers, there may be some surface runoff. Most of this water drains into local basins and then evaporates, but some percolates down to the subsurface water table.

Figure 8.12 (p. 204) shows that at Phoenix, Arizona, moisture surpluses are rare. Where soil conditions are favorable, irrigation agriculture can be very productive, but irrigation water must be obtained from non-desert upland areas where orographic precipitation is much greater than potential evapotranspiration. In Figure 8.12, Blue Canyon is an example of an upland area with a large winter and spring water surplus available for lowland irrigation. Because there is little water available for evapotranspiration in desert areas, most of the net radiation gain goes into heating of the land surface and the lower troposphere. At Phoenix, afternoon summer temperatures typically exceed 40°C (104°F), and soil temperatures in places like California's Death Valley can reach 90°C (194°F).

Chaparral

An almost unique assemblage of vegetation has evolved where winters are mild and rainy and summers are hot and dry. This dry summer subtropical climatic region appears along the western coasts of continents between about 30° and 45° latitude, on both sides of the equator. Both the climate and associated vegetation are often described as Mediterranean, because the Mediterranean coasts comprise the largest area of the dry summer subtropical climate. This climate and the vegetation associated with it are also found in smaller coastal areas of California, Chile, South Africa, and Australia.

The distinctive vegetation of the dry summer subtropical climate is known in North America as *chaparral*. Chaparral consists of an almost impenetrable mat of brush, ranging from knee-high to twice the height of a person. The shrubs are small-leaved and deep-rooted to survive the summer drought period, and are generally evergreen (Figure 10.14). Because summer drought increases the danger of fire, most chaparral species have evolved the capability of resprouting from subsurface roots after being burned off above ground. Damp north-facing slopes may be covered by oak woodlands, and flat areas may be a virtual savanna, with large oaks rising from grass or shrub. As in the case of desert vegetation, chaparral vegetation on different continents and in different hemispheres is remarkably similar in appearance, even though the plants are unrelated. Convergent evolution of plants is especially clear in this distinctive climatic realm.

The chaparral vegetation region in California often receives nationwide attention when late summer fires sweep over mountain slopes. Dry winds from the interior deserts can make fire fighting in chaparral extremely difficult and dangerous. If a heavy winter rain occurs before the burned chaparral has resprouted, rapid erosion sends torrents of mud and boulders rushing into valleys. This repeatedly damages communities along the California coast.

The monthly temperature and precipitation regime for Palermo, Italy, shown in Figure 10.15, is representative of dry summer subtropical climatic regions. Winter rainfall is associated with midlatitude cyclones. The mild temperatures at this location are due mostly to the protection from cold polar continental air offered by neighboring mountain barriers. Be-

Figure 10.14 This is a characteristic view of *chaparral* vegetation north of San Francisco, California. The drought-adapted pines (*Pinus sabiniana*) that rise above the chaparral shrubs emphasize the low stature of this evergreen sclerophyll (hard leaf) vegetation type, which is adapted to areas with rainless summers and wet winters. The shrubs are evergreen to take immediate advantage of the short period in the spring when soil moisture left over from the winter combines with rising temperatures to produce a flush of new growth. Similar vegetation composed of different species is seen in the Mediterranean region, Chile, South Africa, and southern Australia. (T. M. O.)

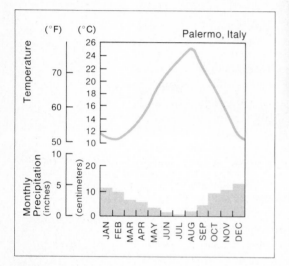

Figure 10.15 Palermo, Italy, has a Mediterranean climate characterized by moderate temperatures and a dry summer (Köppen *Cs* climate). As the graphs show, winter in Palermo is cooler and more moist than summer. The average annual precipitation is nearly 80 cm (31 in.), but for June, July, and August combined the precipitation is only 4 cm (1.5 in.). (Doug Armstrong after H. Nelson, *Climatic Data for Representative Stations of the World*, © 1968, by permission of University of Nebraska Press)

Figure 10.16 The local water budget for San Francisco, California, is characteristic of stations with a Mediterranean climate (Köppen *Cs* climate). A deficiency of moisture persists through the dry summer months, and plants that live through the summer must draw upon fog drip and stored soil moisture. From October through March, when most of the precipitation is received, soil moisture is replenished, and a surplus is generated. (Doug Armstrong after D. Carter and J. Mather, *Publications in Climatology*, vol. 19, 1966)

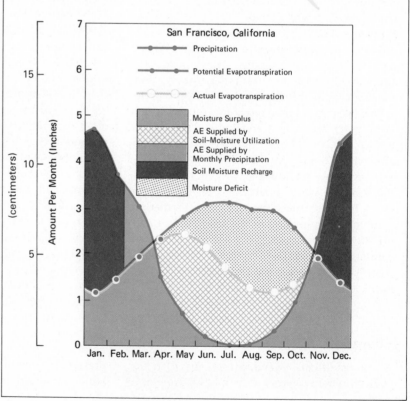

cause of atmospheric subsidence associated with the subtropical highs, summers are hot and dry. In other locations, such as San Francisco, upwelling ocean water along the coast keeps summer temperatures cool.

The average water budget for San Francisco (Figure 10.16) shows a large moisture deficit during summer and fall, and a smaller surplus during spring, after soil moisture storage has been recharged. In most dry summer subtropical climatic regions, local water surpluses are not adequate to sustain irrigation agriculture during the summer. Winter surpluses from mountain regions must be stored in reservoirs to be delivered to irrigated cropland in the summer.

Midlatitude Forests

In the pre-agricultural period, hardwood forests were dominant across the eastern United States, western Europe, Japan, Korea, and eastern China. Similar forests occupied much smaller areas of South America, southern Africa, Australia, and New Zealand. These forests tend to be located within humid continental climatic regions, but they also occur in most of the humid subtropical and marine west coast climatic regions. Most of these regions have been cleared for cropland and pasture, and their upland forests have been cut over for lumber and firewood, so that very little of the original forest remains.

The midlatitude forests on the different continents include different combinations of species. In eastern North America, oak, hickory, maple, and beech each tend to be dominant in various areas. The multi-storied canopy of trees usually rises 30 meters (100 ft) or more above the ground. Some shrubs and shade-tolerant annuals occupy the ground surface, but the forest tends to be relatively open below the canopy. Most of the trees drop their leaves before the onset of winter, so the appearance of the forest changes dramatically through the seasons (see Figure 10.17). In a portion of the region soil water in the root zone of the forests is frozen in the winter and is therefore unavailable

Figure 10.17 **(left)** The mixed broadleaf deciduous forests are alive with color for a few weeks in autumn; this example is from southern Wisconsin. Each species progresses through a sequence of color changes and leaf fall. Variable temperature and moisture conditions cause the date of maximum color to vary by as much as four weeks or more (R. A. Muller)

(right) This hardwood forest in southern Kentucky is seen in the spring before the deciduous trees produce a new set of leaves. Note the amount of sunlight reaching the forest floor. Such bright conditions are never encountered within the world's evergreen forest types. (T. M. O.)

to the trees. With no uptake of moisture from the soil, continued transpiration of moisture by plant leaves would be fatal to the plants. But even before this occurred, the moisture-laden cells composing plant leaves would be ruptured and destroyed by winter freezes, causing the death of the plant. Therefore, broadleaf plants must drop their leaves and remain dormant until the threat of frost is past. Thus the deciduous habit can result either from seasonal drought, as in the mixed tropical forests, or from seasonal cold.

In the southern hemisphere, where winters are mild because of the strong maritime influence, the midlatitude forests are dominated by broadleaf evergreens such as *Eucalyptus* and *Nothofagus* (evergreen beech).

The broadleaf deciduous forests in the eastern United States and western Europe become mixed with coniferous (cone-bearing) needle-leaf evergreens on their northern margins, and with broadleaf evergreens on their southern margins. Pines tend to be dominant on the higher portions of the coastal plain of the southeastern United States, where sandy soils lack nutrients and store only limited amounts of moisture. The southern pines are an important timber resource, but also are more susceptible than deciduous trees to fires that sweep areas of the coastal plain from time to time. Some scientists believe that the southern pines are only an intermediate successional stage toward the broadleaf deciduous forest climax.

Most midlatitude forest regions experience large temperature ranges from winter to summer and a relatively even distribution of precipitation throughout the year. Pittsburgh, Pennsylvania, is near the climatic boundary between Köppen's humid subtropical (*Cfa*) and humid continental (*Dfa*) climates; its monthly temperature and precipitation regimes are displayed in Figure 10.18. In general, summers are hot, but temperatures well below freezing are common during winter months. Monthly precipitation varies little. Klagenfurt, Austria, is representative of midlatitude forests with humid continental climates; the large range of its seasonal temperature is shown in the thermoisopleth diagram in Figure 10.19. This diagram

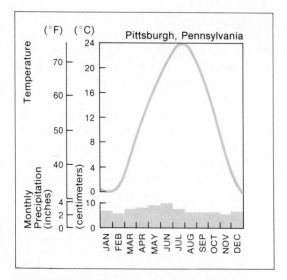

Figure 10.18 Pittsburgh, Pennsylvania, is near the boundary between the humid subtropical and humid continental climates (Köppen *Cfa* and *Dfa* climates). The graph shows the great range of the average monthly temperature between summer and winter. The monthly precipitation is nearly constant through the year, however. (Doug Armstrong after H. Nelson, *Climatic Data for Representative Stations of the World*, © 1968, by permission of University of Nebraska Press)

should be compared with that in Figure 10.7, which illustrates the temperature regime in a tropical rainforest.

The average water budgets of most places within midlatitude forest regions show small summer deficits and relatively large winter and spring surpluses. The changing patterns of the upper air circulation in the middle and higher latitudes, however, tend to produce clusters of wetter or drier, or warmer or colder years in each region (see the Case Study on climatic variation after Chapter 8). Figure 10.20 illustrates some of the hydrologic impacts of climatic variation during an extensive drought across the northeastern United States in the 1960s, as represented by water budget components.

Midlatitude Grasslands

Grasses are the dominant vegetation where precipitation does not meet the needs of trees and

Figure 10.19 Klagenfurt, Austria, has a humid continental climate (Köppen *Dfa* climate). As the thermoisopleth shows, the range of temperature through the year exceeds the range through the average day. In January the noon temperature is approximately −4°C (25°F), whereas in July the noon temperature is 22°C (72°F). Note that in November the temperature through the day varies by only a few degrees, partly because of the moderating effect of cloudy weather. (Doug Armstrong after C. Troll, *World Maps of Climatology*, © 1965, Springer-Verlag Publishing)

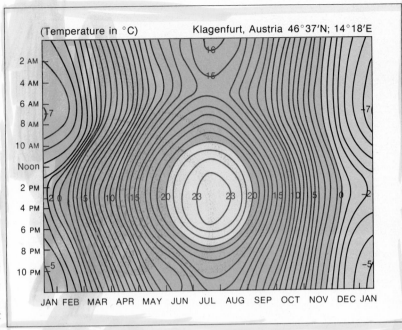

shrubs, or where repeated fires prevent tree regeneration from seedlings. Vast areas of continuous grasslands once extended from Texas to Alberta and Saskatchewan in central North America, and across Eurasia from the Soviet Ukraine to Manchuria. These grasslands included *tall-* and *short-grass prairies* and *steppes*. Similar extensive grasslands also were present in South America, particularly in Argentina and Uruguay. Intense agricultural exploitation has left very few areas of natural grassland in any of these regions.

The midlatitude grasslands contained a large number of plant species that are different from the grasses of the tropical savannas. The dominant midlatitude grasses were usually perennials that lie dormant during the winter and continue their growth in the next growing season. Near the boundary between forest and grassland, where moisture is comparatively abundant, the natural grassland vegetation was usually prairie grass, one to two meters in height. Most tall-grass prairie areas throughout the world are now utilized for intensive agriculture and wheat farming, in which a useful domesticated grass is substituted for the native grass species. Where there is less moisture, the dominant vegetation is short prairie grass, generally less than a meter high. In the driest grassland the grass grows in bunches or tufts, with bare ground often visible between the clumps (Figure 10.21). This is the common Eurasian type of grassland, known as *steppe* grassland. Patches of trees may occur here and there in all grassland types, especially along streams.

Like the tropical savannas, the temperate grasslands of Eurasia and North America were

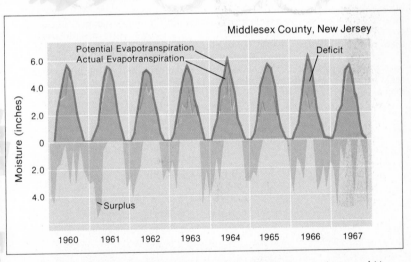

Figure 10.20 These local water budgets for Middlesex County, in central New Jersey, illustrate variability in water budgets in a midlatitude forest at the equatorward margin of the humid continental climatic region. During the drought years of 1962 through early 1966, deficits were large, and winter-spring surpluses were not great enough to meet the water-resource needs of the region. Water restrictions were common, and there was some loss of shrubs and trees in suburban areas. (Doug Armstrong after Robert A. Muller, 1969)

Figure 10.21 This tussocked grassland is in South
Island, New Zealand. Radiocarbon dating of wood
fragments indicates that extensive forests occupied
some of this region 500 years ago. The conversion to
grassland may have occurred as a result of human use
of fire. (G. R. Roberts, Nelson, N. Z.)

once immense natural pastures, supporting
vast numbers of grazing animals. Upwards of
40 million bison and an even larger number of
antelope pastured on the American grasslands—single herds of 100,000 to 2 million animals were seen by explorers and fur traders in
the early 1800s. All modern breeds of horses
appear to have descended from the enormous
herds of native horses that once dominated the
Eurasian grasslands. Native horses and camels
were part of the prehistoric fauna of the North
American grasslands, but were apparently an-
nihilated by the Paleo-Indian hunters who mi-
grated from Asia about 13,000 years ago.

The tall-grass prairies in Illinois and Iowa
are utilized now for corn, small grain, and hog
farming. Many ecologists believe that the climate of this region would have supported forests, were it not for the fires that repeatedly
swept over the open plains. The average annual
rainfall is as great as in forested areas such as
New York and Pennsylvania. Farther west in
Saskatchewan, the Dakotas, Nebraska, Kansas,
and Oklahoma, wheat crops have replaced me-
dium-height prairie grasses.

Beyond the wheat belt, on the high plains
of Alberta, Montana, Wyoming, Colorado, and
Texas, short-grass prairies are still present.
Here tree growth is precluded (except along
stream courses) by complete drying out of the
subsoil during periodic droughts. This phenomenon is very clear in the morphology of the
soils developed in this area, as will be seen in
Chapter 11. The area is too dry for unirrigated
agriculture, and its predominant use is as unf-
enced pasture for beef cattle. The medium-
height prairie grass grades into short grass
through a broad transition zone centered just
east of the 100th meridian from central Texas
to the Dakotas, with a westward swing into
Saskatchewan and Alberta. This transition zone
is tempting but dangerous for agriculture.

Figure 10.22 This graph shows the annual ratio of actual evapotranspiration to
potential evapotranspiration (*AE/PE*) in percent for Dodge City, Kansas, which is
located near the drier western margin of the midlatitude grasslands. The clusters of drought years in the 1930s and 1950s, "dust bowl" years, are remarkable
features of the climate. In 1934, for example, *AE* amounted to only 33 percent
of *PE*, but in 1944, the ratio climbed to 94 percent. These climatic fluctuations
have been characteristic of the Great Plains and most other subhumid climatic
regions, and they are likely to have impacts on most natural systems and human activity.

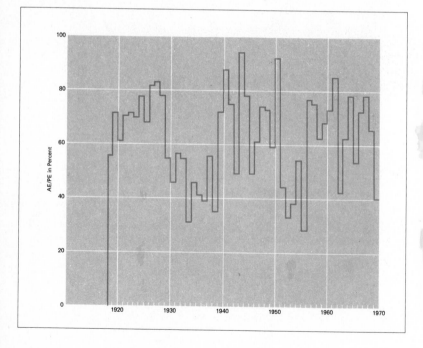

Figure 10.23 Huron, which is located in the prairie grasslands of South Dakota, has a humid continental climate with moderate moisture and cold winters (Köppen *Df* region) and a subhumid climate, according to Thornthwaite. The local water budget for Huron shows that precipitation occurs in all months, and is most plentiful during early summer. However, high summer temperatures cause potential evapotranspiration to be high then as well, and moisture deficits normally occur. The soil moisture is recharged during the cooler months, when the vegetation's demand for moisture is small, but winter precipitation is not large enough, on the average, to generate surpluses for runoff. (Doug Armstrong after D. Carter and J. Mather, *Publications in Climatology*, vol. 19, 1966)

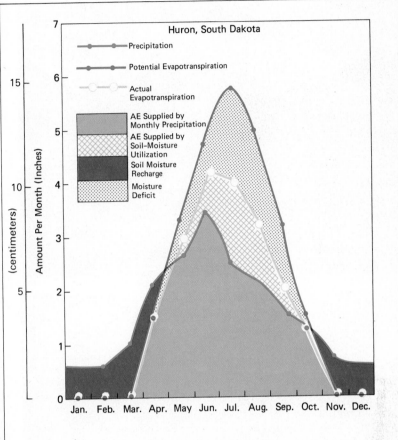

Cycles of wet years repeatedly lure wheat farmers onto the short-grass prairies only to meet disaster when the inevitable dry years follow, which transform the ploughed land into a "dust bowl" (Figure 10.22).

Most of the world's grasslands are located where average precipitation barely equals potential evapotranspiration. In the grasslands of continental interiors in the northern hemisphere, summers tend to be hot, with periodic thunderstorms. Winters are very cold and dry.

The average water budget for Huron, South Dakota, located in the transition zone between the tall and short grasslands, is shown in Figure 10.23. Despite a summer rainfall maximum, there is still a relatively large summer moisture deficit, and precipitation is not great enough during winter to generate a significant surplus. In addition, summer rainfall from convective thunderstorms is highly variable from year to year. Some climatologists and ecologists believe that clusters of drier than normal years prevent the establishment of forests within the wetter margins of the subhumid climatic regions.

Northern Coniferous Forests

Coniferous (cone-bearing) needleleaf evergreen forests dominated by spruce, fir, and pine extend in a broad band across North America, Europe, and Asia between about 50° and 65°N latitude in the subarctic climatic region. These forests, also known as *taiga* or *boreal forests*, endure the largest annual temperature ranges encountered on the earth. Winters are bitterly cold and very dry, with only light snowfalls. These areas are dominated by the Canadian and Siberian highs and are the source regions for continental polar air masses. Summer is brief, but daylight periods are long and temperatures are mild, even warm, occasionally exceeding 25°C (77°F).

The conifers of the boreal forests are tall, slim, and tapered, as shown in Figure 10.24. Most conifers are evergreen and do not lose their leaves during winter. Their small needle-shaped leaves and thick bark resist moisture losses during the long cold winters, and their conical crowns may be adapted to intercept the oblique rays of a sun that in their latitude is never high in the sky. Since they are not required to manufacture new leaves before they can begin photosynthesis and growth in the spring, they conserve energy and are able to

(a)

(b)

Figure 10.24   (a) This Douglas fir forest in Washington's Cascade Range shows how a small number of tree species dominate in most coniferous forests. In this area evergreen coniferous forests are a response to occasional summer droughts that exclude broadleaf forests. Farther poleward, where a short growing season and deeply frozen subsoils are a limiting factor, spruce and pine are the dominant trees. Mt. Rainier is in the background.

(b) Pure stands of large trees that are useful for lumber have caused the coniferous forests of the middle latitudes to be intensively exploited by loggers. Here, close to Mt. Rainier, we see steep slopes recently stripped of their forest cover. Clear-cutting in such terrain greatly increases water erosion and soil slippage. The small forest remnant may have been spared to reseed the cleared area. (T. M. O.)

commence growth as soon as temperatures rise sufficiently. This is a clear advantage where the growing season is very short.

Coniferous forests contain a comparatively small number of plant species. Low sun angles and dense foliage allow little light to reach the ground, and cool temperatures also limit plant growth. Spruce is common in the coniferous forests of North America. Larch, which drops its needle-like leaves in the winter, is dominant in eastern Siberia, where winter conditions may be too severe even for needleleaf evergreens. January temperatures in the latter region plummet to −40° to −50°C (−40° to −60°F). On the southern margins of the coniferous forests, the trees are tall and densely packed. Farther poleward, the trees are smaller and the forest more open.

Figure 10.25 shows the monthly regimes of temperature and precipitation at Moose Factory in central Canada. Monthly mean temperatures range from −20°C (−35°F) in winter to 16°C (60°F) in summer. Much of the annual precipitation falls as rain during summer, when potential evapotranspiration is nearly as high as in midlatitude regions. Hence, most summer rainfall is utilized for evapotranspiration. Local water budgets show that the only moisture surplus is in the late spring at the time of snowmelt. Nearly rainless periods occasionally result in small deficits. Some summers are dry enough for the danger of forest fires to be serious.

In North America the coniferous forests of the subarctic climatic region spread southeastward from the McKenzie River valley in northwestern Canada across the Canadian border into sections of Michigan, New York, and New England. Along the northwestern coast of North America the coniferous forest extends southward into Washington, Oregon, and northern California. Here winters are wet and summers are very dry. The virgin forests of the Pacific Coast are often composed of giant trees, and the species differ from those of the taiga. Redwoods (*Sequoia*) and Douglas fir are dominant in the northern California Coast Ranges, and fir, spruce, cedar, and hemlock prevail in the forests of Oregon, Washington, and British Columbia. Moisture stress during the warm season has favored the survival of an evergreen

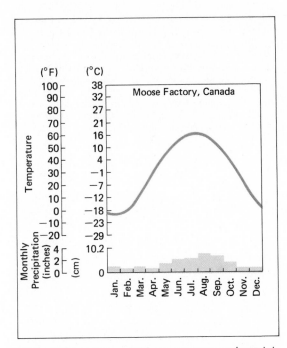

Figure 10.25 Mean monthly temperature and precipitation regimes at Moose Factory, in central Canada. The range between winter and summer temperatures in the subarctic climate is the largest of all global environments. The northern coniferous forest is well adapted to short but warm summers, when much of the annual precipitation falls. (Vantage Art, Inc. after Glenn Trewartha, *Introduction to Climate*, 4th ed., 1968, McGraw-Hill Book Co.)

coniferous forest in this midlatitude region of summer drought. The immense size of many trees causes the biomass in portions of the evergreen forests of the Pacific Coast to be even greater than that in tropical rain forests.

Tundra

Trees cannot survive unless the average temperature during the growing season exceeds 10°C (50°F) for a period of two to three months. Near the Arctic Ocean, the trees of the northern coniferous forest give way to low shrubs, grasses, and flowering herbs, with mosses and lichens on rock surfaces (Figure 10.26). This vegetation formation is known as *tundra*.

Although winter temperatures in the tundra regions do not reach the extremes experienced in the more continental taiga of eastern Siberia,

they are nevertheless very low, and are frequently accompanied by gale-like winds from which there is no shelter. At −18°C (0°F), a 30-km-per-hour (20 mph) wind produces the equivalent of a temperature of −40°C (−40°F). This depression of effective temperature is called the *windchill factor*. At −30°C (−24°F) the same wind lowers the equivalent temperature to −55°C (−68°F). The combination of wind and low temperature causes extreme moisture stress and danger to water-bearing plant tissues, because the freezing of water ruptures plant cells and evaporation induced by the wind cannot be offset by water intake from the frozen ground. As a consequence, tundra plants are small-leaved, like desert plants, and low-growing so that they will be blanketed by snow when icy winter winds sweep over the land surface.

Figure 10.27 is a thermoisopleth diagram for Sagastyr, at latitude 73°N, in the tundra of northern Siberia. The nearly vertical pattern of isotherms shows that the daily variation of temperature is only a few degrees because of the long daylight periods in summer and the absence of sunlight during midwinter. During a year, however, the average mid-day temperatures vary from 6°C (43°F) in July to −40°C (−40°F) in February.

In tundra regions precipitation tends to be low throughout the year. Winters are long, and

Figure 10.26 Low shrubs, grasses, and flowering herbs are the dominant vegetation types in tundra regions, as shown here in Norway. The tundra is wet during the period of thaw because water cannot drain through the permanently frozen ground below the surface. (Brian Hawkes/Carl Ostman Agency)

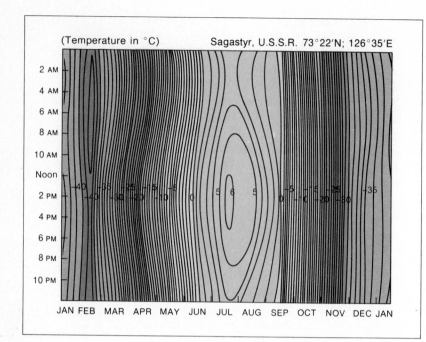

(Temperature in °C)   Sagastyr, U.S.S.R. 73°22'N; 126°35'E

Figure 10.27   Sagastyr, U.S.S.R., is a station in the tundra region of Siberia north of the Arctic Circle (Köppen *ET* climate). The temperature during the year varies through an extreme range because of the lack of sun near the time of winter solstice and the continual daylight at summer solstice. The nearly vertical temperature contour lines imply that the temperature during any given day is essentially constant. (Doug Armstrong after C. Troll, *World Maps of Climatology*, © 1965, Springer-Verlag Publishing)

the tundra is thinly covered by snow for 6 to 8 months. Most of the sparse precipitation falls as rain in the brief summer. Despite the relatively low summer precipitation, the tundra is usually moist and waterlogged during the warm months, when the surface layer of soil thaws but cannot drain downward because the subsoil is permanently frozen. This situation has important consequences for landform development and also for human use of the tundra environment, as will be seen in Chapter 16.

Highland Vegetation

Under average conditions, temperatures decrease with increasing elevation. Thus, one can progress through different climatic and vegetation zones while climbing upward in highland regions.

Figure 10.28 shows that in humid climatic regions the sequence of vegetation types encountered with increasing elevation is similar to the sequence met in traveling from the equator to the poles. In dry climates the increase in precipitation with elevation results in spectacular vegetation changes within short horizontal distances. In the San Francisco Peaks region of northern Arizona, where the vertical sequence of North American life-zones was first studied, the upward progression begins with desert shrubs at an elevation of 1,200 meters (4,000 ft) and ends in a spruce–fir forest at 3,000 meters (10,000 ft) that is not unlike the spruce–fir forests of subarctic Canada. Normally growing thousands of miles apart, the two completely dissimilar vegetation associations exist here within a distance of a few miles. In tropical uplands the diversity of contrasting vegetation is even greater. In the Peruvian Andes, which rise to glacier-clad summits from rainforests on the east and deserts on the west, nearly all the world's latitudinal temperature and vegetation zones can be recognized in vertical succession (Figure 10.29).

Local climates and vegetation in highland areas are very complex, however. Adjacent slopes with different orientations differ dramatically in terms of solar radiation, temperature, and precipitation. North-facing slopes of deep valleys in the middle latitudes receive little direct sunlight over the year, while nearby south-facing slopes may bake in many hours of sunshine each day. Leeward slopes may be

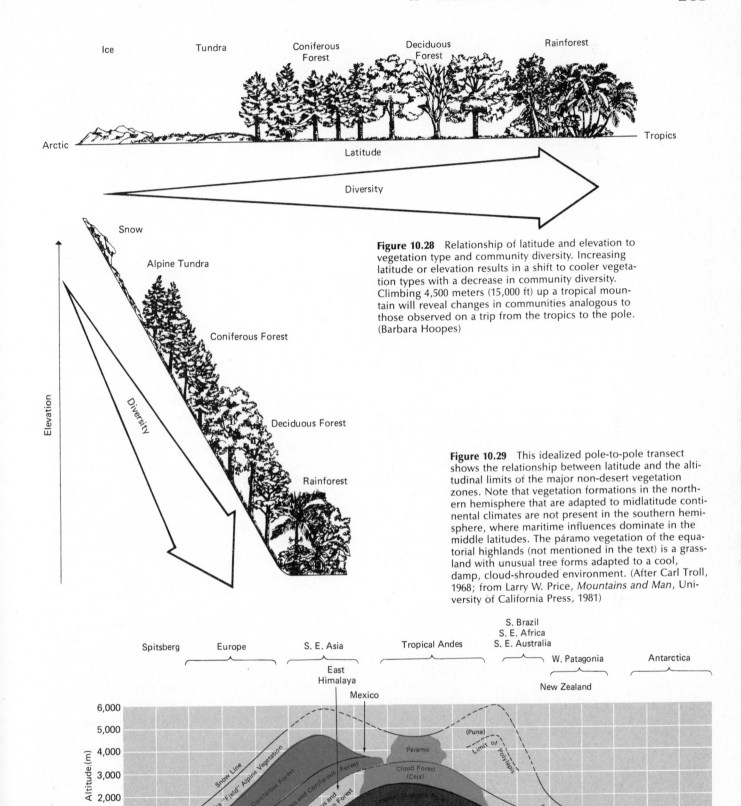

Figure 10.28 Relationship of latitude and elevation to vegetation type and community diversity. Increasing latitude or elevation results in a shift to cooler vegetation types with a decrease in community diversity. Climbing 4,500 meters (15,000 ft) up a tropical mountain will reveal changes in communities analogous to those observed on a trip from the tropics to the pole. (Barbara Hoopes)

Figure 10.29 This idealized pole-to-pole transect shows the relationship between latitude and the altitudinal limits of the major non-desert vegetation zones. Note that vegetation formations in the northern hemisphere that are adapted to midlatitude continental climates are not present in the southern hemisphere, where maritime influences dominate in the middle latitudes. The páramo vegetation of the equatorial highlands (not mentioned in the text) is a grassland with unusual tree forms adapted to a cool, damp, cloud-shrouded environment. (After Carl Troll, 1968; from Larry W. Price, *Mountains and Man*, University of California Press, 1981)

rather dry, but precipitation on windward slopes tends to increase sharply with elevation.

For at least the first 1 or 2 kilometers of elevation, upland areas produce much larger water surpluses than surrounding lowlands. A flourishing vegetative cover in upland regions retards water runoff, reducing soil erosion on steep slopes and decreasing flood hazards in adjacent lowlands. At the same time, forested uplands release water slowly to streams, sustaining their flow through rainless periods. Thus preservation or human management of upland vegetation, the control of soil erosion, and the stabilization of stream flow all go hand-in-hand. This is a matter of critical concern to humans who live within the highland regions and far beyond them on the great river floodplains of the earth.

SUMMARY

Solar energy and moisture determine the general type of vegetation that can grow in a particular location. Normally the vegetation of any place is a mosaic of various plant species that form a plant community. When disturbance of natural vegetation occurs, or when agricultural land is abandoned, the area experiences a largely predictable succession of plant communities. Eventually a mature or climax vegetation may be attained. The climax vegetation is assumed to be in equilibrium with the environment and does not change further. The climax vegetation is determined by soil conditions and the availability of moisture and energy, but climatic variability and various disturbance factors also have a significant impact.

In this chapter the natural vegetation formations have been classified into nine global types. Forests are restricted to the humid climatic realm, but various types have evolved differently in response to seasonal energy and moisture regimes. Thus there are tropical rainforests, tropical mixed forests, midlatitude forests, and northern coniferous forests. Grasslands and specialized xerophytic vegetation dominate the dry climatic realm. Savannas occur where there is winter drought and summer rain, whereas evergreen chaparral vegetation is present where there is winter rain and summer drought. Energy availability is so restricted toward the polar margins of the humid climatic realm that forests give way to low-growing forms of vegetation collectively called tundra. In highland areas the distributions of local climates and vegetation are complex, but the vertical sequence of plant communities shows similarities to the poleward gradations of climate and vegetation.

REVIEW QUESTIONS

1. Illustrate some specific plant adaptations to varying environmental constraints.
2. In what general ways do plant communities usually change during succession toward a climax vegetation?
3. How does tropical rain forest vegetation reflect the climatic characteristics of the equatorial zone?
4. What is a thermoisopleth diagram?
5. How do forests in areas of the tropics that experience a dry season differ from the rain forests of the equatorial zone?
6. What edaphic condition is widespread in savanna regions?
7. What are the two broad categories of desert vegetation? How do these two dissimilar vegetation forms survive in areas of extreme moisture deficiency?

8. What unusual problems do Mediterranean environments pose for vegetation? What is the nature of the vegetation that has evolved in such areas?

9. In what other areas of the world does one encounter forests like those originally covering the eastern United States? How do these forests reflect environmental stress?

10. What general climatic characteristic is associated with the midlatitude grasslands? Are the locations of grasslands strictly controlled by climatic parameters?

11. Why are needleleaf evergreen trees better adapted than broadleaf deciduous trees to severe winters and a short growing season?

12. How does low stature benefit tundra plants?

13. Where, over a short distance, can one view vegetation changes similar to those occurring with changes in geographic latitude?

APPLICATIONS

1. In the vicinity of your campus there is probably land, once disturbed or cleared, that fairly recently has been allowed to revert to a more natural or uncontrolled vegetation cover. Locate two such sites and compare the composition of their vegetation. Is there any good explanation for the differences you can see?

2. Analyze the map of global vegetation (Figure 10.4) in comparison with the map of global climate (Figure 8.3). Make a list of the climates of the natural vegetation types. Now make a final list of global regions where the general associations of climate and vegetation do not seem to fit, and suggest some possible reasons.

3. Are there any preserved remnants of the prehistoric natural vegetation of your region? Was the prehistoric vegetation actually regarded as a climax type, or was it part of a plant succession after some natural disturbance? In what areas of North America would you expect the prehistoric vegetation to have been other than a climax form?

4. How does the size of plants appear to be related to the availability of energy and moisture?

5. There are similarities in the appearance of the vegetation in desert and tundra regions. What are the resemblances? Is this a result of the same or different climatic stresses?

6. One peculiar aspect of the climatic regime of tundra regions is also seen in the tropical rainforests. This climatic characteristic is especially evident in a comparison of thermo-isopleth diagrams from different vegetation regions. What is the similarity? Are its causes the same in the two regions?

FURTHER READING

Billings, W. D. *Plants, Man, and the Ecosystem,* 2nd ed. Belmont, Calif.: Wadsworth (1970), 160 pp. This brief paperback, part of the Fundamentals of Botany series, includes many succinct sections that supplement this chapter.

Eyre, S. R. *Vegetation and Soils: A World Picture,* 2nd ed. Chicago: Aldine (1968), 328 pp. This book, written from the perspective of the British Isles, focuses on relationships between vegetation and soils. It is organized by global vegetation types and is especially useful for Chapters 9 and 10 of this text.

Shelford, Victor E. *The Ecology of North America.* Urbana: University of Illinois Press (1963), 610 pp. Shelford's book is a detailed account of North American biomes. It is packed with information on plant and animal interactions, and is a valuable reference book, available in a paperback edition.

Tivy, Joy. *Biogeography.* New York: Longman (1971), 394 pp. Plant distribution and vegetation formations are explained in terms of environmental factors.

Vankat, John L. *The Natural Vegetation of North America: An Introduction.* New York: Wiley (1979), 261 pp. This very informative paperback presents the ecological basis of the major vegetation formations of North America from tundra to tropical rainforest.

Walter, Heinrich. *Vegetation of the Earth in Relation to Climate and the Eco-Physiological Conditions.* Translation of 2nd German ed. New York: Springer-Verlag (1973), 237 pp. This work stresses relationships between climate and vegetation and also includes short descriptions of the vegetation formations of each natural vegetation type.

The Amazonian Forest: An Endangered Ecosystem

Although the vast tropical rain forest of the Amazon River Basin occupies one-twentieth of the earth's land area, it remains the earth's least known environment. Centered in Brazil, the Amazonian forest reaches far into the neighboring countries of Bolivia, Peru, Ecuador, Colombia, and Venezuela, covering more than 3 million square km (2 million sq miles). Offering optimum conditions for plant growth, the great wet forest—oppressively humid and infested with parasites, insects, and reptiles—has been described by outsiders as "green hell."

There is no question that the largest diversity of organisms of every type is present here, though only a fraction of them have been described scientifically. More than twenty percent of all known higher plants and birds on our planet evolved in the Amazon Basin. Better knowledge of the region would no doubt put the figure still higher. As many as 235 different tree species have been identified on a single acre. The Amazonian forest is a seemingly inexhaustible storehouse of edible fruit, nuts and berries; of medicinal plants; of hardwood; and of fish equal in numbers to those in entire oceans. But pessimistic ecologists predict that all of this, as well as the remaining rain forests elsewhere in the world, may vanish in less than 40 years.

The wet tropical forests have long been used by humans, but in a limited and ecologically unthreatening manner. The traditional forest occupants were primitive hunters and collectors and groups of people en-

gaged in shifting "slash-and-burn" agriculture. Despite its unpleasant sound, the slash-and-burn system, practiced in Southeast Asia, equatorial Africa, and large portions of Central and South America, was a stable and ecologically sound use of the rain forest. In this system, also known as *swidden, ladang,* and *milpa* agriculture, plots of a few acres were cleared of the smaller trees, the debris of which was burned. The ash added enough nutrients to the impoverished tropical soils (see Chapter 11) to permit a variety of crops to be grown for two or three years, after which the crops failed and the fields became congested with weeds. The plot was abandoned and a new area was cleared, cropped, and then abandoned in turn. The crops involved included maize, manioc, squash, bananas, guava, rice, cucumbers, sugarcane, and cassava. Fields reoccupied by secondary forest could be cleared again several decades later.

The slash-and-burn method can support only small groups of people using large areas of forest in a subsistence economy that requires no surplus production for trade. It was undisturbed by external groups and involved no intensive commercial exploitation of the forest, although some forest products were collected and marketed to outside groups. In the modern world many governments therefore see these forests as undeveloped "voids" in their national territories and have actively encouraged their commercial use or complete replacement by cropland. This has become particularly true

where governments are beset by severe economic problems and seek foreign investments, export commodities, and new agricultural land for exploding populations.

In 1970 the Brazilian government announced a program to resettle some 3 million landless peasants from the drought-stricken Northeast of Brazil on 100-hectare (250-acre) homesteads in the Amazonian forest. The settlers were to become rice farmers. The wedge was the proposed Transamazon Highway, to be bulldozed westward through the forest for 3,300 kilometers (2,000 miles) from the Northeast to the Peruvian border. The road was opened in 1975, and the emigrants began to arrive. But by the end of 1981 less than 10,000 families—990,000 short of the target for 1980—had been settled.

Clearly, the plan had misfired. While the population in the Northeast continued to explode, adding 6 million more people in the crisis area, the costly Transamazon Highway failed to have any remedial effect. The ancillary benefits—an outlet for the timber and mineral wealth of the Amazon Basin—have also receded. The Transamazon Highway project was launched when the world price of petroleum was but $2 a barrel. By the time the road opened, fuel costs had begun their upward spiral. A few small sawmills appeared, exploiting mahogany and cedar, but they seem defeated by the cost of transport and the absence of dense stands of commercial trees. Large-scale mineral extraction is still a hope for the future, frus-

CASE STUDY

Bulldozers clearing the Amazonian forest in northern Brazil. Forest removal of this type is carried out to create new cropland for colonists, pasture land for beef cattle, and tree plantations for timber and pulpwood. Thus far, such projects have achieved little success in Amazonia. (Rolf and Mari Wesche)

trated at present by high costs that would require further borrowing by the country with one of the world's largest foreign debts.

Every scheme for Amazonian development seems to have backfired. The varieties of rice selected were clearly inappropriate to Amazonian conditions. By the third year, crops were starved for nutrients and succumbed completely, while erosion of the cleared soil ran rampant, causing annual losses as high as 1,000 tons per hectare. Faced with disaster, the Brazilian government in 1973 switched emphasis from small homesteads to corporate cattle ranches of vast size, threatening the forest still more. The ranches were to produce cheap beef for export. Some three and a half times the acreage reserved for homesteads was assigned to more than 300 ranches, with individual holdings covering as much as a million hectares (250 million acres). But where the forest was converted to pasture, inedible weeds succeeded the imported grasses within a few years and the cattle starved, or sickened and died. Where the forest was replaced by single-species tree plantations for timber and wood pulp—another encouraging prospect—insect pests and plant pathogens soon took control. Even the fishing industry anticipated as a food source for settlers failed. Too late it was discovered that the fish of forest streams fed heavily on forest fruits and seeds. Destruction of the forest starved the local aquatic fauna.

Meanwhile, agricultural pests and plant diseases followed the new road as it penetrated the forest, as did the diseases of alien humans. The latter are rapidly extinguishing the original Indian populations inhabiting the forest, just as the Indians of the Caribbean were destroyed centuries earlier. Malaria has become a severe problem as a consequence of highway construction. Under natural conditions the black waters of rainforest streams and swamps are too acid and in shade too deep for mosquito larvae. But the construction of an elevated roadbed has blocked natural drainage lines, creating hundreds of small sunlit lakes. These receive nutrients from erosion of bare soil, reducing the acidity of the impounded water to a level favorable for the larvae of the malaria-transmitting mosquitoes.

Despite all this, felling of the rainforest continues in the Amazon Basin and elsewhere. A U.S. National Academy of Sciences report estimates that the tropical forests are being cleared at the rate of 50 hectares (125 acres) a minute—the loss each year of an area the size of the British Isles. Aggravating the situation is the fact that 90 percent of the world population increase in the next twenty years is expected to be in the tropics, which will make forest preservation increasingly difficult.

The great fear of ecologists is that the rainforests will vanish before we begin to know what potential resources they contain and before we truly understand their internal processes and their role in the planetary system. They are already the major source of the world's pharmaceuticals, including the chemicals used in cancer control and contraceptive pills. It is suspected that they contain many unknown beneficial substances—of what importance we can only guess. As a major "sink" for atmospheric CO_2, the tropical rainforests may play a significant role in reducing the buildup of atmospheric CO_2 released by fossil fuel burning. Thus they must have mitigated the much-discussed global temperature increase of the last hundred years. How much we do not know! This question, by itself, should make tropical forest preservation a matter of interest to all of us.

Lime Banks by Andrew Wyeth, 1962. (Collection, Mr. Smith W. Bagley; copyright © by Andrew Wyeth)

*A thin layer of soil cloaks the earth's surface,
softening its contours and supporting its vegetation.
The soils we see today are products of the interactions
of living and nonliving materials with energy and
moisture over thousands of years.*

A vital factor in the natural productivity and human utilization of any locality is the nature of its soils. Soils supply the water and nutrients for land plants, and plants support the food chains that sustain the web of life on earth. Thus it is important for geographers to understand the processes of the soil system. Most people use the term "soil" very imprecisely. The loose "dirt" that one can dig into with a shovel or push around with a bulldozer may or may not be a soil in the technical sense of the word. A true soil, as opposed to mere decomposed rock or a recent deposit of river-borne silt, is the product of a living environment, and is significantly different from the raw material from which it is derived. Where there is no life, as on the moon, there may be much loose rock material, but there is no true soil. Since our planet teems with life, soils of some type are present almost everywhere on the earth.

In this chapter we shall examine the most important characteristics of soils, the reasons for geographical variations in soil types, the current system of soil description and classification, and problems of soil management.

11
The
Soil System

Soil-Forming Processes
Weathering
Translocation
Organic Activity

Properties of Soils
Soil Texture
Soil Structure
Soil Chemistry
Soil Color
Soil Profiles

Factors Affecting Soil Development
Parent Material
Climate
Site
Organisms
Time

Major Pedogenic Regimes
Laterization
Podzolization
Calcification
Salinization
Gleization

Classification of Soils

Soil Management

SOIL-FORMING PROCESSES

For soil to develop, two things must happen. First, water percolating down through loose rock material must cause physical and chemical modifications in the original material. Second, the activities of living organisms must bring about further changes. The less water and organic activity present, the weaker the development of the soil. The more strongly developed the soil, the more it is apparent that percolating water and active organisms have caused the original material to develop visible layers with varying physical and chemical characteristics. The presence of these layers, produced by soil-forming processes, identifies a true soil.

Soils develop on rock material that has already been reduced to fine fragments. This material may be either a mass of decomposed rock,

273

(a)

(b)

(c)

or sediment that has been transported and deposited by an agent of erosion, such as running water, wind, or glacial ice. Thus, the story of soil development begins with the initial fragmentation of the solid rock of the earth's crust, a process known as *weathering.*

Weathering

Solid rock can be fragmented by both mechanical and chemical processes. All rock masses are brittle; they break to form organized systems of deeply penetrating cracks, called *joints,* as a result of slow stretching or twisting motions of the earth's crust (Figure 11.1). When molten lava cools and solidifies, it breaks into prismatic forms known as columnar jointing as a result of contraction (see Figure 12.14, p. 322). Rock masses also expand somewhat when erosion of the overlying land reduces the downward pressure that confines them. The release of pressure allows massive, sparsely jointed volumes of rock to dilate by the separation of sheets that are parallel to the land surface in a process called *exfoliation* (Figure 11.1). Joint production and exfoliation seem to be the only rock fragmentation processes that do not in some way involve water.

Water enters rock masses by way of joint systems, as well as through tiny pores between the mineral grains composing most rocks (see Chapter 12, p. 316). When water freezes in

Figure 11.1 (a) Granite rock usually breaks into blocklike masses along joints when it is exposed at the earth's surface. The joints in this granite at Pike's Peak, Colorado, have been enlarged by physical and chemical weathering processes that slowly transform such rock into sand and clay. (Ward's Natural Science Establishment)

(b) The reduction of confining pressure due to erosion into masses of granitic rock causes the rock to expand by breaking into parallel sheets. These are often curved and resemble an onion structure. This process, called *exfoliation,* affects only sparsely joined rock, as here in California's Yosemite region. (T. M. O.)

(c) This photo illustrates the effect of frost weathering in cold climates in high latitudes and at high altitudes. The angularity of the forms produced here in the Teton Range of Wyoming is characteristic of frost weathering. (T. M. O.)

joints and pore spaces, it splits rocks into smaller fragments, just as freezing water bursts pipes when a heating system fails in winter. This is because the phase change from liquid water to ice produces an increase in volume of nearly 10 percent. Wherever winters are severe, this so-called *frost weathering* is an important agent of rock disintegration (Figure 11.1). The prying action of plant roots and the burrowing activities of animals also help to widen the joints in rocks.

In dry climates the evaporation of water carrying dissolved salts results in the growth of crystals of evaporite minerals on rock surfaces and in pore spaces between rock particles. As the crystals grow they wedge apart the mineral grains composing the rock. Subsequently, expansion and contraction of the evaporite crystals due to daily changes in temperature and humidity help to loosen adjacent particles.

The products of all these forms of mechanical weathering are chemically unchanged rock debris of all sizes, from giant blocks and slabs to microscopic silt particles.

Except in very cold climates, chemical weathering is more important in the fragmentation of rock than is mechanical weathering. The positively charged hydrogen ions and negatively charged hydroxyl (OH) ions in water are quick to react with the chemical compounds (minerals) of which rocks are composed (see Chapter 12, p. 316). Chemical weathering is assisted by plants, whose root secretions and decomposing litter release carbon dioxide and organic substances that modify downward-percolating water into a weak acid. This acidic soil moisture is more reactive with most rocks than is natural rainwater. Some of the products of the chemical reactions between soil moisture and subsurface rock material also are seen at the surface, such as the orange rust that forms on iron, and the green coating that appears on copper roofs and leaky copper plumbing. Both of these are products of the chemical reaction known as *oxidation*, in which oxygen ions combine with ions of rock-forming material, producing a weaker, swollen microstructure. Most soils and decayed rocks are shades of brown, due to the oxidation of iron-bearing minerals. In general, chemical reactions cause rock dis-

integration by swelling and softening some minerals and altering others to new types.

Chemical weathering is most effective where moisture is plentiful and high temperatures accelerate chemical reactions. The result is a *weathered mantle* composed of fine particles of chemically altered material overlying the solid bedrock (Figure 11.2). This mantle, which may be several meters deep, often has a high clay content. Although only certain minerals decompose to clay, these minerals are abundant in many rock types. The amount and specific types of clay formed are indications of the duration of weathering as well as the temperature, moisture, and chemical characteristics of the environment in which the weathering occurred.

Translocation

The mantle of fine particles produced by chemical and mechanical weathering is not yet a true

Figure 11.2 Chemical and physical weathering processes break down massive rock into the small particles that form the inorganic component of soil. In warm, moist climates, weathering proceeds actively along the exposed surfaces of jointed rock. Some of the mineral nutrients supplied to the soil by the parent rock are incorporated into growing vegetation. When plant litter decays, nutrients are returned to the surface of the soil and rainwater subsequently washes them downward into the soil. (John Dawson after Arthur N. Strahler, *Physical Geography*, 3rd ed., © 1960 by John Wiley & Sons, by permission)

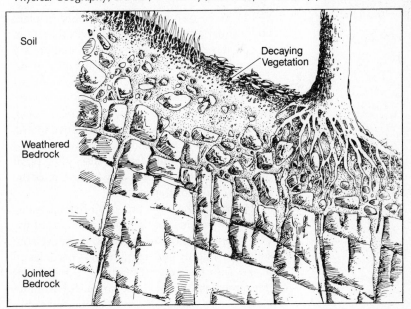

soil. Additional changes are required. Some of these result from the process of *translocation,* in which both solid and dissolved materials are moved downward by water sinking into the ground. Translocation involves both the loss of material from the upper part of the soil and the arrival of translocated substances in the lower part of the soil. Translocation of material is one of two processes that cause well-developed soils to become differentiated into layers of varying density, color, texture, and chemical composition. These layers are known as *soil horizons.*

Downward flushing of both solid and dissolved (or *leached*) matter is known as *eluviation.* Every soil has a somewhat porous *eluvial horizon* (or layer) from which translocated material has been lost. The deposition of translocated material at a lower level in the soil is known as *illuviation.* The fine particles arriving in the *illuvial horizon* make this layer of the soil denser than the eluvial horizon above it. The illuvial layer sometimes acquires a yellow or orange color due to the arrival of oxides of iron and aluminum leached from the eluvial layer by acidic water. In dry regions calcium carbonate ($CaCO_3$) is translocated into the illuvial horizon, producing whitish flecks or veins of lime.

Organic Activity

Soil horizons also reveal the influence of plant and animal activity. Organic activity is crucial to soil development, and organic matter is a vital component of soils. Several types of organisms play a role in soil formation: plants contribute vegetative matter such as leaves, twigs, flowers, seed pods, and dead roots; fungi and bacteria reduce this vegetative matter to *humus;* and ants, termites, earthworms, and other soil inhabitants (which also alter plant detritus) mix the humus downward into the mineral matter of the soil.

Humus is the dark brown to black organic substance that makes the upper portions of the best soils so much darker than the deeper subsoils. Organically generated humus becomes so mixed with the mineral matter of the soil that it is almost impossible to separate. It has many beneficial properties. It increases the soil's ability to retain both moisture and soluble plant nutrients. It is an important source of the phosphorus and nitrogen required by plants. And it maintains a soil structure that is neither too compact nor too porous for plant growth.

Aside from releasing nutrients and producing humus from vegetative debris, soil bacteria are vital in *nitrogen fixation*—the conversion of gaseous nitrogen to forms that can be utilized by plants. Nitrogen, which is essential to plant growth, is present in the air that occupies pore spaces in the soil, but it cannot be taken up directly by plants in gaseous form. Usable nitrogen is made available to plants by the action of bacteria that inhabit the soil or are parasitic on the roots of the large family of plants known as *legumes,* which include peas, soybeans, clover, and alfalfa, as well as many uncultivated plants. Progressive farmers rotate their other crops with legumes to add extractable nitrogen to the soil. It is estimated that in the eastern United States organic activity fixes more than 135 kilograms of nitrogen per hectare (or 120 lbs per acre) each year.

The abundance of organisms in the soil is far greater than one would imagine. There may be a million earthworms and 25 million insects in one hectare (2.47 acres) of pasture land. As many as a million bacteria can inhabit one cubic centimeter of soil. All are functioning components of the soil system.

PROPERTIES OF SOILS

It is possible to compare soils of various types by focusing on several properties, including texture, structure, chemical characteristics, color, and profile development. These properties reflect the environment of a soil and the parent material from which it was derived. They are the products of a number of soil-forming factors to be discussed subsequently.

Soil Texture

Soil *texture* refers to the size distribution of the mineral particles composing the soil. Soil par-

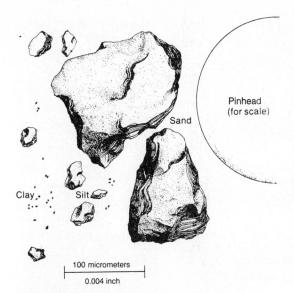

100 micrometers

0.004 inch

Figure 11.3 The particles that constitute the inorganic component of soil are classified according to size. Particles smaller than 2 micrometers in diameter are called clay, particles from 2 to 50 micrometers in diameter are called silt, and larger particles are considered sand or gravel. (John Dawson)

Table 11.1   Particle Size Classification

	Diameter	
Particle Class	MILLIMETERS	MICROMETERS
Gravel	greater than 2.0	greater than 2,000
Very Coarse Sand	2.0–1.0	2,000–1,000
Coarse Sand	1.0–0.5	1,000–500
Medium Sand	0.5–0.25	500–250
Fine Sand	0.25–0.10	250–100
Very Fine Sand	0.10–0.05	100–50
Silt	0.05–0.002	50–2
Clay	less than 0.002	less than 2

Source: U.S. Department of Agriculture. *Soil Taxonomy,* 1975.

Figure 11.4 The texture of a soil is determined by measuring the portions of clay, silt, and sand in the inorganic part of the soil. This is done by sifting the soil sample through a series of screens graded from coarse to fine. The soil texture triangle shown in the figure can be used to classify the texture of a soil sample once the percentages of the components are known. Texture classes are separated by white lines. If a soil sample contains 30 percent clay and 40 percent sand, for example, it would be classified as a clay loam. (Doug Armstrong after E. M. Bridges, *World Soils,* © 1970, Cambridge University Press)

ticle sizes are classified into four general categories: gravel, sand, silt, and clay—with further subdivisions within each category (Figures 11.3 and 11.4). Table 11.1 indicates the size ranges included in the categories and their subdivisions.

The United States Soil Conservation Service has established a classification of soil textures that describes the various mixtures of different particle sizes. In dealing with gravelly soils, it is the finer matrix that is classified, with the term "gravelly" or "gravel" appended. Figure 11.4 shows the standard classification scheme. The term *loam* indicates a mixture of sand, silt, and clay, which has a more favorable structure for root growth than does a soil composed of a more uniform particle size, such as sand or clay. Not shown in Figure 11.4 are the gravelly loams. Coarse-grained sandy soils are permeable and absorb water easily, but dry out rapidly. Their permeability also permits rapid leaching of soluble nutrients. Because individual clay particles (called micelles) are so small as to be visible only under an electron microscope, clay soils are very dense, making them

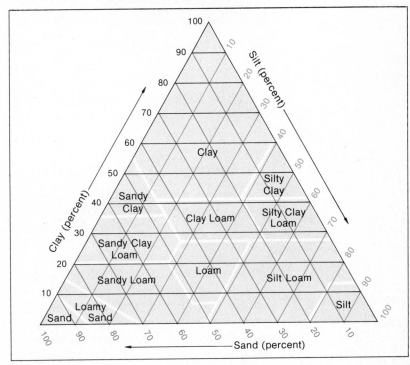

Figure 11.5 Four common soil ped structures are shown here: (a) platy, (b) prismatic, (c) blocky, (d) granular. Scales are in inches. (Roy W. Simonson, Courtesy USDA)

hard to work. They accept moisture very slowly and then hold it tenaciously.

Soil Structure

The particles composing most soils clump together in characteristic small masses called *peds*. The sizes and forms of peds determine the *structure* of the soil. As shown in Figure 11.5, soil peds vary considerably in form. This is a consequence of variations in the type and content of clay, in organic activity, and in both natural and artificial disturbances, including vegetation change, climatic fluctuations, and agricultural tillage. Most plants grow best in loamy soils with a granular or crumb structure,

having peds measuring 1 to 5 mm (0.04 to 0.2 in.) in diameter. Soil structure determines the rate of water absorption and the ease of root penetration.

Another aspect of soil structure is soil *bulk density*—mass per unit of volume, including pore space. Bulk density increases with clay content and is a measure of soil compactness. The greater the bulk density, the less the pore space between soil particles, and the more compact the soil. High bulk density is undesirable because diminished pore space reduces the rate of water acceptance by the soil. Unfortunately, pore space is often reduced unintentionally by the use of heavy agricultural machinery on clay-rich soils. This results in soil compaction and can lead to increases in surface runoff, soil erosion, and gully development.

Soil Chemistry

The chemical behavior, or "fertility," of soil is closely related to the *clay–humus complex* in the soil. The clay–humus complex consists of microscopic particles of humus and clay bound together so that they behave like large molecules (Figure 11.6). The extremely small (less than 2 micrometers) clay and humus particles carry negative electrical charges and remain in suspension in soil moisture. As a result they attract positively charged ions, or *cations*, such as those of the chemical bases calcium, magnesium, potassium, and sodium. These essential plant nutrients are easily dissolved and would be leached from the soil were it not for their retention by the clay-humus complex.

In a fertile soil the clay–humus complex maintains a delicate balance. It must hold nutrients strongly enough to keep them from being leached away, but not so strongly that plants cannot extract them from the soil. Many different types of cations are attracted and held in the clay–humus complex, with various degrees of strength. Weakly bound cations may be replaced by others that are more strongly attracted. Basic cations, such as the sodium ion, are easily replaced by metallic cations, such as iron or aluminum ions, or by hydrogen ions from water. The ability of a soil to absorb and retain exchangeable cations is known as the *cation exchange capacity*, or *CEC*. A high CEC indicates a fertile soil. A soil's CEC is related to the types and amounts of clay and organic matter present.

Not all soils have an active clay-humus complex. Desert soils are rich in soluble nutrients because so little water moves through them; but they contain almost no humus and very little clay. When such soils are watered artificially, the lack of a clay-humus complex permits rapid leaching of their soluble nutrients. This is a major problem of desert irrigation agriculture.

The hydrogen ions in a soil may present a problem, because they can displace plant nutrients. Hydrogen ion concentration, known as *acidity*, is measured in terms of the *pH scale*, ranging from about 3 to 10 in soils. Pure water has a pH of 7, which is regarded as neutral. This means that pure water contains 10^{-7} grams of hydrogen ions per 1,000 cubic centimeters (cc). The lower the pH, the more acidic the soil.

The best agricultural yields are obtained from soils with pH values between 5 and 7 (Figure 11.7). Acidic soils with low pH values

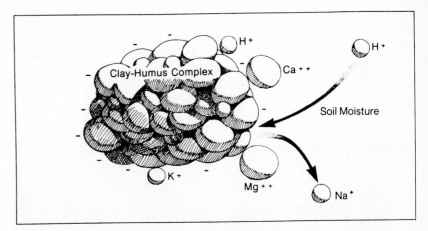

Figure 11.6 The finest inorganic particles and humus in a soil bind together to form the clay-humus complex. On a submicroscopic scale, the particles of the clay-humus complex act like giant molecules with the power to attract cations electrically. The complex performs an important function in a soil by preventing chemical nutrients needed by plants from washing out of the soil. Acidic soil moisture contains an excess of hydrogen ions that can replace basic cations on the surface of the complex, as the figure shows (H^+ replacing Na^+). Hence acidic soil moisture removes basic inorganic nutrients from the soil. (John Dawson)

Figure 11.7 The pH value of a soil, an indication of its hydrogen ion concentration, is one of the measures that can be used to estimate a soil's suitability for agriculture. A low pH indicates that a soil is acidic and may have lost many of its nutrients by exchange with hydrogen ions. A high pH indicates that a soil contains strong alkalis, which may be damaging to plant root tissues. (Doug Armstrong after Lyon and Buckman, *The Nature and Property of Soils*, 4th ed., © Macmillan Co.)

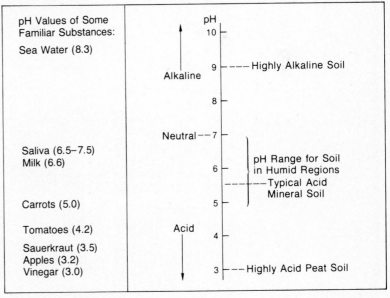

occur in wet areas having abundant partially decayed vegetation on the soil surface. It is possible to raise the pH of acidic soils by adding lime (CaCO₃) to them. However, soils with pH values greater than 7, called *alkaline* soils, can also present problems. They may contain

amounts of sodium that exceed plant tolerances and contribute to high bulk density and thus very poor soil structure. The alkalinity of such soils, which are common in semiarid regions, can be diminished by adding a source of hydrogen, such as ammonium sulfate ([NH₄]₂SO₄).

Soil Color

An obvious characteristic of any soil is its color. Soil color often changes as one digs downward. It also varies from place to place. Soils of the humid tropics commonly are orange or red. In the temperate grasslands soils are dark brown to black. Soils under coniferous forests tend to be gray near the surface, with an orange or yellow zone below.

Soil color results almost entirely from the amounts of organic matter and iron present, and from the chemical state of the iron. Organic matter colors the soil dark brown to black. Iron that has been oxidized (combined with oxygen) contributes reds, yellows, and browns. Where oxygen has been excluded, iron compounds have greenish and gray-blue hues. This occurs most often where waterlogging has kept air from moving through the soil. Other coloring matter is sometimes present, especially the white of calcium carbonate (lime). Soil color identifications are made by comparing soil samples with color charts designed specifically for the purpose of soil description (Figure 11.8b).

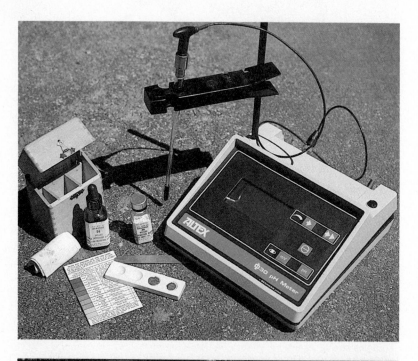

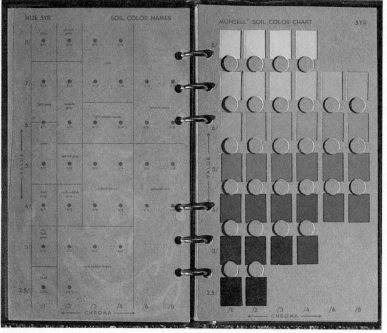

Figure 11.8 (a) Soil pH is measured using various techniques. Here we see a standard pH "kit" of the type used in the field for many decades. The soil pH is indicated by the color produced in a white powder sprinkled over soil moistened with a pH indicator solution. At the right is a newer electronic pH meter that can be used either in the field or in the laboratory.

(b) Soil color is determined by the use of a Munsell Soil Color Chart bound into a standard field reference booklet. The page with the color chips is placed over the soil to be described, so that the soil can be seen through the holes below each of the color chips. The color that best matches the soil is characterized in terms of *hue* (position in the color spectrum), *value* (degree of lightness or darkness), and *chroma* (color saturation, or departure from a gray of the same value). (T. M. O.)

Soil Profiles

Every soil has its own distinctive sequence of layers, or horizons, resulting from the processes of translocation and organic activity. These horizons constitute the *soil profile*. Five separate layers, termed the *O, A, B, C,* and *R* horizons, can often be distinguished. Each of these is further subdivided, as shown in Figure 11.9.

The *O* horizon consists of undecomposed plant litter or raw humus at the soil surface.

Figure 11.9 Standard horizons in soil profiles. No single profile contains all of the horizons shown. Additional subhorizons similar to those indicated for the C horizon include: *B2t*—illuvial clay; *B2ir*—illuvial iron; *B2h*—illuvial humus; *B2m*—strong cementation; *Csi*—cementation by silica; *sa*—enriched by salts; *f*—permanently frozen; *x*—hardpan composed of sand and/or silt. (Vantage Art, Inc. modified from Robert Ruhe, *Geomorphology*, 1975, Houghton Mifflin, and Soil Survey Staff, 1951, 1962)

Below it is the *A* horizon, at the top of which humus is mixed with mineral particles. The presence of organic matter makes the *A* horizon dark in color at the top; farther down the loss of fine mineral and soluble substances by eluviation (Figure 11.10) causes it to be lighter in color and relatively porous and light in texture (sandy or silty).

The *B* horizon receives material translocated from the *A* horizon, giving it a higher clay content and a greater bulk density than the *A* horizon. It is in the *B* horizon that we encounter blocky and prismatic soil structures and sometimes dense clay "hardpans" as much as a meter thick. The *B* horizon may also be vividly colored by iron oxide leached from the *A* horizon. The *O, A,* and *B* horizons collectively are known as the *solum*—the portion of the soil displaying the effects of pedogenic processes.

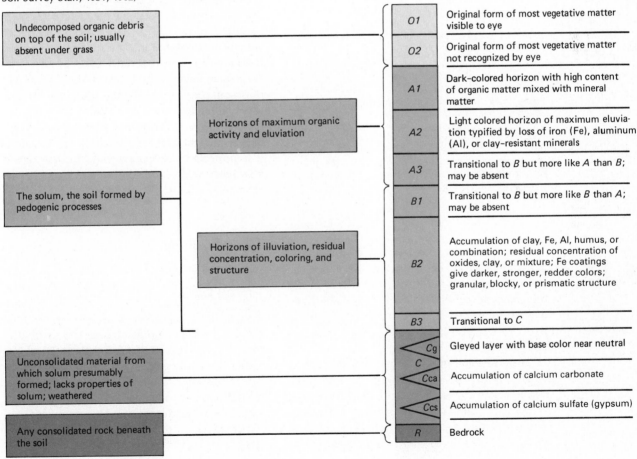

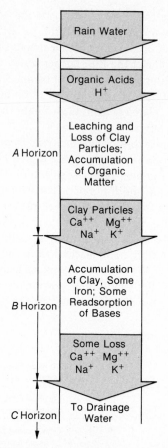

Figure 11.10 In the eluviation process, rainwater, which has been made slightly acidic by dissolved carbon dioxide or organic acids in humus, infiltrates the soil. The hydrogen ions in the acidic water displace basic cations from the clay-humus complex, causing a downward movement of soluble nutrients. (Doug Armstrong after E. M. Bridges, *World Soils* © 1970, Cambridge University Press)

The C horizon is composed of weathered material or loose deposits that have not yet been affected by soil-forming processes. In dry regions some whitish calcium carbonate (lime) may be deposited in the C horizon. The R horizon, where present, consists of unweathered bedrock.

From one soil profile to another the characteristic horizons vary greatly in thickness, depth below the surface, and strength of development. Laboratory procedures are necessary to determine such important aspects of soil profiles as bulk density and the concentrations of organic matter, carbonates, clay, oxides, and other constituents at various levels. Since many soil profiles reveal changes through time in the soil-forming processes, the profiles are indicators of past as well as present environmental conditions. As such they are used by archaeologists to reconstruct the environments of ancient cultures. Soil profile characteristics are more routinely used to identify and map specific soil types for purposes of crop selection, irrigation scheduling, fertilization, and other aspects of land-use planning.

FACTORS AFFECTING SOIL DEVELOPMENT

Our present understanding of soils has developed largely out of work begun by Russian soil scientists more than a hundred years ago. Over the vast area of Russia, soil characteristics seem closely associated with large-scale patterns of climate and vegetation. The studies of the early Russian soil scientists and their more recent successors around the world have indicated that soil profiles show the influences of five separate factors: (1) parent material, (2) climate, (3) site, (4) organisms, and (5) time. Soil scientists today call these the *factors of soil formation*. It is important for geographers to understand how each of these factors influences soil characteristics.

Parent Material

The inorganic material on which a soil develops is called the soil's *parent material*. This can be rock that has decomposed in place, or loose material that has been deposited by streams, glaciers, rockfalls, landslides, or the wind. The parent material determines what chemical elements are initially present in the soil. Some parent materials have an abundant supply of the nutrients most needed by plants; others lack them in varying degrees; and some contain substances that are toxic to certain types of vegetation.

If the parent material is rich in soluble bases—calcium, magnesium, potassium, and sodium—which are easily dissolved by water, it can continually supply these nutrients despite constant leaching from the soil. Limestone and basaltic lava have a high content of soluble bases and in humid areas produce the most fertile soils. Outstanding examples are seen in southeastern Pennsylvania near Lancaster and Harrisburg and in Alabama's Black Prairie, which are areas of limestone bedock; in regions of basaltic rock in Virginia, Maryland, Delaware, New Jersey, and Pennsylvania; and on the calcareous (lime-rich) glacial deposits of the so-called "corn belt," extending from Iowa to Ohio.

If soluble nutrients are not abundant in the parent material, water moving through the soil removes bases and replaces them with hydrogen ions (Figure 11.6). The soil thus becomes increasingly acidic and decreasingly suitable for agriculture. Soils developed on sandstone are often infertile, being poor in nutrients and coarse in texture, which facilitates leaching. Thus the natural vegetation of the sandy Coastal Plain from Delaware to New Jersey consists of pine forest, which has very low nutrient demands. The soils of this region are too poor for any type of agriculture. Similar "pine barrens" occur on sandy soils in the Carolinas, Georgia, and Florida.

In addition to soluble bases and nitrogen, crop plants require iron, phosphorus, and sulfur, and trace amounts of such elements as boron and copper. If these are not available in the soil's parent material, farmers must supply them artificially.

Abrupt changes in natural vegetation commonly reflect a change in soil due to variations in parent material. Even so, soils from dissimilar parent materials may become quite similar with the passage of time if other soil-forming factors are the same.

Climate

On a global scale, major soil types show a close relationship to climatic zones. The energy and moisture delivered by the atmosphere influence many aspects of soil formation. These include translocation, the rates of chemical reactions, and organic activity in the soil. Abundant rainfall aids translocation, and warm and wet conditions favor chemical reactions and organic activity.

Both vegetative production and the activity of soil bacteria and of larger organisms are curtailed in desert areas and in tundra regions at high latitudes or lofty elevations. In the dry desert environment, both plant litter and soil organisms are minimal, and in the tundra, organic litter decays slowly, often forming acidic peat rather than a clay–humus complex. Plant production is at a maximum in the warm and wet tropics, but here the destruction of litter by organisms is so rapid and thorough that the soil is actually poor in organic matter.

Climate affects the chemistry of soil moisture, which, in turn, affects the solubility of various substances in the soil. For example, iron can be removed only by acidic water. Soil water tends to be acidic in cool, wet areas, which are normally covered by coniferous forest. Therefore iron is leached from the eluvial layer in such areas (see p. 276). In dry regions, lime leached from the upper portion of the soil is redeposited at a lower level where the moisture evaporates rather than moving through to the water table.

Many soils contain features formed thousands of years ago under different environmental conditions. Such "relict" features are important indicators of past climates and vegetation. Soil features have given evidence of shifts in the forest–tundra boundary in high latitudes and in the forest–grassland boundary in the midcontinent region. Soils also reflect expansion and contraction of the world's deserts, as well as less severe climatic fluctuations in nearly all parts of the world. The interactions between climate and soil formation will be made clearer in the discussion of major pedogenic regimes later in this chapter.

Site

The specific location of a soil helps determine the soil type. Since water drains downward, soil at the foot of a slope will evolve in a wetter environment than soil on the hillcrest or on the slope itself. The material at the slope foot is finer than that upslope because fine material is washed downslope on the surface, while the greater dampness at the slope foot causes clay formation by chemical weathering within the soil itself.

Generally, soils on slopes are thinner, stonier, lower in organic matter, and less well developed than those on level or low-lying land. On level surfaces the effects of translocation are much stronger, producing an eluviated *A* horizon resting on a more massive clay-rich *B* horizon.

Organisms

The type and intensity of organic activity vary geographically. Since much of this variation is due to climate or microclimate, the climatic and organic factors in soil development are sometimes hard to separate. We have already noted the general influences of organisms in the discussion of soil-forming processes, but there remains one aspect of special importance—the *nutrient cycle*.

Wherever plants and animals exist, there is a constant cycling of material and energy between life forms and their environment. Organisms need the nutrients in soils to carry out their life-sustaining processes. They return these same nutrients to the environment as waste products, or as litter, or in the form of their own bodies when they die. This establishes a nutrient cycle. Without constant uptake and return of nutrients, soluble compounds would soon be leached out of the soils of humid regions. This would cause a steady decrease in the soils' capacity to support life.

Nutrients not used by organisms are indeed gradually flushed from the soil. Which nutrients are lost depends upon the climate and life forms present (Figure 11.11). We have noted previously that pines do not require nutrient-rich soils. Therefore the soils under pine forests gradually lose their soluble nutrients and become acidic. On the other hand, tropical forests have high nutrient demands. Despite the heavy rainfalls of the humid tropics, the forest soils are stabilized by rapid nutrient cycling between the vegetation and the soil. Human removal of such forests, in which the bulk of the nutrients are stored at any time, is followed by rapid soil deterioration. The soil can then be restored for agricultural use only by constant application of costly fertilizers.

Time

Time is required for translocation and organic activity to produce strong horizon development in soils. But do soil horizons continue to strengthen with the passage of time, or does the soil eventually reach an equilibrium with its environment and cease to be altered further? Because the environment and soil-forming factors frequently change, the answer to this question is uncertain.

It has been possible to study the rate of soil formation on parent material that was deposited in historic time. When the volcano Krakatau, between Java and Sumatra, exploded in 1883, large amounts of volcanic ash fell on the surrounding land. Under the moist tropical climate prevailing there, soil development on the ash was rapid—the soil thickened at a rate of about 1 cm (0.4 in.) per year. In drier areas in Central America, it has taken a thousand years to produce a soil about 30 cm (1 ft) thick on volcanic ash.

Many of the world's most valuable soils are fertile because they are young. They have formed too recently to be strongly affected by chemical and mechanical eluviation. The soils of river floodplains are usually very productive. These soils have developed upon *alluvium* composed of sand, silt, and clay deposited by streams during floods. Soils on recent volcanic ash and glacial deposits are also normally fertile because they have not yet been strongly modified by translocation.

By contrast, flat land surfaces of great age tend to have infertile soils. In such environments thin impoverished *A* horizons overlie impermeable *B* horizons composed of dense clays or cemented into a rock-like mass by translocated lime, iron oxide, or silica. Here and there

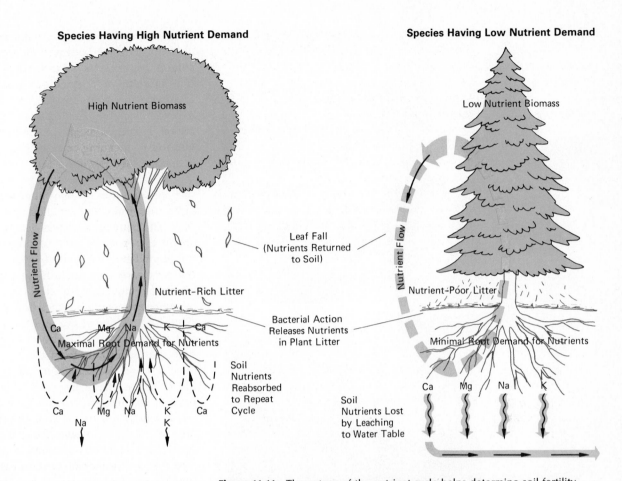

Figure 11.11 The nature of the nutrient cycle helps determine soil fertility.
  (**left**) Plant species with high nutrient demands prevent soluble compounds from being leached from the soil. The plant extracts nutrients and then returns them to the soil in the form of plant litter. The constant two-way exchange between the plant and the soil maintains a high concentration of nutrients in the soil.
  (**right**) Plant species with low nutrient demands permit unused soil chemicals to be leached away, and themselves return few nutrients to the soil. Thus the soil's fertility declines. (Vantage Art, Inc.)

on the plains of Australia, Africa, and South America, and also in the southwestern United States, the land surface has been eroded down to these "hardpans," which produce resistant crusts over the weathered parent material. These so-called *duricrusts* are best developed in areas that have well-defined dry seasons. Figure 11.12 shows a duricrust layer cemented by calcium carbonate.

dogenic regimes produces a soil of a distinctive general type that reflects major geographic variations in energy and moisture budgets.

MAJOR PEDOGENIC REGIMES

Most of the earth's soils have been created by one of a small number of distinctive *pedogenic,* or soil-forming, *regimes*. These are related to climate both directly and indirectly through the influence of vegetation. Each of the major pe-

Laterization

The unique feature of most tropical soils is their high content of iron and aluminum oxides in relation to silica, giving them a brick-red color. Such soils result from intensive chemical weathering and leaching under hot, wet conditions.

(a)

(b)

Figure 11.12 A calcium carbonate duricrust, called calcrete, or *caliche* in the southwestern United States, forms a resistant caprock and produces a bold ledge along the valley of the Virgin River in southwestern Nevada: (a) general view; (b) detail of the calcrete caprock. In tropical savanna regions, ferricrete (laterite) crusts produce similar landforms. Such crusts form within the soil and are exposed by later erosion of the upper part of the soil profile. (Dr. John S. Shelton)

Silica is normally the most abundant mineral in decomposed rock because it is a common constituent of many rock types and is highly resistant to solution. To remove silica from soils requires a very aggressive regime of weathering and soil leaching. This distinctive soil-forming regime is known as *laterization* (from the Latin word *later*, meaning brick or tile). The process is diagrammed in Figure 11.13.

In the humid tropics, decomposition and insect consumption of plant litter are usually too rapid to permit the formation of abundant humus. The paucity of humus decreases the soil's ability to retain the soluble nutrients that are not immediately taken up by plants. When forests are cleared and the nutrient cycle is broken, leaching removes nearly all soil cations, and in extreme cases even the silica from decomposing clays. The residue of the process is quartz sand and oxides and hydroxides of iron and aluminum that eventually form a rock-like red duricrust known as *laterite*, or *ironstone* (Figure 11.14). Occasionally the residue is *bauxite*, the principal commercial ore of aluminum. Laterite crusts have been reported as developing in as little as thirty years when tropical forests are removed. However, such crusts are most common in the seasonally-wet savanna regions fringing the Amazon and Zaire (Congo) basins and in Southeast Asia, sometimes attaining thicknesses of as much as 6 meters (20 feet). The formation of laterite crusts in savanna regions seems to involve oxide concentration by water table fluctuations that help to mobilize and remove cations well below the land surface.

The process of laterization in a less extreme form has created the red soils on old land surfaces in the southeastern United States. These soils were rapidly exhausted by intensive cotton farming in the 1800s. The laterized red clay soils extending from Alabama through the Carolinas are, nevertheless, usable agriculturally with proper fertilization and careful erosion control measures.

Podzolization

The vast coniferous forests of the higher middle latitudes in North America and Eurasia do not

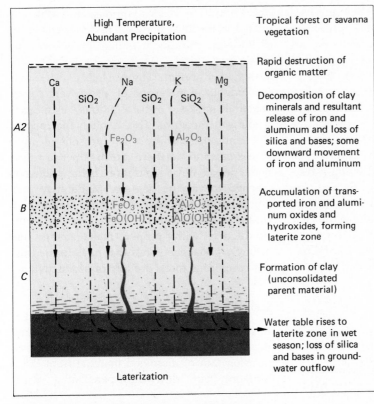

Figure 11.13 (**left**) Diagrammatic illustration of the process of laterization, showing movements of chemical substances. Substances in red in this figure are accumulating in place as residuum or by illuviation. (Vantage Art, Inc.)

(**right**) This laterized soil has developed on deeply weathered rock in central Puerto Rico. The soil is low in nutrients but is permeable to water and is used agriculturally. (Reproduced from Soil Science Society of America, C. F. Marbut Memorial Slide Collection)

Figure 11.14 This photo shows some of the forms taken by laterites. The rounded spherules (0.5 to 1 cm diameter) are iron-rich *pisolites* from Western Australia. These constitute ore in one of the world's largest iron mines. Resting on the pisolites are a more typical slaggy laterite from Brazil and a porcelain-like laterite fragment from Indonesia. The latter two types commonly form ledges similar to that seen in Figure 11.12. (bottom). (T. M. O.)

have high nutrient demands. They produce acidic plant litter and permit soluble bases to be leached from the soil. Since organisms do not thrive in the acidic environment of the soil (which is frozen during the winter), there is very little humus production. As in the moist tropics, the lack of an active clay–humus complex is a factor in the leaching of soluble nutrients, including humus itself. The acidic soil moisture is able to decompose clay and remove iron and aluminum in solution, leaving a residue of silica. The mobilization of iron and aluminum and immobilization of silica are the reverse of the laterization process and produce a light gray ashy-looking A horizon. Thus the general soil-forming process is called *podzolization*, from the Russian words *pod* (under) and

zola (ash). The oxides, humus, and clay eluviated from the A horizon are deposited in the B horizon, coloring it orange or yellow, and sometimes black, in sharp contrast to the bleached A horizon (Figure 11.15).

Podzolization can occur also where parent materials are poor in basic cations, resulting in acidic soil moisture. This is often the case on the sands of coastal plains. The soils of much of eastern and northern North America reflect some degree of podzolization, being acidic and leached to varying degrees. Moderately podzolized soils that have developed under deciduous or mixed forest vegetation can be highly productive with proper management, but strongly podzolized soils that have formed under coniferous forest are low in productivity.

Calcification

In semiarid and arid regions, where little moisture moves through the soil, there may be some translocation of dissovled material from the A horizon, but downward-percolating water

Figure 11.15 **(left)** Diagrammatic illustration of the process of podzolization, showing movements of chemical substances. Substances in red in this figure are accumulating in place as residuum or by illuviation. (Vantage Art, Inc.)
(right) This podzolized soil in New York State has developed on sandy glacial outwash. The bleached eluvial A2 horizon and the iron-enriched illuvial B horizon are shown distinctly. (Reproduced from Soil Science Society of America, C. F. Marbut Memorial Slide Collection)

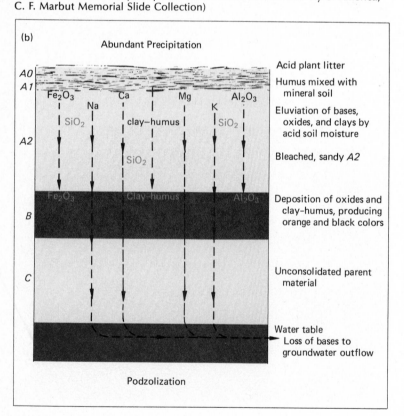

evaporates the *B* or *C* horizon. The most visible translocated mineral is calcium carbonate, which is deposited in various forms in the *B* and *C* horizons, forming a subsidiary *calcic horizon*. This subsoil enrichment process is called *calcification* (Figure 11.16).

Soils produced by the calcification process show varying degrees of horizon development. In midlatitude grasslands, dark brown to black humus-rich *A* horizons normally occur above light-colored *B* horizons that contain only flecks or nodules of calcium carbonate. In semidesert areas there is almost no organic horizon, but a strongly developed *B* horizon containing conspicuous veins of calcium carbonate. The depth of the calcic horizon indicates the depth to which moisture penetrates in the soil before evaporating.

The calcification process produces soils that are rich in nutrients and are especially productive where there is a thick organic horizon. Such

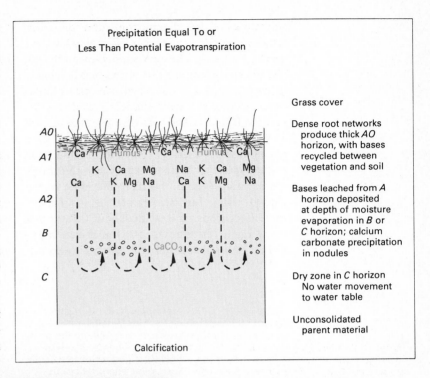

Calcification

Figure 11.16 (**top**) Diagrammatic illustration of the process of calcification, showing movements of chemical substances. Substances in red in this figure are accumulating in place as residuum (humus) or by illuviation (CaCO₃). (Vantage Art, Inc.)

(**bottom left**) This calcified soil has developed under tall-grass prairie in central Iowa and is used for growing crops such as corn and soybeans. The parent material is loess, which is fine dust transported long distances by wind.

(**bottom right**) This calcified soil has developed under short-grass prairie in eastern Colorado. The organic content and the soil depth are much lower than those of the soil in Iowa. This soil is used primarily to grow small grains. (Reproduced from Soil Science Society of America, C. F. Marbut Memorial Slide Collection)

soils were developed in the transition zone from tall-grass prairies to short-grass steppes in the middle latitudes. The outstanding examples are the Russian Ukraine, the plains of Manchuria in northern China, the North American Great Plains east of the 100th meridian, and the Argentine Pampas. The soils of these regions are so valuable that, except for urban sprawl, they are almost all used as cropland. Fertile soils produced by calcification extend into the short-grass prairie west of the 100th meridian in the United States, but here rainfall is insufficient for reliable crops. This is "dust bowl" country—tempting to the farmer, but the scene of repeated droughts resulting in economic and hu-

man disasters. A vast area of very similar short-grass steppe in the Soviet Union's province of Kazakhstan (east of the Caspian Sea) has also been the scene of chronic crop failures since the first introduction of large-scale agriculture into the region in the 1950s.

Salinization

In dry regions, water that runs off hillslopes is often trapped in depressions with no drainage outlet. This may produce ephemeral lakes that soon evaporate. In these depressions the water table can rise close to the level of the land surface. Above the water table there is always a *capillary fringe* in which water rises upward through pore spaces in the soil. If the water table is close to the land surface, capillary action brings water all the way up to the surface, where it evaporates. Salts contained in the water are left behind as a white powder covering the dry soil, as in Figure 11.17. This process is known as *salinization*. The deposits consist of various chlorides, sulfates, and carbonates of calcium, magnesium, and sodium, including common salt (NaCl) and gypsum (CaSO$_4$). Because high salt concentrations are toxic to most plants and soil organisms, salinization causes land to become barren and useless.

Figure 11.17   (a) Excessive application of irrigation water in this location on the Mesopotamian Plain of Iraq has raised the subsurface water table to the point at which capillary rise of moisture brings dissolved salts to the surface, where they are precipitated as the moisture evaporates. Such saline encrustations destroy the land for agriculture. Induced salinization of irrigated cropland has plagued the Mesopotamian area for thousands of years, and is a problem in most irrigated desert areas, including those in the United States. In the distance is a field of irrigated barley.

(b) This trench exposes a cross section of a gleyed histosol in Ireland. The alternating color bands indicate the state of the iron in the soil (red colors indicate oxidation; blue-black colors indicate reduction, with oxygen excluded). The material is being quarried for peat, which, when dried, is used for household fuel and to generate electricity in commercial power plants. (T. M. O.)

(a)

(b)

Unfortunately, salinization is often caused by human use of the land. When desert lowlands are irrigated for agriculture, some of the water percolates through the soil to the water table. This raises the water table, causing waterlogging and artificial salinization of the soil. This has been a problem in irrigated areas since ancient times. We first read of it in 4,000-year-old cuneiform records from Mesopotamia, which mention steady decreases in tax collections because of declining yields and land abandonment attributed to salinization of irrigated cropland. Russian scientists have estimated that every year artificial salinization in dry lands throughout the world ruins as much as 300,000 hectares (741,000 acres) of cropland. The Mesopotamian Plain in Iraq, Egypt's Nile Valley, the Indus Valley in Pakistan, and the San Joaquin and Imperial valleys in California are all seriously affected by artificially-induced salinization.

Gleization

In regions of high rainfall, low-lying areas may be waterlogged naturally. Here the water is not saline, because salts are leached away in the constantly wet environment, and vegetation can grow thickly. The litter from this vegetation cannot be decomposed normally due to the low level of bacterial activity in the oxygen-poor muck. As a result, the upper soil takes the form of dark *peat*, composed of partially rotted plant matter (Figure 11.17). Decay of plant material in the damp environment releases organic acids that react with the iron in the soil. The iron is reduced, rather than oxidized, giving wetland soils a black to bluish-gray color. Acidic wetland soils of such colors are called *gley* soils, and the formative processes involved are called *gleization*.

Gley soils occur throughout the world. In North America they are most often related to depressions resulting from irregular deposition of debris by continental ice sheets. Therefore, they are scattered through the American Midwest north of the Ohio and Missouri rivers, as well as in marshy areas on river floodplains, especially in the Mississippi River Valley.

To make gley soils productive for dryland crops, it is necessary to drain them and raise their pH values by adding lime, thus creating the proper chemical environment for humus-producing microorganisms.

CLASSIFICATION OF SOILS

The classification of soils is even more difficult than the classification of climates, since there are many more variables to take into account. Whereas a climate may be characterized by its temperature and moisture regimes, soils vary in texture, structure, organic content, cation exchange capacity, the details of horizon development, parent material, site, and relative age. Furthermore, soils, unlike climates, can differ markedly from point to point within an area the size of a city block. Finally, human activities can completely alter soils both physically and chemically, either by design or unintentionally.

Soil classification begins with detailed field work that involves the description and mapping of localized soil types. In the United States these form more than 10,000 different *soil series* that are named after the geographic location in which the soil is found. Different soil series having features in common are subsequently grouped into *soil families*. The classification scheme then ascends through a series of steps including about 1,000 U.S. soil *subgroups*, 230 world-wide *great groups*, 47 *suborders*, and 10 *orders*. The higher levels of this classification system were adopted in 1960 by the U.S. Department of Agriculture, superseding an earlier highly generalized classification based on presumed climatic and biotic influences on natural soil characteristics. The current classification system, officially termed the *United States Comprehensive Soil Classification System*, is more often called the *7th Approximation System*, since it was actually the seventh revision of the original proposal submitted to soil scientists for test applications.

In the 7th Approximation System, the focus is on the specific properties of soil as it exists now, rather than on its environment, genesis,

Table 11.2   Nomenclature for Principal Soil Horizons: U.S. Comprehensive System*

Epipedon: Surface horizon darkened by organic matter; includes eluvial horizons.

    Mollic epipedon (from Latin *mollis,* soft; Greek *epi,* outer; Greek *pedon,* earth)

        Dark colored; at least 1 percent organic matter; more than 50 percent base saturation (that is, 50 percent of total exchange capacity saturated by base cations).

    Umbric epipedon (from Latin *umbra,* shade)

        Dark with humus, but less than 50 percent base saturation.

    Ochric epipedon (from Greek *ochros,* pale)

        Light in color and low in organic matter.

Subsurface horizon

    Albic horizon (from Latin *albus,* white)

        Bleached due to removal of oxides and clays.

    Argillic horizon (from Latin *argilla,* clay)

        Enriched by translocated clay.

    Natric horizon (from Arabic *natrun,* sodium carbonate)

        Having columnar or prismatic structure and a high sodium content.

    Spodic horizon (from Greek *spodos,* wood ash)

        Enriched by translocated dark organic matter or iron and aluminum oxides.

    Oxic horizon (from *oxygen*)

        Strongly weathered, with bases leached to leave a residue of hydrated iron and aluminum oxides and permeable clays.

    Cambic horizon (from Latin *cambiare,* to exchange)

        Little changed from weathered parent material, but having minor illuvial carbonate or clay and/or greater coloring by iron and aluminum oxides than higher or lower horizons.

* This table includes only the most common of the many horizons possible.

or natural condition. This avoids the problem of determining the conditions under which particular soils have formed and developed through time. The 7th Approximation System creates an entirely new terminology, in which mainly Greek and Latin roots are combined to form terms that describe a soil's properties precisely. The system includes ten major soil orders. The generalized formative influences on the seven most widely distributed soil orders are illustrated in Figure 11.18. The general relationship between climate and the soil orders is illustrated in Figure 11.19. The nature of the new classification is apparent when we encounter the nomenclature for the distinctive types of soil horizons, which are outlined in Table 11.2.

The names of the orders, suborders, great groups, and subgroups of the new classification are very descriptive, once the formative elements are learned. For instance, a *Plinthaquult* (great group) has a well-developed textural *B* horizon (order *Ulti*sol), including a zone of concentration of iron and aluminum compounds (*plinth*ite), in a wet setting (*aqu*).

Such terms are now familiar to North American soil scientists, but their number and complexity are sufficiently forbidding that at the introductory level it is not practical to go beyond the terminology of the principal soil orders. In the following pages, we shall summarize the nature of the ten soil orders in the 7th Approximation system, whose global distribution is shown in Figure 11.20. The suborders are summarized in Appendix III. The soil orders are described here in order of increasing level of development.

Entisols (from the word "recent") are soils with poor horizonation (Figure 11.21), because of any of several factors: youth of the soil, rapid erosion during soil formation, waterlogging, or human interference, such as plowing. Obviously, these conditions differ considerably, and thus the range of Entisol types is wide. Suborders make the necessary distinctions.

Histosols (from the Greek *histos,* tissue) are composed primarily of plant material (see Figure 11.21). They develop in waterlogged environments, where organic matter decomposes slowly due to the lack of oxygen required by bacteria (Figure 11.17). Suborders specify drainage conditions and the degree of decomposition of the plant material. Histosols are usually, but not always, acidic, and their horizons are based on the degree of compaction and state of decomposition. They can develop almost anywhere, from the high arctic to equatorial forests, but they are most extensive in tundra regions and glaciated landscapes of higher latitudes. Histosols can be very productive when drained; however, the subsequent oxidation and drying of the organic material cause compaction and subsidence of the land surface, and increase the hazards of fire, flooding, and wind erosion.

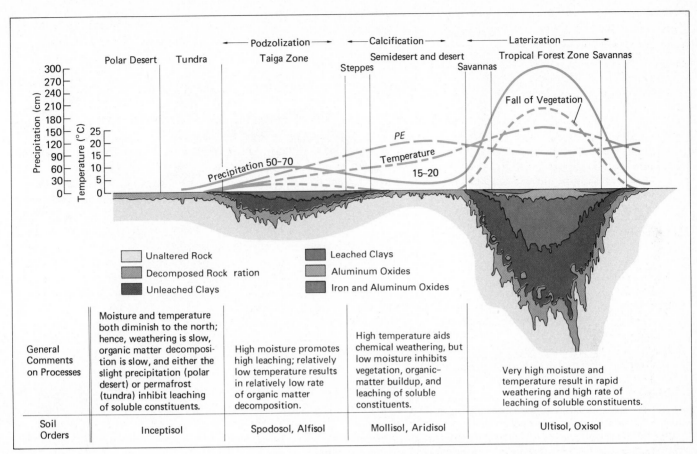

Figure 11.18 Latitudinal distribution of influences on soils. This diagram summarizes how precipitation, energy, potential evapotranspiration (*PE*), and vegetation affect rock weathering and soil formation. Irregular boundaries indicate uneven depths of the zones indicated. (Vantage Art, Inc. modified from N. M. Strakhov, *Principles of Lithogenesis*, vol. 1, 1967, Plenum Publishing Corp., and P. W. Birkeland, *Pedology, Weathering, and Geomorphological Research*, 1974, Oxford University Press)

Vertisols (from the Latin *verto*, to turn) are clay-rich soils in which horizon development is impeded by the churning effects of expansion and contraction due to seasonal wetting and drying. Such soils develop in semiarid regions and are dominated by sodium or magnesium compounds and related clay minerals that have exceptional water-absorbing capacity. When dried, Vertisols shrink, harden, and develop systems of cracks as much as 2.5 cm (1 in.) wide and 50 cm (20 in.) deep. These cracks collect organic debris that falls into them. Wetting causes the soil to swell, closing the cracks and churning sticky soil masses against one another. In much of California, Vertisols of the suborder xererts are the normal soil type and are known as "adobe" from their use in the making of sun-dried brick in the Spanish period of California history. Vertisols of the suborder usterts are ex-

tensive in Texas, Australia, and India. Vertisols are widespread in summer-dry subtropical climates, winter-dry tropical areas overlying base-rich basaltic rocks, and wherever parent materials consist of sodium-rich marine sediments. These soils are rich in plant nutrients but are difficult to use agriculturally because of their poor structure and problems of nutrient availability induced by the chemical effects of excessive sodium.

Inceptisols (from the Latin *inceptum*, beginning) are young soils that have developed in

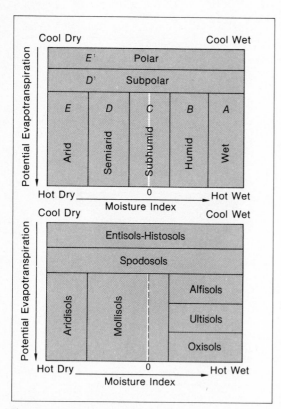

Figure 11.19 (above) Climate is one of the important factors in soil formation. The top diagram shows the principal climate regions according to the Thornthwaite system. Potential evapotranspiration, a measure of energy, increases downward from cool to hot on the vertical scale. The moisture index, a measure of the moisture available for systems, increases from dry to wet along the horizontal scale. The bottom diagram classifies the principal soil types according to the same method.

The similarities between the two diagrams show that some soils form primarily in a particular climatic region. Waterlogged Entisols and Histosols are most common in tundra regions, where subsoils are frozen. Mollisols rich in humus and bases form under grassland in the subhumid climate region, where the moisture index is small or zero. Aridisols with little profile development occur in the arid climatic region. In the humid and wet climate regions, the soils arrange themselves primarily according to energy rather than to moisture, partly because of the important effect of vegetation type on soil development. Thus soils range from thoroughly leached and acidic Spodosols developed under coniferous forest, through more neutral Alfisols, to successively less fertile Ultisols and Oxisols formed under subtropical and tropical humid climates.

Figure 11.20 World-wide distribution of soil orders according to the U.S. Comprehensive Soil Classification System (7th Approximation). (Adapted from USDA, Soil Conservation Service, 1972)

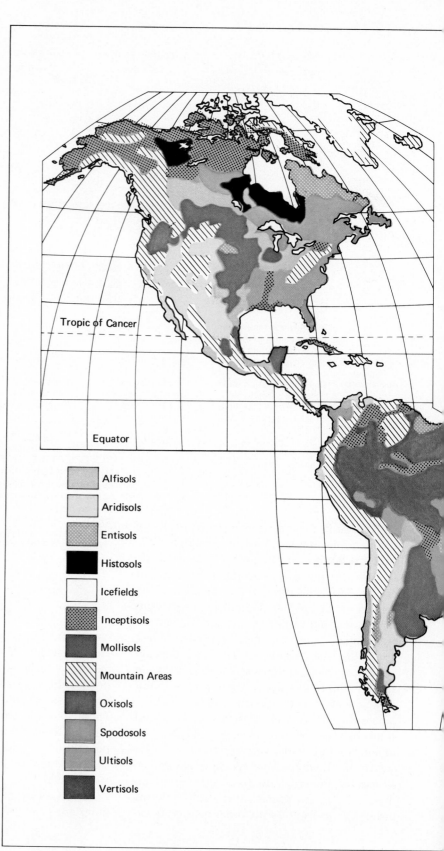

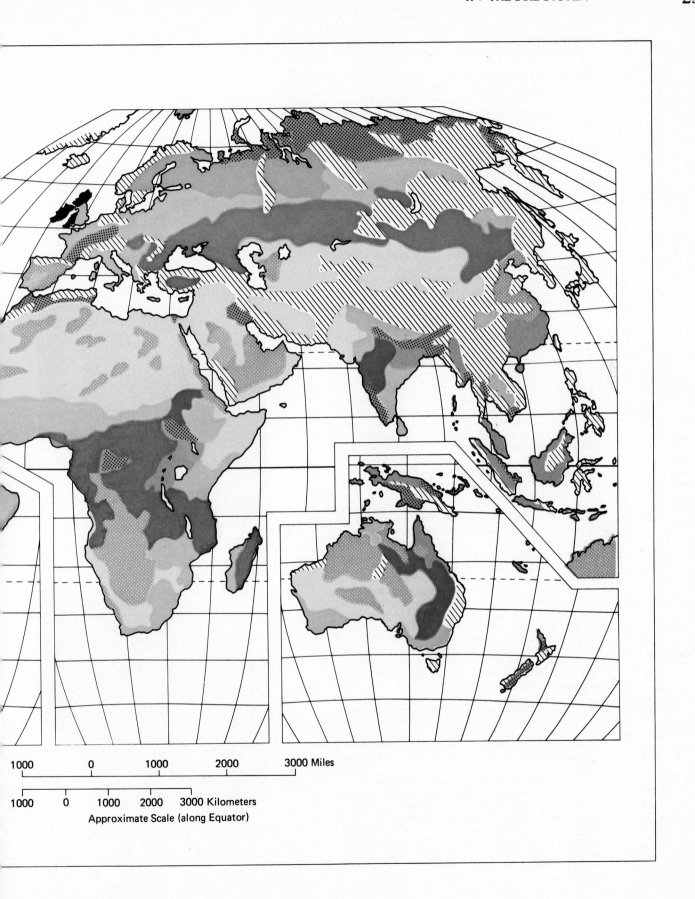

Figure 11.21 Characteristic types of soil are associated with certain climates, vegetation, and drainage conditions, as these illustrations suggest.

(a) *Tundra Region.* Tundra vegetation consists of mosses, lichens, low shrubs, grasses, and flowering herbaceous plants. Tundra soils are developed over ground that is permanently frozen and thus impermeable to water. Repeated disturbances of the soil by freeze and thaw, as well as seasonal waterlogging, are characteristic. Networks of ice wedges are also common in low-lying areas. The soils have poor horizonation.

(b) *Coniferous Forest Region.* Coniferous forests produce acidic litter and do not recycle soluble bases in large amounts. Consequently, the soils beneath such forests are acidic, with clay, humus, bases, and even the oxides of iron and aluminum eluviated by acidic soil moisture. The result is the distinctive podzolized soil type, in which a bleached sandy *A2* horizon overlies an oxide-colored illuvial *B* horizon. Such soils are infertile unless neutralized by the addition of lime.

(c) *Short-Grass Prairie Region.* Grasslands soils are generally high in nutrients due to low precipitation, rapid nutrient cycling, and a rich clay-humus complex. The organic horizon is well developed and dark in color due to its humus. Short-grass prairies have shallower soils than tall-grass prairies because soil moisture does not penetrate as deeply. The soils formed in the transition between short- and tall-grass prairies are exceptionally fertile. Short-grass prairies have a visible zone of lime accumulation in the *B* or *C* horizon due to shallow penetration of moisture.

(d) *Tall-Grass Prairie Region.* Tall-grass prairies receive more precipitation and develop soils that are deeper and even richer in organic matter than the soils of short-grass prairies. They have high base saturation, however, and are somewhat more leached than the short-grass prairie soils. This fact suggests that at times these soils have been occupied by forests, which have affected the soils' development. They are the most fertile soils located within a zone of reliable rainfall.

(e) *Midlatitude Forest Region.* Beneath the broadleaf deciduous forests of the midlatitudes, soils tend to be well supplied with mineral nutrients due to nutrient cycling where deep roots penetrate the parent material and nutrients stored in the vegetation are returned seasonally to the soil in leaf fall. In midlatitudes, bacterial action supplies abundant humus.

(f) *Tropical Forest Region.* Plant litter decomposes so rapidly in the humid tropics that humus formation is weak. Nevertheless, the steady decay of plant litter on the forest floor returns nutrients to the soil, where thay are immediately taken up by vegetation. Thus soil nutrients are largely locked up in the vegetation itself, with the subsoil being eluviated of clays, bases, and even silica. Oxides and hydroxides of iron and aluminum color the soil orange or red.

(g) *Desert Region.* Desert soils contain little clay and organic matter due to the lack of chemical weathering and the sparse plant cover. Thus they lack a clay-humus complex. They are high in nutrients since there is little eluviation, but are easily leached when irrigated. Desert soils commonly include calcrete (caliche) layers.

(h) *Bogs and Meadows.* Decaying plant material makes most waterlogged soils strongly acidic, and the upper horizon may be dark peat. Changes in water level produce color changes as iron is oxidized by contact with air or reduced by the exclusion of air. (John Dawson, adapted from S. R. Eyre, *Vegetation and Soils*, Edward Arnold Publishers)

296

(a) Tundra Region: Entisols, Histosols

(c) Short-grass Prairie Region: Mollisols

(b) Coniferous Forest Region: Spodosols

(e) Midlatitude Forest Region: Alfisols

(f) Tropical Forest Region: Ultisols, Oxisols

(d) Tall-grass Prairie Region: Mollisols

(g) Desert Region: Aridisols

(h) Bogs and Meadows: Histosols

humid regions on recent alluvium, glacial or aeolian (wind-produced) deposits, or volcanic ash. Although some soluble compounds have been removed from the *A* horizon, these soils show no clear illuvial horizon, and can be very productive if not too permeable or stony.

Aridisols (from the Latin *aridus,* dry) are the soils of deserts and semideserts around the world (Figure 11.21), normally having the thinnest profiles of any regional soil type, due to shallow and infrequent penetration of water. They contain a minimum of organic matter, have maximum stoniness, and often include concentrations of calcium carbonate (lime), calcium sulfate (gypsum), and sodium chloride (salt), either in discrete masses or in surface or subsurface layers (see Figure 11.17). Clay hardpans and iron and silica crusts may also be present, the latter inherited from prior pedogenic regimes. Calcification and salinization may be factors locally. Aridisols are usable with irrigation, but are easily leached and require regular fertilization to remain productive.

Mollisols (from the Latin *mollis,* soft) have dark, humus-rich *A* horizons (mollic epipedons) with high base saturation (see Figure 11.16). They develop under midlatitude grasslands in the transition zone between arid and humid climates. The vegetation cover generally maintains a rich clay–humus complex due to the extremely dense mass of roots, which saturates the upper soil with organic decay products. Moisture is adequate to encourage a flourishing soil biota, but not sufficiently abundant to cause vigorous soil leaching. Mollisols show varying degrees and depths of calcification. Unlike Vertisols, which may also be dark and rich in organic matter, Mollisols do not experience major volume changes, nor do they harden when dry. Thus, provided there is sufficient moisture, the best Mollisols are the "cream" of agricultural soils. Their natural cover of wild grasses has been replaced everywhere by crop grasses: wheat, barley, rye, corn, and sorghum. Unfortunately, they extend into areas that experience periodic droughts, leading to crop failures and recurrent "dust bowl" conditions. Mollisols rarely, if ever, form on bedrock. They are derived from alluvium, glacial deposits, and loess (fine material deposited by the wind).

Mollisols vary significantly in character as we move westward from Illinois to the much drier high plains of eastern Colorado. This is reflected in the soil suborders.

Spodosols (from the Greek *spodos,* wood ash) are soils in which a leached and eluviated light-colored *A* horizon (albic horizon) overlies an illuvial *B* horizon (spodic horizon) that is colored by translocated iron or aluminum compounds or related organic carbon (Figure 11.15 and Figure 11.21). An iron-cemented clay hardpan may be present. Spodosols are most common in forested cold-winter areas and wherever the parent material is nutrient-poor sand. The major soil-forming process is podzolization. Spodosols are notoriously infertile, most of their soluble bases having been replaced by hydrogen ions. In addition to their acidity, their sandy texture causes problems of moisture retention. However, some root crops, such as potatoes, thrive in these soils. To be productive for grains, spodosols must be neutralized by treatment with calcined lime (quicklime), which may increase grain yields by as much as one-third and pasture production by 300 percent.

Alfisols (from aluminum, *Al,* and iron, *Fe*) have yellowish-brown *A* horizons (ochric epipedons) that have been partially leached of bases, causing the upper soil to be colored by iron and aluminum compounds (see Figure 11.21 and 11.22). A clay hardpan is usually present in the illuvial zone. Alfisols are transitional between the Mollisols of the lime-accumulating *pedocal* zone and the Spodosols and Ultisols of the leached, iron and aluminum enriched, *pedalfer* regions. In the United States, they are found mainly in the southern Great Lakes area; however, Alfisols are widely distributed, occurring from middle to tropical latitudes. The retention of a significant proportion of bases allows the better Alfisols to be very productive agriculturally, and they support part of the intensively farmed American "corn belt."

Ultisols (from the Latin *ultimos,* ultimate) are similar to Alfisols but more thoroughly leached of bases (Figure 11.21 and 11.22). Typically they are found on land surfaces older than those occupied by Alfisols. Thus Ultisols may be an advanced state of Alfisol development. Ultisols occur in climates that are warmer and wetter than those of Alfisols. They are redder due to a greater proportion of iron and aluminum oxides in the *A* horizon, and they are significantly more leached, poorer in humus, and less productive than Alfisols. The bases present in the upper soil have been brought up by deeply penetrating tree roots and are fed into the *A* horizon by way of plant litter. Forest cutting interrupts this nutrient cycle and results in rapid leaching of the bases remaining in the upper soil. In North America, Ultisols are found throughout the southern Atlantic states and lower Mississippi Valley. These soils were severely damaged by 150 years of intensive cotton farming, which resulted in erosional losses of nearly the entire solum in some localities.

Oxisols (from the word "oxide") are even more thoroughly leached that Ultisols and are the soils characteristic of the wet tropics (Figure 11.13 and Figure 11.21). They do not occur in North America. The diagnostic feature of Oxisols is a subsurface oxic horizon consisting of a residue of clay and iron and aluminum oxides and hydroxides, with virtually all bases removed. Oxisols are the consequence of the laterization process. They are found on long-exposed land surfaces in those portions of the tropics that receive heavy precipitation either seasonally or every month. However, oxisols are more widely distributed than the climates presumably required for their development. Oxisols found in such relatively dry regions as the Australian interior indicate climatic change in these areas. Oxisols are primarily encountered in the equatorial forest, savanna, and scrub forest regions of Central and South America, Africa, and Southeast Asia. As in the case of Ultisols, the bases present in Oxisols are a consequence of nutrient cycling by the natural vegetation. Thus, clearing the natural vegetation results in rapid leaching of soluble plant nutrients, leaving a severely impoverished soil.

Figure 11.22 **(left)** This Alfisol is located in northern Michigan and has developed on sandy glacial outwash. The soil is relatively infertile because of the excessive drainage of water through the sandy parent material. Alfisols developed on other parent materials formerly supporting deciduous forests may be quite productive. (7th Approximation: typic udipsamment)
(right) This Ultisol has developed on sandy coastal plain sediments in central North Carolina. The soil is strongly acidic and over 200 cm (80 in.) thick. (7th Approximation: plinthic paleudult)

SOIL MANAGEMENT

The soils in which plants grow are one of our planet's most priceless resources, equalled in importance only by air and water. Human replacement of natural vegetation with crop plants has seriously affected the earth's soils. As we have seen, the soil is a dynamic system that is in constant interaction with the things living and growing in it. In their natural state many soils seem to reach an equilibrium condition. Over the space of a year, the nutrients taken from the soil by its vegetative cover are returned to the soil as litter in the process of nutrient cycling.

Clearing the land for agriculture interrupts the cycling of nutrients. Since we remove most of the useful products of our croplands, only a portion of the nutrients that plants remove from the soil are recycled into it in the form of organic litter. Thus agriculture without artificial fertilization inevitably results in some loss in soil fertility. At the same time, soil erosion is increased by exposing bare ground to the impact of rain and the force of wind and flowing water, and by disturbing the soil through plowing and cultivating. In the United States, annual losses to erosion in agricultural areas average from 9 to 12 tons per acre, whereas new topsoil forms at a yearly rate of only about 1.5 tons per acre. The U.S. Department of Agriculture estimates that valuable topsoil is currently flushing out of the mouth of the Mississippi River at an average rate of 15 tons per second. Erosion by water

and wind removes the vitally important *A* horizon that contains the humus needed for moisture retention and soil fertility. Loss of the *A* horizon reduces the soil's water-accepting capacity; this increases the proportion of water that runs off on the surface, which results in still more erosion.

In eroded soils, plant roots must seek nutrients in dense subsoils instead of the loose, humus-rich *A* horizons to which field crops are adapted. As soils become thinner and less fertile, plants become increasingly sensitive to periodic moisture variations. Thus many "droughts" are actually normal events that have a disastrous impact because the soil system has been weakened by human activities.

Loss of soil by erosion creates other problems. The erosion of deforested hillslopes has choked streams with sediments, making them shallower. This increases the hazard of flooding on the plains beyond the uplands. Deforestation and soil erosion in upland areas have become an acute problem in the foothills and high valleys of the Andes and Himalaya mountains and in the East African Highlands, due to rapid population growth that is pushing agriculturalists onto ever steeper land. This is increasing flooding in the nearby heavily populated low-

Figure 11.23   (left) Gullying of this cornfield in Missouri resulted from the improper practice of plowing and harrowing up and down the slope. The damage was done in only a few days by several heavy rains. At the time of the photo, this land, which originally had a natural cover of grass, had been used agriculturally for only three years.

(right) This scene in southern Wisconsin illustrates contour strip cropping, which retards runoff and greatly reduces soil erosion. According to the Soil Conservation Service, agricultural yields have doubled where contour strip cropping has been introduced in this area. (U.S. Soil Conservation Service)

lands. Soil erosion also shortens the life of reservoirs, many of which are filling with silt two or three times more rapidly than predicted. As we saw earlier, land deterioration due to salinization is also a very serious problem in dry regions all around the world, possibly affecting as much as 75 percent of all irrigated land.

Of course, measures are being taken to preserve our soil resource. The cost of fertilizers of various types is a major expense on any modern farm. Crop rotation can stave off soil exhaustion. To reduce wind and water erosion, crop stubble and litter are left on the fields, or cover crops are planted after harvesting. In rolling country, plowing along the contour rather than up and down the slope will hold back runoff and reduce soil erosion by as much as two-thirds. Contour plowing can be combined with strip cropping, in which different crops are alternated in bands along the contour, as shown in Figure 11.23. A new erosion-reducing method is "minimum tillage," in which the land is disturbed as little as possible in agricultural operations, with weed control by herbicides, or

natural sod being left between the crop rows. This can reduce erosion to half its usual rate.

In deforested uplands massive plantings of trees can reduce erosion and sediment input into streams. In irrigated areas deep drains and more carefully controlled water applications will prevent the artificial rises in water tables that cause salinization of dryland soils.

All these procedures are being carried out in many parts of the world, but only after centuries of neglect. Unfortunately, in some regions it is already too late, causing an exodus of people from rural areas. Ironically, the deterioration of soils around the world has become evident at the very time that the world's population and food demands are increasing explosively. While crop yields have increased markedly in many areas, the cost of producing such yields has risen even faster, and crop failures have become more frequent. Quite possibly the future of humanity will depend less on world political events than on how we handle the fragile soil resource that sustains life on our planet.

SUMMARY

Soil is a mixture of mineral and organic matter and is the product of a living environment. The mineral matter results from rock weathering, and the organic matter is a product of bacterial decomposition of plant material. Soil development begins with the fragmentation of rock by mechanical and chemical weathering processes. The nature of weathering, which produces the raw material for soil development, varies with changing energy and moisture conditions. Translocation and organic activity convert the decomposed rock to soil.

Soil differs from weathered rock by having a profile consisting of distinct soil horizons. These vary in texture, structure, bulk density, color, and chemical composition. Soil pH is a measure of acidity, or hydrogen ion concentration, and cation exchange capacity expresses the soil's ability to retain soluble bases. The clay–humus complex is of special importance in a soil's ability to hold moisture and essential plant nutrients. The nutrient cycle between

plants and their soils is another factor in the maintenance of soil fertility.

The five principal influences on the formation of soil are parent material, climate, site, organic activity, and time. Climate, together with organic activity, is responsible for three different soil-forming regimes: laterization, podzolization, and calcification. Drainage conditions combine with climate to create two additional soil-forming regimes: salinization and gleization. Because of the great number of variables involved in soil formation and soil character, classification of soils is extremely difficult, as reflected in the complex terminology of the 7th Approximation classification system currently in use in the United States. The broadest divisions in this system are the ten global soil orders, of which nine are represented in North America.

Soil management is an important task. Any interference with the natural soil system disturbs soil equilibrium and usually increases erosion and nutrient losses. Agricultural use of soils demands careful attention to these problems. Several highly effective procedures have been developed to combat soil erosion and soil exhaustion. However, in many areas of the world remedial action has been too late in coming—the most productive portion of the soil has already been lost.

REVIEW QUESTIONS

1. How does water penetrate into rock masses to initiate weathering processes?
2. What are three mechanical processes that aid in the fragmentation of large masses of rock?
3. What specific soil phenomena are a consequence of the process of translocation?
4. What is the clay-humus complex, and how do soils benefit from its presence?
5. How is soil texture determined?
6. What is the relationship involving soil acidity, hydrogen ions, and basic cations in soils?
7. Draw an idealized soil profile and indicate what is happening in each soil horizon.
8. What five factors most strongly influence profile development in undisturbed soils?
9. Why would archaeologists be interested in soil profile characteristics?
10. What is the pedological importance of nutrient cycling?
11. How does the podzolization process differ from laterization?
12. What soil-forming regimen can be induced by human activity?
13. In the 7th Approximation soil classification system, which of the soil orders is related to the podzolization process? Which has the maximal development of a humus-rich A-horizon? Which order is the most thoroughly leached of basic cations?
14. How does soil erosion in upland areas affect surrounding lowlands?
15. Make a list of agricultural practices that have the objective of minimizing soil erosion.

APPLICATIONS

1. Cemeteries are good places to observe the effects of weathering on various types of fresh rock. Go to an old cemetery in your area; staying on the cemetery paths, look at the weathering phenomena on the grave markers. What weathering effects are visible? What types of stone are most resistant to weathering? What types are least resistant? Can you explain the varying responses to weathering of the different types of stone?
2. What soil series are present in the agricultural areas nearest to your campus or your home? How would the soils be classified according to the U.S. Comprehensive System (7th Approximation System)? Do these soils contain the standard O, A, B, and C horizons? Do they contain any special features like Cca horizons?
3. Are the soils in your area derived from varying parent materials? If so, examine the soil texture and pH at varying depths on the different parent materials. In what ways do the soils reflect their parent materials?
4. What are the principal soil management problems in your area? Are there any unusual problems related to the physical or chemical properties of local soils?
5. How can the water budget concept be applied to the various degrees and types of soil development?
6. What changes in soils would occur where a midlatitude forest advances into a former grassland area? where a grassland expands into a former forest area?
7. How is it possible for a calcrete layer 1 m thick to be found 1 m below the surface in a desert region where the heaviest rainfalls only wet the soil to a depth of 0.5 m?
8. How could natural changes in the environment duplicate human effects on soil systems in varying settings?

FURTHER READING

Bennet, Hugh H. *Soil Conservation.* New York: McGraw-Hill (1939), 933 pp. This abundantly illustrated classic, dealing with the processes and results of soil erosion in the 1930s, was written by the first Chief of the U.S. Soil Conservation Service. It is useful for its regional treatment of soil erosion problems in the United States and foreign areas and for its wealth of detail concerning specific localities.

Birkeland, Peter W. *Pedology, Weathering, and Geomorphological Research.* New York: Oxford University Press (1974), 285 pp. This work provides a thorough treatment of weathering, the factors of soil formation, and the study of soils as geological deposits that convey considerable information on recent environmental history. It is not concerned with soil-plant relations or agricultural applications.

Bridges, E. M. *World Soils,* 2nd ed. New York: Cambridge University Press (1978), 128 pp. This brief work is a well-illustrated (with color) outline of soil development and soil types in major natural regions of the world.

Buol, S. W., F. D. Hole, and **R. J. M. McCracken.** *Soil Genesis and Classification.* Ames: Iowa University Press (1973), 360 pp. A thorough treatment of soil-forming factors, processes, and resulting soil types classified according to the U.S. Comprehensive System.

Carter, V. G., and **T. Dale.** *Topsoil and Civilization,* 2nd ed. Norman: University of Oklahoma Press (1974), 292 pp. The authors interpret the rise and fall of great world civilizations as due to human-induced deterioration of their resource bases—particularly the soils that supported their agricultural systems.

Clarke, G. R. *The Study of Soil in the Field,* 5th ed. Oxford: Clarendon Press (1971), 145 pp. This is an excellent manual outlining in very readable fashion the exact techniques used in analyzing soils in the field, with information on soil mapping and the use of aerial photographs.

Eckholm, Erik P. *Losing Ground: Environmental Stress and World Food Prospects.* New York: W. W. Norton (1976), 223 pp. The author presents an ominous view of the problem of current human-induced soil deterioration in varying environments, with special stress on future problems of food production, especially in the Third World. This book should be on every geographer's reading list.

Soil Survey Staff, U.S. Dept. of Agriculture. *Soil Taxonomy: A Basic System of Soil Classification for Making and Interpreting Soil Surveys.* U.S.D.A. Handbook 436. Washington, D.C.: U.S. Government Printing Office (1975), 754 pp. This is the updated official presentation of the U.S. Comprehensive Soil Classification System (7th Approximation), providing the system's rationale and detailed application.

Steila, Donald. *The Geography of Soils.* Englewood Cliffs, N.J.: Prentice-Hall (1976), 222 pp. This valuable paperback is a good introduction to the U.S. Comprehensive Soil Classification System (7th Approximation), including material on the uses and management of soils of the different orders in the classification scheme.

United States Department of Agriculture. *Soils and Men.* Yearbook of Agriculture, 1938. Washington, D.C.: Dept. of Agriculture (1938), 1,232 pp. Although dated, this work contains a wealth of information on soils, their classification, characteristics, uses, and maintenance.

CASE STUDY

The Terrace Builders

For any region to support sizable numbers of people, crops must be grown, requiring extensive areas of farmable soil. The value of arable land, which inhabitants of the benign environments of the middle latitudes accept unquestioningly, is best seen in the harsh lands that lack it. Skellig Michael, a snag of rock off the Irish coast, is a vivid example. Sometime before the ninth century, Skellig Michael's solitude and isolation caused it to be chosen by Irish monks as the site of a monastery. But no food could be produced there. The island was completely barren, exposing naked rock everywhere. Not to be denied, the monks labored to create productive patches by building retaining walls in rock niches and filling them with soil brought by boat from the mainland, 10 km (6 miles) distant. We know no more than this, for the island was abandoned more than 1,000 years ago, probably due to attacks by Viking raiders. But we can still see the rock niches and retaining walls, packed with foreign soil and now covered with wildflowers.

Attempts to create productive cropland on a larger scale are seen in many parts of the world. Steep hillslopes have been reconstructed into staircase-like flights of horizontal terraces where only steep slopes and thin moisture-deficient soils existed naturally.

Agricultural terracing to create deep soils on steep slopes is widespread where population pressure is high. In some areas, such as the Philippine island of Luzon, the landscape has been completely transformed by terracing. Although requiring great labor to construct and maintain, terraces are found in a va-

This extremely steep land in Ifugau Province on the Philippine island of Luzon has been made suitable for agriculture by slope terracing. The stone-faced terraces are 2 to 6 meters high, and support irrigated rice crops. (Robert Reed)

riety of climates, being conspicuous along the Mediterranean coastlands, especially in Spain, Italy, and Lebanon, on the slopes of the Pyrenees and Sierra Nevada of Spain, and on the southern slopes of the Atlas Mountains overlooking the Sahara Desert. They also appear in the Canary Islands, the Andes of South America, the highlands of Yemen, the Himalayan foothills, and the uplands of Sri Lanka (Ceylon), and on the volcanoes of Java and Sumatra. In some regions, terraces contour the hillsides; in others, they are built into the floors of narrow, steeply rising valleys and catch the runoff from adjacent rocky slopes. As at Skellig Michael, the terrace fill

was sometimes carried from a great distance. The objective was not to create mere flat land, but deep, moisture-retentive soils—possibly to be irrigated by elaborate canal systems.

Geographers have long disputed whether the technology involved in agricultural terracing originated in a single location and diffused outward or was developed independently in many widely separated regions. Logic seems to favor the latter since the idea of terracing is a simple one, with natural examples provided by rock ledges and steplike landslide terrain, both of which are frequently seen in the very areas noted for artificial terraces. Terracing in the Andean area goes back at least 2,500 years, in Asia perhaps much further.

In many regions, terraces are irrigated, necessitating careful construction, leveling, and maintenance—along with the intricate canal and aqueduct systems that water them. Even in wet areas of the tropics, irrigation is practiced, for rice is grown in paddies on the terraces. These upland rice paddies are marvels of primitive technology, as were the irrigated terraces for maize built by the Incas in the Peruvian Andes. From the tropics to the middle latitudes, orographic rainfall may be adequate to nourish maize, potatoes, vegetables, tea, coffee, cocoa, sugar cane, vineyards, orchard crops, and cereals. Even unirrigated terraces are gardened rather than farmed, being carefully fertilized with animal manure, human waste, bird or bat droppings (guano), and even fish heads and volcanic ash. Constant attention is devoted to the state of terrace walls often made of well-fitted stone. Two or more crops may be grown in the same field—a tree crop and some lower-growing form.

The creation of agricultural ter-races transforms steep mountain slopes into lush gardens, sometimes studded with villas surrounded by ornamental plants—one of the most attractive of all of the landscapes created by human activity. In areas of frequent strife, fortified villages may be perched on projecting mountain spurs between the green fields.

The effort expended on terracing is extreme, but provides an enduring relationship between the land and its occupiers. Where terracing is not practiced the results are far less satisfactory. Fields may be cleared on 45° slopes where just standing erect is difficult. The rate of erosion is catastrophically high, and the plots are soon abandoned with new ones being sought out every few years. This form of shifting cultivation produces a highly visible patchwork of active and abandoned fields in various stages of reversion to brush or forest. This pattern is conspicuous in parts of Central and South America, East Africa, South and Southeast Asia, and even the southern portions of Europe. Under such treatment, the entire soil resource can be lost in a few decades, with bare rock becoming exposed where forests once prevailed. With the rapid growth of village populations in such settings, forest clearing for fuel often precedes agricultural use of the land. This is certainly the case in portions of the Andes, East African highlands, India, and Nepal. Here the use of the land seems purely exploitive, with little thought of a permanent relationship to a particular plot.

It is difficult to explain the variations in local traditions that lead one group to mine the land and another to tend it as the terrace builders do. Part of the explanation lies in the system of land tenure, with those having secure title finding it to their advantage to improve the land, while those with no enduring claim take what they can and move on. But this by no means accounts for all local variations seen. Everywhere there is a tendency to do things as they have traditionally been done, even if the traditions were established in ignorance of their effect or under circumstances that have changed profoundly.

Cotapaxi, by Frederick Edwin Church. (On loan to the Metropolitan Museum of Art from John Astor)

The large-scale features of the earth's crust—continents, ocean basins, mountain ranges, and plateaus on land and beneath the sea—have been formed by the earth's internal energy, sometimes manifested catastrophically by earthquakes and volcanic eruptions.

To this point we have been speaking of physical processes powered by radiant energy from the sun. But not all change on the earth is due to outside forces. A California winery building is slowly being torn apart by movements of the land under it. The Yellowstone region has risen 0.6 meters (2 feet) in the last 50 years. Earthquakes in Iran, Turkey, Central America, and China repeatedly shatter villages and cities. Volcanoes violently spring to life and then rest for centuries. Islands along the Norwegian coast grow larger each year as the land rises slowly out of the sea. All of these changes require a source of energy within the earth itself.

The solid outer shell of the earth, known as the *lithosphere*, which includes the crust and upper mantle, is affected by energy released by the radioactive decay of unstable elements deeper within the mantle. Without this internal energy and the motions it generates, the earth would be smooth and featureless—all its relief worn away by rain, wind, and running water. But the presence of deep canyons, rough hills, and soaring mountain ranges all tell of an ongoing contest between two sets of forces. One of these, powered by gravity and radiant energy from the sun, tends to wear down and smooth the earth; the other, powered by the earth's internal energy, upheaves the land, creating the irregularities that we call "scenery."

Catastrophic earthquakes and volcanic eruptions have played their role in the formation of some of the earth's grandest scenery. But most relief features associated with the earth's internal forces are the result of small unspectacular movements repeated again and again over millions of years. It is worth remembering that most of the rock masses seen at the earth's highest elevations were formed on ancient ocean floors or many kilometers below the surface of the land.

In this chapter we shall discuss the different scales of relief features found on the earth and the origin of the rock types and geologic structures that comprise these features. Subsequent chapters will treat the gradational processes that modify these structures into the familiar landforms we see around us.

12
The Lithosphere

Orders of Relief

First Order of Relief: Continents and
Ocean Basins
Second Order of Relief: Major Subdivisions
of the Continents and Ocean Basins
Third Order of Relief: Landforms

Minerals and Rocks

Minerals
Igneous Rocks
Sedimentary Rocks
Metamorphic Rocks
The Rock Cycle

Crustal Deformation

Volcanism
Magma Sources and Eruptive Styles
Tensional Faulting
Strike-Slip Faulting
Folding
Compressional Faulting
Broad Warping and Isostacy

ORDERS OF RELIEF

When we speak of relief, we mean differences in elevation between high and low places on a surface. However, relief features have horizontal as well as vertical dimensions. Relief features on the earth's surface range in scale from tiny raindrop imprints to the vast continents and ocean basins. Thus we need some means of distinguishing between relief scales. We can do this by speaking of *orders of relief*. The continents and ocean basins, which are our planet's largest-scale relief features both vertically and horizontally, are regarded as the *first order of relief*.

First Order of Relief: Continents and Ocean Basins

The arrangement of the earth's continents and ocean basins is one of the most important aspects of physical geography. The distribution of land and water masses is a major influence on

Figure 12.1 This photograph, taken from a spacecraft orbiting more than 480 km (300 miles) above the earth's surface, shows a portion of the east coast of Africa, looking toward the Arabian Peninsula, with the Red Sea at the left and the Gulf of Aden at the right. The opening of the Red Sea is believed to have begun 5 to 10 million years ago when the lithospheric plates carrying Africa (bottom) and Arabia (top) began to drift apart. (NASA)

the atmospheric circulation and the distribution of plant and animal species around the world. But the arrangement of the continents and oceans is not permanent. These features have changed in size, form, and position throughout the earth's history, and are continuing to change today (Figure 12.1).

The continents and ocean basins are features of the thin outer layer of the earth known as the *crust* (Chapter 1). The crust is not an unbroken shell. In fact, it is divided into a dozen or more adjoining *lithospheric plates*. These plates, which are 70 to 100 km (40 to 60 miles) thick, include the upper portion of the earth's mantle. The separate continents are embedded in a continuous layer of denser crustal material that is exposed in the ocean basins. The general name for the rocks of the continents is *sial*—indicating that their chemical makeup is dominated by silica (SiO_2) and alumina (Al_2O_3). The rocks of the ocean floors, and the mantle beneath, are termed *sima*—silica-magnesia (MgO) rocks (Figure 12.2). Sial is about four-fifths as dense as sima; therefore the separate continents "float" about four-fifths submerged in the continuous layer of sima (Figure 12.3).

When the first accurate world maps were constructed some 400 years ago, it was immediately noticed that the facing coasts of the continents of Africa and South America could fit together, as in a jigsaw puzzle. This suggested that these continents had broken apart and had somehow moved to their present positions. The hypothesis of "continental drift" was first set forward in a formal way in 1912 by the German meteorologist Alfred Wegener (Figure 12.4). Wegener pointed out that coal beds in temperate regions are composed of tropical vegetation types, suggesting that today's midlatitude areas had once been located near the equator. Likewise, ancient glacial deposits are seen in regions that are now tropical; thus low-latitude landmasses once had polar or subarctic locations. Furthermore, geological structures and rock types along the west coast of Africa and the east coast of Latin America seemed to match. Finally, the fossilized remains of ancient plants clearly indicated former connections among all the widely separated southern hemisphere continents.

For almost half a century most geologists scoffed at Wegener's ideas, for there seemed to be no conceivable force that could move an entire continent. The fact that the sial of the continents projects downward into the denser sima beneath made the possibility of horizontal movement of the continents seem all the more unlikely.

In 1944 the English geologist Arthur Holmes proposed a solution for the puzzle. He suggested that it is the sea floors that are moving—and dragging the continents along with them. Holmes hypothesized that the driving force is a process of convection ("boiling") in the upper part of the earth's mantle. This would cause slow horizontal circulations of molten material below the earth's crust, which would drag the crust this way and that. The only feasible location for the subterranean "flows" is in the asthenosphere (Chapter 1)—the mantle layer having temperatures high enough to keep rock material melted and pressures low enough to allow the molten material to circulate at a rate of centimeters per year.

During World War II, it was discovered that all the oceans are divided by continuous undersea ridge systems. The best example is the Mid-Atlantic Ridge. In addition, trenches deep enough to hold the world's highest mountains almost completely ring the Pacific Ocean. Holmes suggested that volcanic eruptions along the oceanic ridge systems create new areas of ocean floor, while old areas of ocean floor eventually disappear by descending into the oceanic trenches, a process later termed *subduction* ("underflow").

In the next two decades a mass of new evidence emerged that supported Holmes's

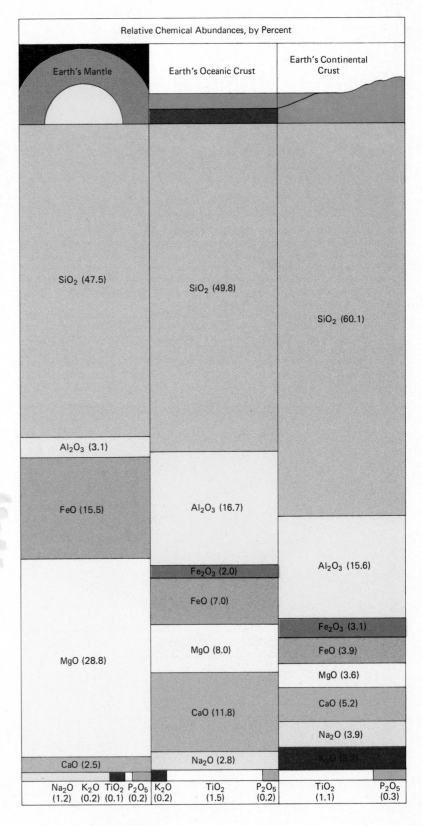

Figure 12.2 The average abundances of various elements change significantly from the earth's mantle upward to the rocks of the continents and ocean floors. Silica (SiO_2) is dominant in all three locations. Compared to the mantle material, continental rocks have a much higher content of silica and alumina (Al_2O_3), so that they are broadly termed *sial*. The mantle contains greater amounts of magnesium (Mg) and iron (Fe), and is known as *sima*. The lavas of the ocean floor, which the diagram shows to be intermediate chemically between continental sial and mantle sima, are distinctive in their high content of calcium.

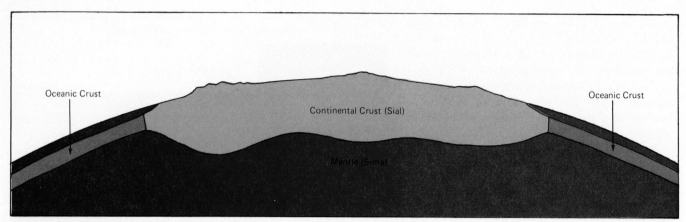

Oceanic Crust

Continental Crust (Sial)

Oceanic Crust

Mantle (Sima)

Figure 12.3 (above) Because the sial of the continents is lighter than the sima of the mantle, the continents float like rafts in the denser mantle material. To be in bouyant equilibrium, the continental crust must project downward into the mantle under large mountain ranges, as indicated here. Each segment of the continental and oceanic crusts may be imagined displacing an equal weight of sima in the upper mantle. The vertical scale of the diagram is greatly exaggerated. The average thicknesses of the oceanic and continental crusts are 10 km (6 miles) and 40 km (25 miles), respectively, with mountain "roots" reaching as deep as 70 km (40 miles).

Figure 12.4 (below) Alfred Wegener suggested in 1912 that about 200 million years ago all of the landmasses of the earth were grouped together, forming a supercontinent, Pangaea. Pangaea subsequently separated into two continents, Laurasia in the northern hemisphere and Gondwanaland in the southern hemisphere. The rifting continued, with crustal plates and associated continents moving in the directions indicated by the arrows. The plate that carries India, for example, separated from Antarctica and drifted northward until it collided with the Eurasian plate. The drift of the continents is now thought of in terms of plate tectonics, seafloor spreading, and subduction. In this figure, subduction zones are shown in black and spreading centers in red. Thin black lines indicate transform faults; purple areas are continental shelves. (*The Atlas of the Earth*, p. 36, © 1971, Mitchell-Beazley, Ltd.)

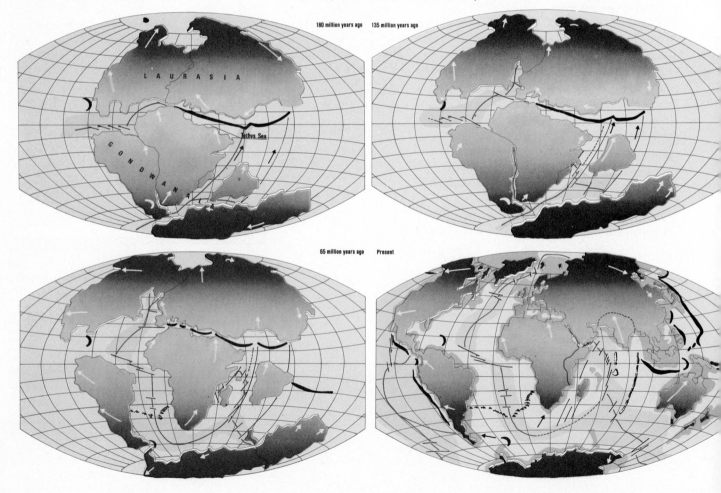

hypothesis. The continents do indeed move, riding passively like rafts on lithospheric plates that seem to move like conveyor belts from the oceanic ridges toward the oceanic trenches in the process now known as *sea-floor spreading*. The continents cannot themselves be subducted downward because of their low density relative to the material beneath them in the crust and mantle. The global pattern of crustal plates, spreading centers, and subduction zones was established by oceanographers in the 1960s and is shown in Figure 12.5. The movement of these crustal plates and the interactions between them are known as *plate tectonics*.

The sea-floor spreading hypothesis has been supported by many findings since the 1960s (Figure 12.6). Scientists have drilled into the floors of the oceans hundreds of times, principally from the research vessel *Glomar Challenger*, recovering cores of sediment and rock for analysis. These have shown that the sea floors are extremely young compared with the continents. Whereas the earth itself is almost 5 billion years old, and 3.8 billion-year-old rocks have been found on the continents, the most ancient portions of the sea floors have an age of only about 170 million years. Furthermore, the thickness and age of oceanic sediments increase with their distance from the oceanic

ridges. This supports the idea that the sea floors are being formed at the ridges and are moving away from them (Figure 12.7).

Some of the most convincing evidence for sea-floor spreading comes from the record of ancient magnetism in the volcanic rocks of the ocean floors (Figure 12.8). When volcanic rock solidifies from a molten condition, tiny grains of the iron-bearing mineral magnetite align with

Figure 12.5 This map indicates the probable boundaries of the lithospheric plates composing the earth's crust, and the relative motion of the plates. The red lines represent spreading centers where new crust is being formed, and the blue lines mark regions where plates are descending into the mantle in the subduction process. Blue arrows are on the upper (overriding) plate and point in the direction of subduction of the descending plate. Red arrows indicate the general relative directions of plate motion. The seven major plates are identified in bold type and several of the smaller plates in lighter type. Note that the Pacific basin is largely rimmed by subduction zones.

Many prominent features of the earth's crust are explained by the motion and interaction of lithospheric plates. Mountain uplift is active at the boundaries of converging plates along the west coasts of North and South America, where the mountains are young. Along the Atlantic coasts, however, there are no plate boundaries and no present mountain-building activity. Earthquake and volcanic activity is associated with converging plates, such as those around the rim of the Pacific Ocean basin. Nonexplosive volcanism occurs at the seafloor spreading centers. (Calvin Woo from John F. Dewey, "Plate Tectonics," *Scientific American*, copyright © 1972 by Scientific American, Inc. All rights reserved)

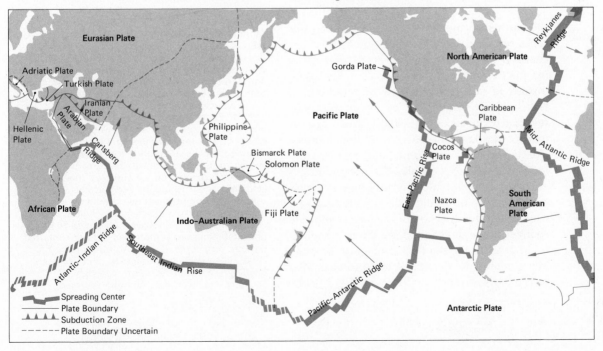

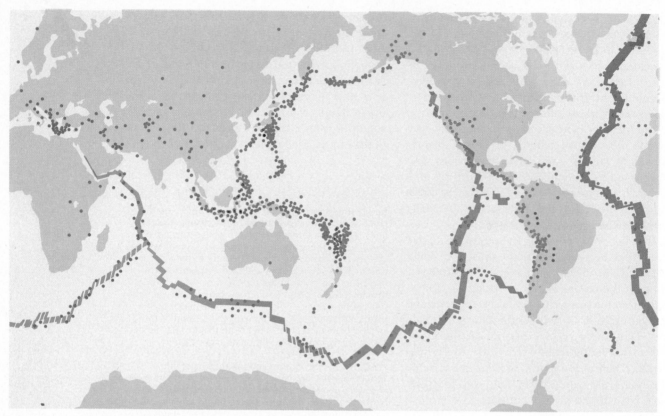

Figure 12.6   The distribution of earthquakes parallels the locations of major features of the earth's surface, including oceanic ridges and trenches, and continental boundaries. Recognition that these bands of seismicity mark the edges of lithospheric plates has been a comparatively recent development. With it has come acceptance of the theory of moving crustal plates that are either diverging from or converging along zones of strong seismic activity.

Figure 12.7 (opposite)   (a) A newly developed spreading center causes continental plates to rift and move apart.

(b) The development of new crust at a spreading center causes a sizable ocean basin to form over tens of millions of years. The sediment deposits are thickest near the continental plates, where the oceanic crust is oldest. This diagram represents the Atlantic Ocean basin. The magnetic stripe pattern indicates the record of the earth's magnetism frozen into new crustal rock as it cools. The pattern is the same on both sides of the spreading center; such patterns occur in all ocean basins.

(c) Deep ocean trenches, such as those in the western Pacific, occur where a plate of oceanic crust plunges downward into the mantle beneath either a continental plate or another oceanic plate (plate convergence). Crustal deformation at a convergent boundary leads to mountain uplift, formation of island arcs, and earthquakes and volcanic activity.

(d) A collision between two continents produces high mountain ranges such as the Himalayas. The continental crust is doubled in thickness and the surficial sediments are compressed to form contorted geologic structures. (Calvin Woo from Robert S. Dietz, "Geosynclines, Mountains, and Continent Building," *Scientific American*, copyright © 1972 by Scientific American, Inc. All rights reserved)

the earth's magnetic field. Studies of volcanic rocks reveal that the global magnetic field has changed in strength and reversed in polarity (so that a compass needle would point south instead of north) at intervals of hundreds of thousands of years. The last major magnetic reversal occurred about 700,000 years ago. A map of the lavas on the ocean floors shows alternating stripes of normal and reversed polarity. These stripes parallel the oceanic ridges and their general pattern is strikingly symmetrical on the opposite sides of the ridges. Changes in the direction of the stripes indicate changes in the direction of sea-floor spreading.

In general, the sea floors appear to be moving laterally 1 to 10 cm (0.4 to 4 in.) each year carrying the continents with them. The Atlantic

Ocean is widening by this process, whereas continents are being forced toward the Pacific Ocean from two sides. The floor of the Pacific Ocean is detached from the adjacent continents and is forced to descend under them in the oceanic trench system that almost completely encircles the Pacific (Figure 12.5).

Second Order of Relief: Major Subdivisions of the Continents and Ocean Basins

The continents themselves contain large-scale relief features in the form of mountain systems like the Alps or Rockies, large depressions on the scale of the Mississippi Valley and the West Siberian Lowland, and great plateaus such as those of Tibet and South Africa. Features on this scale constitute the *second order of relief*. The ocean basins also exhibit a second order of relief in the form of the oceanic ridge system, the submarine trenches, and other large-scale irregularities (Figure 12.9).

All second-order relief features have been produced by either vertical or lateral motions of the earth's crust. All can be explained in terms of plate tectonics. The great continental mountain systems were created by compression and volcanic activity where one crustal plate has

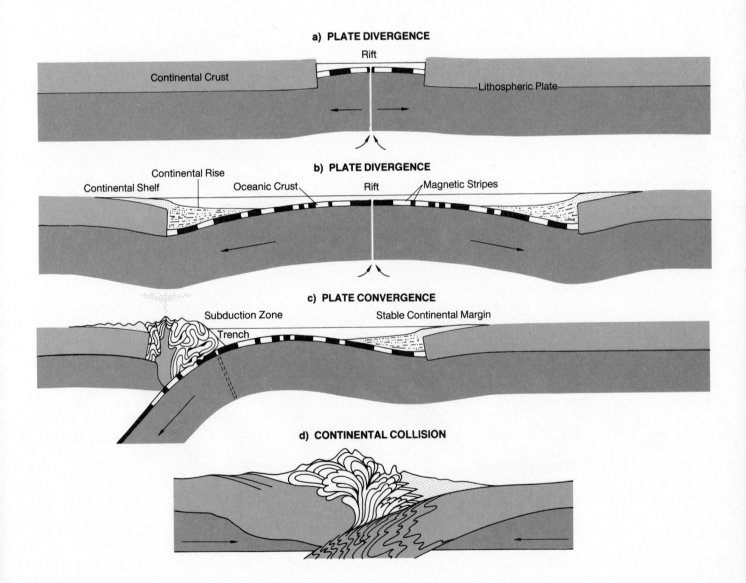

a) **PLATE DIVERGENCE**

Rift

Continental Crust

Lithospheric Plate

b) **PLATE DIVERGENCE**

Continental Rise

Continental Shelf   Oceanic Crust   Rift   Magnetic Stripes

c) **PLATE CONVERGENCE**

Subduction Zone   Stable Continental Margin

Trench

d) **CONTINENTAL COLLISION**

pushed against another. The rugged Andes Mountains bordering the west coast of South America are currently being forced up as the eastward-moving Nazca plate descends under the westward-moving South American plate (Figure 12.5). Similar activity in the past explains the rise of the mountain systems of the Pacific coast of North America. The Himalaya Mountains and Tibetan Plateau were raised when the crustal plate that carries India was forced northward against the Eurasian plate.

Along the Atlantic coasts of the Americas, Europe, and Africa, the continents ride on crustal plates that have been moving away from the Mid-Atlantic Ridge for the past 100 to 200 million years. Since there are no plate boundaries near these coasts—and therefore no compression or volcanic activity—they are low-lying for the most part. However, 200 to 300 million years ago these same coasts were colliding as an older Atlantic Ocean was squeezed closed. The pres-

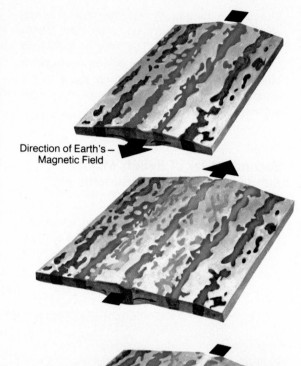

Direction of Earth's Magnetic Field

Figure 12.8 Evidence of seafloor spreading. The Reykjanes Ridge south of Iceland is an active center of seafloor spreading, where reversals in the polarity of the geomagnetic field have been recorded in new seafloor as it was emplaced by volcanic eruptions at the center of the ridge. Each colored stripe maps a band of rock laid down when the earth's magnetic field was aligned in its present direction, and the white stripes outline seafloor formed while the magnetic field was reversed. Note that these patterns are not perfect stripes. The asymmetries suggest uneven production of lava along spreading centers.

ent Atlantic coasts and their borderlands were then the sites of mountain systems like those currently rimming the Pacific. These ancient highlands have been effaced by subsequent erosion.

Between the continents and ocean basins is a series of second-order relief features that are submerged by the seas but are actually portions of the continents. These are the continental shelves, slopes, and rises (Figure 12.10).

The *continental shelves* are the submerged rims of the continents. They descend from coastlines to an average depth of about 130 meters (430 ft). Although they have an average slope of less than a tenth of one degree, they are far from smooth in many places. They were exposed to the atmosphere as recently as 15,000 years ago. At that time, close to 3 percent of the water that presently resides in the seas was held on land in the form of vast ice sheets. The average width of the shelves is 75 km (46 miles), but some areas have shelves far wider than this, while others lack a shelf altogether.

The continental shelves have great economic importance. Because the shallow water allows sunlight to penetrate to the bottom of their landward portions, and because nutrients are constantly supplied by erosion of the land, the shelves are able to support an abundant marine food chain. This has generated controversies among fishing nations over the ownership of the continental shelves. More recently, large deposits of oil have been found in the sediments blanketing the shelves in the Persian Gulf, Gulf of Mexico, North Sea, Yellow Sea, South China Sea, and other places. This has intensified ownership quarrels among maritime nations, and it helped trigger the brief but destructive Falkland Islands war between England and Argentina in 1982.

The continental shelves are succeeded by the slightly steeper (4 to 5 degrees) and more irregular *continental slopes,* which descend to some 3,000 to 3,600 meters (10,000 to 12,000 ft). The less steeply sloping *continental rises* below the continental slopes are embankments of sediment washed from the land and deposited in deep water along the margins of the sialic continental slabs. It is the sediment blankets of the continental shelves, slopes, and rises that are crumpled and upheaved by compression during the collisions of crustal plates, thereby generating the earth's great mountain chains.

Stretching out from the bases of the continental rises at depths of 5,000 meters (16,000 ft) or more are the deep ocean floors. Some portions of these, the so-called *abyssal plains,* are flat and featureless blankets of marine sediments that cover the older volcanic rocks of the sea floors (Figure 12.9). At one time it was assumed that the sea floors are composed almost entirely of abyssal plains. But this is far from true; there is a highly diversified relief beneath the waters of the seas. This includes not only the ocean trenches and ridge systems, but also large and small plateaus, great fault scarps, clusters of undersea mountains, and mysterious erosional canyons (see Chapter 18).

Third Order of Relief: Landforms

The individual hills, valleys, cliffs, and plains of the earth's surface are the *third order of relief.* These sculptural details of the second-order relief features are called *landforms.* A landform is a feature that can be seen in its entirety in a single view. The largest third-order relief features are plains resulting from erosion or the deposition of sediments, individual mountains within a mountain range, and particular valleys within a valley system. There is no lower limit to the size of a landform.

Most landforms have been produced by erosion or deposition of material rather than by motions of the earth's crust. Chapters 13 through 18 focus on the origins of landforms, which are the stage upon which the various physical systems interact, and which influence these interactions in many ways. Their slopes provide gravitational energy; their heights are sources of water and sediment; they channel or block movement; their form and composition provide problems or possibilities for human activity. But before we look more closely at landforms, we shall examine the raw material for landform development—the materials composing the earth's crust.

Figure 12.9 The ocean basins, which are first-order relief features, contain second-order relief features, such as deep trenches and extensive undersea mountain ranges. This physiographic diagram of the North Pacific Ocean basin shows the plains, trenches, and seamounts characteristic of the ocean floor. Heights above and below sea level are given in feet. Note in particular the island arcs along the continental side of the great trenches. Such arcs are located over subduction zones where oceanic crustal plates are plunging downward into the mantle, and they are sites of volcanic and earthquake activity. The corrugations on the seafloor west of California are an exaggerated depiction of the actual relief, and merely signify lavas of varying ages and magnetic polarities. (National Geographic Society)

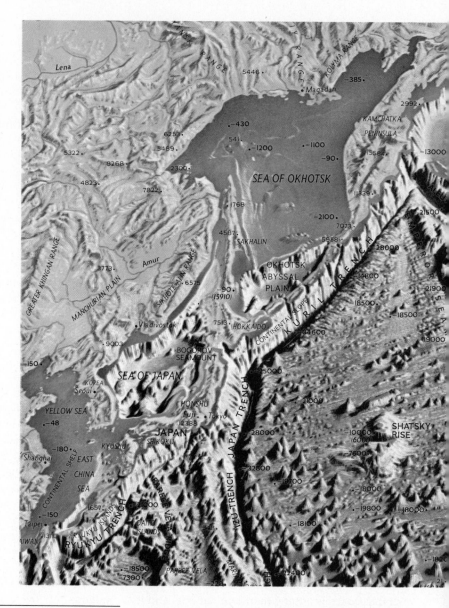

MINERALS AND ROCKS

We inhabit the surface of the lithosphere, or "sphere of stone"—the outer shell of the earth that is composed of either solid or fragmented rock material. Rock masses differ greatly in origin and composition. Because of this, they vary in physical strength, chemical stability, and permeability to water. The weathering of different rock types produces different end-products. Furthermore, rock masses are arranged in many different ways: in layers that are horizontal, wrinkled, or broken and offset, or in seemingly uniform masses. All these variations are influences on landform development. Thus, to understand the appearance of the earth's surface, it is necessary to understand the origin and characteristics of rocks.

Minerals

When one looks at ordinary rocks under a microscope, it becomes apparent that they are composed of much smaller particles. These either interlock or are cemented together by some

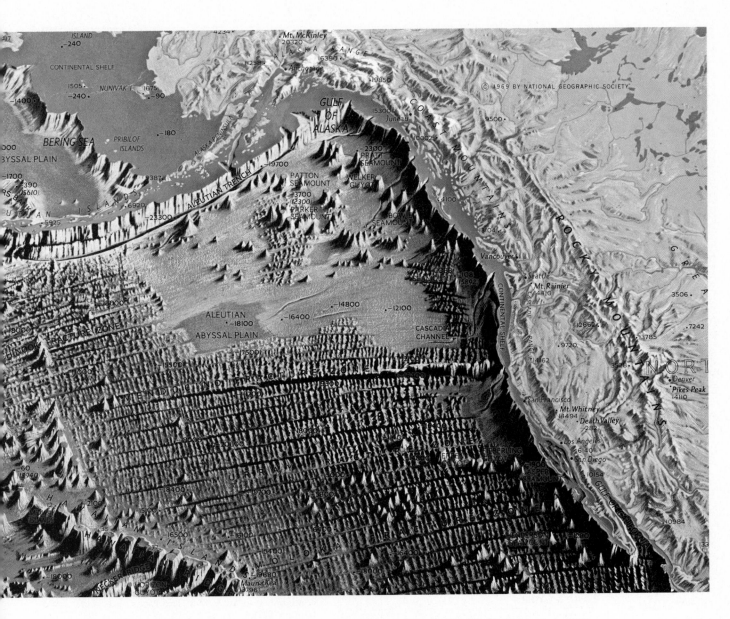

other substance. Each of these particles is one of about 2,000 naturally occurring inorganic substances called *minerals*.

Minerals are combinations of chemical elements in solid form. The atoms of each mineral have a characteristic three-dimensional arrangement that gives its crystals a certain geometric form, as well as a particular luster, color, hardness, and specific gravity (mass relative to that of an equal volume of water). Many minerals are extremely beautiful, and perfect specimens are sought by collectors (Figure 12.11). Certain minerals are quite familiar, such as *halite* (com-

mon salt, NaCl), *ice* (solid H_2O), and *geothite* (iron rust, FeO[OH]).

The high-density metals, such as iron, aluminum, copper, manganese, lead, zinc, nickel, tin, and silver, are rarely found free as separate minerals. They are usually combined with other elements to form minerals that must be artificially decomposed, or "refined," to extract the metal. Such minerals are known as the *ores* of the metals. Gold and platinum are unusual in that they normally occur in pure form. Occasionally, "native" (uncombined) copper is present in a natural deposit, such as that on the

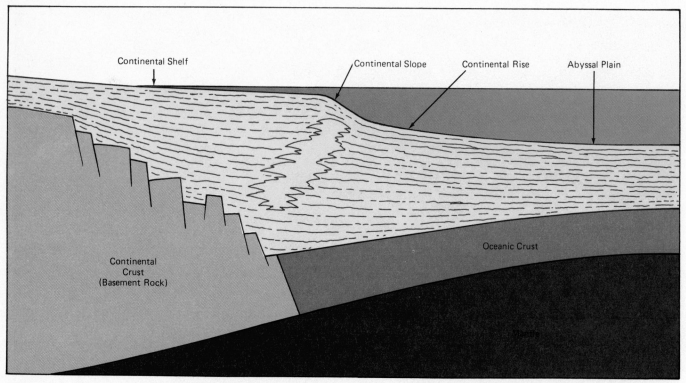

Continental Shelf

Continental Slope   Continental Rise   Abyssal Plain

Continental
Crust
(Basement Rock)

Oceanic Crust

Mantle

Figure 12.10 This is a generalized cross-section of an Atlantic-Coast type of continental margin, where the continental shelf is broad and no ocean trench is present. The continental shelf, slope, and rise are composed of sediments lithified into rock (yellow) that covers the continental basement (pink), which usually consists of ancient plutonic and metamorphic rock types. The continental edge preserves the fractures created by the rifting that formed the Atlantic Ocean. The irregular configuration within the sedimentary mass consists of massive reef structures built outward from the continent in warm seas. When lithospheric plates converge, the sediments shown in yellow are crumpled into folds and overthrusts of the type seen in major mountain systems, and slivers of the oceanic crust may be driven onto the adjacent continent. (After P. B. King, and Pacific Rise Study Group, 1981)

Keweenaw Peninsula of Michigan, which was worked by Indians in the pre-European period. Iron meteorites have also been a source of metal for non-industrial peoples.

The elements combined in minerals are held together by electrical bonds. Thus all multi-element minerals are composed of combinations of positive and negative ions. The most abun-

dant ions in minerals are those of silicon (Si), which has a positive charge, and oxygen (O), which carries a negative charge (Table 12.1). Most rock-forming minerals are silicates, in which positively-charged ions of the hard and soft metals are combined with silica (SiO_2). About 92 percent of the earth's crust is composed of silicate minerals (Table 12.2). Oxides, which form another large family of minerals, result from the chemical decay of silicate min-

Table 12.1   Major Elements of the Earth's Crust

Element	Weight (Percent)	Volume* (Percent)
Oxygen (O)	46.60	93.77
Silicon (Si)	27.72	0.86
Aluminum (Al)	8.13	0.47
Iron (Fe)	5.00	0.43
Calcium (Ca)	3.63	1.03
Sodium (Na)	2.83	1.32
Potassium (K)	2.59	1.83
Magnesium (Mg)	2.09	0.29
Totals	98.59	100.00

*Computed as 100 percent, hence approximate. After Brian Mason, *Principles of Geochemistry*, John Wiley & Sons, Inc., 3rd ed., 1966.

Figure 12.11 These minerals are important ores. Pyrite, goethite, and hematite are all iron ores. Chalcopyrite and malachite are copper ores. Galena is the major ore of lead, sphalerite of zinc, cinnabar of mercury, and cassiterite of tin.

erals, which releases metallic ions that combine with the oxygen ions in water. Since oxygen is much more abundant than silicon in the silicate minerals, and is also an important constituent of the carbonate minerals (which combine metallic ions with negatively-charged CO_3), almost 94 percent of the volume of the earth's crust consists of oxygen ions (Table 12.1). Of the vast number of different minerals known to occur on the earth, only those few silicates and carbonates shown in Table 12.2 are common constituents of unaltered rocks.

Igneous Rocks

Much of the rock of both the continents and the ocean basins has solidified from a molten condition. Molten rock-forming material, or *magma*, is present everywhere on earth below a depth of about 70 km (40 miles). Occasionally this fluid material, with a temperature of 900°

Table 12.2 Major Rock-forming Minerals

Mineral Group	Mineral	Generalized Chemical Composition	
		POSITIVE IONS	NEGATIVE GROUP
Silicates	Olivine	Mg, Fe	(SiO_4)
	Garnets	Mg, Al, Ca, Fe	
	Pyroxenes	Na, Mg, Al, Ca, Fe	(SiO_3)
	Amphiboles	Na, Mg, Al, Ca, Fe	(SiO_{11}), (OH)
	Micas	Mg, Al, K, Fe	(Si_2O_5), (OH)
	Clay Minerals	Al, K	
	Plagioclase Feldspar	Na, Al, Ca	(SiO_2)
	Orthoclase Feldspar	Al, K	
	Quartz	Si	O
Carbonates	Calcite	Ca	(CO_3)
	Dolomite	Ca, Mg	

After A. Lee McAlester, *The Earth,* Englewood Cliffs, N.J.: Prentice Hall, 1973.

Figure 12.12 Red-hot lava erupted at the earth's surface at temperatures of 900° to 1,200°C (1,600° to 2,200°F) rapidly congeals into extrusive igneous rock (foreground). "Fire fountains" such as this one seen in Hawaii commonly accompany extrusions of basaltic lava and may erupt for periods of several minutes at a time, rising and falling, and often being intermittently active for several days and occasionally even months. (Warren Hamilton/U.S.G.S.)

to 1,200°C (1,600° to 2,200°F), forces its way through the crust and spills out at the surface as red-hot *lava* (Figure 12.12), which cools rapidly to form volcanic rock. Volcanic eruptions also hurl rock particles and bits of lava into the air. This *pyroclastic* ("fire-broken") material rains down to form loose deposits of *volcanic ash,* also known as *tephra*. Extremely violent eruptions vent great volumes of fine glowing particles that are welded together by heat when they settle to the ground. This produces the rock known as *tuff*. Much larger volumes of magma solidify slowly below the surface, deep within the crust. In all cases the product is *igneous rock* (from the Latin *ignis,* fire). Igneous rocks that have solidified below the land surface are termed *intrusive*, while those that solidify at the surface are *extrusive*.

As fluid magma cools, various silicate minerals "precipitate" out in crystal form at successively lower temperatures. The mineral compositions of the more common igneous rocks are shown in Figure 12.13. Magma that crystallizes in the high-temperature environment beneath the earth's surface cools much more slowly than magma that pushes closer to the surface. The greater the time required to solid-

ify, the larger the mineral crystals composing the resulting rock. Igneous rock formed by cooling over periods of thousands of years at a depth of many kilometers is coarse-grained *plutonic rock* (after Pluto, the Roman god of the underworld). The rock mass itself is called a *pluton*. Where many individual plutons have formed a large volume of igneous rock surrounded by other rock types, the result is a *batholith,* such as the Idaho batholith or the Sierra Nevada batholith. Deep-seated plutonic rocks, such as granite, become exposed at the surface by crustal upheaval and the erosion of covering rock masses.

When magma forces its way to the land surface, the result is a volcanic eruption. These eruptions vary in nature (Chapter 15) but produce both lava and tuff. Lava is fluid or semifluid magma that flows out at the surface, where it normally solidifies into rock in hours

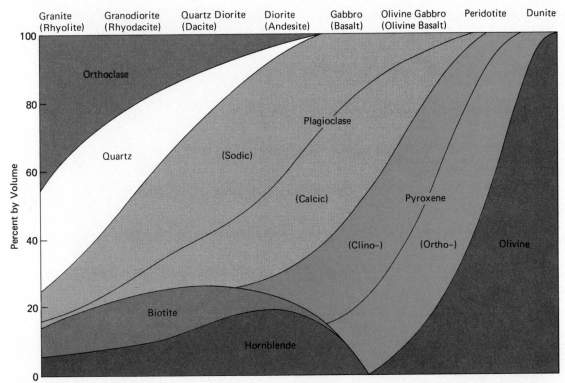

Figure 12.13 Igneous rocks are classified by the proportions of their constituent minerals. Here the approximate mineralogical composition of the more common igneous rocks is shown in terms of the percentage of the total volume occupied by each mineral. Thus a granite should contain approximately 40 percent orthoclase, 30 percent quartz, 15 percent plagioclase, 10 percent biotite, and 5 percent hornblende. Intrusive rock types are labeled across the top, with their extrusive equivalents in parentheses. Peridotite and dunite are rare, probably being formed only in the earth's upper mantle. (Doug Armstrong)

or days. The rapid cooling of lava results in much smaller mineral crystals than those of plutonic rock. Most lavas contain a scattering of 1- to 2-mm-long crystals (phenocrysts) imbedded in a fine-grained matrix (groundmass) of other crystals that are too small to be visible to the naked eye. Tuff consists of materials ranging in size from microscopic crystals to giant blocks (e.g., Figure 1.9).

Almost every type of intrusive igneous rock has an extrusive equivalent, as shown in Table 12.3. This table also shows that a limited number of silicate minerals, including quartz, feldspars, micas (biotite and muscovite), hornblende, pyroxene, and olivine, dominate the compositions of igneous rocks. The most common igneous rocks are the extrusive *basalts* (Figure 12.14), which are dark lavas that compose the sea floors and are also widespread on the land; extrusive *andesites*, which are lighter colored lavas that form the bulk of the earth's large volcanic cones; and coarse-grained intrusive *granites* (Figure 12.15), which form the cores of the continents and are often exposed where the crust has been upheaved in mountain systems.

Widespread tuffs, such as those of the Yellowstone region, are generally composed of light-colored *rhyolite*, the extrusive equivalent of granite.

Although igneous rocks form about 80 percent of the earth's crust, on both the continents and sea floors they are commonly hidden by a veneer of sedimentary rock. As was pointed out in Chapter 1, there is a fundamental difference between the rocks of the continents and those of the ocean basins. The igneous rocks of the continents are mainly low-density plutonic types, whereas those of the ocean floors are higher-density basaltic lavas that erupted along the oceanic ridge system. We saw previously

Table 12.3 Mineral Composition of Common Igneous Rocks

Intrusive Rock	Mineral Composition		Extrusive Rock
	ABUNDANT	LESS ABUNDANT	
Granite	Quartz, K- and Na-feldspars (orthoclase, plagioclase)	Biotite, hornblende, and muscovite	Rhyolite
Diorite	Na- and Ca-feldspars (plagioclase) and hornblende	Biotite and pyroxenes; quartz usually absent	Andesite
Gabbro	Ca-feldspar (plagioclase), pyroxenes, and olivine	Hornblende	Basalt
Peridotite	Olivine and pyroxenes	Oxides of iron; feldspars usually absent	No extrusive equivalent

(a)

(b)

(c)

Figure 12.14 *Basalt* is common extrusive volcanic rock (or lava) and forms the floors of the ocean basins.

(a) This ten-times enlargement of a thin section shows basalt to consist largely of three different silicate minerals and volcanic glass. (M. E. Bickford, University of Kansas)

(b) This solidified volcanic lava on Hawaii shows the contrasting surfaces of *pahoehoe* lava (shiny) and *aa* lava (rough). The hotter pahoehoe flow partially covered a previous aa flow. (T. M. O.)

(c) The Devil's Postpile in California's Sierra Nevada consists of prismatic columns of *andesite*, an extrusive rock intermediate in chemical composition between granite and basalt. The columns were caused by thermal contraction as the rock solidified. Columnar lava always produces a ramp-like deposit of debris, or *talus*, such as that in the foreground. (T. M. O.)

that the basaltic lavas of the ocean basins are considerably younger than most of the plutonic rocks of the continents.

Sedimentary Rocks

Fragments of rocks broken by natural processes are known as *clasts*. Rock weathering (Chapter 11) constantly produces fragmented rock material known as *clastic debris*. This material is moved from higher to lower elevations by the

(a)

(c)

(b)

Figure 12.15 *Granite* is the most common intrusive igneous rock on the continents.

(a) The thin section (enlarged about ten times) shows that granite is a coarse-grained rock consisting of many different minerals that have an interlocking structure. Silicate minerals are the main constituents of granite; when granite weathers, it first decomposes into sand. Further chemical alteration transforms some minerals to clay. (M. E. Bickford, University of Kansas)

(b) This road cut near Prescott, Arizona reveals a typical weathering profile in granitic rock. Chemical weathering gradually converts masses of jointed granite to isolated *corestones* surrounded by decomposed rock. If these corestones do not disintegrate first, they may in time be exposed at the surface as piles of boulders. (T. M. O.)

(c) The Sierra Nevada range in California, shown here in the vicinity of Mount Whitney, consists largely of granite. The boulders in the foreground were separated from massive parent rock by chemical weathering along the joints of the granite, while the vertical slabs in the background were produced by frost weathering. (U.S. Forest Service)

Table 12.4 Common Sedimentary Rocks

UNCONSOLIDATED SEDIMENT	GRAIN SIZE	LITHIFIED ROCK
Angular boulders, cobbles, pebbles	> 2 mm	Breccia
Rounded boulders, cobbles, pebbles	> 2 mm	Conglomerate
Sand	0.02–2.0 mm	Sandstone
Silt	0.002–0.02 mm	Siltstone (Mudstone)
Clay	< 0.002 mm	Shale

Clastic Origin brackets the above table.

UNCONSOLIDATED SEDIMENT	MINERAL COMPOSITION	LITHIFIED ROCK
Calcareous parts of marine organisms and direct calcium carbonate precipitates	Calcite ($CaCO_3$)	Limestone
Magnesium replacement of calcium and direct magnesium carbonate precipitates	Dolomite ($CaMg(CO_3)_2$)	Dolomite
Amorphous silica	Chalcedony, Quartz (SiO_2)	Chert (Flint)
Compacted plant remains	Carbon (C)	Bituminous Coal
Salt left by evaporation of sea or saline lake water	Halite (NaCl)	Rock Salt
Gypsum left by evaporation of sulfate-laden water	Gypsum ($CaSo_4 \cdot 2H_2O$)	Gypsum

Chemical Origin brackets the above table.

energy of running water, wind, waves, and glacial ice, and by the direct pull of gravity. All of these agents of erosion eventually lose some of their energy, resulting in deposition of some or all of the debris they are carrying. Streams transport clastic debris to lakes, inland seas, and the oceans, where sediment accumulates in layers. Sediment is also deposited on the land—on river floodplains, in crustal depressions, along the margins of glaciers, at the bases of cliffs, and wherever sand-carrying winds lose velocity.

Clastic sediments deposited on the continents consist mainly of boulders, cobbles, gravel, sand, and silt. Marine sediments formed on the continental shelves close to shorelines include sand, silt, and clay; farther from the shore on the outer shelves we find only clays, chemically precipitated calcium carbonate, and the limy and siliceous remains of tiny marine animals. Under the deeper water of the continental rise areas we once more encounter sands and silts, emplaced by great submarine slides and flows of sediment down the continental slopes. When sediments of any type become deeply buried under subsequent deposits, compaction and the deposition of cementing substances cause them to become *lithified*, or con-

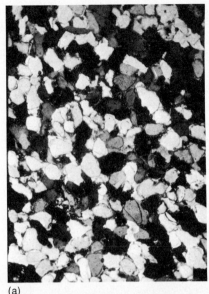

(a)

(b)

(c)

Figure 12.16 Sedimentary rocks consist of fragments of rock debris or organic material compacted and cemented together by various chemical substances, most commonly silica, calcite, and iron oxide.

(a) *Sandstone,* one of the most common *sedimentary rocks,* is usually composed of grains of silicate minerals cemented together by other minerals, as shown in this enlarged thin section. In general, sandstone is any rock composed of fragments that are in the size range of sand grains. When sandstone weathers, the cementing material disintegrates, releasing the individual grains. Sandstone is formed by the consolidation of beds of sand deposited by wind or water both on the land and in the sea. (M. E. Bickford, University of Kansas)

(b) This photograph illustrates the laminated nature of sedimentary rocks, which form *strata* that vary in thickness and physical and chemical characteristics. (Warren B. Hamilton/U.S.G.S.)

(c) This thick series of sedimentary rocks exposed above the Colorado River near Moab, Utah, produces a landscape of cliffs and ledges unobscured by soil and vegetation. The ledge-forming strata are sandstone, with weaker shale producing the slopes between successive ledges. In eastern North America the same rocks would produce a much smoother landscape that would be covered by soil held in place by forest vegetation. (T. M. O.)

verted into *sedimentary rock* of either clastic or chemical origin (Table 12.4).

Sedimentary rocks form in layers (Figure 12.16). These may be millimeters to hundreds of meters thick. Each distinct layer, called a *bed* or *stratum,* indicates a period of sediment deposition. The separations between strata are *bedding planes,* which indicate a period of no deposition at that location. Frequently the strata above and below a bedding plane are dissimilar, indicating a change in the conditions of sediment delivery. Individual beds may also vary laterally due to differences in energy conditions and distance from the original source of the sediments. A single layer deposited in a certain

time interval can change in lateral succession from *conglomerate* (cemented gravel) to *sandstone* (cemented sand), *siltstone* (cemented silt), *mudstone* (cemented silty clay), *shale* (cemented clay), and finally *limestone,* which is formed either as a chemical deposit or as an accumulation of the skeletal remains of tiny marine animals.

Table 12.5   Structure and Composition of Metamorphic Rocks

	METAMORPHIC ROCK	TEXTURE	MINERAL COMPOSITION	DERIVED FROM
Foliated	Slate	Fine-grained; smooth, slaty cleavage; separate grains not visible	Clay minerals, chlorite, and minor micas	Shale
	Schist	Medium-grained; separate grains visible	Various platy minerals, such as micas, graphite, and talc, plus quartz and sodium plagioclase feldspar	Shale, basalt
	Gneiss	Medium- to course-grained; alternating bands of light and dark minerals	Quartz, feldspars, garnet, micas, amphiboles, occasionally pyroxenes	Granite
Nonfoliated	Quartzite	Medium-grained	Recrystallized quartz, feldspars, and occasionally minor muscovite	Sandstone
	Marble	Medium- to course-grained	Recrystallized quartz or dolomite plus minor calcium silicate minerals	Limestone or dolomite

Limestone, which is composed of calcium carbonate, has considerable economic value. From it is made the cement used to construct highways, buildings, sidewalks, patios, and swimming pools. *Dolomite* is a calcium-magnesium carbonate rock that forms as a chemical precipitate or by chemical alteration of limestone. Limestone is peculiar among common rock types in that it dissolves completely where there is abundant moisture and vegetation. This creates very distinctive landscapes, as we shall see in Chapter 15. Dolomite too is soluble, but much less so than limestone.

Marine organic deposits that have an exceptionally high carbon content form *hydrocarbons*—the principal constituents of petroleum and natural gas. Due to their low density, petroleum and natural gas migrate upward to fill openings in the more porous rocks, especially sandstone. These are the "reservoir rocks" in oil and gas fields.

Economically, the most valuable of all sedimentary rocks is *coal*, which originates as luxuriant vegetation growing in freshwater lagoons and swamps. To be preserved, this organic material must accumulate in a stagnant-water environment and be acted upon by bacteria that can thrive without oxygen. To be transformed into coal, the resulting organic complex must be compressed by deep burial. The first stages of coal formation are occurring today in the swamps of the southeastern United States.

Two other sedimentary rock types of economic value are *rock salt* and *gypsum*. These are chemical deposits formed on the beds of evaporating lakes and inland seas in dry regions. The uses of salt are too numerous to mention. Gypsum likewise has many uses, among the most important being the manufacture of plaster of Paris and gypsum board, which is the standard material used to sheath the inner walls of houses.

Submarine volcanic activity can saturate sea water with silica ions, resulting in the chemical precipitation of silica as thin beds of *chert*. Pod-

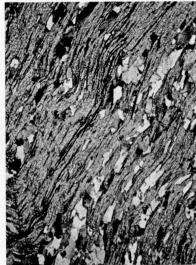

Figure 12.17 Metamorphic rocks form by the transformation of preexisting rocks under conditions of heat and high pressure.

(a) This outcrop of *gneiss* exhibits the swirled patterns often seen in metamorphic rock. (M. E. Bickford, University of Kansas)

(b) *Schist* is a crystalline rock dominated by a layered arrangement of platy minerals, as seen in this thin section enlarged ten times. Because weathering tends to be most effective between the sheets, schists tend to break into thin flakes. (Warren Hamilton/U.S.G.S.)

like masses of chert, called "flint," are frequently found in limestone and, due to their hardness and ability to retain sharp edges, were used in the manufacture of stone tools by primitive peoples.

Metamorphic Rocks

Crustal motions related to plate tectonics sometimes cause rock masses to be forced deep down into the lowest portions of the earth's crust. Here pressures and temperatures are hundreds of times those at the earth's surface. This causes the rock material to deform and flow in a plastic manner or to melt and recrystallize in different minerals. This process of *metamorphism* of rock material transforms limestone to *marble*, shale to *slate*, sandstone to *quartzite*, granite to *gneiss*, and lava to *schist* (Table 12.5). Metamorphism most often occurs in or near subduction zones, where crustal plates converge.

In many metamorphic rocks the minerals are "smeared out," or oriented along visible planes of flow (Figure 12.17). Where the original rock contained a mixture of minerals, as in granite, metamorphism may segregate them in wavy bands of contrasting color. The result is gneiss. Complete melting of the original rock generates new fluid magma, which may force its way upward in the crust to become a pluton of igneous rock. The process of metamorphism, then, is transitional to complete melting.

Near the centers of all the continents are rigid areas of very ancient *crystalline rocks*—rocks clearly displaying mineral crystals—principally, granite, gneiss, and schist. These ancient rocks may be exposed at the surface, as around Hudson Bay in Canada and in New York's Adirondack Mountains, or they may be "basement rocks" covered by younger sedimentary rocks. These oldest portions of the continents are known as *crystalline shields*. Most of the rocks of the crystalline shields were subjected to extreme metamorphism in pre-Cambrian time (Chapter 1), and many of the plutonic rocks appear to have been created by the melting of even more ancient sedimentary rocks. The crystalline shields represent the deep roots of ancient mountain systems that were erased long ago by the processes of erosion. They contain the earth's oldest known rocks (about 3.8 billion years in age), all of which are metamorphic types exposed where erosion has

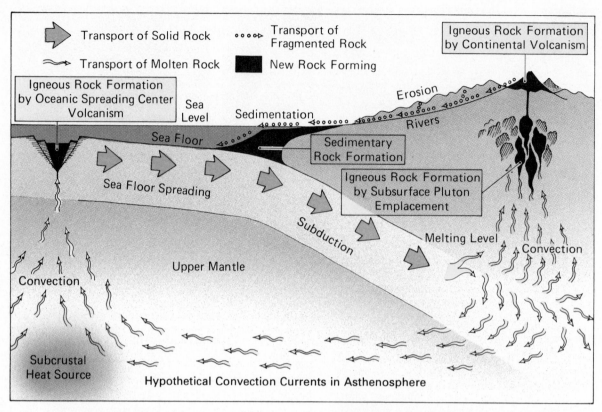

Figure 12.18 This diagram displays the flow of material in the rock cycle. The diagram represents a vertical plane cutting through the earth from its surface to the asthenosphere in the upper mantle. The circulation is driven by subcrustal heat sources and gravity. Arrows show the transport of rock material in different forms. Dark areas represent locations in which new rock masses are formed. Metamorphic rocks are produced by partial melting and recrystallization of older rocks in subduction zones and regions of pluton formation. (T. M. Oberlander)

removed thicknesses of tens of kilometers of overlying younger rocks.

The Rock Cycle

During the course of time, rock materials pass from one form to another in the *rock cycle* (Figure 12.18). Our understanding of the rock cycle has been greatly advanced by the discovery of sea-floor spreading and the crustal subduction process. We have long known that most of the rocks exposed at the earth's surface are sedimentary types composed of the debris of older rocks of all types. We know too that despite the nearly 5-billion-year history of the earth, all the ocean floors have ages of less than 200 million years. Thus in the rock cycle the older oceanic crust is swallowed by subduction, descending deep into the lithosphere where it becomes metamorphosed. At depths of 70 to 90 km (45 to 55 miles), it begins to melt into new magma, which rises into the continental crust in the form of igneous intrusions. Some of this igneous material erupts onto the surface as volcanic ash and lava. Erosion of the surface ash,

lava, sedimentary rocks, and the deeper metamorphic and plutonic basement rocks generates new sediments. These sediments are gradually transformed into new sedimentary rock, which in time may be subducted, metamorphosed, melted, and recycled as new igneous rock. In this rock cycle the same atoms that have been part of the earth since its formation are used over and over in successive generations of rock material.

Differences in the rock types resulting from the rock cycle play a very important role in the appearance of landscapes. We shall explore this topic in Chapter 15. But first it is necessary to look at the geological structures created by the deformation of rocks as crustal plates move and interact.

CRUSTAL DEFORMATION

In addition to rock type, the "architecture" of rock masses plays an essential role in the appearance of landscapes. The varying structural configurations of rock bodies are the result of past tectonic activity, including volcanism and crustal compression and extension.

Volcanism

Volcanism refers to the intrusion of magma into the earth's crust and the extrusion of volcanic gases and molten material at the earth's surface. Although volcanism's greatest effect is in the oceans, the floors of which are composed of basaltic lava, its more visible displays on the land are among the earth's most awe-inspiring phenomena.

Large-scale volcanism is related to lithospheric plate boundaries (Figure 12.19). Three types of plate boundaries exist: those where plates are pulling apart, those where plates are sliding past one another horizontally, and those where plates are pushing together with one

being subducted under the other. The first two are zones of nonexplosive outpourings of lava; the third is the location of violent tephra-producing volcanic eruptions.

Magma Sources and Eruptive Styles

Eruptions of the earth's most active volcano, Hawaii's Kilauea, can be viewed from close at hand in complete safety. Other eruptions devastate large areas with amazing suddenness and spare only those eye-witnesses who are far away. This fact was reaffirmed by the May 1980

Figure 12.19 This map shows the locations of the known active volcanoes of the world; for clarity, some volcanoes have been omitted from regions where many are present. The map also shows the principal lithospheric plate boundaries, oceanic spreading centers, and subduction zones. Most regions of important volcanic activity are located near active plate boundaries. Note the large number of active volcanoes around the rim of the Pacific Ocean, which is ringed by subduction zones. The volcanoes of Iceland and the Atlantic Ocean are associated with plate divergence from along the Mid-Atlantic Ridge. Volcanoes in the Caribbean, in East Africa, and in the eastern Mediterranean are associated with plate boundaries as well. Conversely, the east coasts of North and South America, where there are no plate boundaries, are devoid of volcanic activity. Isolated volcanic regions far from plate boundaries, such as the Hawaiian Islands, are due either to "hot spots" in the earth's mantle or fractures in the lithospheric plates, which allow molten rock from the earth's interior to reach the surface. (Andy Lucas after F. M. Bullard, *Volcanoes*, 1962, University of Texas Press)

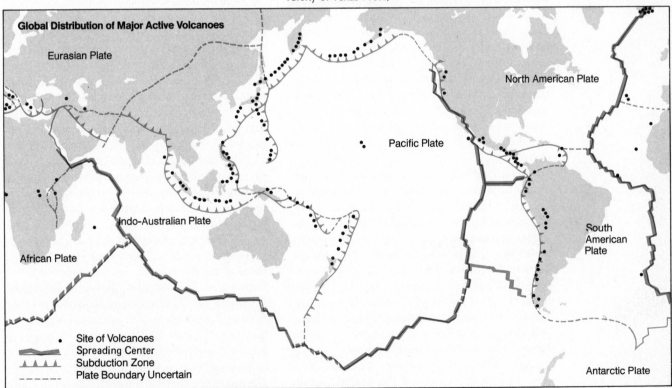

Global Distribution of Major Active Volcanoes

Eurasian Plate

North American Plate

Pacific Plate

Indo-Australian Plate

South American Plate

African Plate

• Site of Volcanoes
  Spreading Center
  Subduction Zone
  Plate Boundary Uncertain

Antarctic Plate

eruption of Mt. St. Helens in the state of Washington, which took more than 60 human lives in a matter of minutes. Volcanoes take many different forms, and so do volcanic eruptions. This is a consequence of variations in the source of volcanic energy—the magma feeding the volcano.

There are two general sources of magma. One is the portion of the mantle that lies below the rigid crust of the earth. The permanent layer of fluid rock material here is chemically basic sima, having a silica content of 50 percent or less. When simatic magma reaches the surface, the magmatic gases are still dissolved in the melt, making it very fluid. It pours out freely, with little explosive activity, despite the presence of impressive steam clouds.

The other source of magma is the melting of rock masses that have descended into the earth's furnace-like interior in the process of subduction. Magma produced by the melting of older rocks contains more than 65 percent silica and is "cooler" than basic magma—about 900°C (1,600°F) versus 1,200°C (2,200°F). It has a lower density than basic magma, and its gases have separated out below the surface, where they are under enormous pressure. Therefore, when it is able to break through to the surface, silicic magma always erupts violently. The first eruption of gases and steam reduces the pressure confining the magma, which allows more magma to flash into gas, causing more explosions, more pressure release, more gas, and so on, in a chain reaction that may continue for hours or even days. We shall have more to say about different types of volcanic eruptions in Chapter 15.

Silicic volcanism generates much smaller amounts of lava than does basaltic volcanism, but ejects great volumes of coarse and fine pyroclastic tephra, in explosive eruptions. The tephra (volcanic ash) and lava are principally andesitic to rhyolitic in composition. Explosive volcanism occurs at the continental edges, areas of crustal plate convergence, and in volcanic *island arcs*, such as the West Indies and the Japanese, Philippine, and Indonesian archipelagos (island chains)—all of which lie above presently (or recently) active subduction zones.

As we shall see in Chapter 15, basaltic lava, which forms the floors of the oceans, sometimes floods out in great volumes on the land, as in the states of Washington, Oregon, and Idaho. Such a phenomenon requires special circumstances, to be discussed in the later chapter.

Tensional Faulting

The earth's crust is subject to many different types of stresses as lithospheric plates collide, pull apart, and scrape past one another. These stresses can always be resolved into either tension or compression, both of which cause the deformation of rock masses (Figure 12.20). Rocks are strained by tension where the crust has been stretched or arched upward, and also where two adjacent sections of the crust move parallel to one another in opposite directions, which is known as *shear*. Tension causes rock masses to rupture and be displaced along fracture planes called *faults*. Individual fault traces may extend tens to a thousand or more kilometers across the earth's surface and commonly penetrate tens of kilometers down into the earth's crust. Sudden slippage of rock masses in contact along faults produces shock waves that are felt as earthquakes.

Faults related to crustal extension allow segments of the crust to tilt or sink. Sometimes enormous areas are affected. The Great Basin, centered in Nevada, offers one of the best displays of the effects of faulting to be seen on our planet. A viewer of a nineteenth-century map of this vast region remarked that it resembled "an army of caterpillars crawling north." The topography consists of parallel mountain ranges separated by flat-floored basins (Figure 12.21). The basins are blocks of the earth's crust that have gradually sunk or tilted downward, leaving neighboring blocks standing as mountain ranges. This is clearly a region in which tension has stretched out the earth's crust and ruptured it on a massive scale.

Faults that result from crustal extension, like those of the Great Basin, are called *normal faults*. Such a fault forms a plane that extends into the earth at a steep angle that is known as the *fault dip*. Fault dip is measured as the angle from the horizontal (Figure 12.22). The map direction of a fault is termed the *strike*, which is measured as an angle east or west of true north or south. The angle measured is the projection of the fault

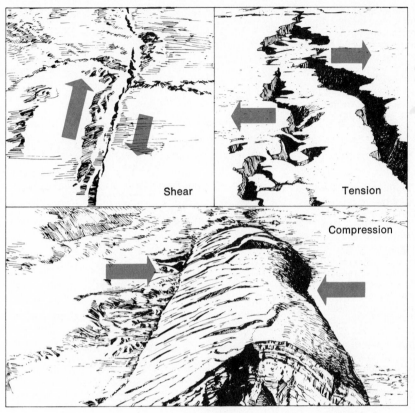

Shear

Tension

Compression

Figure 12.20 Shear, tensional, and compressional forces on the earth's crust. These three drawings show some of the possible landforms that can result from these forces or combinations of them.

(**top left**) Shear produces horizontal displacement, causing offset stream valleys, interruption of ground-water flow, and demarcations in vegetation.

(**top right**) Tension has produced a down-dropped block, forming a steep-sided *graben*.

(**bottom**) Compression has folded sedimentary rocks into a ridge. (John Dawson)

Figure 12.21 The scenery of Nevada is dominated by linear mountain ranges separated by flat-floored basins. These basins are strips that subsided along faults as the crust was stretched and fractured by large-scale geologic movements beginning about 25 million years ago. The white patches in the basins are salt flats. (T. M. O.)

Figure 12.22 The attitude of any inclined plane, such as a fault or a tilted layer of rock, may be described precisely in terms of the *strike* and *dip* of the plane. The strike is the compass bearing of the line of intersection between the inclined plane and a horizontal plane. It is read in degrees east or west of a north-south direction. The dip is the angle the inclined plane makes with a horizontal plane; it is given in degrees together with a compass direction. In this diagram, the strike is about N 30° E and the dip is to the northwest at about 50°. (Vantage Art, Inc.)

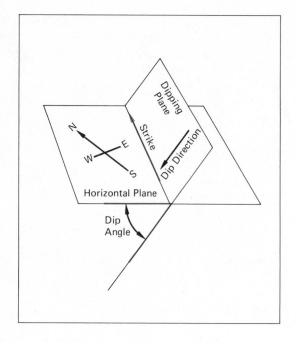

Figure 12.23 Faults are the result of the rupture of rocks due to crustal tension, compression, and shearing. Fault motion is described by the relative displacement, or *slip*, of the rock units in contact along the fault. Motion may be parallel to either the dip or the strike of the fault plane, or oblique to both.

In the center of the diagram is a *normal fault*, in which the type of motion is *dip slip*, with the block resting on the fault plane slipping down relative to the adjacent block. This type of motion results from tension and produces crustal extension. The faults on the right of the diagram result from compression and produce shortening of the crust. In both *overthrusts* and *reverse* faults, the rock mass above the fault plane moves up and over the rock mass below the fault plane. In all three faults the motion is parallel to the dip of the fault plane, but in the latter two the direction is the reverse of normal dip-slip motion so the type of motion is *reverse dip slip*.

Shearing stress involves side-by-side forces exerted in opposite directions. This results in horizontal motion parallel to the fault strike, or *strike slip*. The direction of the strike slip is designated in terms of the movement of the block on the far side of the fault. In this case, looking across the fault from either side, we see that the far block has moved to the right, so the motion is *right lateral strike slip*.

The final fault on the left shows displacement oblique to both the strike and dip of the fault plane. Both strike slip and dip slip are involved in *oblique slip*. Whereas strike-slip faults are always vertical and dip-slip faults are always inclined, oblique slip can occur on both vertical and inclined faults. (Vantage Art, Inc.)

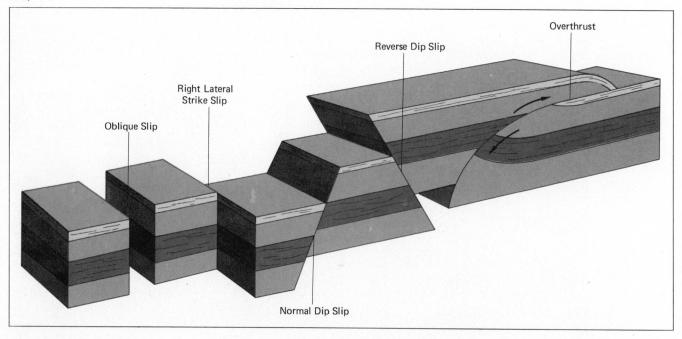

to a horizontal plane (Figure 12.22). In normal faulting the block resting on the inclined fault plane slips downward with respect to the adjacent block (Figure 12.23). Movement down the dip of the fault is *dip-slip* motion. Normal dip-slip faulting has formed most of the mountain ranges in the Great Basin.

Strike-Slip Faulting

Some of the earth's most important faults have been produced by motion in which adjacent segments of the crust move past one another horizontally rather than vertically. Faults of this type cannot produce mountains, but they are major sources of destructive earthquakes. Faults on which the motion is parallel to the strike rather than the dip of the fault plane are called *strike-slip* faults (Figure 12.23). One of the best known of these is the San Andreas fault, which extends 1,000 km (650 miles) northwestward through southern and central California. It was a sudden 3 to 6 meter (10 to 20 ft) offset on this fault that produced the famous San Francisco earthquake of 1906.

Strike-slip faulting seems to be necessary to permit sea-floor spreading to continue. The velocity of motion away from oceanic spreading centers cannot be uniform throughout a large crustal plate because of the spherical shape of the earth, and because of varying rates of lava generation along the oceanic ridges. In order for portions of crustal plates to move at different velocities, the plates must break into segments that can slip past one another. These strike-slip faults are called *transform faults* because they transform sea-floor spreading into fault motion. The San Andreas fault seems to be a transform fault that has broken through the edge of the North American continent. Similar large transform faults are found in Turkey, Iran, China, and New Zealand, as well as throughout the oceans.

Folding

Where crustal plates are moving toward one another, rocks are compressed. Compression causes sedimentary rocks to be deformed into wrinkles called *folds*, which are a clear indica-

tion of crustal shortening (Figure 12.20). Strong compression also can produce faults by causing rock masses to break into slabs that overlap one another like roof shingles.

Everyone is aware that rock is brittle and breaks under sudden impact. But large volumes of rock, yielding slowly under compressive stress, can undergo amazing deformation without breaking. In the geologic process of folding, rock can be deformed in whatever ways a tablecloth can be rumpled when you push your hands over it.

Despite the complexity of form that is possible, some basic elements are recognizable in any system of folds in sedimentary rocks. Where rock strata have been forced into wrinkles, the ridges are called *anticlines* and the troughs are *synclines* (Figure 12.24). The inclined layers between the anticlinal crests and the synclinal troughs are the fold *limbs*. These terms apply only to geologic structure, not to the visible surface landforms, which are created by erosion and may reflect geologic structure in varying ways, as will be seen in Chapter 15.

Regions of anticlines and synclines very often have considerable economic importance. Most sedimentary rocks that are compressed into simple folds were originally marine deposits formed on continental shelves. Many of these sediments contain petroleum and natural gas produced by the decay of marine microorganisms. When the sediments are folded, the hydrocarbons, which are lighter than water, migrate toward the anticlinal crests where they form "pools" in porous rocks, especially sandstones. These anticlinal structures are the reservoirs from which we derive our major energy supplies. The theory of plate tectonics helps account for the fact that the Middle East is the world's greatest supplier of oil. In this case the Arabian crustal plate pushed into the Iranian plate, rumpling the continental shelf sediments between the two. This produced ideal reservoir structures for petroleum accumulation. Other types of reservoir structures involve faults that trap subsurface hydrocarbons against impermeable rocks, and salt intrusions that dome and pierce sediments, creating hydrocarbon traps in upturned layers.

It should be pointed out that folding may

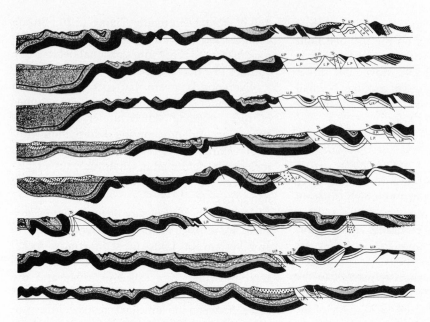

Figure 12.24 The Zagros Mountains of western Iran offer one of the world's finest displays of simple fold structures formed by crustal compression (see Figure 15.28). In this group of structure cross sections, the arches with descending fold limbs are *anticlines*, and the U-shaped troughs with rising limbs are *synclines*. The eastern (right) portions of the cross sections show *reverse faults* (thrust faults) also resulting from lateral compression.

The folding in this region was produced by the spreading of the Red Sea (Figure 12.1), which has pushed the Arabian Peninsula away from Africa and against the portion of Eurasia represented by central Iran. (From the British Petroleum Co.)

only be a superficial phenomenon that affects masses of sediments deposited over a rigid "basement" of older rocks. During such folding the sediments become detached from the basement rocks, which are themselves not folded. However, the basement rocks are often faulted by the compressive forces, a process discussed in the following section.

Compressional Faulting

Under certain conditions rocks fracture rather than wrinkle when the earth's crust is shortened by strong compression. The resulting fault planes vary from oblique to nearly horizontal. The block resting on the fault plane rides up and over the adjacent block. Such faults are known as *reverse faults* or *thrust faults*. Where

compression is especially powerful, the fault planes flatten to parallel the earth's surface, permitting vast sheets of rock to slide or be pushed tens of kilometers horizontally. Such far-traveled slabs are called *overthrusts* (Figure 12.23). Thrust faults and overthrusts are normal features of large mountain ranges formed by compression in regions where crustal plates have collided.

In many areas of the European Alps, three or four overthrusts, each having the area of a state like Massachusetts or Connecticut, are piled atop one another. The incredible force required to produce such crustal deformation is vivid evidence of the enormous energy within the earth. The dynamic ever-changing nature of the earth also is evident in the fact that the normal faults of Nevada, recently produced by crustal extension, cut through thrust faults caused earlier by strong crustal compression.

Overthrusts are attracting increasing attention as possible areas of petroleum reserves, because some overthrust masses have pushed across oil-bearing fold structures. In the United States two overthrust belts are being explored for petroleum. One lies west of the Blue Ridge Mountains of the Appalachian region, where ancient basement rocks and younger sedimentaries were sheared off and forced westward. The other zone is in the Northern Rocky Mountains in Wyoming, Montana, and Alberta, where the overthrusts have moved eastward across oil-bearing folds.

Broad Warping and Isostacy

In addition to folding and faulting, the earth's crust experiences broad vertical motions. Although these do not create striking geological structures, they are important because they help control erosion and deposition. Rising areas are attacked by stream erosion, forming regions of hills or plateaus gouged by canyons. Sinking areas are usually regions of low relief in which sediments are deposited. Where hilly or mountainous coastal areas have sunk, they have been transformed into islands and peninsulas, as in Greece, Yugoslavia, and western Turkey, and on a smaller scale in the San Francisco Bay area of California.

Broad uplift may be a result of localized crustal heating due to mantle convection, or it may reflect the interaction of converging crustal plates. Smaller areas often rise because of a large subsurface intrusion of magma. Such an event may be occurring at present near Mono Lake in California, where recent earthquake activity and local changes in elevation have given rise to predictions of a volcanic eruption within this decade.

Some areas are rising today because of removal of load from the land surface. A load that has come and gone several times over the past 2 or 3 million years is glacial ice, which has been more than a kilometer thick over areas of millions of square kilometers (see Chapter 17). Both Scandinavia and eastern Canada are presently rising at rates up to 20 meters (65 ft) per 1,000 years because continental ice sheets have withdrawn from them within the last 10,000 years.

The response of the earth's crust to changes in load is related to the fact that the lithosphere appears to be floating in buoyant equilibrium in the denser material of the underlying mantle. Each segment of a sialic (continental) area, with an approximate density of 2.8, displaces an equal weight of sima in the upper mantle,

which has a density of about 3.3. Accordingly, the continents float about four-fifths submerged in the denser material of the mantle, just as a slab of wood floats in water. This concept is known as *isostacy* (from the Greek *isos*—equal, and *stasia*—standing).

For isostatic equilibrium to prevail, any large-scale upward projection of the earth's crust must be compensated by a downward projection of crustal material into the earth's mantle. Thus all continental mountain systems have a deep subcrustal "root," a fact verified by seismic and gravity surveys. Conversely, major depressions on the continents are areas of thin crust, with the "Moho" between the crust and the mantle (Chapter 1) rising to a higher than average level.

Just as the addition of the weight of a continental glacier disturbs buoyant equilibrium and depresses the crust, the removal of weight creates an imbalance that causes the crust to rise. This must be accomplished by lateral flow of sima in the asthenosphere. Accordingly, the transfer of material by the processes of erosion causes the eroding area to rise in order to maintain isostatic equilibrium, whereas the area receiving the weight of the eroded sediment sinks for the same reason. To permanently remove a mountain range by erosion, much more time is required and a far greater volume of sediment must be transferred than would be necessary without isostatic compensation. Isostatic uplift in response to erosion triggers more erosion, which in turn causes more isostatic uplift—a vivid example of a positive feedback relationship, and an important factor in landform development.

Geologic structures such as domes, folds, fault blocks, and overthrusts are hardly ever seen intact; they are attacked by erosion as soon as they begin to form. But it is important to understand them because they control the development of many types of landforms as well as the location of all mineral deposits. After a general discussion of the factors influencing landforms in Chapters 13 and 14, we shall look at the detailed effects of geologic structure on landforms in Chapter 15.

SUMMARY

Without disturbances by forces within the planet, the earth's surface would soon be worn smooth by the vigorous processes of erosion. Our planet's internal energy is most evident in the form of earthquakes and volcanic eruptions. However, the major differences in elevation on the earth are produced by very small crustal movements repeated over periods of millions of years.

The earth's relief features may be divided into three scales of magnitude. The largest-scale features, the first order of relief, are the continents and ocean basins. These are not permanent, but change in form and position through geologic time. The active elements are the sea floors, which are pushing outward from the oceanic ridge systems where new ocean crust is generated by volcanic activity. The continents are embedded in the denser material of the earth's mantle and are carried along by the lateral motion of this material. This motion is thought to be produced by convection in the earth's mantle. The second order of relief consists of major continental and oceanic mountain systems and related large-scale relief features. The motions of the earth's crust related to sea-floor spreading and the subduction of crustal plates produce these features. The third order of relief consists of individual landforms that are generally related to processes of erosion and deposition. These landforms provide the stage for all human activity and exert many influences on other physical systems.

Landforms are themselves composed of rock or rock debris. All rocks are composed of minerals, which are natural combinations of chemical elements in solid form. The most common rock-forming minerals are the silicates in which silicon and oxygen combine with other chemical elements.

Rock originates in several ways. Igneous rock crystallizes from a molten condition. It is generated as fluid magma far below the earth's surface and is either intruded into the crust or extruded at the surface. Most sedimentary rock is formed by the lithification of deposits of loose rock debris and organic matter, both on the land and on the sea floor, although some limestones and cherts are inorganic chemical precipitates. Metamorphic rock is a product of the recrystallization of older rocks that have been subjected to great heat and pressure deep within the earth's crust. The rock cycle is an endless sequence of material transformations, in which the same atoms are used over and over to form new rock material from older rocks.

The earth's crust has been affected by various types of motions, from gentle warping to violent deformation, by forces resulting from the interaction of moving lithospheric plates. The rise of molten material from the earth's mantle and from areas of rock melting in subduction zones has created surface volcanic activity as well as crustal uplift. Volcanism is associated with the margins of lithospheric plates. Oceanic volcanism is generally nonviolent, being produced by very fluid basaltic magma that is erupted at a high temperature. Volcanism related to subduction zones is far more dangerous, as andesitic and other more silicious types of lava are erupted explosively due to the pressure of gases that have separated from the magma.

Where the earth's crust is being stretched by tensional forces, it is collapsing by the development of normal dip-slip faults. The Great Basin region of western North America is the world's best example of such a process. Strike-slip faults transform sea-floor spreading into faulting in which masses of the earth's crust move horizontally past one another. Crustal compression can produce either folding, which creates anticlinal and synclinal structures, or thrust faulting and overthrusting, which shorten and thicken the crust. Petroleum reserves may be present in anticlinal fold structures covered by large-scale overthrusts. Broad warping of the crust affects landscapes by stimulating stream erosion in rising areas and causing the deposition of sediments in crustal depressions. Isostatic adjustments, by which the earth's crust maintains bouyant equilibrium, causes eroding areas to rise, which significantly retards the erosional destruction of large positive relief features.

REVIEW QUESTIONS

1. How are sima, sial, and isostacy interrelated?

2. What is the difference between the current plate tectonic theory and the continental drift hypothesis advocated earlier by Alfred Wegener?

3. What was the evidence for continental drift adduced by Wegener? What is the evidence for sea-floor spreading adduced by proponents of this new paradigm?

4. In plate tectonic terms, what is the significant difference between the margins of the Atlantic Ocean and those of the Pacific Ocean?

5. Indicate five second-order relief features of the oceanic basins.

6. Is the barrel of a ball point pen a mineral? Are there any minerals in your view at this moment? What, exactly, is a "mineral"?

7. How can you distinguish a plutonic rock from an extrusive volcanic rock? How could either be distinguished from a coarse-grained sandstone?

8. Which would be heavier, a piece of rhyolite or a piece of basalt of the same size? Why?

9. In what location(s) is rock metamorphism most likely to occur?

10. Why is it that the hottest magma produces the least explosive volcanism?

11. Volcanism seems to occur both where crustal plates are pulling apart and where they are pushing together. Explain this apparent contradiction.

12. What is a crystalline shield?

13. Describe the "rock cycle."

14. What different types of faults can you name, describe, and trace to a particular mode of origin?

15. What is the economic importance of subsurface anticlinal structures? Can you imagine a circumstance in which a synclinal structure might be valuable?

16. What factors could be responsible for broad uplift of the land surface?

APPLICATIONS

1. Poorly informed persons periodically assert that California will someday "fall off" during an earthquake. Why is this impossible?

2. On an outline map of the world indicate the locations of offshore oil fields on the continental shelves.

3. Make a collection of the rocks present in the region in which your campus is located. What does each rock specimen tell you about the geological environment in which it was formed? Do any of the rock types create distinctive landforms? If your campus is in an area of glacial deposition, see how many different rock types you can collect and determine the relative proportions of local and far-traveled rocks. Be aware that an occasional diamond has been found in glacial deposits in the midwestern United States!

4. What kinds of geological materials are quarried or mined in your county or state? In a state almanac or annual geological survey report, look up the annual value of your state's mined or quarried products. Are you surprised about the values of any of the various materials?

5. Geological studies show that in some areas of lithospheric plate convergence, volcanic island arcs (like Japan and the Philippines) have been "welded" onto a continental margin. Draw a series of cross sections through an island arc resulting from subduction, showing how an island arc distant from any continent could become welded onto a continent.

6. How could you distinguish a volcanic tuff from a specimen of sandstone or siltstone?

FURTHER READING

Decker, Robert, and **Barbara Decker.** *Volcanoes.* San Francisco: Freeman (1981), 244 pp. Extremely lively account of modern volcanic phenomena, with less attention to the classic events of the past. Clearly relates volcanism to plate tectonic theories.

Glen, William. *Continental Drift and Plate Tectonics.* Columbus, Ohio: Merrill (1975), 188 pp. A concise and well-illustrated account of the essentials of plate tectonic theory.

Hamblin, W. K. *The Earth's Dynamic Systems.* Minneapolis: Burgess (1978), 470 pp. Extremely well-illustrated treatment of rocks, geological structures, and landforms. Notable for use of LANDSAT images.

Hurlbut, Cornelius S., Jr. *Minerals and Man.* New York: Random House (1970), 304 pp. Hurlbut's informative book on the occurrence, characteristics, and uses of minerals is renowned for its superb color photographs of mineral specimens.

King, Philip B. *The Evolution of North America.* Princeton, N.J.: Princeton University Press (1977), 197 pp. A spirited overview of the geologic structure and history of North America by a leading authority. King offers many sidelights and personal observations that enliven the subject matter.

McPhee, John. *Basin and Range.* New York: Farrar, Straus, Giroux (1981), 216 pp. McPhee is one of the most evocative of current writers. Here he focuses on the earth's crust and makes it and its investigators come alive. Highly recommended for pure stimulation.

Sheets, Payson D., and **Donald K. Grayson.** *Volcanic Activity and Human Ecology.* New York: Academic Press (1979), 672 pp. This collection of authoritative articles deals with volcanic hazards and volcanic effects on environments and human ecology from prehistoric to modern time. An impressive variety of topics is covered in depth.

Sorrell, Charles A. *Rocks and Minerals: A Guide to Field Interpretation.* New York: Golden Press (1973), 280 pp. An extremely handy guide to rock and mineral identification. The first seventy pages are an excellent introduction to minerals and rocks. Exceptional illustrations.

Sullivan, Walter. *Continents in Motion.* New York: McGraw-Hill (1974), 399 pp. The story of the development of the theory of plate tectonics and sea-floor spreading, presented as the unraveling of a scientific mystery rather than in textbook fashion.

Wilson, J. Tuzo. *Continents Adrift and Continents Aground.* San Francisco: W. H. Freeman (1976), 230 pp. This is a collection of seventeen superbly illustrated articles on plate tectonics originally published in *Scientific American.*

Windley, Brian F. *The Evolving Continents.* New York: Wiley (1977), 385 pp. This advanced treatment outlines the geological development of all the continents in terms of plate tectonic theories. Very well illustrated with diagrams and cross sections that demonstrate the great variety of lithospheric plate interactions.

Wyllie, Peter. *The Way the Earth Works.* New York: Wiley (1976), 296 pp. Excellent, thoughtful account of the earth's geological environment. Notable for its text rather than its illustrations.

Earthquakes

A little after 02:00 [2 A.M.] on December 16, the inhabitants of the region were awakened suddenly by the groaning, creaking, and cracking of the timbers of their houses and cabins, the sounds of furniture being thrown down, and the crashing of falling chimneys. In fear and trembling, they hurriedly groped their way from their houses to escape the falling debris. The repeated shocks during the night kept them from returning to their weakened and tottering dwellings until morning. Daylight brought little improvement to their situation, for early in the morning another shock, preceded by a low rumbling and fully as severe as the first, was experienced. The ground rose and fell as earth waves, like the long, low swell of the sea, passed across the surface, bending the trees until their branches interlocked and opening the soil in deep cracks. Landslides swept down the steeper bluffs and hillsides; considerable areas were uplifted; and still larger areas sank and became covered with water emerging from below through fissures or craterlets, or accumulating from the obstruction of the surface drainage. On the river, great waves were created which overwhelmed many boats and washed others high upon the shore, the returning current breaking off thousands of trees and carrying them into the river. High banks caved and were precipitated into the river; sandbars and points of islands gave way; and whole islands disappeared. (Earthquake History of the United States, 1973)

Obviously, this report must be from fault-shattered California or some other part of the tectonically-active mountainous west of North America. Or is it? In fact, the site of this description was New Madrid, Missouri—then a community of log cabins next to the Mississippi River—the date: December 1811. Further shocks occurred there in January and February of 1812. Reports of similar events are also available from Quebec (1638) and Charleston, South Carolina (1886), with only somewhat less catastrophic effects having occurred again in Missouri (1895), western Texas (1931), western Ohio (1937), and southern Illinois (1968), to name only a few normally placid geologic settings that have experienced sudden seismic shocks.

Unlike volcanic eruptions, which have restricted geographic distributions, earthquakes can occur anywhere. Furthermore, unlike most other environmental hazards, they occur without warning. By their direct and indirect effects, including fires, flooding, landslides, and so-called "tidal waves," earthquakes have caused more destruction and

taken more lives than any other single type of natural catastrophe.

What are earthquakes? The slow distortion of rock masses by tectonic stress causes a build-up of potential energy in the rock much as in a compressed or extended spring. When the deformation, or strain, surpasses the elastic limit of the rock, it ruptures suddenly or slips along preexisting faults, releasing the stored energy in the form of seismic shock waves that speed outward from the point of rupture. Several types of shock waves are involved, including rapidly moving body waves of varying character that pass through the mass of the earth and slower surface waves that follow the earth's outer skin. The surface waves create the ground motion we call an earthquake. The point below the earth's surface at which the energy is released is the earthquake *focus*, or *hypocenter*. The point on the earth's surface directly above the focus is the earthquake *epicenter*. It is at the epicenter that shock waves are most strongly felt.

The type of ground motion produced and the damage occurring

during an earthquake vary according to the depth of the focus, distance from the epicenter, and nature of the local geologic material. In a small tremor, the sensation produced ranges from a gentle horizontal shaking to one or more sharp jolts. Near the epicenter of a large earthquake, one has the feeling of being in a rough sea in a small boat, with the ground visibly rising and falling in moving crests and troughs similar to ocean waves. Normally, the shaking lasts no more than a few minutes. However, smaller aftershocks, sometimes numbering in the hundreds, may continue over a period of days to months, indicating continuing subsurface rock displacements.

The violence of an earthquake is measured on two different scales that specify the magnitude and surface effect of the energy released. The *magnitude scale* devised in 1935 by Charles F. Richter of the California Institute of Technology is a logarithmic one in which each successively higher unit represents about 32 times the energy release of the preceding unit. Thus the difference in energy released by earthquakes of magnitudes 4 and 8 on the Richter scale is not 2 times, but more than a million times. The largest earthquakes felt since the invention of *seismographs* (instruments that record earthquake intensity) have attained Richter magnitudes of about 8.9. Any earthquake of magnitude 8 or above is considered catastrophic.

The earthquake *intensity scale*, which measures the surface effects of earthquakes, was developed by the Italian geophysicist G. Mercalli in 1905. The Modified Mercalli scale of 1931, which is now in general

CASE STUDY

use, rates these effects from I to XII (see table). There is no correspondence between the Richter and Mercalli scales because the surface effects of an earthquake can vary from point to point with no change in earthquake magnitude, due to the manner in which the local geologic material transmits seismic shock waves.

There is normally less ground motion on solid rock near the earthquake epicenter than on unconsolidated alluvium or artificial land fill some kilometers from the epicenter. Unconsolidated deposits magnify the ground shaking and often collapse, in severe earthquakes behaving almost as fluid. This phenomenon was conspicuous at Lisbon, Portugal, in 1755, San Francisco in 1906, Anchorage, Alaska, in 1964, and in the New Madrid earthquakes of 1811–1812, in which there was catastrophic shaking and rupturing of the ground on the Mississippi River floodplain but little effect in the adjacent bedrock hills. Artificial land fills are a ubiquitous part of our expanding urban scene, especially in hilly and coastal regions. Fortunately, North America has experienced few severe earthquakes since the explosive expansion of cities and their suburbs in the present century. Unfortunately, every day brings us closer to the first great earthquake that will strike one of our modern urban centers, as has recently befallen Managua, Nicaragua (1972), Beijing (Peking), Tangshan, and Tianjin (Tientsin), China (1976), and Guatemala City, Guatemala (1976).

Earthquakes can be anticipated in areas overlying subduction zones and lithospheric plate boundaries, and also along the oceanic ridge system. In all of these locations, great masses of rock are being moved past one another as part of the processes of sea-floor spreading. We may expect repeated news of earthquake disasters from Peru, Chile, Central America, California, Alaska, Japan, the Philippines, New Zealand, Indonesia, the Mediterranean, Turkey, Iran, the Himalayan foothills, Burma, and China.

But some of history's most devastating earthquakes cannot be explained in terms of the gross motions of lithospheric plates—for example, those of the Mississippi Valley, Charleston, and Quebec, all apparently in the middle of the North American plate. And there is the greatest earthquake in European history, the Lisbon quake of 1755, which in 6 minutes took hundreds of thousands of lives and destroyed castles, fortresses, cathedrals, and mosques throughout Portugal, southern Spain, Morocco, and Algeria. The epicenter of this earthquake was apparently offshore in the Atlantic, in a region where there is no major tectonic feature to explain such a geologic cataclysm.

In time, geologists will uncover the reasons for earthquakes in such areas. For example, recent evidence suggests that the Mississippi Valley, a midcontinental area of frequent seismic activity, may be an ancient continental rift similar to those presently containing the Red Sea and the East African lake system. Several types of evidence suggest that this billion-year-old rift, which is now buried by later sediment accumulations, has been reactivated by the intrusion of mantle material at depth, followed by isostatic adjust-

---

Modified Mercalli (MM) Intensity Scale of 1931

I Not felt except by a very few under especially favorable circumstances.

II Felt only by a few persons at rest. Delicately suspended objects may swing.

III Felt quite noticeably indoors. Standing motor cars may rock slightly. Vibration like passing of truck.

IV Felt indoors by many, outdoors by few. Dishes, windows, doors disturbed; walls make cracking sound. Sensation like heavy truck striking building.

V Felt by nearly everyone. Some dishes, windows, etc., broken; a few instances of cracked plaster; unstable objects overturned. Disturbances of trees, poles, and other tall objects sometimes noticed.

VI Felt by all, many frightened and run outdoors. Some heavy furniture moved; a few instances of fallen plaster or damaged chimneys.

VII Everybody runs outdoors. Damage slight to moderate in well-built ordinary structures; considerable in poorly built or designed structures; some chimneys broken.

VIII Damage considerable in ordinary substantial buildings, with partial collapse; great in poorly built structures. Fall of chimneys, factory stacks, columns, monuments, walls. Heavy furniture overturned. Sand and mud ejected in small amounts.

IX Damage great in substantial buildings, with partial collapse. Buildings shifted off foundations. Ground cracked conspicuously. Underground pipes broken.

X Some well-built wooden structures destroyed; most masonry and frame structures destroyed; ground badly cracked. Rails bent. Landslides considerable from river banks and steep slopes. Shifted sand and mud. Water splashed (slopped) over banks.

XI Few, if any, masonry structures remain standing. Bridges destroyed. Broad fissures in ground. Underground pipelines completely out of service. Earth slumps and land slips in soft ground. Rails bent greatly.

XII Nearly all structures of every type damaged greatly or destroyed. Waves seen on ground surface. Objects thrown upward into the air.

Source: Bolt, Bruce, et al. 1975. *Geological Hazards*. New York: Springer-Verlag, p. 9.

ments that may be triggering earthquakes.

While ground motion during earthquakes causes enormous damage, subsidiary effects may be equally dangerous. The enormous fires that are triggered by earthquakes—as in San Francisco in 1906 and Tokyo in 1923—may be more destructive than the ground shaking. The New Madrid earthquakes caused changes in elevation of 2 to 6 meters (6 to 20 ft) over an area of 100,000 sq km (40,000 sq miles), creating temporary waterfalls in the Mississippi River and resulting in permanent submergence of 600 sq km (250 sq miles) of virgin forest by groundwater outflow and surface stream flooding. In coastal and undersea regions, earthquakes resulting from sudden movement along faults produce seismic sea waves, or *tsunamis* (see Chapter 16), popularly called "tidal waves," which have swept over populous coastal regions, causing enormous loss of life. Earthquake tremors may set up resonance effects in large and small water bodies, causing them to slosh

back and forth rhythmically, often with destructive effects on docks and ships at anchor. During the Lisbon earthquake of 1755, these resonance effects, termed *seiches*, were observed as far away as Switzerland, the British Isles, and Scandinavia.

Almost all earthquakes trigger landslides, ranging from gigantic rockslides in mountain regions to stream bank cavings along flat floodplains. While the latter may create destructive waves in rivers, the former have often obliterated entire communities (see the Case Study following Chapter 13).

Landslides, devastating in themselves, often create still further dangers. In 1959 an earthquake sent a great rockslide crashing into the Madison River Canyon in Montana, just outside Yellowstone National Park. While this slide snuffed out the lives of 26 campers, an even greater hazard was created, as the slide blocked the Madison River, causing it to back up in a rapidly rising lake. Had the lake risen to the top of the landslide dam and spilled over, it would have washed away the

dam, producing a flood of cataclysmic proportions in the valley beyond. Men and heavy earth-moving equipment were rushed to the scene to labor around the clock to excavate a controlled spillway for rising Earthquake Lake. Fortunately, the work was finished in the nick of time, and the second disaster was averted. The highest flood crest ever recorded resulted from the washout in 1895 of a landslide dam 274 meters (900 ft) high in the Himalayan foothills in India; the resulting flood flow was 73 meters (240 ft) deep.

It would be no comfort to victims of earthquakes, volcanic eruptions, hurricanes, and tornadoes to be told that each of these phenomena is the result of processes that help to maintain equilibrium in the earth's intertwining physical systems, but such is indeed the case. Some disequilibrium is building even as you read this, and before many months have passed, it will be compensated by a seemingly violent event that will take lives, destroy property, and be recorded in newspaper headlines as another natural catastrophe.

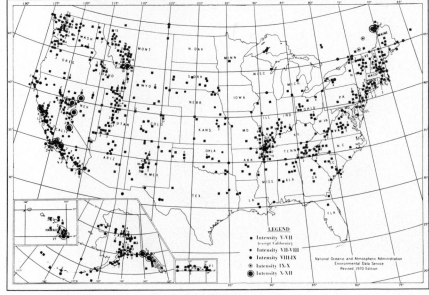

Earthquakes (intensity V and above) in the United States through 1970.

Historic earthquake epicenters in the United States are recorded on this map according to their intensities on the Modified Mercalli scale. It can be seen that no region of the United States is safe from earthquake damage. When looking at such a map, one must imagine a halo of destruction around each epicenter, often covering thousands of square kilometers. The Mississippi Valley earthquakes of 1811-1812 were felt over some 2,600,000 sq km (1,000,000 sq miles) to distances as remote as Washington, D.C., with chimneys being toppled as far away as Pennsylvania, Ohio, Georgia, and South Carolina. (Vantage Art, Inc., from Earthquake History of the United States, *1973, U.S. Dept. of Commerce, National Oceanic and Atmospheric Administration)*

341

Winter Landscape from the Ch'ing Dynasty. (Courtesy of the Smithsonian Institution, Freer Gallery of Art, Washington, D.C.)

In addition to having oceans, unique climates, and a living biosphere, the earth is unlike all other planets in our solar system with respect to surface form. No known planet has scenery like the earth's because no other planet in our solar system has a similar atmosphere, or hydrosphere, or such a constantly active interior. No other planet has been intricately sculptured by erosional processes, or has the variety of rock types, geologic structures, or surface forms seen on the earth. No other planet has had its early history so thoroughly effaced by continuing dynamic processes that constantly change the planetary surface.

The form of the land surface is the most visible and certainly one of the most influential of all aspects of the human environment. Thus it is an important area of study within physical geography. The variable form of the earth's surface, like other aspects of our environment, can be explained in terms of energy, materials, and system thresholds. As we have seen in earlier chapters, the materials are rocks and minerals of various types and water in all its forms. The energy sources that affect these materials are solar radiation, gravitational force, and the heat produced by radioactive decay of unstable elements deep within the earth. System thresholds determine when and where change will occur as energy is applied to the materials exposed at the earth's surface.

There is unmistakable evidence that the form of the earth's surface changes constantly. Sometimes the changes are easily seen, as when a severe flood alters a river channel or a landslide tears away a hillside. Other changes are too slow to be visible over a human lifetime. Valleys deepen and expand. Ocean waves gradually gnaw away the edges of the land. Our hills seem everlasting, but after each spring thaw or heavy rain, streams are brown, yellow, or red with sediment eroded from the land. The

13
Landforms

The Differentiation of Landforms

Geologic Structure
Tectonic Activity
Gradational Processes
Time

Slopes: The Basic Element of Landforms

Mass Wasting
Water Erosion on Slopes

Models of Landform Development

The Cycle of Erosion Concept
The Equilibrium Theory of
    Landform Development

The Tempo of Geomorphic Change

Towering pillars of limestone in southern China are an example of a unique landform produced by a particular combination of weathering processes and rock structures. Many different factors give landforms their characteristic shapes.

Figure 13.1 Landforms of the type shown here are third-order relief features produced by a variety of erosional and depositional processes.

The valley occupied by the stream was created by river erosion and enlarged by glacier scouring. In the right background is the summit of Mt. Rainier, a volcano in the Cascade Range in Washington—a constructional landform that is composed of lava and volcanic ash. The peak to the left is on a rim created by the collapse of an earlier summit of the volcano. In the middle distance are moraines, ridgelike deposits left by the Emmons Glacier when it was much larger than at present. In the foreground are cobbles washed out of the glacier by meltwater. The larger boulders and some of the cobbles were brought here by a rockslide in 1963. The rockslide covers the surface of the Emmons Glacier in the center of the photo. (T. M. O.)

hills, large and small, are being washed to the sea—slowly, grain by grain.

The remainder of this book focuses on the form of the land surface and the processes that create individual relief features of the third order, which we call *landforms* (Figure 13.1). The study of landforms, known as *geomorphology,* draws on many of the facts presented in the earlier discussions of energy, system thresholds, the hydrologic cycle, climate, vegetation, and soils—all of which influence landform development. In turn, terrain features affect other physical systems. They determine surface elevation, thereby affecting temperature, causing orographic rainfall, and producing rainshadows. They influence the nature and amount of surface and subsurface water supplies. They affect the local character of vegetation and soils. Thus landforms are an active part of the physical environment.

This chapter considers the factors that cause landforms to differ, discusses the processes affecting slopes, and concludes with some general theories of landform development. Subsequent chapters will analyze the landforms created by specific geomorphic processes and geologic structures.

THE DIFFERENTIATION OF LANDFORMS

The limestone hills accurately portrayed in Chinese paintings, like the one at the beginning of this chapter, certainly are unlike hills in Kentucky or Missouri made of the same rock. Why is this? And why are landforms in widely separated parts of the world similar to one another but different from features only kilometers away? Four factors are involved: (1) the *geologic structure*, which includes the type and arrangement of the material composing the landform; (2) *tectonic activity*, or the presence or absence of local movements of the earth's crust; (3) the current *gradational processes* or those active in the past; and (4) the span of *time* during which

the tectonic and gradational processes have operated in a particular way.

Geologic Structure

Geologic structure refers to the physical nature and geometrical arrangement of the rock materials from which landforms are built up or in which they are sculptured by erosion (Figure 13.2). In Chapter 12 we saw that geologic materials are extremely diverse, ranging from blan-

Figure 13.2 This oblique aerial photograph taken over western Iran shows a complex landscape that is thoroughly dominated by geological structure. All of the ridges are composed of resistant limestone strata, while valleys are eroded in weak shales. Strong folding has inclined the strata so that variations in rock resistance can be exploited by erosion. (Aerofilms and Aero Pictorial Ltd.)

kets of dust that have settled from the atmosphere to enormous volumes of hard rock that has crystallized from subsurface magma. Some materials are chemically and physically simple; others are extremely complex. They may be arranged in layers or in seemingly structureless masses. Sedimentary rocks may be horizontal, tilted, or folded, and together with igneous and metamorphic rocks may be divided by joints, broken by faults, or thrust vast distances horizontally. All of these varying conditions are reflected in local and regional landforms. Strong deformation of rock masses producing complex geologic structures is usually associated with present or past margins of interacting lithospheric plates and results from the sea-floor spreading process. The diverse effects of geologic structures on landforms are important enough to merit separate treatment. We will discuss these effects in detail in Chapter 15.

Tectonic Activity

The disturbances of the earth's crust that result from the earth's internal energy are known as *tectonic activity*, or simply *tectonism* (from the Greek word *tekton*, meaning builder). Tectonic activity is a reflection of the restlessness of the earth's interior, which is manifested in the sea-floor spreading process and related deformational and volcanic effects. Tectonism influences scenery in two ways: by broadly elevating and depressing the earth's crust, and by producing the crustal extension, compression, and volcanism that create the geologic structures that influence the form of the landscape. Because the geologic structures resulting from tectonic activity are attacked by erosion even as they are being formed, these structures are rarely seen intact. But it is tectonic uplift, including isostatic adjustment, that provides the potential energy that stimulates erosion. Most areas that are rough or mountainous are experiencing tectonic uplift at the present time. Areas that have little relief are either sinking and being filled with sediments or have been free of vertical motion for many millions of years.

Gradational Processes

Whereas tectonic activity tends to cause the earth's surface to be uneven, atmospheric phenomena and gravity tend to smooth the surface. Solar radiation drives the atmospheric circulation, which generates rain, snow, wind, and ocean waves. Surface water and ice, moving downslope under the influence of gravity, pick up and transport rock fragments produced by weathering. The result is a constant shifting of fragmented rock material from high to low places on the earth's surface. This tendency for high places to be worn down and low places to be filled in is known as *gradation*. Gradation includes the fragmentation of rocks by weathering, the detachment and removal of rock fragments by the agents of erosion, and the deposition of rock debris at lower elevations. If our planet had no internal energy, gradation would eventually produce a smooth earth. But energy within the earth constantly disturbs the surface—creating new highs and lows, and new work for the gradational processes.

The chief physical agent of gradation is water (Figure 13.3). It is the main factor in the break-up of rock by chemical and mechanical weathering, and the principal mechanism by which rock debris is moved from high to low elevations. Although channeled flows of running water are the most widespread single agent of erosion and deposition, other important gradational agents are the direct force of gravity, wind, glaciers, and waves and currents in lakes and seas. Any one process may be dominant over vast areas, producing a distinctive imprint upon the landscape. Often different processes work jointly or alternate in dominance over varying spans of time.

Usually climate determines which of the gradational processes is most active in an area. Changes in the type or intensity of gradational processes can generally be traced to a climatic change. An example is the periodic cutting and filling of gullies in the arid southwestern United States, widely thought to reflect fluctuations in rainfall intensities over periods of years. A regional gully-cutting episode after about 1200 A.D. is thought to have destroyed the agricultural system of the Pueblo Indians, who inhabited the impressive cliff dwellings of Arizona, New Mexico, and southwestern Colorado, causing the Indians to abandon their homes permanently. This particular episode is but one in a continuing alternation of gully cutting and filling in the region, where the erosional systems appear to be very sensitive to climatic fluctuations.

Climate also affects gradation through the influence it exerts on the vegetative cover (Figure 13.4). Vegetation is the dominant factor controlling the rates of erosion by water and wind. A dense vegetative cover holds soil in place, breaks the impact of wind and rain, and maintains a surface that absorbs rainfall rather than allowing it to run off and erode the land.

On hillslopes that are not protected by vegetation, loose material produced by weathering is periodically flushed away by water runoff. This prevents soil development and generates a gullied, rocky landscape. The erosional loss from bare soil is 50 to 100 times greater than that from a grass-covered area. Similarly, hillslopes covered by grass are eroded many times faster than hillslopes covered by temperate forest vegetation. But in the humid tropics the intensity of rainfall overcomes the protective effect of a dense vegetative cover, and the rate of hillslope erosion increases (Figure 13.5). Human activity has increased erosion rates in all environments, mainly as a result of direct or indirect modification of vegetation (Figure 13.5).

The relationship between climate, gradational processes, and landforms is complex. In certain regions with distinctive climates, the landforms also seem distinctive. Whether or not the gradational processes in such places are truly unique has been a matter of argument among geomorphologists. We shall consider the specific geomorphic features associated with various climates in more detail in Chapter 16.

Time

Where vast areas of hard rocks have been eroded to very low relief, as in portions of Af-

rica and Australia, it is clear that the passage of time is reflected in the landscape. In the complete absence of tectonic activity, the effect of time would be to permit gradation to reduce landscapes to monotonous plains. However, there are few places on the earth where such landscapes have been identified.

We may also think of the effect of time in terms of response times and relaxation times, as described in Chapter 2. Some types of landforms respond very quickly to changes in geomorphic processes. River channels and beaches are examples. River channels enlarge rapidly during floods and then return to their pre-flood form when floodwaters recede. Beaches become steeper and narrower when a storm sends large waves against the shore and they rebuild quickly after the storm subsides. Such adjustments, or responses, are always in a direction tending to reestablish equilibrium between the landform and the changed process affecting it. The time required for restoration of process/form equilibrium is the relaxation time. In many quick-response landform systems, the inputs of energy and materials fluctuate continuously, so the thresholds for erosion and deposition are crossed frequently. As a result, form changes are more or less continuous, with equilibrium being approached but never actually attained.

In cases where the response and relaxation times are very long, landforms may preserve "leftover" features resulting from processes that are no longer active. These are called *relict* forms. The clearest examples are the landforms produced in high mountains throughout the world by glacial erosion and deposition that ended more than 10,000 years ago (Chapter 17). To erase the evidence of glacial modification of landscapes will require perhaps a million years of gradation by other processes.

SLOPES: THE BASIC ELEMENT OF LANDFORMS

The chief problem in landform analysis is to explain the conditions that produce the various

Figure 13.3 Water acting as an agent of erosion and deposition affecting the varied materials and structures of the earth's crust has generated widely different landforms, as these examples show.

(a) This valley in New Zealand was carved by the Motueka river, then filled with sediments washed out of a melting glacier. The present sinuous river slowly changes its course by lateral migration, alternately eroding and depositing a veneer of sediment and producing a wide floodplain between the valley walls. (G. R. Roberts, Nelson, N. Z.)

(b) Water acting on these sedimentary rocks in Red Rock Canyon, California, has cut a variety of shapes and patterns in layers of varying resistance to erosion. (Florence Fujimoto)

(c) Surface runoff and streamflow are carving a dendritic, or branched, drainage pattern into this landscape north of Christchurch, New Zealand. The relatively uniform resistance of underlying rock favors the development of a treelike drainage pattern as each of the side tributary branches extends by headward erosion away from the main stream channels. (G. R. Roberts, Nelson, N. Z.)

(d) The canyon of the Gunnison River in western Colorado, more than 600 meters (2,000 feet) deep, is a splendid example of the power of channeled flows of water incise in hard rock. The ancient metamorphic rock of the canyon walls is the same as that exposed in the deepest portions of the Grand Canyon of the Colorado River in Arizona. (T. M. O.)

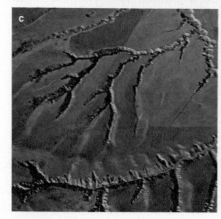

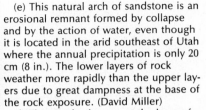

(e) This natural arch of sandstone is an erosional remnant formed by collapse and by the action of water, even though it is located in the arid southeast of Utah where the annual precipitation is only 20 cm (8 in.). The lower layers of rock weather more rapidly than the upper layers due to great dampness at the base of the rock exposure. (David Miller)

(f) The attack of waves at the base of these limestone hills near Dorset, England, has removed the support for the overlying rock. The entire face has become a nearly vertical cliff because of removal of rock from the foot. (G. R. Roberts, Nelson, N. Z.)

(g) The rock spires in Bryce Canyon National Park, Utah, are products of water erosion of weakly consolidated sedimentary rocks. They exhibit a step-like structure because the rock layers have different resistances to erosion. (David Miller)

(h) The view from Zabriskie Point in California's Death Valley reveals that although there is little rainfall in desert regions, the effect of water erosion is extremely conspicuous due to the absence of a protective cover of vegetation. (T. M. O.)

Figure 13.4   These views illustrate variation in slope form in physically similar interbedded shales and sandstones in two different climatic regions.

In California's Coast Ranges, the grass cover developed in response to an annual rainfall of about 50 cm (20 in.) protects a thin soil, increases moisture infiltration into the ground, and reduces erosive surface runoff.

In southern Utah, a rainfall of about 25 cm (10 in.) is insufficient to produce vegetative protection of slopes, resulting in intricate gully erosion and exposure of bare rock ledges that shed coarse waste onto the slopes. The nature of the rainfall regimes in the two areas is also a factor. In California, most of the precipitation falls in low-intensity drizzles associated with the passage of winter frontal systems. In southern Utah, summer thunderstorm activity produces brief episodes of high-intensity precipitation on bare ground. (T. M. O.)

Figure 13.5   This diagram illustrates how the average rate of surface lowering by erosion, known as the rate of denudation, is affected by precipitation and vegetation. The example considered is a 10° slope in a warm region where the average annual temperature is 25°C (77°F). Soil wash by surface runoff is the most important process of denudation. The rate of lowering is least where the annual precipitation is about 100 cm (40 in.) because in such regions vegetative protection of the soil surface reduces the erosive impact of rainfall and the increase in soil permeability due to organic activity allows little surface runoff to be generated. The rate of denudation is greater in drier regions because the vegetation is sparse and surface runoff readily erodes the exposed bare ground. Rainsplash on bare ground also plays a role. In very moist regions, more moisture is available than the soil can absorb, and large moisture surpluses are generated. The large volume of water moving through the soil and weathered mantle accelerates the rate of subsurface chemical erosion, so that the abundant runoff of such regions carries away much material in solution as well as by soil wash. (Doug Armstrong after M. G. Wolman)

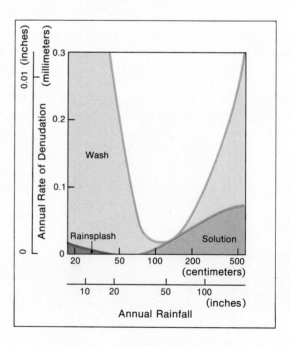

types of slopes and flat surfaces that make up the earth's landscapes. Figure 13.6 provides a sample of the diversity of slope types present in different regions.

Most of the slopes we see around us were initiated by downward erosion by channeled flows of running water. The same process that cuts gullies in a sloping cornfield like that in Figure 11.23 (p. 300) has excavated the Grand Canyon of the Colorado River. Stream incision provides new vertical surfaces that are quickly transformed into slopes by other erosional processes. These processes loosen material and move it down the slopes and into streams, ever widening the excavation made by vertical stream erosion. Rock weathering, mass wasting, and water-assisted erosion all play a part in the development of slopes.

Mass Wasting

Gravity provides the energy for all slope-forming processes. Gravity itself can cause the material composing hillslopes to move downward without the assistance of any moving fluid in the process known as *mass wasting*. Such movements range from sudden catastrophic rockslides to the slow downslope creep of soil and rock fragments over hundreds of years.

Solid rock is held together by the attractive forces between neighboring atoms, but slopes composed of soil and rock fragments remain in place because of the friction between solid particles. Without these frictional and cohesive forces, hillsides would collapse under the pull of gravity. Weathering of the rock forming a slope, or water saturation of the weathered rock, may occasionally reduce friction and cohesion to the point where rocks break loose or a portion of the hillside avalanches downward.

For any loose material on a slope, whether soil or a layer of rock debris, there is a maximum angle of inclination, known as the *angle of repose*, that the material can maintain without slipping downward. At this threshold angle, gravitational stress is just balanced by the cohesion or friction of the material on the slope. If new material, or a large volume of water, is added to a slope that is already at the angle of

repose, a portion of the slope may fail and slide downward. Soil that has become water-saturated is most likely to slide, because the absorbed moisture lessens the friction between soil particles at the same time that it increases the bulk weight of the soil.

At the bases of rock cliffs there are normally cones of debris consisting of loose rock fragments that have fallen from the face of the cliff (Figure 13.7). This material, called *talus* (large chunks) or *scree* (small particles that shift underfoot), accumulates at an angle of repose of 34 to 39 degrees. Talus slopes resulting from rock falls are present wherever there are rock cliffs but are most common in alpine areas and dry regions where bare rock exposures are most plentiful.

In most areas, where there are no rock cliffs, mass wasting acts slowly and invisibly. The soil of every sloping pasture or forested hillside moves downhill a fraction of a centimeter per year (Figure 13.6c). This form of mass wasting is aptly known as *soil creep*. Soil creep causes fence posts and tombstones on sloping ground to tilt conspicuously.

The creep process is related to expansion and contraction of the soil by wetting and drying and by freeze and thaw. Figure 13.8 shows how changes in soil volume always result in slight downslope displacement of the soil mass. The rate of displacement is related to the slope angle but is generally less than one centimeter a year. Since creep velocity is a function of the slope angle, the transported material (known as *colluvium*) accumulates at the base of the slope, thus gradually elevating the slope foot and causing the slope angle to become less steep with the passage of time (Figure 13.9). As creep-transported colluvium moves into depressions, the slope also becomes progressively smoother.

A distinctive type of mass wasting is seen in subarctic and highland tundra regions where soils that are silty and moisture-retentive cover ground that remains frozen throughout the year. We have seen that in many tundra regions only the upper meter or so of the soil thaws during the spring and summer. The water released by thawing cannot drain downward through the still frozen subsoil. For a few days

Figure 13.6 These photographs show the variety of slope types that result from varying geologic materials and climatically controlled slope-forming processes.

(a) Stone Mountain, Georgia, is a core of unjointed granitic rock that rises boldly above jointed, chemically weathered granite in the Piedmont region of the Appalachians. (Warren Hamilton/U.S.G.S.)

(b) Chemical weathering in jointed granite in arid eastern California creates a jumble of exposed rock. In the distance is the steep wall of the Sierra Nevada, its crest splintered by frost weathering; Mt. Whitney is in the center. (T. M. O.)

(c) Smooth rolling slopes in Devon, England, result from chemical weathering of limestone, accompanied by slow downward transfer, or "creep," of the resulting mantle of soil. (G. R. Roberts, Nelson, N. Z.)

(d) Precipitous slopes have been cut in basaltic lavas in the upland areas of the Hawaiian Islands. Strong vertical incision by torrential streams during heavy rains is accompanied by rapid sliding of the water-saturated soils on steep slopes (scar at left). (Butch Higgins)

Figure 13.7 Frost weathering of well-jointed granite rock at elevations above 3,500 meters (11,000 ft) in California's Sierra Nevada loosens rock masses and produces rock falls, particularly during the spring, when ice is melting. This creates cones of rock rubble, known as *talus*, at the base of frost-shattered cliffs. (T. M. O.)

the water-saturated soil loses its cohesion. When the saturation reaches a threshold value it activates the soil, which begins to sag slowly downslope. Different sections move at different rates, which makes the slope appear to be covered with overlapping scales or lobes of soil (Figure 13.10). This process is known as *solifluction* (soil flowage). Despite their active appearance, solifluction lobes move only a few centimeters each year. Movement ceases when the water content declines below the threshold value for soil activation. Like soil creep, solifluction fills in depressions and tends to smooth the landscape.

Slopes covered by silt or clay soils that have an unusual capacity to absorb rainwater often collapse suddenly in the form of *slumps*. Like solifluction lobes, slumps result from loss of cohesion and increase of weight in water-saturated soils. A slump is a rapid rotational movement in which a mass of soil slips downward

and tilts backward, leaving a conspicuous scar on the slope. A hillside slump usually terminates in an *earthflow*, consisting of a bulge of the collapsed soil that is pushed out and down the slope (Figure 13.11). Slumps and earthflows are often destructive. They can be triggered by human activity, such as lawn watering or slope steepening during construction of roads or buildings. Slumping also occurs along stream banks, especially during the declining stages of floods when the banks are saturated with water.

The most awesome of all gravitational transfers of material are the immense *rockslides* that occur in mountainous country (Figure 13.12). In these slides as much as several cubic kilometers of rock may fall and "splash" outward surprising distances with incredible speed and destructive force. Nearly all rockslides are triggered by unusual wetting of weathered rock or by the sudden shock of an earthquake. Mountains are periodically drenched with moisture due to

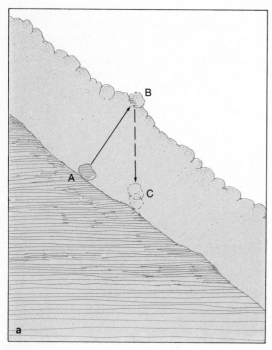

Figure 13.8 (a) Soil tends to creep downslope under the influence of gravity. This schematic diagram illustrates the mechanism responsible for soil creep. When soil (A) becomes moist or freezes, it expands (B) at right angles to the slope, but when the soil dries and contracts, the soil particles tend to move vertically downward (C), which results in a net downslope migration of the soil surface. (John Dawson)

(b) The "needle ice" shown in this photograph, at a much enlarged scale, is a mechanism that moves soil downslope. When water in damp ground freezes, needles of ice grow outward from the surface, carrying surface soil particles with them. When the needle ice melts, soil particles tend to fall vertically downward or forward, resulting in a net downslope displacement of the particles. This increases the effect shown in (a). (David Cavagnaro)

Figure 13.9 This upslope view of an artificial roadcut reveals the contact between chert bedrock (on the right) and the mantle of *colluvium* (to the left) that is moved downslope by slow gravitational transfer. In this instance, the colluvium has filled an erosional depression, causing the hillslope to become smoother. The fragments of bedrock in the colluvium are angular rather than rounded as in stream-deposited *alluvium*. (T. M. O.)

heavy orographic rainfall and spring melting of deep snow. This may be enough to trigger damaging earthflows or rockslides, as occurred in the mountains of central Utah and western Nevada in the spring of 1983. Mountain regions are also particularly prone to earthquakes, which are an expression of the geological movements that raised the mountains. If an earthquake occurs when weak rocks are saturated with water, rockslides are to be expected. In such an event, two different thresholds are involved. First, build-up of strain in the rocks of the earth's crust reaches a threshold that permits sudden movement of large subsurface rock volumes. This sends out shock waves (the earthquake) that momentarily reduce rock cohesion to the second threshold, which permits collapse of the slope. Large rockslides have occurred in historic time in all the earth's high mountain regions, and slide scars and deposits are a part of all high mountain landscapes.

Figure 13.10 Solifluction lobes, shown here on a slope in Canada's Yukon Territory, develop where silty, water-retentive soils cover permanently frozen ground. The lobes exhibit irregular margins because the rate of downslope movement of the thawed soil varies from point to point. (Larry W. Price)

Water Erosion on Slopes

The effectiveness of gravity as an agent of landscape change is greatly multiplied when it sets water in motion. Slope erosion by running water probably removes 50 to 100 times as much material as mass wasting processes.

Slope erosion commences with *rainsplash*. When large raindrops strike bare soil, their impact can splash soil particles several millimeters to centimeters laterally (Figure 13.13). On slopes, more than half the particles splashed into the air fall downslope from their previous location, thereby causing a net downslope shift of soil particles. Because vegetation and plant litter intercept rainfall, rainsplash erosion is most effective in dry regions where bare soil is widely exposed.

Where there is a cover of vegetation and soil, average rainfall events usually do not produce surface runoff of water. Runoff normally occurs where rain falls on areas with poorly developed or nonexistent vegetation and soils and, occasionally, where the downpours on vegetated surfaces are torrential. In such instances, water builds up on the surface and begins to move down any available slope, initiating *overland*

flow. The initial overland flow is in the form of a slowly moving sheet that has no erosive effect. However, as it gains speed, the sheet passes a depth/velocity threshold that causes it to break into turbulent threads, or *rills*. In turbulent flow (see Chapter 14), there are velocity vectors and pressure gradients in all directions, including upward. Turbulent flow in rills thus lifts soil or rock particles into the flow, initiating erosion by the process of *slopewash*. Slopewash is the main process of slope erosion in regions lacking a vegetation cover.

On rare occasions during extremely heavy rains, flows of water receive so much soil and rock debris from slopes that a viscous stream known as a *mudflow* or *debris flow* results (Figure 13.14). These flows are intermediate between water flows and earthflows; they are much higher in density than the former, but far more fluid and faster moving than the latter. Mudflows are composed of fine particles, whereas debris flows contain more coarse material, including large boulders. Since both types of flows result from exceptionally heavy rains on unprotected soils, they occur where the vegetation is poorly developed, as in arid or alpine

(b)

(a)

Figure 13.11   (a) Weathered rock that is saturated with water has little cohesive strength, and sometimes large bodies of rock debris break loose and flow rapidly downslope. The enormous earthflow in the photograph occurred in Idaho, near the headwaters of the Pahsimeroi River, visible in the foreground. It has a characteristic lobed appearance, and spread into the valley before coming to rest. Slope failures are frequently found in semiarid regions where sudden heavy rains saturate the soil and where the vegetation cover is too sparse to bind the soil. (John S. Shelton)

(b) A smaller, more typical slump and earthflow about 50 years old in California's Coast Ranges. Slump and earthflow phenomena are widespread in central California due to the presence of mechanically weak rock and heavy clay soils with high water-absorbing capacity. (T. M. O.)

(c) A destructive slump and earthflow in Oakland, California, destroyed 14 homes. The structures visible have moved downslope from street level. (T. M. O.)

(d) The hillside scar shown in the photograph was produced in 1955 by a rockslide that moved tons of soil, rock, and vegetation into Emerald Bay on Lake Tahoe at the border of California and Nevada. (Warren Hamilton/U.S.G.S.)

(c)

regions. They also are a hazard where fire has recently destroyed the vegetation. Although these flows subside quickly, they may travel distances of several kilometers and can destroy roads, bridges, buildings, and even whole towns.

Slopes that are vegetated, and even some bare slopes, show little evidence of rill erosion. Such slopes are able to absorb most rainfalls, which percolate into the soil. This soil moisture then moves downslope between soil particles as subsurface *throughflow*. Throughflow is capable of removing soluble substances as well as fine particles that can filter between the coarser particles composing the soil. In this way well-vegetated slopes in humid areas can gradually be eroded with little or no runoff on the surface itself.

Figure 13.12 The hillside scar shown in the photograph was produced in 1955 by a rockslide that moved tons of soil, rock, and vegetation into Emerald Bay on Lake Tahoe at the border of California and Nevada. (Warren Hamilton/U.S.G.S.)

MODELS OF LANDFORM DEVELOPMENT

The interplay of diverse gradational processes, earth materials, and geologic structures results in an enormous variety of individual landforms. These combine to produce a great number of general landscape types. To unify the study of landforms, geographers have sought some common principle in their development. Two quite dissimilar models of landform development stand out as being widely applicable. They focus, respectively, on the effects of time and of the equilibrium tendency in the development of landforms.

The Cycle of Erosion Concept

By the late 1800s, the theory of organic evolution presented by Charles Darwin in 1859 was finally gaining acceptance. At this appropriate moment the American geographer William Morris Davis wrote a series of essays in which he proposed that landforms, like organisms, could be described and analyzed in terms of their evolutionary development through time. Thus Davis classified landforms according to their stage in a theoretical cycle of development, using the terms "youth," "maturity," and "old age." He then inferred the processes by which the landforms of each stage would slowly evolve into those of the next stage.

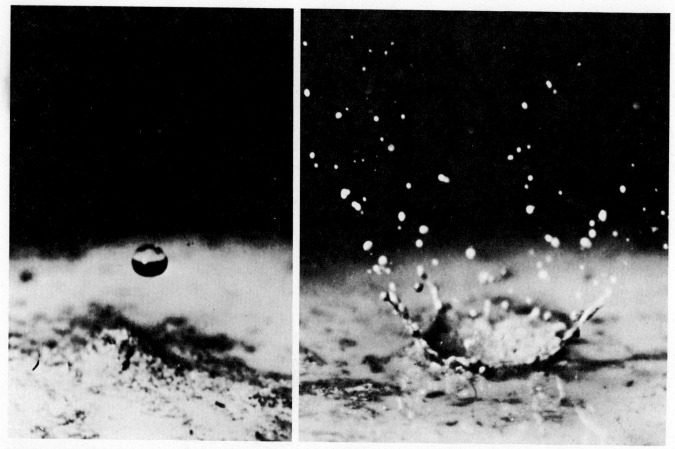

Figure 13.13   These close-up, high-speed photographs of a raindrop splashing into bare soil show graphically the work that can be done by water.
   (left) The raindrop is 3 mm in diameter and is traveling at a speed of 11 meters per second.
   (right) When the drop strikes the ground, particles of soil are thrown several centimeters by the impact. The impact of raindrops on a bare slope tends to produce a net movement of soil particles down the slope. (U.S. Navy Office of Information)

Davis focused first on the evolution of landscapes dominated by the effects of stream erosion. To explain his *cycle of erosion* in the simplest way, he visualized a sudden uplift of the land surface from a lower to a higher elevation, after which erosion begins to affect the uplifted mass. Rainfall and runoff on the raised area initiate erosion as streams flow down the newly created slope toward the sea. At first these streams rapidly erode their beds downward, cutting narrow valleys that are V-shaped in cross section. As long as areas of the original uplifted land surface remain visible between the new valleys, the landscape would be in the *youthful* stage of landform development (Figure 13.15). In this stage, settlement and human activity take place on the plateau-like surfaces between the narrow, steep-sided valleys.

As streams deepen their valleys, they decrease their altitude above sea level. This decreases their potential energy and, therefore, the kinetic energy available to do erosional work. Finally, in "late youth" the streams become "graded," with no excess energy to convert to the work of downward erosion. At this stage, valley deepening ceases. This permits slope gradational processes and lateral stream erosion to widen the valleys and create flat floodplains. In the stage Davis called *maturity*, the original uplifted land surface is converted entirely into hillslopes leading down to flat-floored valleys (Figure 13.15). In the valleys, streams meander back and forth over continuous floodplains veneered with stream-deposited alluvium. Since the only level land is in the valleys, human activity is concentrated there.

Further evolution from maturity to *old age* is assumed to proceed very slowly, requiring tens of millions of years. In the hypothetical old age stage envisaged by Davis, the uplifted surface has been worn down almost to sea level, producing a lowland with faint relief, called a *peneplain* ("almost a plain"). Isolated areas of higher ground that are remote from the larger valleys are called *monadnocks* (Figure 13.15), after solitary Mount Monadnock in New Hampshire. Davis stressed that his simple model could include variations, such as further uplift during any of the intermediate stages in the cycle of erosion.

Davis and his students devised other cycles of erosion for landscapes in which stream erosion was not the dominant process. There were cycles of evolution for deserts, coasts, glaciated areas, and regions of soluble limestone bedrock. In general, Davis viewed landforms as products of *geologic structure, geomorphic process,* and *stage of evolution.* Stage did not imply any fixed amount of time, but referred only to the development of forms in a sequence. Form evolution is slow where rocks are resistant or processes weak, and rapid where materials are weak and processes are vigorous.

Davis's work is important because it provided an easily understood basis for the organization and classification of landforms and caused geographers to begin to focus on the relationships between different landforms. However, the cycle of erosion concept originated at a time when the processes by which landscapes are transformed had only begun to be investigated. Today the Davisian cycle of erosion is regarded as being of value mainly as a way of introducing students to the concept of landscape change. The terms "youth," "maturity," and "old age," introduced into geomorphology by Davis, have been retained, but only as descriptive terms for stream-dissected landscapes. Even this involves problems, as the youthful valley form is common in maturely

(a)

(b)

Figure 13.14 Mudflows and debris flows are common in areas where steep slopes have little vegetative protection as a consequence of fire, aridity, or a short growing season. Violent rainfalls of short duration cause surface rocks and finer particles to be flushed into valleys where they form viscous flows into adjacent lowlands.

(a) Mudflows, such as this example seen in the California desert, often transport large boulders many kilometers. In the center of the view the mudflow deposit has been eroded by later flows of water carrying gravel.

(b) Debris flows transport a higher proportion of coarse material than do mudflows; consequently, they are more viscous and less free-flowing. This debris cone in the California desert has been built by the gradual accumulation of angular rock fragments carried out of a canyon by thousands of years of debris flow activity. (T. M. O.)

(a) Initial surface

(b) Youth

(c) Maturity

(d) Old age

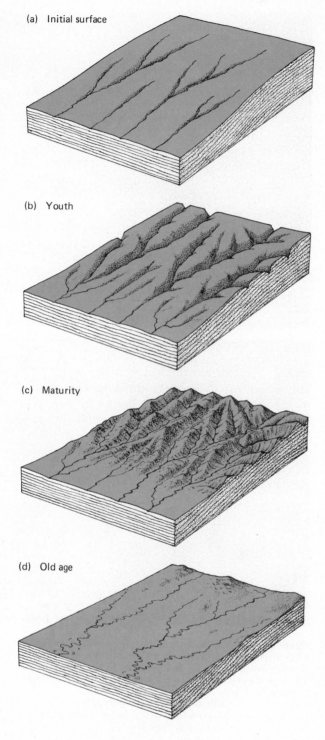

Figure 13.15 This sequence of diagrams illustrates stages of the Davisian cycle of landscape evolution. This cycle applies to moist regions where erosion is accomplished primarily by flowing water. The diagrams assume that the underlying rock is uniform and exerts no controls over landform development.

(a) The initial surface is a landscape of low relief. After uplift of the region, streams have energy to begin cutting downward.

(b) In the stage of youth, the streams have cut narrow steep-sided valleys, and much of the initial surface is preserved between the valleys.

(c) In the stage of maturity, the uplifted area has been eroded into a mass of hills. The streams have stopped incising and have widened their valleys. Little or no trace of the initial plain remains.

(d) In old age, mass wasting and flowing water have eroded the region to a plain of low relief (*peneplain*) with isolated hills, or *monadnocks*. (T. M. O.)

dissected ("ridge and ravine") landscapes, and mature valleys are not uncommon where the dissection is still in a youthful stage. Although the cycle of erosion concept is a useful introduction to landforms, and is no doubt valid over very long periods of time, it does little to increase our understanding of geomorphic processes and the forms they create.

The Equilibrium Theory of Landform Development

Since about 1950 geomorphologists have increasingly emphasized the actual mechanics of landform development. This approach focuses directly on the detailed relationships between form and process in the landscape. It emphasizes the general tendency for the form of the land surface to equate the energy of the erosional processes to the resistance of the material they affect. This equalizes the expenditure of energy throughout the erosional system and the landscape it affects.

Two forces cause material to be put in motion on a slope. One is *downslope gravitational stress*, which is proportional to the slope angle. If you tilt a table, an object on it slides off when the downslope component of the pull of gravity exceeds the friction between the object and the table top. More specifically, the object starts to slide at the threshold when the component of gravitational force parallel to the slope exceeds the component of gravity normal (perpendicular) to the slope, plus a frictional component related to the roughness of the slope and the object (Figure 13.16).

A second force that causes material to move is *shear stress*. This is the oblique downward and forward force exerted by one material rubbing against or flowing over another material. Shear stress is proportional to the density and velocity of the moving material. Water, ice, or wind moving over any surface exerts shear stress on that surface and may cause loose particles on it to move. The shear stress exerted by water or ice is also related to slope angle, because water and ice are usually moving due to downslope gravitational stress. The magnitudes of both of these stresses, as well as the resistance of the material affected, can be expressed numerically and used in equations that explain or predict geomorphic processes and forms.

If erosion reduces the angle of a slope, it has the effect of reducing the downslope gravitational stress on the material forming the slope. It also reduces the shear stress exerted by moving erosional agents. Then how far can erosion reduce a slope angle? Certainly not below the minimum that gives the erosional agent enough energy to continue to remove material from the slope.

On the other hand, any increase in the slope angle increases both the downslope gravitational force and the shear stress imposed by an erosional agent. This results in more vigorous erosion, which tends to reduce the slope angle. Here is an example of *negative feedback*, in which disturbance of a system that is in equilibrium triggers changes that tend to restore the original system (Figure 13.17). Thus, when a natural hillslope is made steeper due to highway construction or some other artificial modification, a destructive slump may follow. This is the way the hillslope returns to its original equilibrium angle. Negative feedback is the principal means of maintaining equilibrium in physical systems and is a normal feature of many process/form systems. Negative feedback causes many landform systems to be self-regulating, so that they tend to maintain a steady state with the passage of time. In such a system, changes in form occur only when there is a change in the material or in the nature or intensity of the gradational processes.

Wherever the relief is high and the materials composing a slope vary in resistance to fragmentation and removal, there are corresponding variations in the slope angle (Figure 13.18).

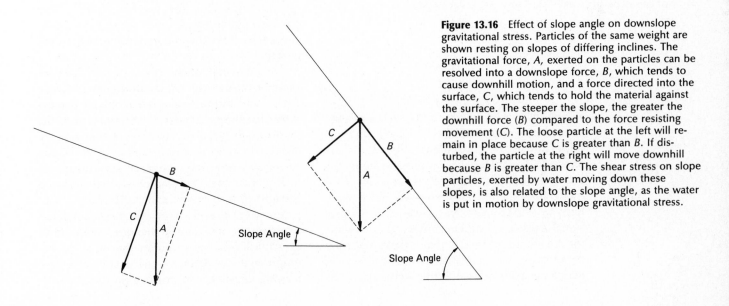

Figure 13.16 Effect of slope angle on downslope gravitational stress. Particles of the same weight are shown resting on slopes of differing inclines. The gravitational force, *A*, exerted on the particles can be resolved into a downslope force, *B*, which tends to cause downhill motion, and a force directed into the surface, *C*, which tends to hold the material against the surface. The steeper the slope, the greater the downhill force (*B*) compared to the force resisting movement (*C*). The loose particle at the left will remain in place because *C* is greater than *B*. If disturbed, the particle at the right will move downhill because *B* is greater than *C*. The shear stress on slope particles, exerted by water moving down these slopes, is also related to the slope angle, as the water is put in motion by downslope gravitational stress.

Slope Angle

Slope Angle

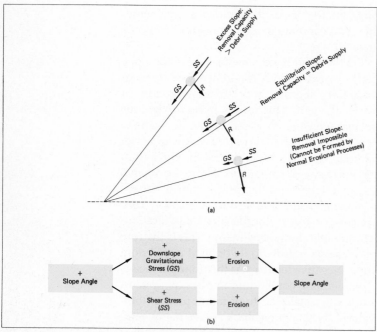

Figure 13.17   Principles of equilibrium slopes.
(a) Hypothetical slope angles indicate downslope gravitational stress (*GS*), shear stress (*SS*) exerted by the gravitationally-driven erosional agent (such as slopewash), and resistance to detachment of slope particles (*R*). Resistance increases with decreasing slope angles (see Figure 13.16), while *GS* and *SS* diminish with decreased slope. The steepest slope is unstable, as the sum of *GS* and *SS* far exceeds *R*, causing rapid erosional removal. The intermediate slope is stable, as the sum of *GS* and *SS* just balances *R*, so that removal can occur without changing the slope angle. The lowest slope cannot be produced by the stresses shown, as particle resistance *(R)* exceeds erosional stresses, so that erosional removal is impossible.
(b) Diagrammatic representation of a negative feedback relationship that causes slopes to be self-adjusting toward angles that equate stress to resistance. Any increase in slope increases the erosional stresses, which in turn decrease the slope angle until equilibrium is reestablished. (Vantage Art, Inc.)

Where the rock is massive and resistant, slopes created by erosion will be steep, which maximizes downslope gravitational stress and shear stress. Where material is easily removed by the active agents of erosion, slopes are gentle, thus moderating the erosional stresses. Since variations in slope angles tend to equalize the erosional stresses on different rocks, it is possible for the whole landscape, tough rocks and fragile ones, to wear away at the same rate over very long spans of time. If this were not so, the

local relief in all landscapes composed of varying rock types would become progressively greater with the passage of time, which is not the case.

The delicate adjustment of surface forms to local materials and gradational processes is, indeed, the most important thing to understand about landforms. It is impossible to modify either landforms or geomorphic processes artificially without impinging upon process or form thresholds that trigger reactions in the natural system. The reaction will be one that tends to restore an equilibrium between process and form. Too often, however, this response is a costly or destructive event in human terms. We cannot make a significant change in a natural system and expect the system to remain passive; sooner or later it will respond with a change of its own to absorb the effect of the artificial change.

For example, artificially straightening a river for purposes of navigation or to prevent flooding also shortens the river and increases its slope in the shortened section. Rivers often respond to this type of disturbance by cutting downward in the straightened section. This restores the original slope in the disturbed section but leaves the upstream continuation of the channel "hanging," forcing adjustments there as well. The stream's downcutting produces sediment that may clog the channel downstream, increasing the flooding there. Often such inevitable adjustments are unforeseen by those who make the initial change in the natural system.

A somewhat different type of equilibrium pertains to depositional landforms that involve inputs and outputs of material. Many depositional landforms persist only because periodic losses of material by erosion are balanced by arrivals of new material (Figure 13.19). Talus cones, beaches, river deltas, and volcanoes are a few examples of landforms that may be thought of in terms of material budgets. Such features enlarge or shrink as the balance between input and output changes. Any natural or artificial change in either the input or the output will result in a quick response in the feature, causing it to grow or shrink. Growth

(a)

Figure 13.18 Differential weathering and erosion due to variations in rock type.

(a) Variations in rock resistance are especially clear in arid regions. Here in Monument Valley, Utah and Arizona, massive sandstone forms vertical walls. The lower layers of thinly bedded sandstones and shales produce much gentler slopes, although even here the more resistant layers make small cliffs. (T. M. O.)

(b) In this view over the Watauga River in forested eastern Tennessee, the steep relief on the left and in the background rises high above gentler slopes on the right due to variations in rock resistance to weathering and erosion. Steep slopes are required to effect removal of the resistant Lower Cambrian sedimentary rock on the left, whereas younger and less resistant sedimentary rock in front of the bold escarpment can be removed even on subdued slopes. Such variations in slope angle make it possible for all parts of the landscape to be eroded at similar rates over long periods of time. (Warren Hamilton/ U.S.G.S.)

(b)

in one place due to reduced output at constant input means shrinkage somewhere else, for an output that is reduced always reduces the input to some neighboring system. The relationship between river sediment and beach maintenance has already been noted in this regard.

The value of the equilibrium concept of landform development is its focus on the exact relationship between geomorphic processes and surface forms. It is especially useful in assessing the impact of natural and artificial changes in landform systems. Finally, it can be described

Figure 13.19 The black sand beach at Kalapana on the island of Hawaii is an example of an ephemeral landform resulting from an unbalanced materials budget. The black volcanic sand of the beach was created when molten lava fragmented upon contact with seawater. The supply of sand is not being renewed from any present source. Meanwhile, storm waves periodically remove some of the sand from the beach, and the wind transfers dry beach sand inland. Consequently, the beach has been shrinking for about 30 years. (T. M. O.)

and analyzed mathematically. Consequently, the equilibrium model has largely replaced the cycle of erosion concept in landform analysis.

THE TEMPO OF GEOMORPHIC CHANGE

At this point it is appropriate to ask: How rapidly does the landscape change due to tectonic and gradational processes? In general, the max-

imum rates of crustal uplift greatly exceed the most rapid rates of lowering of the land surface by erosion. The latter, measured as the average depth of removal over the entire area concerned, is known as the *denudation rate*. Repeated precise surveys have shown that the rate of uplift in regions of mountain growth is about 5 to 10 meters (15 to 30 ft) per 1,000 years. In mountain regions with steep slopes, the rate of sedimentation in reservoirs indicates that the denudation rate rarely exceeds 1 meter (3 ft) per 1,000 years. Over very large areas of variable relief, the rate is much less. The denudation rate over the entire area drained by the Mississippi River and its tributaries is only about 5 cm (2 in.) per 1,000 years. Even this is higher than the natural rate, for erosion in this region has been greatly augmented by human activities.

The landscapes we see around us are produced for the most part by small changes that occur frequently over vast spans of time. Mountain ranges are lifted a meter or less in a century or more. Though small in amount, and rare on the human time scale, mountain-building movements are frequent on the geologic time scale, in which we routinely speak in terms of millions of years. An uplift rate of only 1 cm per year sustained for 1 million years is more than adequate to raise a fragment of the submarine continental shelf to the height of Mount Everest (Figure 1.7, p. 14).

Some types of erosion and deposition do proceed on a human time scale. Downward and lateral erosion by streams takes place mainly during floods that occur once every year or two. Measurements of the sediment loads carried by streams indicate that the majority of the sediment is transported during the five to ten days of greatest flood discharge each year. Only in the exceptional floods that occur at intervals of many years can streams clear out the sediment in their channels completely and erode downward into bedrock. Such events change the landscape slowly, but they are so uncommon that a greater quantity of work is probably accomplished by smaller events that occur with greater frequency. It is suspected that this is a general principle in landform development.

SUMMARY

No other known planet has landforms similar to those of the earth. This is largely because the other planets in the solar system lack the plate tectonic processes that create complex geologic structures and the liquid water that is the main agent of landscape sculpture on the earth.

The diversity of landforms on the earth's surface is a consequence of four general factors: geologic structure, tectonic activity, gradational processes, and the passage of time. Geologic structure involves both rock type and the geometry of rock masses. Tectonic activity produces geologic structures and causes the earth's crust to be elevated or depressed locally at varying rates. Gradation is the constant transfer of rock material from high to low places on the earth's surface. Gradation includes mechanical and chemical weathering of rock and the effects of agents of erosion that are set in motion by gravitational force.

Water, in its various forms, is the chief agent of gradation. The dominant gradational process in any region is largely controlled by climate, and its intensity is often related to the vegetative cover. Where tectonic activity has ceased, landscapes can be worn to very low relief over long periods of time. Response time refers to the disturbance of a landform system by a change in geomorphic processes, and relaxation time pertains to the subsequent restoration of equilibrium in the disturbed landform system.

Slopes of varying origins are the basic element of landforms; therefore slope development is the major problem of landform analysis. Mass wasting is the transfer of material down slopes by gravitational force, unassisted by any moving agent of erosion. It includes soil creep, slumps and earthflows, solifluction, and rock-slides. Slopes lacking vegetative protection are eroded by rainsplash and the overland flow of water, which initiates slopewash. Slopes protected by vegetation experience little surface erosion but they lose material in the process of subsurface throughflow of water in the soil.

Two models of landform development have achieved popularity. The cycle of erosion concept originated by W. M. Davis is a descriptive model suggesting that landforms develop from stage to stage in an evolutionary cycle related to rock resistance and the passage of time. The terms "youth," "maturity," and "old age" are used to describe erosional landscapes of differing appearance. The more useful equilibrium theory of landform development is an explanatory model stressing the specific relationships between geomorphic processes and resulting forms. The emphasis is on the tendency of landforms to equate erosional stress to material resistance to minimize and evenly distribute the work (or expenditure of energy) in the geomorphic system. Many landform systems are self-adjusting through a negative feedback process in which changes imposed on the system tend to be canceled out by the system's response after some threshold value of instability is attained. Artificial disturbances of landform systems cause imbalances between processes and forms that trigger responses that are predictable but often come as a costly surprise.

Strong crustal motion is localized, but its rate can greatly exceed regional rates of erosion. This accounts for the earth's varied relief. Most landscape changes are gradual and the result of events of moderate magnitude repeated over and over through vast spans of time.

REVIEW QUESTIONS

1. Why is it impossible for any other planet in the solar system to have a surface configuration like that of the earth?

2. What four general factors are most influential in causing terrestrial landforms to differ from place to place?

3. In what ways does tectonic activity influence landforms?

4. In what two ways is water the chief agent of gradation?

5. What are the other agents of gradation? What determines which one will be dominant in an area?

6. There are several ways in which time is of importance in landform studies. Explain.

7. How is the "angle of repose" relevant to the concept of geomorphic thresholds?

8. What is the difference between "creep" and solifluction?

9. How are large rockslides usually triggered?

10. Outline the sequence of events leading to slope erosion by surface water.

11. What conditions favor the occurrence of mudflows or debris flows?

12. Characterize the landscape (valleys and interfluves) in each stage of the erosional cycle proposed by William Morris Davis.

13. Why is the Davisian cycle of erosion concept in disfavor among contemporary geomorphologists?

14. In terms of the equilibrium theory of slope development, why do resistant rocks normally form steeper slopes than easily eroded material?

15. What problem commonly arises when landforms or geomorphic processes are artificially modified by human activity?

16. How do magnitudes of local and regional rates of denudation compare with local rates of uplift in tectonically active regions?

APPLICATIONS

1. What are the landforms of your area on the macro, meso, and micro scales? Are there any unusual landforms or "textbook examples" of particular landforms in your vicinity?

2. Look at the slope forms and slope angles in your area. What do you think explains the differences from place to place?

3. What energy transformations occur (a) in the case of a rockslide that blocks a valley, damming the stream in the valley? (b) in the case of a wave that removes enough material from the base of a cliff to leave the higher part of the cliff without support?

4. It has been noted that for a beach to persist, particle input must be equivalent to erosional loss. What landforms in your region can be thought of in terms of a similar material budget?

5. W. M. Davis's cycle-of-erosion concept has been criticized on grounds that a full cycle from youth to old age could rarely be completed. What different factors could disturb the course of an erosion cycle?

6. Evaluate the relative roles of geologic structure, tectonic activity, gradational process, and time in the creation of the scenery of your region. Name other regions in which each of these factors is dominant.

FURTHER READING

Bloom, Arthur L. *The Surface of the Earth.* Englewood Cliffs, N.J.: Prentice-Hall (1969), 152 pp. This well-written paperback outlines the processes of landform development—brief but unusually good.

Bradshaw, Michael J., A. J. Abbott, and **A. P. Gelsthorpe.** *The Earth's Changing Surface.* New York: Wiley (1978), 336 pp. An abundantly illustrated account of landform development; unusual and interesting.

Brunsden, Denys, and **John Doornkamp,** eds. *The Unquiet Landscape.* Bloomington: Indiana University Press (1974), 171 pp. This is a magnificently illustrated collection of articles on various aspects of landform development, from a series appearing in the British periodical *The Geographical Magazine.*

Butzer, Karl W. *Geomorphology from the Earth.* New York: Harper & Row (1976), 463 pp. Different climatic regions are discussed in this textbook on landform development. It is not highly technical and uses a geographical approach.

Hunt, C. B. *Natural Regions of the United States and Canada.* San Francisco: W.H. Freeman (1974), 725 pp. This is a well-illustrated introduction to the regional landforms of North America. Fairly complete, but nontechnical.

Lobeck, A. K. *Geomorphology: An Introduction to the Study of Landscapes.* New York: McGraw-Hill (1939), 731 pp. Although dated, this text is included here because of its excellent photographs, maps, and diagrammatic illustrations of landforms of all types.

Ritter, Dale F. *Process Geomorphology.* Dubuque, Iowa: W. C. Brown (1978), 603 pp. As the title suggests, this is a modern process-oriented text. Illustrations consist largely of graphs and diagrams, and some portions are at a relatively advanced level.

Shelton, John S. *Geology Illustrated.* San Francisco: W.H. Freeman (1966), 434 pp. This book is unsurpassed for crisp photographic illustrations of landforms. The text and organization are more imaginative than most—highly recommended.

Twidale, Raoul C. *Analysis of Landforms.* New York: Wiley (1976), 572 pp. The best illustrated treatment of landforms of all types; by a leading Australian geomorphologist.

Utgard, R. O., G. D. McKenzie, and **D. Foley.** *Geology in the Urban Environment.* Minneapolis: Burgess (1978), 355 pp. This paperback is a collection of articles concerning the significance of landforms and geomorphic processes in urban settings, demonstrating that landforms are not merely a "rural" topic.

CASE STUDY

When Mountains Fall

Rapid gravitational transfers of soil and rock on slopes occur on a wide range of scales, from the sliding of a pebble disturbed by a gust of wind, to the sudden collapse of a mountainside during an earthquake. But the larger the sliding mass, the more puzzling its behavior. It is important to learn how rockslides do behave to assess the hazard they present to settled valleys in mountain regions—particularly those that are earthquake prone.

The largest slope detachment for which evidence presently exists is in the valley of the Karkheh River in western Iran, an area of anticlinal mountains produced by simple folding. Here, some 11,000 years ago, a 300-meter-thick slab of limestone on the flank of an anticlinal arch peeled away over a horizontal distance of 15 kilometers (9 miles), and thundered downward about 1,000 meters, blocking the Karkheh River. The magnitude of the detachment is remarkable, but even more so is the fact that the sliding mass, which descended only 1 kilometer over a slope of about 20 degrees, somehow traveled a horizontal distance of 20 kilometers before coming to rest. To do this it had to rise up and over a 600-meter-high ridge midway in its course. The rockslide deposit covers an area of about 270 square kilometers and has a volume estimated at 30 cu km (7.2 cubic miles).

How does such an immense mass actually move? How could it rise half the height it fell, and do so after traveling 7 kilometers horizontally?

Eyewitness accounts of the movement of major rockslides vary considerably, and do not always accord with physical evidence. Most observers of large rockslides describe them as being turbulent flows involving huge amounts of dust. Often the slide mass appears to be preceded by a blast of air powerful enough to level forest trees *before* the hurtling rock debris reaches them. In sinuous valleys, the onrushing debris stream appears to slosh from side to side, riding up on the outsides of curves, a fact easily verified where vegetation has been stripped from valley walls by the speeding mass.

To understand the Karkheh River slide we must look at phenomena observed in other parts of the world. One of the most dangerous of all the world's mountains is Huascarán, which rises to 6,768 meters (22,200 feet) in the Andes Mountains of Peru. The particular hazard at Huascarán is the immense accumulation of ice on the summit, which lies only 9 degrees south of the equator, in the area affected by heavy ITC precipitation. In 1962 a huge mass of ice detached and fell from the peak. Frictional heating caused the ice to flash into water producing an instantaneous debris flood that killed some 5,000 people in towns and villages in the valley below. In 1970 an earthquake offshore in the Peruvian trench—an active subduction zone—caused Huascaran's massive ice cornice to collapse again. Ice and rock fell vertically almost 1,000 meters, then pulverized on impact, once more producing a cataract of water, mud, and giant boulders. Measured from the moment of the earthquake, the flow moved 16 kilometers (10 miles) in slightly less than 2 minutes—a speed of 133 meters (436 feet) per second, or about 480 km (300 miles) per hour. Approximately 80,000 people were killed in that 2-minute catastrophe as many villages, one entire town, and a large portion of another were engulfed.

Aside from its tragic consequences, the notable aspects of the Huascarán disaster of 1970 were the tremendous speed of the debris mass, and the fact that, despite an apparent depth of at least 150 meters, it appeared to be separated from the ground, passing *over* some obstacles without disturbing them. The Huascarán example may shed some light on the Karkheh River rockslide and other massive slides known to have traveled surprising horizontal distances. Based on study of a large prehistoric rockslide in southern California, Ronald Shreve of the California Institute of Technology proposed that the transport mechanism was a cushion of air compressed under the moving mass of rock. Air trapped under rapidly moving rock masses buoys them up and minimizes friction against the ground. Slides can therefore attain extremely high velocities and can even surge strongly uphill instead of banking up against obstacles. Consequently they can travel much farther than predicted by the usual equations pertaining to sliding masses. To trap the air most effectively, the slide must "splash" outward after its initial fall, especially from a topographic platform of some kind. Supporting the air cushion hypothesis is Shreve's observation that rockslide deposits frequently preserve the original sequence of rock layers involved in the detachment. This seems to imply a low degree of turbulence during transport.

Although the air cushion hypothe-

A portion of the great rockslide into the Karkheh River Valley in western Iran. The detachment is 15 km (9 miles) wide, and extends the full width of this aerial photograph. In the center is a portion of the rockslide deposit. In the foreground is the anticlinal mountain that was "leapfrogged" at high velocity by the slide mass. The abrupt edge of the slide mass is visible on the crest of this mountain at the extreme right. The rockslide traveled another 8 km (5 miles) after jumping this ridge. (Aerofilms and Aero Pictorial Ltd.)

sis appears to resolve difficult questions about rockslides, it has not been accepted universally. Kenneth Hsu of the Geological Institute in Zurich, Switzerland, has pointed out that meteorite impacts on the moon have caused enormous masses of lunar rock to be ejected horizontally for distances of hundreds of kilometers, forming conspicuous features on photographs of the lunar surface. Similar blankets of *ejecta* are associated with impact craters on the planet Mars. Lunar gravity is only one sixth that of the earth, facilitating long-distance transport, but since the moon is devoid of a gaseous atmosphere, an air cushion cannot be invoked as a transport mechanism there. Martian gravity is about 38 percent as strong as earth gravity, and the surface atmospheric pressure on Mars is only 7 thousandths that of the earth, once more making an air cushion an unlikely transport mechanism. Could it be that large meteorite impacts volatilize so much material that a cushioning gas is generated to convey the ejecta blankets on the moon and Mars?

Hsu does not believe so, stressing the turbulent flow characteristics observed within moving rockslides by eyewitnesses. Instead, Hsu believes that constant collisions between the particles in the turbulent mass progressively accelerate particle velocity. This and the buoying effect of dust are the impelling forces for horizontal rockslide transport in his view. Hsu prefers to refer to large rockslides as "debris streams" to emphasize that large detached masses often flow rather than slide. In contrast to Shreve's assertion of lack of ground contact, Hsu cites examples of ground scoured or grooved by passing debris streams.

Since there appears to be evidence in support of both theories of debris transport in large rockslides, it seems probable that the transport mechanism varies with local circumstances. Gliding transport on an air cushion may occur when the rapidly moving debris mass is launched horizontally over a body of air, as at an abrupt steepening of bed slope. Large detachments that cannot be launched over a sheet of air probably move in a turbulent manner,

with a steep front—much as the water in a flash flood, but with a depth and velocity an order of magnitude greater. In either case, large rockslides are much more mobile and have a far greater destructive potential than would be predicted from their mass, vertical fall, and the slope and roughness of the ground they traverse.

While their speed is unexpected, rockslides stop with amazing abruptness when the transport mechanism fails, whether by escape of supporting air or decrease below a critical value of energy transfer by particle collisions. Enormous volumes of rock moving at velocities of hundreds of meters per second stop within a minute or less. The deposits of large rockslides characteristically have wall-like margins that rise steeply above the surrounding topography. None has been permitted an unobstructed view of the sudden "freezing" of the motion of a catastrophic rockslide. As the dust clears, one merely sees that all is still, and is grateful to be a survivor.

369

The Pass Through the Mountains by Fukaye Roshu. (The Cleveland Museum of Art, John L. Severance Fund)

From 100 miles above the earth's surface, the organized patterns created by human activity are almost invisible. However, only from such an altitude can one begin to appreciate the organization of the earth's natural features. Two kinds of phenomena in particular exhibit remarkable regularity. One is erosional, the other depositional. First, and far the more widespread, are the systems of valleys carved by water runoff from the land surfaces. Second, and much more localized, are the patterns of sand dunes built by wind in deserts that seldom experience rainfall or runoff.

Valley systems and dune systems are related. Both result from the friction of a fluid (something that "flows") passing over the material of the earth's surface. The flow of water, however, is confined to channels, whereas the flow of air that concentrates sand and creates dunes has the whole landscape for its bed. The waves of sand raised by the wind as it passes over the land sometimes resemble water waves raised by the wind as it blows across the sea. Despite the difference in the natures of the "channels" for running water and for wind, the two fluids erode, transport, and deposit material in similar ways. For this reason we shall consider them together in this chapter. Since the domain of running water greatly exceeds that of wind erosion and deposition, and is much more the realm of human activity, our attention will center there.

Processes related to channeled flows of water are known as *fluvial* processes, from the Latin *fluvius*, meaning river. Flowing water makes gradation possible by providing an effective transportation system connecting high and low places on the earth's surface. From a high altitude the effects of fluvial processes are conspicuous over nearly all ice-free land areas.

14
Fluvial and Aeolian Landforms

Stream Organization

The Mechanics of Channeled Flow

Stream Discharge
Flow Turbulence
Scour and Fill
Stream Load
Channel Equilibrium
Stream Gradient and the Longitudinal Profile
Profile Knickpoints
Lateral Migration and Channel Patterns

Fluvial Landforms

Valley Development
Floodplains
Fluvial Terraces
Alluvial Fans
Deltas

Aeolian Processes and Landforms

Wind as a Geomorphic Agent
Loess
Dune Landscapes

Flowing water is one of the most important agents of landscape change, even in arid regions. The energy in moving water empowers it to entrain, transport, and redeposit rock waste. The landscape expresses the contest between gradation by flowing water and uplift caused by geologic processes.

Figure 14.1 This side-looking radar image of a 32-km- (20-mile-) wide portion of eastern Kentucky portrays the nature of stream dissection of the land surface where there is little variation in rock type or geological structure. Note the large dendritic stream systems, the smaller tributaries, and the uniform density of channels. The northeast to southwest "grain" shows the influence of bedrock joints on stream development. Radar images like this penetrate clouds and haze, and often give better definition of surface form than do conventional photographs. (Raytheon Company and U.S. Army Engineering Topographic Laboratories)

The only exceptions are the areas covered by sand dunes. From space one can see intricate valley systems carved by running water, as in Figure 14.1; large streams collecting the water and sediment delivered by smaller streams; and the mouths of the earth's great rivers, issuing plumes of sediment that discolor the sea. Clearly, rivers are the essential disposal system in the process of gradation.

In this chapter we shall look first at the way streams are organized in drainage networks. By *streams* we mean channeled flows of all sizes, from great rivers to small creeks. Then we shall

examine the mechanics of channeled flows of water. The vertical and horizontal patterns that channels make and the landforms that result from fluvial processes will be considered next. Finally, we shall look briefly at the quite different effects of wind on landforms.

STREAM ORGANIZATION

Only the few streams that cross very dry regions, such as the Nile River in Egypt, flow far without being joined by other streams. All natural streams in a region are part of a *drainage network* that removes surface runoff from a *drainage basin*, which is the area drained by a system of connected stream channels. Every stream has its own drainage basin, or contributary area (also called its *catchment*), which can range in size from a fraction of a square kilometer to a sizeable portion of a continent (Figure 14.2).

The specific geometrical arrangement of streams within a drainage network is the *drainage pattern*, which can take many forms. These are usually related to the local geological framework. The most common drainage patterns are illustrated in Figure 14.3, which also demonstrates the remarkable uniformity of the drainage pattern in an area undergoing dissection by fluvial processes.

In all drainage networks, small streams feed into successively larger streams. The characteristics of drainage networks can best be analyzed by using the idea of *stream order*. The smallest streams, which have no tributaries feeding into them, are *first order* streams. Where two first order streams join, the resulting larger single channel is a *second order* stream. Similarly, *third order* streams begin at the junction of two second order streams, and so on, as shown in Figure 14.4. A stream of higher order is formed only when two streams of the next lower order join, but not every time a tributary of any size enters. Counting the streams of different orders in any drainage network, one finds that the number of streams in a particular order is from 3 to 5 times the number in the next higher order.

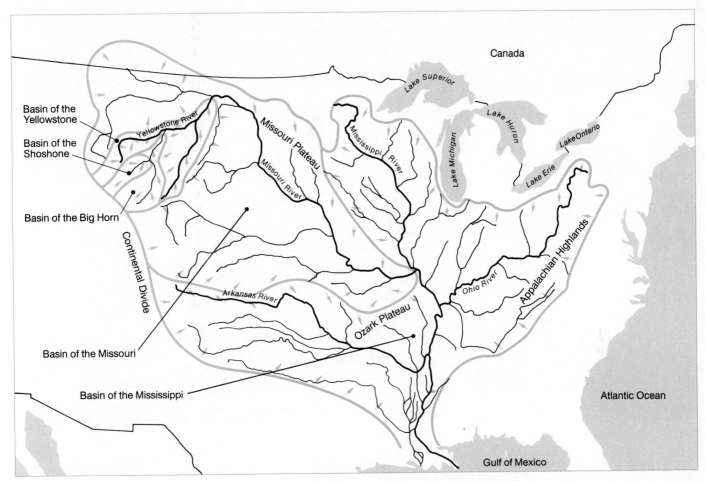

Figure 14.2 The drainage basin of a large river such as the Mississippi contains a hierarchy of smaller nested drainage basins, outlined in blue. In this figure, the basin of the Shoshone River is part of the basin of the Big Horn, which is part of the basin of the Yellowstone, which feeds into the Missouri, which is the major tributary of the Mississippi. The arrows indicate the downhill direction of water flow into the various basins. The Continental Divide along the Rocky Mountains separates basins draining eastward toward the Gulf of Mexico and the Atlantic Ocean from those draining westward toward the Pacific Ocean. (Doug Armstrong)

Thus a drainage network is a pyramid-like phenomenon, supported by many low order streams whose waters eventually funnel into the single highest order stream at the basin outlet.

Within any drainage network the streams of each order usually have a characteristic length, slope, and drainage basin area. The average values for these parameters, like the number of streams in each successive order, change in a very regular manner as the stream order increases, as indicated in Figure 14.4 (b,c,d). Studies of drainage networks suggest that the nature of their development is remarkably consistent from place to place. They seem to have evolved in a way that tends to maximize system efficiency, with a minimum expenditure of energy both in the formation and operation of the system.

THE MECHANICS OF CHANNELED FLOW

To understand fluvial processes and the landforms they create, it is necessary to understand how water flows in channels. Take the example of water in a trough. The water does not flow unless one end of the trough is lifted higher than the other end. This permits the potential energy of the water at the high end of the trough to be converted to kinetic energy as the

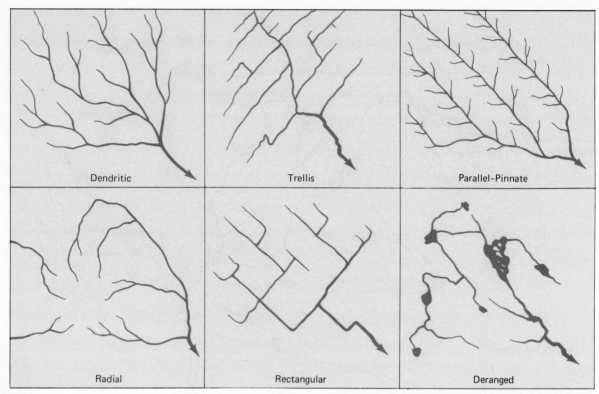

Figure 14.3 These are six of the most frequently encountered drainage patterns. *Dendritic* patterns are found in areas that lack strong contrasts in bedrock resistance, such as flat-lying sedimentary rock or massive crystalline rock that is deeply weathered. *Trellis* patterns develop where inclined layers of sedimentary rock of varying resistance to erosion are exposed at the surface. The parallel segments develop along the outcrops of the erodible layer of rock. *Parallel-pinnate* drainage reflects a topography of long parallel ridges. The short segments drain the flanks of the ridges, and the long segments drain the troughs between. *Radial* drainage indicates the presence of an isolated high mountain area and is common on individual volcanoes or dome-shaped mountainous uplifts. *Rectangular* patterns reflect strong jointing of resistant bedrock, with streams incising along the joint planes. *Deranged* drainage shows no geometrical pattern or constant direction and usually includes numbers of lakes; this pattern indicates destruction of prior drainage by the erosive effects of continental ice sheets. Such areas lack true valleys developed by fluvial erosion. (Vantage Art, Inc.)

water flows to the low end of the trough. But not all potential energy is converted to energy of motion. Some energy is lost in overcoming friction between the water and the trough walls as well as friction within the flow itself. Similarly, water flowing in a natural channel encounters friction with the stream bed and banks. This friction has a significant retarding effect, and causes the velocity of flow near the channel margins to be less rapid than in the center of the flow (Figure 14.5). Boaters heading upstream know that near the banks the downstream current is slower and easier to overcome.

The way water flows in its channel determines the amount of energy it has for erosional and depositional work—meaning the deepening of valleys, the transport of solid and dissolved matter, and the deposition of sediments to create flat valley floors or floodplains. Stream discharge and turbulence, as we shall see, play important roles in providing this energy.

Stream Discharge

The volume of water a stream carries past a given point during a specific time interval is called the stream *discharge*. Stream discharge is measured in cubic meters (or cubic feet) per second, and is equal to the cross-sectional area of the flow times the flow velocity (distance traveled per unit of time). The United States Geological Survey maintains more than 6,000 stream gauging stations to measure stream discharges in the United States (Figure 7.9). The data are used to forecast floods, to assess irrigation water supplies, to plan sewage disposal, and for engineering purposes, such as the design of dams and bridges. Stream discharge at a given location is usually inferred from the height of the stream's waters, using a previously determined relationship known as a *rating curve* (Figure 14.6).

Stream energy is closely related to stream discharge, because discharge influences the flow velocity. Flow velocity in turn determines the stream's capacity to do work in the form of erosion and sediment transport. The velocity of any segment of a stream reflects the balance between the downslope gravitational stress, which is determined by stream bed slope, and the energy lost in overcoming friction at the boundaries of the flow. The larger the flow and the smoother the channel, the less the energy lost to friction, and the greater the stream ve-

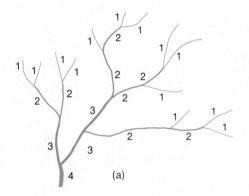

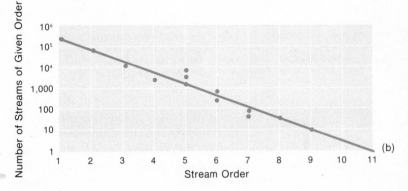

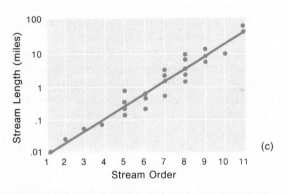

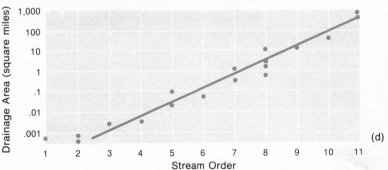

Figure 14.4 (a) This diagram illustrates the system of stream ordering devised by the engineer Robert Horton and simplified by the geomorphologist Arthur Strahler. A second-order stream arises at the junction of two first-order streams; a third-order stream is produced by the junction of two second-order streams, and so on.

(b) (c) (d) These three graphs represent a Horton analysis of a drainage basin near Santa Fe, New Mexico. The graphs show the regularities that are typical features of drainage development, as revealed by a Horton analysis. (b) The numbers of streams of given order decrease regularly with increased stream order. (c) The average length of streams increases regularly with increased order. (d) The average area drained by a stream of a given order increases regularly as stream order increases. (Doug Armstrong after *Fluvial Processes in Geomorphology* by Luna B. Leopold, M. Gordon Wolman, and John P. Miller. W. H. Freeman and Company. Copyright © 1964)

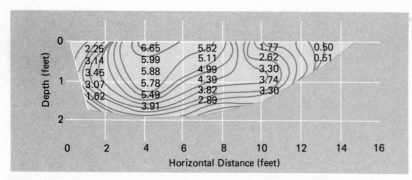

Figure 14.5 This diagram shows the average measured stream velocities, in feet per second, at various points in a cross section of Baldwin Creek, Wyoming. Note that the velocities tend to be lowest near the sides of the channel and highest near the center of the channel. The stream is therefore able to carry material suspended in its waters. Moderate velocities near the stream bed enable some material to be transported over the bed. (Doug Armstrong after *Fluvial Processes in Geomorphology* by Luna B. Leopold, M. Gordon Wolman, and John P. Miller. W. H. Freeman and Company. Copyright © 1964)

locity. Streams normally increase in size (width, depth, and discharge) in the downstream direction as tributaries and groundwater flow into them. Other things being equal, this would cause streams to increase in velocity and energy in the same direction. However, as streams grow larger, their downstream slope (or rate of descent) tends to decrease. This counterbalancing effect prevents a progressive build-up of energy in the downstream direction, and produces a more uniform distribution of stream energy. Even during floods, streams seldom have a flow velocity that exceeds 3 to 5 meters per second (7 to 11 miles per hour).

Flow Turbulence

At very low velocities water can move as a smooth sheet, like the top card in a deck of cards that is pushed forward over the lower cards in the deck. Such flow is *laminar* (Figure 14.7). Flows of this type lack the kinetic energy to put loose particles in motion. If the flow velocity increases, friction within the flow and at its boundaries soon causes the flow to break into separate currents that are no longer parallel. This irregular *turbulent flow* is normal in streams. In turbulent flow, the speed and direction of motion vary continuously, with currents in every direction, including upward. Nevertheless, the average motion is in the downslope direction. The constantly moving eddies that make quiet streams so fascinating to watch are evidence of turbulent flow, as are the boiling rapids of steeply descending streams.

In turbulent flow, most stream energy is consumed in overcoming the friction at the channel boundaries and between adjacent eddies and currents. Only a small portion of any stream's energy is available for the work of picking up rock debris and moving it through the channel. Even so, turbulent flows can be highly erosive. Rapid pressure variations near the stream bed and forceful upward-moving currents cause solid material to be lifted into turbulent flows and carried away by them. This removal of material, in both solid and dissolved form, is one of the most significant aspects of fluvial systems, as we shall now see.

Scour and Fill

A rapidly flowing stream may detach particles from its bed in the process of stream bed *scour*. In scouring, specific amounts of energy are re-

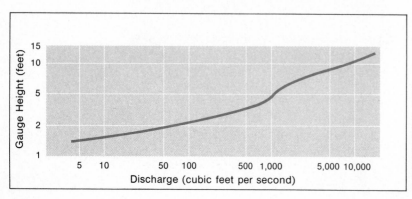

Figure 14.6 This *rating curve* for Seneca Creek, Maryland, allows one to infer the stream discharge from a simple measurement of stream height at a gauging station. The average discharge of this stream is 100 cu ft per second, but discharges more than 50 times as great have been observed. (Doug Armstrong after *Fluvial Processes in Geomorphology* by Luna B. Leopold, M. Gordon Wolman, and John P. Miller. W. H. Freeman and Company. Copyright © 1964)

Figure 14.7 These cross sections compare laminar and turbulent flow, with overall motion forward from AA' to BB'. In laminar flow (left), the fluid moves like a deck of cards being deformed by internal shearing. The movement is slow and in a single direction. Under these conditions, sediment cannot be incorporated upward into the flow or supported within it. Thus the flow is nonerosive. Such flow is rare in streams but is common in slowly moving groundwater and glacial ice. Turbulent flow (right) is characterized by instantaneous velocities in all directions, which allows sediment to be picked up and supported within the flow. (Vantage Art, Inc.)

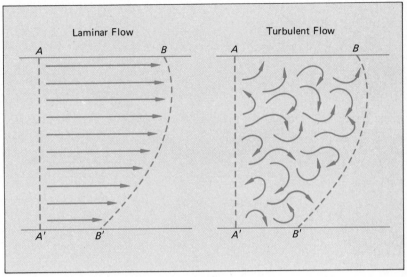

quired to put materials of different sizes and shapes in motion. The most easily scoured particles are those of sand size, with diameters between 0.05 and 1.0 millimeters. The flow velocity must be 15 to 30 centimeters per second (0.3 to 0.7 miles per hour) to put such particles in motion. Figure 14.8 shows that smaller clay or silt particles resist detachment more than sand. This is due to the stronger molecular bonds between the smaller particles. Particles larger than sand size, including gravel, cobbles, and boulders, also resist motion more than sand because of their greater weight.

As Figure 14.8 reveals, it takes more energy to put a particle in motion that it does to keep it in motion, particularly in the smaller size ranges. Although a 0.1 mm particle requires a current velocity of about 20 cm/sec to be displaced, it will subsequently continue to move at a velocity as low as 1 cm/sec. Nevertheless, a continued decrease in stream velocity allows progressively finer particles to settle out of the flow and become part of the deposited sediment in the process of stream bed *filling,* or fluvial *aggradation.*

Stream Load

The effectiveness of streams as sediment carriers is one of the obvious facts of nature.

China's Hwang Ho (Yellow River) earned its name because of the color of its sediment-laden waters, as did the Colorado (Red) River of the southwestern United States, and the White River that drains the Badlands of South Dakota. The Mississippi River carries nearly 300 million metric tons of solid sediment to the sea each year, plus another 150 million tons of dissolved matter. To transport this much material by rail would require a daily train of 24,600 boxcars. Table 14.1 (p. 379) compares the sediment loads of a variety of large and small streams around the world.

Streams move their load of rock debris in three ways. Part of the transported material is dissolved in the water, forming the *dissolved load* (or *chemical load*) of the stream. Fine particles of clay and silt are carried in suspension within the flow of water as the *suspended load.* It is the suspended load that gives many streams their color and opaque appearance. Particles too heavy to be carried in suspension are bounced and rolled along the channel bottom as the stream's *bed load.* The bed load and suspended load together constitute a stream's *solid load.*

The dissolved load of a stream is contributed largely by groundwater inflow into the stream. The amount of dissolved matter in groundwater varies according to the composition of the rocks and soils, and the climate, weathering pro-

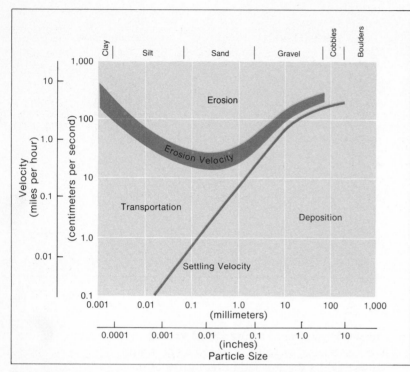

Figure 14.8 The ability of a stream to remove and transport material depends on the local velocity of flow, on the size of the particles, and to some extent on the shape of the particles. This diagram for uniform material relates particle size and flow velocity to mechanisms of erosion and deposition. Combinations of size and velocity located to the right of the settling velocity curve represent the regime of deposition; large particles settle out of a slow-moving stream and become deposited on the bed. Particles with velocities and sizes to the left of the settling velocity curve can be transported either by saltation (skipping or bouncing) along the bed or by suspension in the stream. The erosion velocity represents the minimum water velocity at which particles are carried away from the stream bed. Erosion velocity is shown as a band because it depends on particle shape, bed material, and other factors. The graph shows that once a particle becomes entrained in the flow of water it can be transported by water moving more slowly than the erosion velocity. (After Morisawa, 1968, from Hjulstrom, 1935)

cesses, and vegetation cover in the local area. Human pollution of streams has added greatly to their chemical load in many places, especially near industrial cities and where agriculture or mining are active. The average yearly rate of removal of dissolved matter over the entire United States is about 40 tons/km², or about 100 tons/mile². Although this is largely an invisible process, the weight of dissolved material removed from the U.S. as a whole is more than half the average annual rate of removal of solid matter (71 tons/km², or about 185 tons/mile²). Thus chemical erosion of the land is a major (though largely unseen) aspect of gradation.

To transport its solid load a stream must have forward and upward velocities greater than the constant downward pull of gravity on the particles carried in the flow. Material carried in suspension glides along irregular paths within the flow, while heavier material skips or rolls along the stream bed. Bouncing or skipping transport is known as *saltation,* and rolling or sliding transport is called *traction* (Figure 14.9).

As the particles of the bed load move along they collide with one another and with the solid rock of the stream bed. Impacts against the channel floor break loose new fragments, slowly lowering the stream bed. Constant battering causes the bed load particles to become smaller and more rounded as they progress downstream—cobbles are reduced to pebbles, and pebbles are fragmented into sand and silt. Sand, silt, and the clay washed directly into streams by overland flow are the only solid particles carried far downstream by large rivers.

Any decrease in flow velocity reduces the forces that support the suspended load and propel bed load movement. This causes some of the solid load to drop out of the flow. First to settle out are the heaviest of the particles moved by saltation and traction. These come to rest in a new deposit that may be either temporary or permanent. Deposition of the finest of the particles carried in suspension occurs only when the water is almost still, as in a lake or marsh.

The transporting power of a large stream is enormous. In 1933, a flash flood in California's Tehachapi Mountains caught a train crossing a trestle bridge. The locomotive and tender were carried a kilometer downstream as part of the stream's bed load and were so thoroughly buried by gravel that a metal detector had to be used to find them.

Channel Equilibrium

A stream is a sensitive dynamic system with the ability to adjust the form of its channel in a matter of hours in response to changes in inputs of energy and material. By scouring and filling, a stream adjusts the slope of its bed and

Table 14.1 Characteristics of Selected Rivers

River and Location	Average Discharge at Mouth		Length, Head to Mouth		Area of Drainage Basin		Average Annual Suspended Load (millions of metric tons)	Average Annual Suspended Load per sq km of Basin (metric tons)
	M³/SEC	FT³/SEC	KM	MILES	KM²	MILES²		
Amazon (Brazil)	180	6,400	6,300	3,900	5,800	2,200	360	63
Congo (Congo)	39	1,400	4,700	2,900	3,700	1,400		
Yangtze (China)	22	800	5,800	3,600	1,900	700	500	260
Mississippi (U.S.)	18	650	6,000	3,700	3,300	1,300	296	91
Yenisei (U.S.S.R.)	17	600	4,500	2,800	2,100	800		
Irrawaddy (Burma)	14	500	2,300	1,400	430	170	300	700
Brahmaputra (Bangladesh)	12	415	2,900	1,800	670	260	730	1,100
Ganges (India)	12	415	2,500	1,600	960	370	1,450	1,520
Mekong (Vietnam)	11	390	4,200	2,600	800	300	170	210
Nile (Egypt)	2.8	100	6,700	4,200	3,000	1,200	110	37
Missouri (U.S.)	2.0	70	4,100	2,500	1,370	530	220	160
Colorado (U.S.)	0.2	6	2,300	1,400	640	250	140	210
Ching (China)	0.06	2	320	200	57	22	410	7,200

Sources: Holeman, John N. 1968. "The Sediment Yield of Major Rivers of the World," Water Resources Research, 4 (August): 737–747.
Fairbridge, Rhodes W. (ed.) 1968. The Encyclopedia of Geomorphology. Vol. III, Encyclopedia of Earth Sciences Series. New York: Reinhold.
Espenshade, Edward B. (ed.) 1970. Goode's World Atlas. 13th ed. Chicago: Rand McNally.
Curtis. W. F., Culbertson, J. K., and Chase, E. B. 1973. "Fluvial-Sediment Discharge to the Oceans from the Coterminous United States," U.S. Geological Survey Circular 670. Washington, D.C.

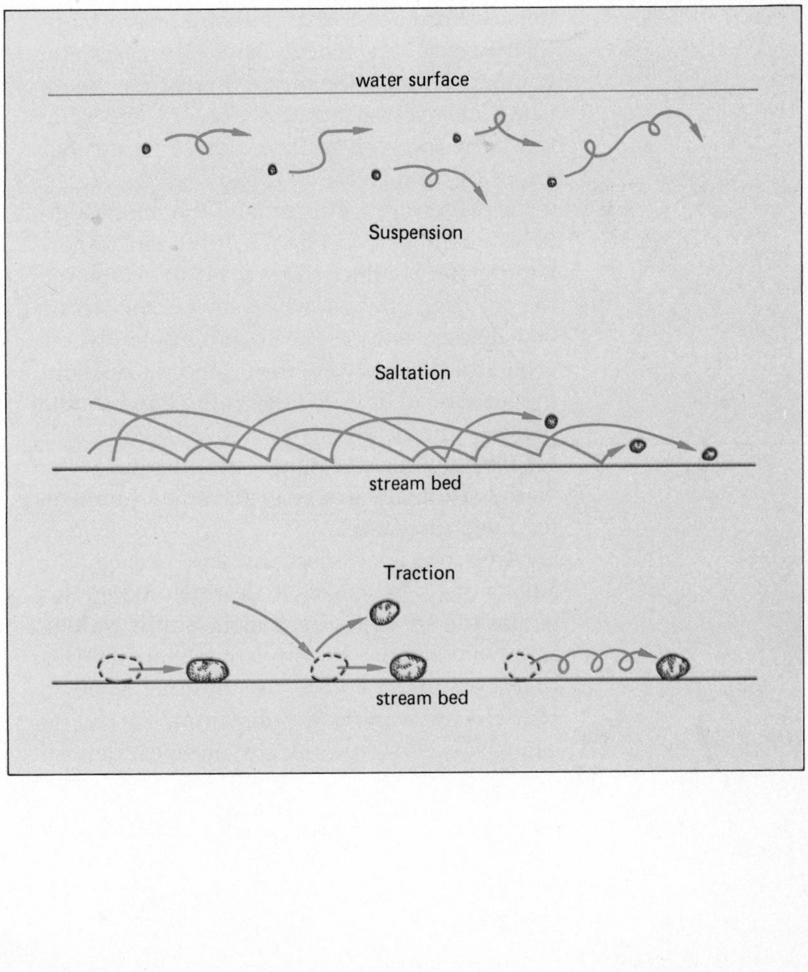

Figure 14.9 Solid material is transported in streams as suspended load and bed load. The finest particles are supported in suspension within the flow (top), while the bed load moves by both saltation and traction, in which particles are dragged or rolled by fluid shear stress, or are knocked forward by a saltating particle. (Vantage Art, Inc.)

the shape of its channel so that stream energy remains in balance with the work of sediment transport. If a stream lacks the velocity to transport the sediment fed into it, some of the sediment is deposited in the channel. This elevates the stream bed. Any elevation of the stream bed increases the channel slope in the downstream direction, which increases the stream velocity at that point. Thus, by depositing sediment the stream increases its energy and ability to move future arrivals of sediment.

Conversely, where stream velocity exceeds that necessary for sediment transport, the stream can reduce its energy by flattening its slope. It can do this either by scouring its bed downward or by lengthening its horizontal distance per unit of vertical descent. The latter increases the stream's *sinuosity* (measured as the stream distance divided by the length of the stream valley).

In both cases negative feedback maintains equilibrium in the fluvial system. The initial disturbance triggers a reaction that either supplies or removes energy to bring the system into balance with the changed conditions.

During a flood, each section of a stream must transmit a discharge that is much larger than normal. As the discharge increases, the volume of water becomes too great for the existing channel size and shape. At the same time, the increase in flow volume is not balanced by an increase in friction, so the stream velocity increases. This enables the floodwaters to scour the bed and banks of the stream, enlarging the channel (Figure 14.10). However, the scouring process, which lowers the stream bed slightly, reduces the stream's potential energy and downstream slope, and so restrains the increase in flow velocity. If the flood should persist, and if the stream bed were easily eroded, the stream slope and channel form would eventually evolve to an equilibrium with the flood discharge.

After the flood discharge has peaked, the stream discharge begins to decrease. When this occurs the stream is once more out of equilibrium. Its channel is now larger than the discharge requires, and it has too little slope to transmit the water and sediment of the diminishing flow. As flow velocity declines, deposi-

tion of sediment begins. This elevates the stream bed, increases the stream slope, and reduces the size of the channel. These adjustments sustain the velocity of the diminishing flow so that subsequent arrivals of sediment can be carried through, thereby reestablishing stream equilibrium.

The sediment underlying the beds of most large streams is an important element in the equilibrium process. It permits channel enlargement and slope adjustment during floods and the restoration of normal channel characteristics after floodwaters pass. However, complete equilibrium is not possible. Because flow conditions change constantly, most streams are in a state of perpetual readjustment, or "quasi-equilibrium."

Whereas stream channels formed in erodible materials can adjust to discharge changes quite rapidly, channels incised in hard rock cannot be adjusted easily. Consequently bedrock channels may not achieve an equilibrium condition until they buffer themselves with *alluvium* (stream-deposited sediments).

Any portion or "reach" of a stream that is in a quasi-equilibrium condition—neither progressively lowering its bed by scour nor raising it by sediment deposition—is said to be *graded*. A graded reach maintains a rate of descent and a channel form that give it just enough energy to transmit its fluctuating input of water and sediment. Since absolute equilibrium is impossible, a graded condition really implies that no long-term change is occurring despite repeated short-term adjustments.

Stream Gradient and the Longitudinal Profile

The rate at which a stream channel descends from higher to lower elevations is the *stream gradient*. This is measured as the vertical fall per unit of horizontal distance, expressed in like units (such as meters per meter), or as meters per kilometer or feet per mile. Gradient changes are a means of balancing stream energy to the work of water and sediment delivery. A graphic portrayal of a stream's gradient from its source to its mouth is called the stream's *longitudinal profile*.

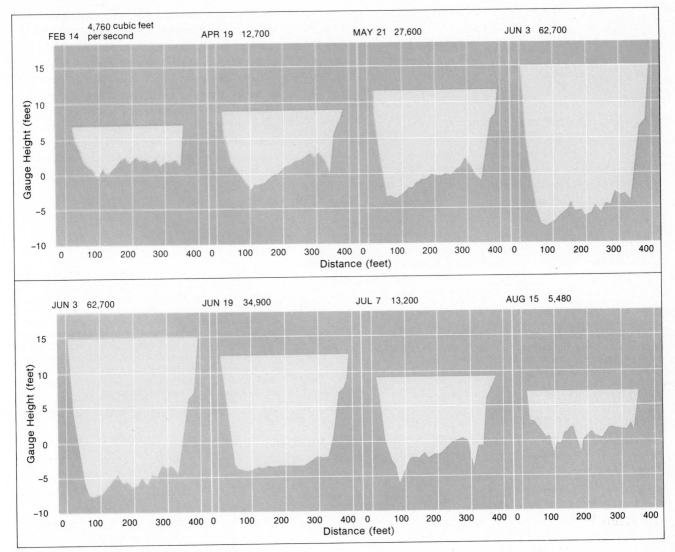

Figure 14.10 These channel cross sections for the Colorado River at Lees Ferry, Arizona, show the scouring and subsequent filling that occurred during a high-water period in 1956. Discharge is in cubic feet per second. The marked enlargement of the channel by scouring was accompanied by an increase in flow velocity to accommodate the greatly increased discharge. In August, when the discharge had decreased and was approaching its initial February value, the cross section of the channel also returned to almost its original size. (Doug Armstrong after *Fluvial Processes in Geomorphology* by Luna B. Leopold, M. Gordon Wolman, and John P. Miller. W. H. Freeman and Company. Copyright © 1964)

The profiles of most streams that are hundreds to thousands of kilometers long are concave upward—steep near their sources and almost flat near their mouths, as shown in Figure 14.11. The initial steepness provides the energy to move coarse bed loads in shallow channels. The flatter downstream gradients indicate increased discharges and larger channels, as well as a reduction in bed load particle size. Figure 14.11 also indicates that short channels commonly have nearly straight profiles. Point-to-point variations in stream profiles are a consequence of changes in discharge, sediment load, channel roughness (friction), and channel cross

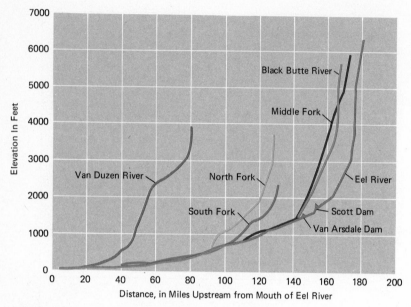

Figure 14.11 Longitudinal profiles of the Eel River and its major tributaries, which drain the Pacific Coast Ranges in northern California. Note the flattening of stream gradients in the downstream direction, resulting in concave-upward stream profiles. The longer the stream, the greater the tendency toward a concave-upward profile. (U. S. Geological Survey)

section (size and shape). An outstanding example is the Missouri River, which steepens its gradient from 0.8 to 1.2 feet per mile below the entry of the much smaller Platte River near Omaha, Nebraska. This large gradient increase is necessary to give the Missouri the energy required to transmit the enormous bed load of sand brought to it by the Platte. One must follow the Missouri 2,240 km (1,400 miles) upstream into its headwaters to find such a steep gradient again.

Figure 14.12 Examples of stream profile knickpoints resulting from different causes. The knickpoints on the Potomac River in Virginia and Maryland (top) result from variations in bedrock resistance and gradually falling sea level in Pliocene and Pleistocene time. The knickpoints on Yosemite Creek in California's Sierra Nevada range are thought to be a consequence of tectonic rejuvenation followed by glacial deepening of the valley of the Merced River (see also Figure 17.17, p. 493). (Somerville, U. S. Geological Survey, unpublished, 1929; and F. E. Matthes, U. S. G. S. Professional Paper 160, 1930)

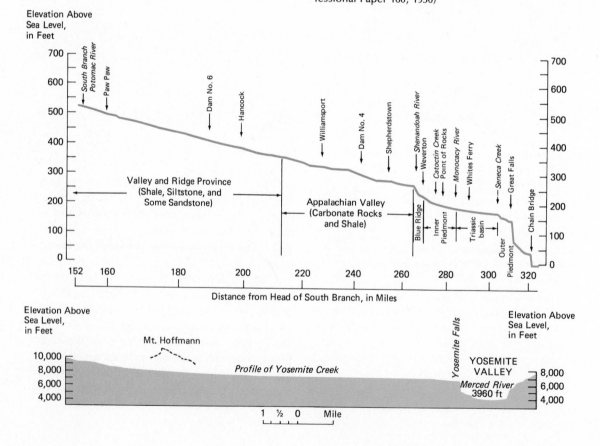

Profile Knickpoints

Stream profiles are sometimes interrupted by exceptionally steep reaches where rapids, cascades, or even waterfalls are present. Such profile interruptions are called *knickpoints*. Knickpoints in stream profiles have various origins. Some result from abrupt changes in bedrock resistance. Others may be related to disturbances of stream equilibrium by crustal movements that have caused bed scouring and channel incision to work headward (upstream) in the drainage system.

Knickpoints in stream profiles are especially common in areas that have experienced severe erosion by glaciers. The many waterfalls and cascades seen in high mountain country usually indicate that glacial erosion has deepened the major valleys, leaving smaller tributary valleys "hanging" above them, such as California's Yosemite Falls (Figure 17.17), and the many small waterfalls in New York's "Finger Lakes" region (see Chapter 17). Glacial erosion may also create steps in the main valleys themselves, each the site of a waterfall or cascade, as in the case of Nevada and Vernal falls in California's Yosemite Valley.

The famous "Fall Line" of the Atlantic coastal region of the United States refers to rapids and cascades developed where streams pass out of the resistant crystalline rocks of the Piedmont region and into the weak sedimentary rocks of the Atlantic Coastal Plain (Figure 14.13). In colonial times the rapids in large streams at the Fall Line provided water power to turn the grindstones of flour mills. Later, the water power was used to generate electricity. The falls and rapids also necessitated the landing of ships' cargoes and the change to overland transportation. This stimulated the growth of such cities as Trenton (New Jersey), Philadelphia (Pennsylvania), Baltimore (Maryland), Richmond (Virginia), Raleigh (North Carolina), Columbia (South Carolina), and Augusta and Macon (Georgia). Their growth was assisted by the fact that the rivers were narrower and easier to cross in the hard rocks above the Fall Line.

Increased kinetic energy in the steeper section of a stream's profile gradually eliminates most knickpoints by channel scouring. However, some knickpoints wear back in an up-stream direction a considerable distance before disappearing (Figure 14.14a). This has been the case with Niagara Falls (Figure 14.14b), which has retreated 11 km (7 miles) since the Niagara River established its present course about 100,000 years ago. Niagara Falls has a complex history related to glacial modification of the landscape, having been covered by glacial ice sheets several times since its inception.

As we shall see in Chapter 16, knickpoints are very conspicuous in streams in seasonally wet tropical regions. The tropical forest and savanna regions of Africa and South America are noted for their spectacular cascades and waterfalls, which are present even well downstream on large rivers. This seems to be a consequence of the inability of tropical streams to cut canyons in certain types of rocks, because of factors to be explained in the later discussion of tropical landforms.

Figure 14.13   The Great Falls of the Potomac River, just outside the city of Washington, D.C., are a striking example of the Fall Line marking the head of navigation on streams flowing from the Appalachian Highland to the Atlantic Ocean. The rivers cascade through ancient crystalline rocks of the Piedmont region, and remain ungraded probably due to gradually falling sea level during the past several million years. (T. M. O.)

Figure 14.14 (a) Streams generally act in such a way as to reduce knickpoints and attain a smooth longitudinal profile. The diagram shows a knickpoint (1) that was formed on Cabin Creek, Montana, by an earthquake. The stream began to scour its channel above the knickpoint and to fill below the knickpoint. Because the material forming the stream bed was easy to erode, the knickpoint was significantly reduced in only a few months (2). Three years later (3) the profile showed no trace of the original knickpoint. (John Dawson after Marie Morisawa, *Streams: Their Dynamics and Morphology*, edited by P. Hurley, © 1968, McGraw-Hill Book Company)

(b) Since the Niagara River began flowing during the Pleistocene era, it has not been able to cut vertically into the caprock of hard dolomite forming Niagara Falls. Instead, the river has cut a long gorge headward by undermining weak shale layers at the base of the falls, causing the collapse of the overlying layers. (From G. K. Gilbert, from O. D. Von Engeln, *Geomorphology*, 1942, Macmillan Company)

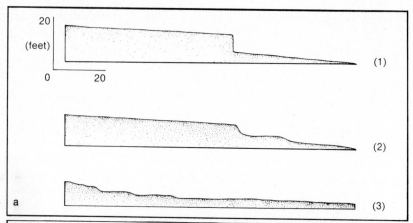

Lateral Migration and Channel Patterns

It is well known that streams can shift position horizontally, as well as scouring and filling their beds. Near Needles, California, the Colorado River has moved laterally as much as 244 meters (800 feet) in a single year, and at Peru, Nebraska, the Missouri River has shifted position from 15 to 150 meters (50 to 500 feet) each year for more than 30 years. Where stream banks are erodible, channels may migrate distances of several kilometers over a period of years. This creates significant problems: on one side of the stream property is gnawed away, while on the other side it expands; political boundaries move with the stream, when it shifts slowly, but stay

in place when it changes course suddenly. Mississippi's boundaries with Arkansas and Louisiana show these effects clearly, as portions of each state have been cut off from the remainder of the state by movements of the Mississippi River (Figure 14.15).

When viewed from above, stream channels display three general patterns: meandering, braided, and straight or nonmeandering (Figure 14.16). The normal pattern is a *meandering* one, in which the stream swings back and forth in either smooth or sharp curves. Meanders occur on streams of every size, from tiny rills on bare ground to kilometer-wide rivers. Regardless of stream size there is a relatively consistent relationship between the width of the stream and

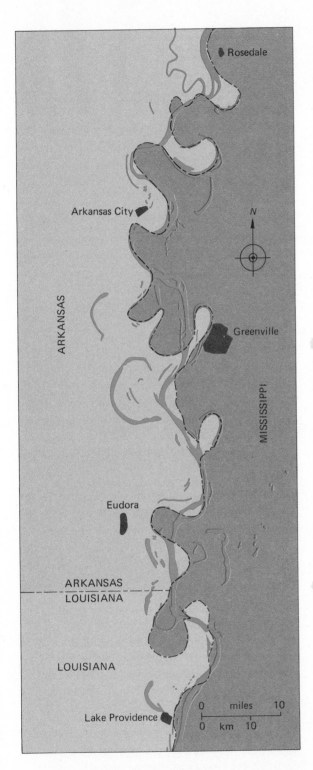

Figure 14.15 This map of the Mississippi River from Rosedale, Mississippi, to Lake Providence, Louisiana, a distance of about 130 km (80 mi), shows how meander cutoffs during floods leave visible evidence in the form of detached political units. The state boundaries were originally fixed along the center of the Mississippi River. The boundaries shift with the slow migration of the channel, but stay in place when there are sudden channel shifts, or *avulsions*, during floods. The areas inside meander cutoffs thus become isolated from the rest of their state. The three cutoffs closest to Greenville were made artificially by the Corps of Engineers in the early 1930s to shorten the river for navigation purposes. Note the *oxbow lakes* that reveal old meander cutoffs.

the *meander wavelength*, or distance between similar points on successive meanders (Figure 14.17). Meander wavelength is normally from 7 to 15 times the channel width.

Even streams that lack true meanders seem to show a succession of deeper *pools* and shallow-water *riffles* whose spacing is related to stream width (Figure 14.18). In fact, laboratory experiments with artificial channels in noncohesive sand or silt show that pool and riffle sequences in straight channels will in time evolve into sequences of meanders. Each pool becomes the site of an outward-pushing channel bend, or meander. Thus the explanation of meander geometry lies in the nature of the flow that produces the initial pools and riffles.

The stream *thalweg*, which is the line that follows the deepest part of the stream channel, swings back and forth across the channel from pool to pool, moving toward the outer bank of each successive curve. Maximum flow velocity occurs at the stream surface, above the line of the thalweg, where the stream is deepest and friction is least. Scouring is most vigorous at the outer (concave) banks where the flow is deepest and where stream velocity accelerates. Deposition of sediment is most evident at the inner (convex) banks, or "points," where flow depth and velocity are least.

The crescent-shaped deposits of sand or gravel formed on the successive points (Figures 14.18 and 14.19) are called *point bars*. Point bars commonly are composed of concentric ridges and depressions known as "bar and swale topography." The ridges are the bed load deposits

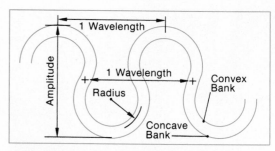

Figure 14.17 The diagram shows how the wavelength, amplitude, and radius of a meander are defined. The radius around a meander is not constant, however, because the form of a meander is more complex than a simple circular arc. The meander wavelength is usually 7 to 15 times the width of the channel. (After Dury, 1969)

Figure 14.16 **(top)** The sinuous Animus River near Durango, Colorado, meanders over a broad floodplain produced by fluvial aggradation. (John S. Shelton)

(bottom) This braided tributary of the Yukon River in northwestern Canada carries outwash from the margin of a receding glacier that is visible in front of the St. Elias Mountains in the distance. Braided channels develop where stream banks are gravelly and collapse easily due to lack of cohesion, and where streams receive excessive inputs of sand or gravel. (Larry W. Price)

accreted during the falling stages of floods, and are composed of coarse sand or gravel. Vegetation may take root in the finer material of the swales, giving the deposit a "banded" appearance. The bars and swales give a clear picture of the movement of meanders through time. The process of undercutting and collapse at the outer (concave) bank and point bar deposition directly across the channel causes the meander loops to shift outward and downstream in the process of *lateral planation*. Most scouring occurs during floods, when stream energy is at its peak. The greatest deposition of sediment occurs during the falling stages of floods, when stream banks collapse at the same time that stream energy declines.

Lateral planation increases the stream length for each unit loss in stream elevation. Therefore its effect is similar to that of vertical stream incision: both produce a flatter stream gradient, which reduces stream energy. An energetic stream that does not have the tools (bed load) to cut vertically into bedrock can achieve equilibrium by moving laterally instead. That is why meandering channels occur where the bulk of the stream load is fine material that is carried in suspension.

A stream load dominated by larger particles that move only by traction and saltation produces a different channel pattern—the *braided* type—which consists of many intertwining shallow channels, as seen in Figures 14.16 and 14.19. In shallow channels the maximum flow velocity is close to the stream bed, where it can thrust against the bed load. To develop the velocity needed to move their bed loads such channels must compensate for their shallow depth by having a steep gradient. This is achieved initially by aggradation.

Braided streams form when bed load particles are deposited in lenticular (lens-shaped) bars in channels abundantly supplied with sand, gravel, and cobble-sized material. These bars divide a stream into many shallow channels and increase the downstream slope by elevating the stream bed. This provides the steeper gradient required to overcome the friction in shallow channels, permitting flow velocity to be high at the stream bed.

Braided channels develop where stream discharge fluctuates widely from day to day or season to season, as in arid and semi-arid regions. Wide fluctuations lead to bank collapse,

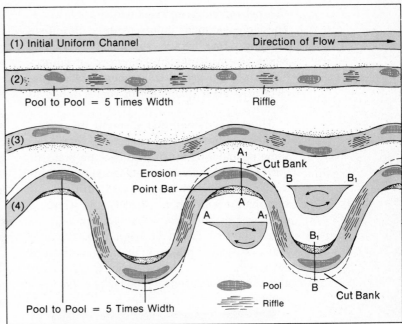

Figure 14.19 (left) Point bars are well developed along this meandering reach of the Rio Grande, which forms the boundary between Texas and Mexico. These point bars consist of the sandy bed load of the Rio Grande, but elsewhere point bars may be composed of gravel or cobbles. All are built up by deposition during the falling stages of flood discharges.

(right) The White River, which issues from the Emmons glacier on the flank of Mount Rainier in the Cascade Range, is an example of a braided stream. Here the bars separating the braids are composed of large cobbles. On other braided streams the channel bars may consist of gravel or sand. (T. M. O.)

Figure 14.18 Streamflow in an erodible straight channel of uniform cross section (1) tends to develop a sequence of alternate deep pools and shallow bars, or *riffles* (2). The straight channel becomes sinuous, with pools forming toward the outer concave banks (3). In time, a meandering channel may be developed (4). A sinuous or meandering stream tends to scour its outer concave banks and to deposit sediment against its inner convex banks; the speed of flow is greatest near the concave banks and least near the convex banks. There may also be a lateral circulation of water, as the cross-sectional diagrams A-A₁ and B-B₁ indicate. The lateral flow is thought to move sediment from the cut bank to the point bar. (Doug Armstrong after G. H. Dury from *Water, Earth, and Man*, edited by R. J. Chorley, © 1969, Methuen and Company)

channel widening, and sediment accumulation. The large sediment inputs required to initiate braiding occur where vegetation is sparse, where stream banks are composed of non-cohesive sand or gravel, and where melting glaciers feed great quantities of coarse waste to streams.

Human activities that increase the movement of sediment into streams have caused some meandering streams to become braided. Such activities include certain agricultural, mining, and logging practices, as well as urban and suburban construction projects. By creating shallower channels, these changes generally increase the frequency and magnitude of floods. This is an acute problem in many rainy tropical areas, where the sediment input to streams is enormously increased by removal of the natural vegetation.

The third channel pattern is the straight, or nonmeandering, type. Straight stream channels are relatively uncommon and seldom extend far. Where present, they indicate that the channel is controlled by the underlying geological structure. Streams incised in resistant bedrock are occasionally channeled by fractures in the rock. Some meander patterns are composed of straight segments that meet at sharp angles, indicating that joints control the channel. The channels of many river deltas are straight because cohesive clays have prevented lateral shifting of the stream beds (Figure 14.25, p. 395).

FLUVIAL LANDFORMS

Valley Development

When a stream has more kinetic energy than it requires to transport its sediment load, the stream reacts to bring its energy into balance with the work it is required to perform. For example, streams flowing down a surface recently created by motions of the earth's crust, or by the construction of a volcano, commonly have excess energy as a consequence of their steep initial gradients. Streams in such settings will cut downward, flattening their gradients

until their energy is in balance with their work of sediment transport. Streams that have achieved quasi-equilibrium will begin scouring when environmental changes increase their discharge, giving them extra energy. Renewed incision to flatten stream gradients and reduce potential energy following a period of stability has been called stream "rejuvenation." The causes of stream rejuvenation are deformation of the land surface, drops in sea level, increases in stream discharge, and decreases in sediment load.

Tectonic uplift of the land surface, which increases the potential energy of streams, is probably the chief natural cause of stream rejuvenation and valley development. This rejuvenation causes valleys to deepen as stream profiles are regraded to restore equilibrium (Figure 14.20). Uplift that is continuous or more rapid than stream incision produces hilly or mountainous landscapes resulting from steady stream downcutting. Uplift in separate pulses results in stream terraces (discussed later in this chapter) and multi-story valleys, in which newer valleys are cut into the floors of successively older and higher valleys.

Lowering of sea level, or uplift of the land relative to the sea, causes valley deepening near coasts. The effects of world-wide sea level changes are discussed in Chapters 17 and 18.

Increases in stream discharges have been an important cause of stream rejuvenation at various times. Increased discharges can be produced by regional changes in climate and vegetation or by local "stream piracy," in which an expanding drainage network invades and "captures" the headwaters of a neighboring stream system. The lateral planation process can also cause stream captures, as a stream meander occasionally pushes laterally into the valley of an adjacent smaller stream that flows at a slightly higher elevation. This diverts the flow of the higher channel into that of the lower one.

Flood discharges, during which valley deepening occurs, have been increased by such human activities as forest removal and urbanization that covers vast areas with impermeable concrete and asphalt. A final cause of stream rejuvenation is decreased sediment load—also frequently the result of human activity, specif-

Figure 14.20   This image taken from a altitude of about 200 km (125 miles) shows the Colorado River's great canyon through the Colorado Plateau in northern Arizona (upper right). The river exits from its canyon at the escarpment of the Grand Wash Cliffs (see arrow), created by uplift along a major fault. Uplift of the Colorado Plateau relative to the area to the west has caused vertical incision of the river to a depth exceeding 1,500 meters (5,000 ft). Healthy vegetation produces the red colors on computer-generated LANDSAT images such as this. Lake Mead appears in the left (west) part of the image. (Department of the Interior, U. S. Geological Survey)

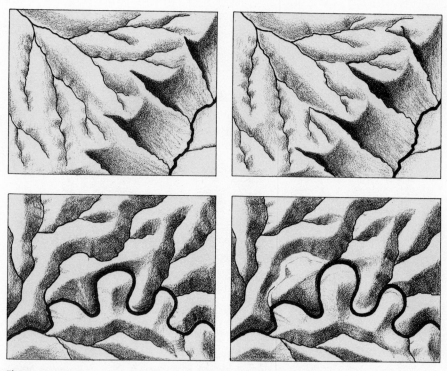

Figure 14.21 Stream captures can be produced by headward extension of valleys or by lateral movements of stream channels. The two examples illustrated here are commonly encountered in areas of hilly relief.

At the top left, the headwaters of two drainage systems meet at an asymmetrical divide, with a gentle descent to the northwest and a much steeper slope to the southeast. The short streams on the steeper eastern slope have much more energy than the low gradient streams on the west side of the divide. Accordingly, the steep-gradient streams deepen and aggressively extend their valleys in the form of ravines. At the top right, the deeply incised ravines have undercut the shallow northwest-draining valleys, "beheading" them by capturing their headwaters. This produces a "fishhook" drainage pattern that includes several "elbows of capture".

At the bottom left, a meander of a large stream pushes close to one of its own tributaries, narrowing and lowering the divide between the two. At the right, the main stream has worn through the divide, capturing the upper portion of the tributary. This leaves the lower portion of the tributary valley beheaded and abandoned. The result is an open lowland lacking a stream channel. (T. M. Oberlander)

ically dam construction, which traps both the bed load and suspended load of streams in artificial reservoirs. Immediately downstream from large dams, streams have excess energy since they have no sediment to transport. Such streams scour downward when large volumes of water are released from the reservoirs. This is occurring now along the Colorado River below the Glen Canyon Dam (completed in 1964) and along the Egyptian Nile below the Aswan High Dam (completed in 1971).

Floodplains

When streams attain a condition of equilibrium and stop deepening their valleys, the valley floors gradually become wider. Valley widening is caused both by lateral stream planation and by normal processes of erosion on valley walls. As meander planation trims back valley walls and sediment is deposited on point bars, a continuous flat-floored trough evolves. The stream swings back and forth across this open lowland

(Figure 14.22). Since the stream channel develops in such a way that it can contain most of the flows through it, but not the rare large flows that occur at intervals of a year or two, this lowland is occasionally flooded and is known as a *floodplain*.

Not all floodplains are produced by stream erosion. Many of the earth's largest rivers flow across vast lowlands created by crustal movements. Examples are the Amazon River of Brazil, the Ganges River of India, the Sacramento and San Joaquin rivers of California and the lower Mississippi River. These streams carry sediment into the structural basins that they cross, which are not true valleys. Many other floodplains are products of stream aggradation that has filled deep V-shaped valleys with sediment. Such floodplains are especially common in coastal regions.

Floodplains are hazardous environments, but they have great economic value. Individual floods may leave a floodplain covered by a depth of several centimeters of fresh silt. These overbank flood deposits cover the older sand and gravel point bar deposits, and create nearly level agricultural land. This land is usually very

fertile since its nutrients are renewed periodically by fresh deposition. However, where human disturbance has accelerated erosion, floods may deposit coarse sand over previously fertile areas, reducing their utility.

River floodplains can contain a variety of fluvial landforms. When overbank flooding occurs, the largest amount of deposition is just outside the stream channel where the flow velocity slows due to reduced depth and increased friction. The thickness of the overbank flood deposit diminishes with distance from the channel. This wedge-like deposition produces paired *natural levees* that slope gently away from the stream on both sides (Figure 14.22). Often there are low-lying *back swamps* between the natural levees and the valley walls that rise above the floodplain. Low ridges or cones of alluvium often extend from the levee into the back swamp. These form where a stream of sediment-laden floodwater has broken across the levee crest. The levee break is called a *crevasse*, and the resulting alluvial deposit is a *crevasse deposit*. In Louisiana, the Mississippi River's natural levees rise 5 to 6 meters (15 to 20 ft) above the back swamps. In the disastrous

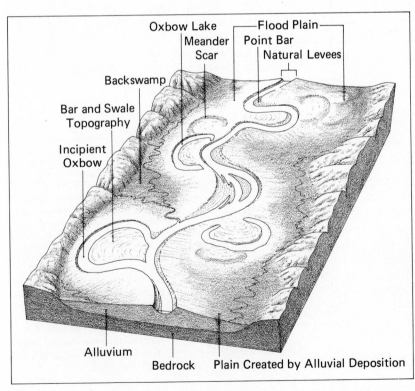

Figure 14.22 Floodplains may result from either fluvial aggradation or lateral stream planation that creates bar and swale topography and wears back the valley walls. Oxbow lakes are the remnants of recently abandoned meanders. When the lake is eventually filled with sediment, a *meander scar* remains. In flood, the overflowing river deposits new sediment on its floodplain and builds natural levees by deposition of silt close to the river channel. (T. M. O.)

1973 flood, the Mississippi River levees received as much as a meter (over 3 ft) of new sediment.

Natural levees are prime locations for agricultural settlement due to their fertile soils and good drainage characteristics. On large rivers they have become industrial sites, accessible by ocean-going ships. To protect developments on such sites, artificial levees are commonly built atop natural levees. When confined in this way, river floods of a given discharge rise higher than under natural conditions when they could spread laterally. If they break through the artificial levees, such high flows can be even more destructive than natural flooding, because of the greater force of the flows.

Where streams have considerable excess energy and the floodplain is composed of easily eroded sand rather than silt, meander shifting may be very active. In such places the meanders often expand into flaring "gooseneck" loops that press close against one another. Flood flows sometimes break through the narrow necks of land that separate adjacent loops, causing the meander to become cut off from the stream, forming an *oxbow lake* (Figure 14.22). Hundreds of present and past oxbow lakes are visible along the Mississippi River from Cairo, Illinois to Baton Rouge, Louisiana (Figure 14.15). Many of these were created at the same time during major floods. Oxbow lakes eventually become marshes, and finally fill with silt and clay, but their traces remain clear long afterward.

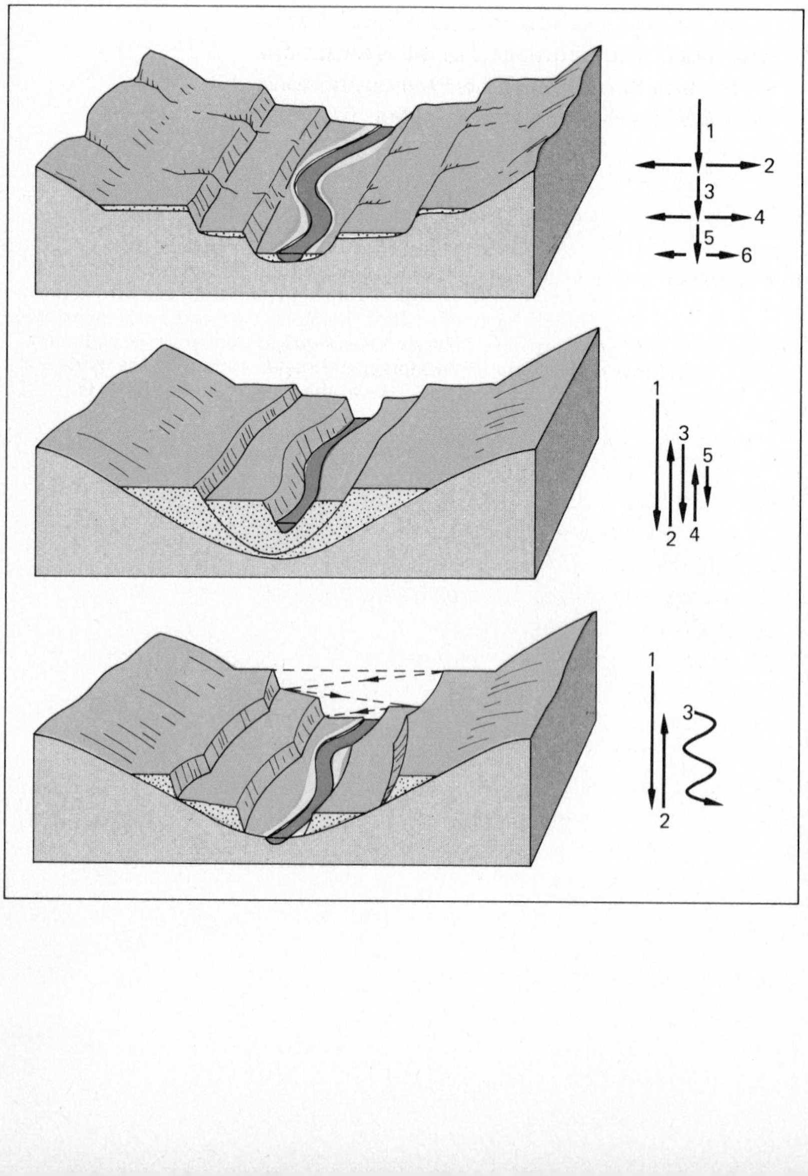

Fluvial Terraces

Fluvial landforms often show that graded streams periodically adjust their profiles by either raising or lowering the stream bed. Where a stream cuts downward into a broad valley floor, the floodplain no longer acts as an overflow channel. Instead, it is left standing well above the level of the highest floodwaters to form a *fluvial terrace* (Figure 14.23).

A temporary phase of aggradation can also produce fluvial terraces. In such a case decreased stream energy or increased sediment input causes deposition that raises the stream bed, forming an *alluvial fill*. Restoration of the preceding conditions causes the stream to cut back through the fill to its previous level. This

Figure 14.23 Stream terraces vary significantly in origin, as these diagrams illustrate. To the right of each diagram is an indication of the nature of the river movements necessary to create the associated terraces: downward arrows indicate stream incision; upward arrows, aggradation; horizontal arrows, lateral planation; sinuous arrows, simultaneous incision and lateral planation. **(top)** Terraces produced by periods of downward stream incision (1, 3, 5) separated by intervals of valley widening by lateral planation and slope erosion (2, 4, 6). These terraces are erosional and are said to be *paired* since those on opposite sides of the stream match in elevation. **(center)** Paired fill terraces produced by two separate phases of alluvial aggradation (2, 4) followed by renewed valley excavation (3, 5). **(bottom)** Valley deepening (1) is followed by aggradation (2) producing an alluvial fill. Subsequently the stream cuts downward at the same time that it is planing laterally in the fill (3), producing *unpaired* terraces. (Vantage Art, Inc.)

leaves a terrace composed of the alluvial fill. Fill terraces are common where streams were temporarily aggraded by sediment washed from glaciers during the ice ages, and also where tectonic motions and major climatic fluctuations have occurred.

River terraces are important indicators of environmental change. They reveal that tectonic, climatic, or hydrologic alterations have forced streams to change their behavior to regain equilibrium. Some rivers are bordered by great flights of terraces, indicating repeated disturbances of equilibrium. The rivers of New Zealand are famed for their magnificent terraces, which reveal a history of strong tectonic activity and complex environmental changes.

Alluvial Fans

Where a steep-gradient mountain stream in a bedrock canyon flows out onto a flat plain, it loses much of its kinetic energy. This occurs because of the spreading of waters that were previously confined to a narrow channel. Where the flow spreads, the bed load is dropped, producing a fan-like deposit issuing from the canyon mouth and spreading over the plain. This deposit, shown in Figure 14.24, is an *alluvial fan*.

Alluvial fans are most common in arid regions. The lack of vegetative protection permits the infrequent heavy rains to flush large quantities of rock debris from slopes. Desert stream floods, which may last only a few hours, are highly charged with sediment. Where a stream flows out of its canyon, the coarsest part of the bed load is dropped close to the canyon mouth. In this way the apex of the alluvial fan is raised much higher than the fan's margins. Many fans are composed largely of the deposits of mudflows and debris flows. Since the density of these flows is many times that of water, they can transport large boulders, which are often seen scattered over fan surfaces.

In mountainous deserts, neighboring fans merge to form continuous ramps of sand and gravel known as *alluvial aprons*, or *bajadas*. In California's Death Valley (Figure 14.24) these debris aprons rise as much as 600 meters (2,000

ft) over a horizontal distance of 8 km (5 miles). The Death Valley fans, like many others, have had a long and complex history of erosion and deposition, and exhibit surfaces of widely varying ages.

There is often abundant groundwater present at the bases of alluvial fans in arid regions, which makes them favored locations for settlement and agricultural development. An outstanding example is Salt Lake City, which sprawls over the surface of a large fan adjacent to Utah's Wasatch Mountains. The Mormon settlers who founded Salt Lake City also established several other settlements in similar settings, including Las Vegas in Nevada and San Bernardino in California.

Deltas

When a river flows into a lake or the sea, its velocity is checked and it loses its load-transporting ability. Its bed load and much of its suspended load is dropped at the river mouth in the form of a *delta*. The term was first used some 2,500 years ago by the Greek historian Herodotus, who noted the similarity of the depositional form to the Greek letter Δ (delta). However, delta configurations are extremely variable. The Mississippi River delta, with its bird's-foot form (Figure 14.25), is quite unlike the classic examples of the Nile and Niger river deltas in Egypt and Nigeria. Variations in delta form result from differences in the rate of sediment supply, the vigor of wave action and coastal currents, and the rate at which the alluvial deposit subsides as a result of compaction or the sinking of the sea floor under it.

When rivers enter deltas, their discharge usually becomes divided among several *distributary* channels. These are created by floodwaters that spill out across low natural levees to produce new channels. The unusual form of the Mississippi delta is primarily the result of slow sinking of the sea floor, which leaves only the crests of the stream's natural levees projecting above sea level. The present bird's-foot configuration developed after a change of the river's course that occurred near New Orleans only about 500 years ago.

Figure 14.24   This alluvial fan is located on the east side of Death Valley, California; its size can be judged from the road crossing it. Rains are infrequent in this region, but torrential storms occur in the mountains from time to time, carrying down large quantities of coarse sediment that is deposited close to canyon mouths. (John S. Shelton)

Alluvial fans commonly join along the base of a desert mountain range to form an alluvial apron or *bajada*, such as this example seen fringing the Panamint Mountains on the west side of California's Death Valley. (T. M. O.)

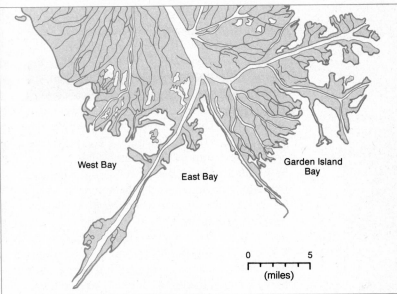

West Bay

East Bay

Garden Island
Bay

0   5
(miles)

Figure 14.25   (left) The delta of the Nile River in Egypt has a different shape from the delta of the Mississippi because of more vigorous erosion by currents of the Mediterranean Sea and the fact that the delta is not subsiding. The construction of the Aswan Dam has changed the sedimentation pattern of the Nile, and its delta is in danger of accelerated erosion. (NASA)

(right) The delta of the Mississippi River in Louisiana resembles a bird's foot because the delta is subsiding, leaving only the crests of its natural levees above sea level. The subsidence is caused by compaction of sediments and also by general tectonic sinking of the floor of the Gulf of Mexico. The bird's foot form was more pronounced before Garden Island Bay was partly filled by a new crevasse deposit. (Doug Armstrong after S. M. Gagliano et al., "Hydrologic and Geologic Studies of Coastal Louisiana," Department of the Army, 1970)

Like floodplains, deltas are fertile and often densely populated. The Nile delta is the home of nearly 30 million people—two-thirds of Egypt's population. Loss of life and property can be heavy in deltas during floods. Some of the worst disasters occur when hurricanes cause flooding along low-lying deltaic coasts. This has been extremely costly to the dense populations occupying the combined deltas of the Ganges and Brahmaputra rivers of India and Bangladesh. It also is a hazard to the inhabitants of the Mississippi delta and the Gulf Coast from Florida to Texas.

AEOLIAN PROCESSES AND LANDFORMS

The greater part of this chapter has been devoted to fluvial processes and landforms because they are essential aspects of the environments in which most people live. Now we shall look briefly at the role of wind in erosion and deposition. Although the effects of wind are distinctive and significant wherever they occur, their most spectacular effects are found in certain sparsely inhabited regions of the world. But the wind can modify the land surface anywhere

in which the vegetation cover is weak or absent. Human activities, by destroying the natural vegetation, can cause wind to become an effective geomorphic agent in regions where its influence would be insignificant under natural conditions.

Wind as a Geomorphic Agent

Processes related to wind action are known as *aeolian* processes, after the Greek god of the winds *Aiolis* (Latin, *Aeolus*). Aeolian erosion and deposition are natural occurrences wherever loose, unprotected sediments are present, as along shorelines and in areas of recent glacial or fluvial deposition. The largest-scale aeolian effects are seen in the world's desert regions, where lack of moisture results in vast expanses of bare dry soils.

A bare surface over which the wind sweeps may be thought of as a vast stream bed. Particles put into motion by the wind behave like those moved by flowing water. The smallest particles are carried in suspension as wind-blown dust. Heavier particles move by saltation and traction, as in the streamflow depicted in Figure 14.9. The velocity of wind, which can greatly exceed that of flowing water, gives it the energy needed to move material. Fine dust can be whirled up into the air by a slight breeze. Dry sand, having particle diameters of about 0.1 mm, begins to move when the wind velocity near the ground reaches 16 km (10 miles) per hour. Coarse sand (2.0 mm) is put in motion by a 50 km (30 miles) per hour wind. But, although wind can achieve far greater velocities than flowing water, its density is much less, so it has less buoyant force. Thus, the wind cannot move rock particles larger than sand, and in fact rarely lifts sand more than a meter above the ground.

Aeolian processes mainly affect loose particles; it is uncommon for wind action to wear away solid rock, even in deserts. Wind erosion of rock essentially means sandblasting. This requires an exceptionally strong wind to blow abrasive quartz particles against soft rock. Wind erosion much more often consists of the removal, or *deflation*, of fine rock debris provided by other gradational processes.

Aeolian deflation usually leaves no spectacular landforms. Since deflation slowly lowers large areas, there may be little left by which to gauge its magnitude. Often the principal evidence of aeolian deflation is a tree or bush left standing on a pedestal of soil, all the surrounding soil having been blown away. In a few extremely arid locations, strong winds have actually eroded parallel grooves in the silts of old lake beds, creating distinctive streamlined forms called *yardangs*. Similar features develop on a much smaller scale on individual rocks, where *ventifacts* (see Chapter 16) are fashioned by natural sand and dust blasting.

Loess

Aeolian deflation is visible chiefly in dust storms (Figure 14.26). These are usually pro-duced by winds related to strong pressure gradients in arid and semi-arid regions. The dust, composed of silt-sized particles, may be lifted thousands of meters into the atmosphere. It settles out far from its source, to form distinctive silty deposits known as *loess*.

Loess, which can be centimeters to hundreds of meters thick, offers excellent agricultural opportunities. It is usually rich in calcium, and soils developed on it are young and unleached. Loess is common in the American Midwest and Mississippi Valley, and in a strip extending across Europe from northern France to the Russian Ukraine. Loess also has accumulated along the northern edge of the highlands of Central and Eastern Asia. The North American and Eurasian loess deposits originated from deflation of Pleistocene glacial deposits. Northern China has extremely thick loess deposits that reflect deflation in the Gobi Desert of Mongolia and northwest China. In northern China loess produces a spectacular landscape that is deeply trenched by gully erosion. Hundreds of generations of humans have lived there in caves carved into the soft material.

Dune Landscapes

The most impressive results of aeolian energy are the great sand seas of desert regions. It is a mistake, however, to think of deserts in general as areas of sand dunes. Most desert terrain consists of gravelly plains, eroded rock surfaces, and stream-dissected relief. Nevertheless, deserts are distinctive in having large areas of dune topography as well (Figure 14.27).

In the deserts of North Africa and Arabia, individual sand seas called *ergs*, with dunes resembling frozen waves, cover tens of thousands of square kilometers. Somewhat less spectacular dune landscapes also cover vast areas in India, western China, and Australia, and small ergs are present in every desert region. An ancient sand sea, comparable to those of North Africa but now grass covered, occupies much of western Nebraska.

Ergs form where winds lose velocity after carrying sand hundreds of kilometers. Thus, ergs are related to large-scale wind patterns that

Figure 14.26 These two views illustrate erosion and deposition by the wind. On the left is a severe dust storm, photographed in Colorado in 1935. Here very strong winds associated with frontal disturbances are seen eroding bare agricultural land, stripping away the most fertile part of the soil—its A-horizon. This hazard exists wherever the natural vegetation has been removed from soil that is periodically dry. On the right is a 6-meter (20-ft) bank of *loess* in Louisiana. Loess is composed of dust-size mineral particles that settle out of the atmosphere downwind from an area of aeolian deflation. This loess is typical in that its cohesiveness causes it to stand in near-vertical banks and to be gullied where vegetative protection is absent. The depth of loess in the Mississippi Valley region varies from centimeters to as much as 15 meters (50 ft). Most Mississippi Valley loess was derived from aeolian deflation of barren late Pleistocene glacial deposits. (USDA Soil Conservation Service and T. M. O.)

Figure 14.27 Sand ripples (foreground) and dunes in Great Sand Dunes National Monument, Colorado. Human figures in the middle distance to the right provide scale. The sand is produced by weathering, carried a short distance by floods of water, and then moved along by strong winds. Sand accumulates in dunes when the transporting wind loses power because of loss of velocity. (Grant Heilman Photography)

produce aeolian sand flows involving vast areas. Once an aeolian sand deposit is initiated, it is self-enhancing. It traps arriving sand, which does not saltate as well over sand as it does over rock or hard-packed soil. Ergs contain dunes of many types, reflecting variations in the sand supply and the nature of the sand-transporting winds, which may change direction seasonally. The most general distinction made among dunes is that between transverse and longitudinal types.

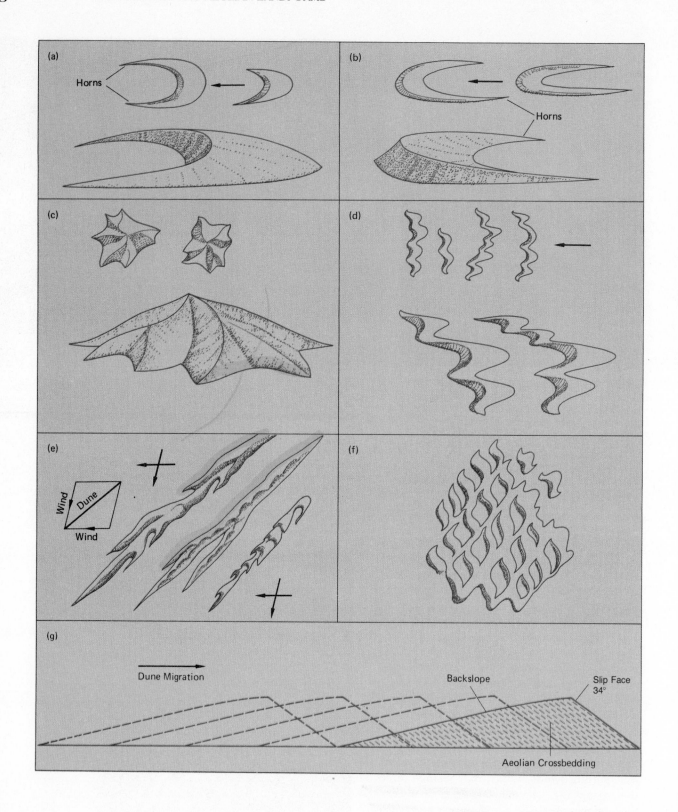

Figure 14.28 **(opposite)** Morphology of sand dunes, with arrows showing the directions of the effective winds. In (a), (b), (c), and (d), typical map patterns of the dune types are shown at the top, with oblique views below; (e) and (f) show the map pattern only.

(a) Individual dunes with elongated tips, or *horns*, pointing downwind are known as *crescentic* or *barchan* dunes. This type of dune is found in dry, vegetation-free environments, and is usually seen migrating across a nonsandy surface of gravel or clay. Barchan dunes may move several tens of meters each year. Symmetrical barchans indicate a nearly constant wind direction.

(b) Single dunes whose horns point upwind are known as *parabolic* dunes. Often they assume a hairpinlike form. Such dunes form where sand blows into a moist area that has a vegetative cover. Vegetation or dampness in the lower part of the dune retards motion there, so the dry crest pushes ahead of the base, causing the horns to ''drag'' behind. Parabolic dunes are found along coastlines and often push into forests, engulfing and killing the forest trees. Similar dunes also occur where older, vegetation-covered, sand accumulations become active again due to natural or artificial destruction of the vegetation. In such instances, the newly active parabolic dunes are called *blowouts*.

(c) Large erg areas sometimes include regularly spaced ''sand mountains'' as much as 200 meters (650 ft) high; these are called *star dunes*. They seem to be fixed in position, and their formation, which is not well understood, seems to require winds from several directions. They are seen in Arabia and the Sahara.

(d) Ergs commonly contain dune ridges transverse to the wind direction. These appear to be composed of linked barchans. They frequently incorporate parabolic segments between the linked barchans. Most transverse dune systems result from effective winds from a single direction.

(e) Where the sand supply is especially abundant, *longitudinal* dunes are often encountered. Some of these are hundreds of kilometers long and as much as 200 meters (650 ft) high. These dunes are thought to show the effect of multidirectional winds. Their morphology varies from smooth *whalebacks* to knife-edged *seif* dunes. Many have a fairly complex topography, including variously oriented subsidiary crests and slip faces.

(f) In some ergs, transverse and longitudinal dune systems appear to cross, producing *grid* dunes. Either seasonal or long-term changes in wind direction may be responsible for such patterns.

(g) This diagram shows the migration of a dune as sand is transferred from the backslope to the slip face, causing the dune to be displaced in the direction of the slip face. The internal structure of a dune consists of steeply inclined laminations of sand produced by accretion on the slip face. The result is *aeolian crossbedding*, which can be seen in many ancient nonmarine sandstones. (Vantage Art, Inc.)

Figure 14.29 Barchan dunes are characteristically crescent shaped, with the horns of the crescent pointing downwind. They form in areas where the supply of movable sediment is limited and vegetation is sparse. Barchans migrate downwind but retain their arcuate shape because the points of the dune migrate faster than the main body.

Transverse dunes form where sand-carrying winds blow mainly from one direction. The highest dune crests are at right angles to the direction of sand-carrying winds (Figure 14.28a, b, d). The beds of sand-carrying streams are often composed of small dunes of the transverse type, formed by flows of water rather than air. Over a period of years small transverse dunes may migrate hundreds of meters in the downwind direction, as in Figure 14.29.

Longitudinal dunes form where the sand-carrying winds come from more than one direction. The crests of individual longitudinal dunes may extend hundreds of kilometers and are believed to represent the sum of the vectors of the various sand-carrying winds, as shown in

Figure 14.28e. There are dune types that have characteristics of both transverse and longitudinal dunes, and some that are complex nonlinear mountains of sand (Figure 14.28c). Images obtained from orbiting satellites make it clear that in any one region dune types are similar, reflecting the local wind patterns and sand supply.

The most recent direction of sand flow in a dune area can be determined by looking at the smallest sand accumulations—the closely spaced *sand ripples* that crawl over the dunes. The migration of sand ripples brings sand up the dune *backslope* (Figure 14.28g) to the dune crest. At the crest the sand falls onto the steep *slip face* of the dune, which always has a slope of about 34°. Sand ripples move in different directions from day to day, often approaching the dune crest obliquely. The ripples are asymmetric, with their steep faces on the downwind side in the direction of ripple travel. Stream beds likewise commonly display ripples, showing that wind and water move particles in much the same way.

Ancient ergs of vast size that are now covered with vegetation, or whose dunes rise from marshes, are graphic evidence of the oscillations in climate that occur through time. Relict ergs similar to the Nebraska Sandhills are widespread south of the Sahara in the vicinity of Timbuktu and Lake Chad, as well as on the margins of other deserts.

SUMMARY

Fluvial and aeolian processes are closely related. Both result from the shear stress of a fluid moving over a surface composed of movable particles. Flowing water forms streams that compose drainage networks. These evolve systematically in a way that seems to minimize energy expenditure. Stream energy results from the conversion of potential energy to kinetic energy. Stream discharge and turbulence play important roles in providing this energy. Although most stream energy is used to overcome friction, there is enough excess energy to cause streams to move solid particles by suspension, saltation, and traction, and to scour their beds during floods. Streams also transport vast quantities of dissolved matter.

Streams tend toward an equilibrium condition through negative feedback processes. Insufficient stream energy causes deposition that raises the stream bed, which increases stream energy. Excessive energy causes channel incision or lateral planation that decreases stream slope and reduces stream energy. These adjustments can be seen in the course of a single stream flood. A graded section, or reach, of a stream has just enough energy to pass along all of the sediment it receives.

The longitudinal profile of a stream is a portrayal of its gradient, or rate of descent, from source to mouth. Knickpoints, or reaches with unusually steep gradients in the form of waterfalls or cascades, sometimes interrupt stream profiles but tend to be eliminated with time. Knickpoints may indicate changes in rock type, tectonic activity, or a "wave of erosion" moving headward through the stream system.

In general, stream flow tends to produce meandering channels. Meandering appears to be a way of equating stream energy to the work of sediment transport. Meanders develop out of pool and riffle sequences that are maintained in the stream channel. Where a coarse bed load is transported, shallow, braided channels evolve. Deposition of sediment in the channel gives braided streams steep gradients that maintain high flow velocities despite energy losses to friction. Straight channels are controlled by underlying geological structures.

The major fluvial landforms are valleys created by stream incision, floodplains produced by lateral stream planation or fluvial aggradation, natural levees built up by overbank flooding, terraces reflecting disturbances of equilibrium, alluvial fans at the mouths of canyons, and deltas where streams empty into lakes or the sea.

Landscape modification by the wind occurs only in areas of sparse vegetation and dry soils. Wind erosion is seen mainly in the form of deflation, which wears down surfaces composed of loose fine-grained material. Material removed by deflation sifts down elsewhere to form deposits of fertile loess. The vast ergs of certain desert regions form where sand-carrying winds lose velocity; they are composed of varying types of transverse and longitudinal sand dunes whose configuration reflects the wind patterns and the sand supply. Relict ergs are a conspicuous indication of major climatic changes through time.

REVIEW QUESTIONS

1. What measurable properties of streams appear to vary in a regular manner related to stream "order"?
2. What are the forces involved in stream flow?
3. What are the relationships between stream energy, stream velocity, and stream discharge?
4. In what three ways are the particles of a stream's solid load moved? Where does a stream's dissolved load come from?
5. How can a sediment-laden stream increase its own energy at a particular point? What happens when a stream has more energy than required to move its sediment load?
6. What is the meaning of "quasi-equilibrium" in reference to streams?
7. What three factors are most often responsible for knickpoints in stream profiles?
8. How is pool-and-riffle spacing related to the wavelength of stream meanders?
9. How can human activities cause a meandering stream channel to develop a braided pattern instead?
10. What is the most common landform created by fluvial processes?
11. What fluvial landforms could you expect to see on a floodplain?
12. In general, what do fluvial terraces indicate about stream behavior?
13. Where are distributary stream channels most often encountered?
14. In what specific ways are fluvial and aeolian processes similar? In what ways do they differ?
15. What chain of events is necessary to produce thick deposits of loess?
16. How are sand ripples, dunes, and ergs related?

APPLICATIONS

1. Most urban areas are near a permanent stream. Look at the largest stream close to your campus. How has it been modified by human activity? What were its natural characteristics in the prehistoric period?
2. Looking at the same stream, where are current areas of bed or bank erosion? of deposition? Can you make any general statement about stream behavior: (a) upstream from your city? (b) in the urban area? (c) downstream from the urban area?
3. How much have lower order streams that enter the main stream (in question 1) changed throughout historic time? Can you see evidence of recent progressive change? What explains the tendencies observed?
4. To maintain constant stream velocity (and therefore transporting power), a decrease in stream depth must be offset by a change in one of the other factors affecting flow velocity. What sorts of changes could compensate for a decrease in stream depth to keep flow velocity constant?
5. Which requires a greater expenditure of en-

ergy—a gradient decrease produced by stream incision or a gradient decrease produced by increased stream sinuosity (increased lateral planation)?

6. The lower Mississippi River has followed several different courses to the Gulf of Mexico. Although the old channels have been silted in, the different locations of the ancient streams are clearly evident. What would be the principal evidence of the Mississippi River's former courses?

7. Why are the boulders found on some alluvial fans much larger than those of alluvial fill terraces?

8. Under what conditions would river deltas grow? shrink? be in a steady state?

FURTHER READING

Belt, C. B., Jr. "The 1973 Flood and Man's Constriction of the Mississippi River." *Science,* 189:4204 (August 1975): 681–684. This article analyzes the effects of human modification of the Mississippi River channel, which caused a flow of only moderately high magnitude to produce a flood of catastrophic proportions.

Chorley, Richard J., ed. *Water, Earth, and Man.* London: Methuen & Co. (1969), 588 pp. There are several chapters on channeled flows, river regimes, and floods in this collection of articles on water in all its forms of occurrence, written by experts on the various topics.

Crickmay, C. H. *The Work of the River: A Critical Study of the Central Aspects of Geomorphogeny.* London: Macmillan Co. (1974), 271 pp. This is an unusually interesting analysis of fluvial and slope processes, written in an engaging manner. The treatment is advanced but nonmathematical.

Gregory, K. J., and **D. E. Walling.** *Drainage Basin Form and Process.* London: Arnold (1973), 456 pp. This is the most detailed current treatment of fluvial processes, reflecting English work in the field and updating Leopold, Wolman, and Miller. Most examples are from the British Isles.

Kemper, J. P. *Rebellious River.* Boston: Bruce Humphries, Inc. (1949), 279 pp. The author recounts the story of early attempts to control the Mississippi River, spotlighting the errors made from a lack of understanding of a complex physical phenomenon.

Leopold, Luna B. *Water: A Primer.* San Francisco: W. H. Freeman & Co. (1974), 172 pp. This is a simple but technically sound introduction to the general principles of hydrology and the action of water on slopes and in channels, by the former chief hydrologist of the U.S. Geological Survey.

———, **M. Gordon Wolman,** and **John P. Miller.** *Fluvial Processes in Geomorphology.* San Francisco: W. H. Freeman & Co. (1964), 522 pp. A landmark when first published, the book presents the findings of several decades of research on fluvial processes by U.S. Geological Survey field workers. It is somewhat technical but highly readable and most informative.

Morisawa, Marie. *Streams, Their Dynamics and Morphology.* New York: McGraw-Hill Book Co., Earth and Planetary Science Series (1968), 175 pp. This brief paperback presents most of the fundamentals of fluvial processes in nontechnical terms in a concise format.

———, **ed.** *Fluvial Geomorphology.* Binghamton State University of New York: Publications in Geomorphology (1974), 314 pp. This is a collection of articles on topics of current interest assembled for a 1973 symposium on fluvial geomorphology. The subject matter is diverse, including an interesting contribution on the greatest flood known, which created Washington's Channelled Scablands.

Russell, Richard J. *River Plains and Sea Coasts.* Berkeley and Los Angeles: University of California Press (1967), 173 pp. Many interesting facets of stream behavior are explored, with most examples being drawn from observation of the Mississippi River—especially good on floodplains and deltas.

Schumm, Stanley A. *River Morphology.* Strouds-burg, Pa.: Dowden, Hutchinson and Ross (1972), 429 pp. This is a collection of influential articles on fluvial processes published between 1850 and 1971. Some are technical, others are in the explanatory-descriptive vein. Each constitutes a major contribution to fluvial studies.

————. *The Fluvial System.* New York: Wiley-Interscience (1977), 338 pp. An advanced treatment of fluvial processes by a leading researcher; offers many original insights.

CASE STUDY

Water Management on the Mississippi

Historically, the best way to avoid the floods of the Mississippi River was to move to high ground. But as the valley became more settled, people were no longer content to adjust their affairs to the natural workings of the river. So the goal became to eliminate nearly all floods.

Until recently the principal defense against floods was the construction of artificial levees atop the natural levees formed by the river. However, confining the river and preventing its spread onto its normal floodplain during peak discharge raises the river height. A break in the levee under those conditions can produce a flood of disastrous proportions.

In the 1930s, the opening of the Bonnet Carré spillway just north of New Orleans, Louisiana, inaugurated a new approach to flood control on the lower Mississippi. The spillway is a controllable outlet 3 km wide and 7 km long leading from the Mississippi above New Orleans to the lower level of Lake Ponchartrain. During periods of high discharge, river water can be diverted to the lake, which eases the pressure downriver at New Orleans and in the lower delta. The Bonnet Carré spillway has been used several times since the 1930s, averting floods on each occasion.

By 1950 hydrologists and engineers reported that increasing volumes of flood water were flowing through the Old River channel into the Red and Atchafalaya rivers. The Atchafalaya River represents a much shorter and steeper route to the Gulf, and it threatened to divert much, if not all, of the Mississippi River into its own channel. So the Old River Diversion Control was built in the late 1950s, allowing about one-third of the Mississippi to flow into the Atchafalaya at any time. However, during periods of very high water, not all of the flow could be led past New Orleans safely, so the Morganza floodway was also built to divert more floodwater from the Mississippi into the Atchafalaya basin.

In 1973 residents along the river knew high waters were coming months in advance. The question was: How high would the water rise? The flood story of 1973 actually began in late 1972. Rainfall had been particularly high in the central Mississippi Valley during the summer and fall. Farther south, rainfall became heavy in the late fall and early winter. By winter, the lower Mississippi was unusually high and there was already talk of possible spillway openings. During the spring, excessive rain continued to fall over the entire basin. The water came dangerously close to the tops of levees. In February the Bonnet Carré spillway was opened, and in mid-April plans were made to open the Morganza floodway for the first time.

There is no perfect flood control system—relieving pressure in one area means adding pressure to another. Opening the Morganza floodway appeared to be the way to protect the greatest number of lives and property, but meant more water downstream in the Atchafalaya River. There was concern that sediment brought by silt-laden Mississippi River waters might fill in the shallow lakes and swamps of the Atchafalaya basin. Several hundred people, as well as livestock, were evacuated from the Morgan City area in the Atchafalaya basin. Oil and gas wells were flooded and their production was temporarily stopped. Farmland was under water, so crops of cotton and corn could not be planted.

Opening the Bonnet Carré spillway affected the ecology of Lake Ponchartrain, for a time at least, by adding sediment and fresh water that changed the salinity and damaged the oyster population. However, there was some speculation that over time the ecology of the lake would be improved by the flushing action of the water, because it washed out debris and water-clogging vegetation and brought in a fresh supply of nutrients.

The opening of floodways such as the Morganza and the Bonnet Carré prevented the main river from flooding. But water in the main stream was so high that the Mississippi's tributaries were unable to flow into it. Consequently, they overflowed locally, forming "inland seas" of backwater flooding, mainly in Mississippi and Louisiana. Farther north, the flooding was worse, the heaviest occurring around St. Louis, Missouri. In some places, levees broke or washed out, and the water poured through them; in other places, the levees held but the water cascaded over them. Near St. Louis, people fled their homes, moved back to their sodden, muddy houses when the rivers dropped, and then had to leave again, and yet again, as new crests came down the river.

The rains and the flooding finally stopped in early summer. The flooding had lasted months, and four separate flood crests had flowed down the Mississippi. Parts of Kansas, Iowa, Illinois, Missouri, Kentucky, Tennessee, Arkansas, Mississippi, and Louisiana had been declared disaster areas.

Because the waters subsided so slowly, it was too late that season to plant a cotton crop in the fertile delta area of Mississippi. Soybean, corn, and sugarcane crops were also affected because planting was delayed. The heavy losses in food crops and livestock led to higher food prices throughout the country. Thousands of homes had to be rebuilt, roads and levees repaired, and massive clean-up projects undertaken.

But even with the great losses, many engineers believed that the flood control measures proved as effective as could be expected. Most of the flood warnings came early enough that people could be evacuated, and loss of life attributed to the flood was relatively small.

Lower Mississippi River Flood Control Plan. (Vantage Art, Inc. from U.S. Army Corps of Engineers, Department of the Army, Lower Mississippi Valley Division)

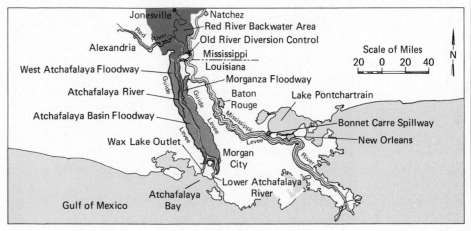

Poster by Ed Mell, 1982. (Art Resources Gallery, Denver, Colorado)

From the window of an airplane, one can hardly fail to be impressed by the geometrical patterns made by landforms in many regions. Seen from a cloud-free flight across the United States, the linear mountains and basins of Nevada, the layered tablelands and sinuous lines of cliffs of central Utah, and the seemingly endless winding ridges of Pennsylvania stand out as highly distinctive forms. The scenic character of such areas results from their peculiar geologic structures. Resistant rocks project boldly above their surroundings, and weaker rocks have been worn to low relief. Jointing often controls drainage patterns and relief forms. Faults and folds produce particular types of landforms that are different from those developed in undisturbed rocks, and sedimentary rocks that are flat-lying display landforms unlike those seen where rock strata have been deformed by tectonic activity.

The distinctive "personalities" of many local landforms, as well as of entire regional landscapes, are largely results of the types and arrangement of the rock masses present. The varying architectural elements of the earth's crust were introduced in Chapter 12. Now we must explore their specific effects in real landscapes.

15
Geologic Structure and Landforms

Differential Weathering and Erosion
Shale
Sandstone
Limestone
Igneous and Metamorphic Rocks

Volcanic Landforms
Basaltic Volcanism
Andesitic Volcanism
Volcanism and Humanity

Crustal Deformation and Landforms
Broad Crustal Warping
Mountain Building
Fault-Controlled Landforms
Fold Structures and Landforms

DIFFERENTIAL WEATHERING AND EROSION

Figure 15.1 illustrates the complexity of the terrain in a small portion of the eastern United States. The variety of relief features seen in this region results from *differential weathering and ero-*

Geologic structure influences landscapes to varying degrees, depending upon rock type, the geometry of rock masses, and the dominant gradational processes. Where vegetation and soil are poorly developed, as in the arid west of the United States, the effect of geologic structure becomes paramount, determining even the minor details of slope form.

PLATE 11

Figure 15.1 This *physiographic diagram* by the master cartographer Erwin Raisz reveals how geologic structure affects the scenery of the lower Hudson River region. North is at the top. In the upstream area, the cross section shows that gently dipping resistant sandstones form the Catskill Mountains, with metamorphic rocks underlying the Hudson River Valley. In the cross section farther south, the highlands are shown to be produced by granite rocks (darkest symbol), with basaltic lavas making ridges in the lowland west of the river. (From the Report of the International Geological Congress, Washington, D.C., 1933)

sion, which refers to the varying responses of different rock types to the weathering processes that fragment rocks and the erosional processes that remove the fragmented material. In the lower Hudson River Valley portrayed in Figure 15.1, differential weathering and erosion are possible because various disturbances of the earth's crust have created complex geological structures, allowing many different rock types to be exposed. Farther west, in the Catskill Mountains, the rock strata are continuous and almost horizontal, so the landscape has a more uniform character. Figure 13.18 (page 363) shows differential weathering and erosion on a smaller scale. In arid regions, even minor variations in rock resistance are revealed, for the processes of weathering and erosion are even more selective where there is no cover of vegetation or soil.

The layering of sedimentary rocks—principally shale, sandstone, conglomerate, and limestone—provides the best opportunity for differential weathering and erosion, as each rock stratum has its own physical and chemical characteristics that may be quite different from those of adjacent strata.

Shale

Shale is mechanically weak, tending to fragment into thin flakes. When chemically weathered, it produces clay soils that are slow to accept water. The resulting high runoff during rainstorms produces an above average number of stream channels per unit of area. Shale is, in fact, the most erodible of all rock types. Because of this, it generally evolves into lowlands or

Figure 15.2 Where shale is interbedded with limestone, sandstone, or conglomerate, the shale erodes to a gentler slope below cliffs of the more resistant rock type. Erosion of the shale slope undermines the resistant caprock, causing its edge to retreat by repeated small-scale collapses. This sandstone cuesta promontory, with underlying shale, is seen on the west flank of the San Rafael Swell in eastern Utah (see Figure 15.23, p. 425). (T. M. O.)

dissected **slopes fringing higher hills that are composed of other material (Figure 15.2). In dry regions, the erosion of weakly cemented shales often creates a maze of closely spaced ridges and ravines known as** *badland* **topography. An excellent example of such a landscape has been set aside as South Dakota's Badlands National Monument, but similar topography occurs in many places west of the Rocky Mountains.**

Sandstone

Sandstone can be physically weak or strong depending upon the substance that cements its grains together. It is often porous, permitting water to soak into it or into the sandy soils that form on it. This reduces runoff and erosion. Since it erodes slowly, sandstone frequently forms ledges or cliffs (Figure 15.3) or hills rising above areas of shale or limestone. Conglomerates, which are composed of cemented gravel or cobbles, are the least common of the clastic sedimentary rocks, but when well-cemented

are even more resistant than sandstone and almost always give rise to bold ledges.

Differential weathering and erosion in areas of sandstone and shale bedrock often produce a series of cliffs and slopes. These vary in form depending upon the tectonically produced tilt, or *dip*, of the strata, as shown in Figure 15.4. Where the dip is gentle, a *cuesta* is formed. This is a plateau capped by a thick layer of resistant sedimentary rock that gradually dips below the land surface. Every cuesta has a steep erosional *scarp* and a gently dipping *backslope* capped by the resistant layer of rock. The cuesta scarp retreats as a consequence of erosion of the weak rock (usually shale) under the caprock; this leads to collapse of the edge of the caprock, due to lack of support.

Flat-topped *mesas* (Figure 15.3) may be detached from the cuesta as its scarp retreats in the direction of bedrock dip. Subsequent erosion eventually reduces mesas to smaller *buttes* that no longer preserve extensive flat summits. Where the rock strata dip steeply, cuestas are replaced by *hogback ridges* (Figure 15.4 and 15.5). The steep slopes that meet at the sharp crest of a hogback are similar in angle, but one slope

Figure 15.3 Sandstone mesas and buttes dominate the landscape in Monument Valley in southeastern Utah. Erosion of shales under the sandstone caprock causes collapse of the caprock, maintaining the vertical cliffs. Where the shale is not exposed at the land surface, the sandstone is eroded into rounded forms and no vertical cliffs are present. Dunes of sand in the center of the photo are derived from breakdown of the sandstone, which fragments during rock falls. (John S. Shelton)

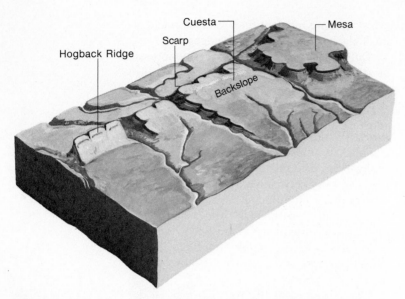

Figure 15.4 *Mesas, cuestas,* and *hogback* ridges are typical landforms that can be produced by differential erosion in regions where a resistant caprock overlies less resistant layers. A flat-topped mesa is formed where the rock strata are horizontal. A cuesta is formed where the strata dip at an angle of a few degrees, so that a cuesta possesses a steep face, or *scarp,* and a gently dipping *backslope.* A hogback ridge is formed where the strata dip steeply. One slope consists of the dipping caprock, and the other is the erosional scarp that cuts across the caprock and the underlying layers. (John Dawson)

(a)

cuts across the edges of rock layers while the other is the top of the dipping bed that creates the ridge.

The arid Colorado Plateau region of the western United States, including parts of Colorado, Utah, Arizona, and New Mexico, is famous for its sharply defined cuestas, mesas, buttes, and hogbacks of naked rock. Forms that are similar, but softened by a forest cover, are present in the eastern United States. In Kentucky and Tennessee, cuestas, buttes, and mesas are important landscape elements, and in large areas of Pennsylvania, Virginia, and Arkansas, hogbacks dominate the scenery. Any cuesta or linear ridge formed by a dipping layer of rock is known as a *homoclinal* (inclined in one direction) structure.

Figure 15.5 (a) This series of homoclinal ridges in eastern Pennsylvania has been developed by the erosion of anticlinal and synclinal folds. The resistant sandstone layers form *hogback* ridges rising above lowlands eroded in shale and limestone. At the left is a *watergap* in which the Susquehanna River cuts through a steeply dipping layer of sandstone. (John S. Shelton)

(b) In more arid environments, where a protective cover of soil and forest vegetation are absent, homoclinal ridges assume a more serrate form. This sandstone hogback north of the San Gabriel Mountains in southern California resembles a row of shark's teeth. (T. M. O.)

(b)

Limestone

Limestone is peculiar in that its dominant mineral, calcite ($CaCO_3$), dissolves in slightly acidic water. Soil moisture contains enough dissolved CO_2 from the atmosphere and from plant decay to become a weak carbonic acid solution, and it also absorbs organic acids released by plants. Moving ground water in the phreatic zone is also slightly acid and therefore corrosive to limestone. The solution (dissolving) of limestone is concentrated along joints and bedding planes, where water can circulate freely and be replenished before it becomes saturated with calcium bicarbonate, the product of limestone solution. Where joints intersect, the solution process is especially rapid, resulting in underground voids of some size. These may enlarge upward to the surface, or the surface rocks may collapse into them. In either case, a surface depression known as a *sinkhole,* or merely a "sink", results. In the scientific literature these are termed *dolines,* a word borrowed from the Serbo-Croatian language, reflecting the importance of limestone landforms throughout Yugoslavia.

Some areas of Indiana have as many as 300 sinkholes (dolines) per sq km (Figure 15.6). Portions of Kentucky and Florida are likewise thoroughly pitted with dolines, which often divert small surface streams to underground routes. Dolines may hold small ponds or marshes, but where the development of dolines is well advanced, surface water can be quite scarce, especially if a dry season occurs.

The most unusual feature created by limestone solution is the *underground cavern.* Vast subsurface voids can form wherever a thick mass of limestone is present at or below the water table in an area of high local relief. Where there is significant topographic relief the water table slopes downward toward the major streams, resulting in active lateral flow of groundwater. This prevents saturation of groundwater with calcium bicarbonate and is necessary for large-scale solution to continue. Accessible limestone caverns are formed in two stages: a period of major solution at or below the water table, and a second period after the water table is lowered, allowing air to enter the cavern.

Of the 48 conterminous states, all but Rhode Island, Vermont, New Hampshire, and Louisiana contain limestone caverns, and some states have hundreds. Some caverns have more than 100 km (60 miles) of passages, including rooms tens of meters high and thousands of square meters in area. Many limestone caverns occupy several levels vertically, and are beautifully decorated with hanging icicle-like *stalactites* and more massive upthrusting *stalagmites,* both of which are composed of calcium carbonate (Figure 15.7). These and a variety of other types of cave deposits, known collectively as *speleothems,* are formed where downward percolating water, enriched in both carbon dioxide and calcium bicarbonate, seeps into the cavern. Some of the CO_2 gas escapes from the water, and this decreases the water's ability to retain calcium bicarbonate, which is precipitated as either calcite or aragonite, two slightly different crystalline forms.

Variations in limestone solubility and the effects of different climates and geologic structures cause limestone landscapes to vary widely in appearance. Nevertheless, all landforms produced by solution of limestone are known as *karst* landforms. The name is derived from the Karst region of northern Yugoslavia, where solution features were first described in detail in 1893.

Because limestone varies in strength and massiveness (degree of jointing and bedding) it can be either a resistant or a weak rock. In humid areas it may dissolve to form lowlands surrounded by higher land on other rock types. Hard limestones can create cliffs and even mountain peaks, such as those in the Alps and Himalayas. In arid regions there is little water available to dissolve limestone, which then acts as a resistant rock, forming cuestas, mesas, and hogbacks. In the humid tropics, where organic acids are abundant, aggressive solution of limestone along joints produces *cockpit karst,* consisting of crater-like dolines (cockpits) encircled by sharp-crested ridges; *cone karst,* composed of clusters of beehive-shaped hills; and *tower karst,* the vertical-sided pillars seen in stylized form in Chinese and Japanese paintings. Some of these forms are shown in Figure 15.8.

The free circulation of groundwater through

(a)

(b)

(c)

Figure 15.6 (a) Limestone is susceptible to chemical erosion by slightly acidic water. In humid regions where the bedrock consists of limestone, subsurface chemical erosion creates surface sinkholes. This landscape in southern Indiana is dotted with sinkholes, some of which contain small ponds. No surface streams are present in such areas; all drainage is underground. (John S. Shelton)

(b) Occasionally artificial road cuts in limestone areas intersect a clay-filled sinkhole or *doline*, as at this location south of Louisville, Kentucky. Note the small-scale solution effects also visible on the limestone outcrop. (T. M. O.)

(c) In karst areas, sizeable streams occasionally issue from underground passages developed by limestone solution. This stream issues in Cedar Sink, a collapsed portion of a cavern in the Mammoth Cave area of Kentucky. The stream is visible for only a hundred or so meters before it resumes an underground course. (T. M. O.)

Figure 15.7 This narrow passageway in New Mexico's Carlsbad Caverns shows the classic features produced by solution in thick beds of jointed limestone. The open passage was dissolved out along the plane of a vertical joint by CO_2-laden groundwater when the water table was much higher than at present. When the water table fell below the level of this passage, air entered and soil moisture dripped from the cavern roof. This moisture carries calcium bicarbonate dissolved from the rocks above. Loss of CO_2 from water slowly dripping from the cavern ceiling causes precipitation of some calcite, which accumulates in downward-growing, icicle-like *stalactites*. Water splashing on the cavern floor precipitates more calcite, producing upward-growing, columnar *stalagmites*. In time the stalactites and stalagmites shown may join to form columns that reach from the floor to the ceiling of the cavern. (T. M. O.)

Figure 15.8 This scene in the south of China epitomizes the beauty of karst landscapes in limestone areas of strong relief in a humid tropical climate. This is an example of *tower karst*. (From *Karst in China*, Beijing)

Figure 15.9 This funnel-shaped sinkhole in Winter Park, Florida developed by rapid collapse over a week-long period in 1981. The dimensions of the crater are about 115 by 100 meters (380 by 330 feet). The collapse involved unconsolidated material atop cavity-riddled limestone, and was triggered by a fall in the water table as a consequence of groundwater pumping and a sustained drought. As the gradual decline in the water table produced a sudden collapse, the phenomenon epitomizes a geomorphic threshold. (George Remaine, UPI)

fer texturally and chemically, so that the entry of water and the nature of chemical reactions vary widely. Because there are no bedding planes in plutonic and gneissic rock, jointing becomes the dominant factor in surface form. Rock masses with high joint density are weathered and eroded downward, whereas masses broken by few joints remain high-standing. Sometimes this leaves great spires or domes of unjointed granitic or gneissic rock jutting above surrounding jointed rock that has been deeply weathered and eroded. Such a landscape is displayed around Rio de Janeiro, Brazil (Figure 15.10).

The decay of granitic rock along joints may temporarily isolate less rotted *corestones* below the soil surface (Figure 11.1). Downward erosion by surface wash sometimes exposes these corestones as scatterings of rounded boulders, or as boulder heaps known as *tors*. Hillsides from Los Angeles to San Diego in California and on into Mexico are mantled with granitic bounders, as are many desert landscapes in the southern portions of both California and Arizona (Figure 15.11). Gneissic rock often forms

the voids remaining when limestone is dissolved give rise to peculiar hydrologic problems. Springs are active or inactive periodically; dolines diverting surface streams to underground paths sometimes become clogged with debris, causing local flooding; subsurface drainage routes change inexplicably, complicating liquid waste disposal and the supply of household water; artificial reservoirs fail to fill as impounded water leaks away through solution passages; and collapses of the ground surface occur with no warning, sometimes in urban areas (Figure 15.9).

Igneous and Metamorphic Rocks

Differential weathering and erosion are less conspicuous in igneous and metamorphic rock than in sedimentary rock. Nonetheless, there are significant variations in resistance in these rock types. Igneous and metamorphic rocks dif-

Figure 15.10 The spectacular rock cones, or "sugarloaves," of Rio de Janeiro, Brazil, result from weathering and erosion in gneissic rock with variable jointing. Unjointed masses resist subsurface chemical weathering and are left standing above surrounding decayed and eroded rock that is densely jointed. (© Louis Villota/The Image Bank)

Figure 15.11 The jointing of granitic rock often causes it to weather into masses of giant boulders that can dominate the scenery in arid and semiarid regions, as in this view in the Mojave Desert of southern California. In a more humid environment such boulders would be present below ground level, obscured by a cover of soil and thoroughly decomposed rock. (T. M. O.)

pinnacles, but neither it nor any other non-granitic rock types forms masses of loose boulders that blanket large areas.

In addition to rock type, the physical arrangement of rock masses plays an essential role in the appearance of landscapes. The varying configurations of rock bodies is the result of past tectonic activity, including volcanism and crustal compression and extension. We shall turn now to this aspect of structure as a control of landform development.

VOLCANIC LANDFORMS

The relationship of volcanism to lithospheric plates was outlined in Chapter 12. There we saw that two dissimilar types of volcanic activity are easily distinguished—non-explosive basaltic eruptions where crustal plates are pulling apart, and explosive eruptions of silicic tephra (volcanic ash) and lava along lines of plate convergence and subduction. Different landforms

are associated with these two contrasting eruptive styles, and, in fact, indicate the degree of volcanic hazard in the locations where they occur.

Basaltic Volcanism

Where lithospheric plates are being pulled apart, basic magma from the earth's mantle rises to fill the opening between them. This erupts at the surface as basaltic lava. The rising magma creates the oceanic ridges, which occasionally push above the sea surface, as in Iceland. Basic magma also erupts through sea-floor fractures, where portions of oceanic plates are sliding past one another. The ocean floor is studded with volcanoes that are now inactive. Many of these appear to have been carried away from the ocean ridge systems by sea-floor spreading. A peculiar form known as the *guyot* is discussed in Chapter 18.

Not all active oceanic volcanoes are associated with plate boundaries. Some lie at the ends of linear chains of submarine mountains that can be followed for a thousand kilometers or more. Such volcanoes mark the position of so-called "hot spots" in the earth's mantle, where rising plumes of magma seem to remain fixed in place while lithospheric plates move over them. The extinct volcanoes that form the majority of such mountain chains formed above and then moved beyond the hot spot, so that the trend of the submarine mountain chain records the direction of plate motion. The Hawaiian chain is the best known example. Its volcanic rocks increase in age with distance from the hot spot presently situated under the southeastern portion of the island of Hawaii itself (Figure 15.12).

Volcanic eruptions on the sea floors build *shield volcanoes*, which have gentle slopes and broad rounded summits (Figure 15.13). Hawaii's Mauna Kea and Mauna Loa are the world's largest mountains, rising some 9,000 meters (30,000 ft) above the sea floor. Their slopes rarely exceed an angle of 10 degrees. All shield volcanoes, on the land and in the sea, are composed of basaltic lava. The Hawaiian shield volcanoes clearly display two contrasting

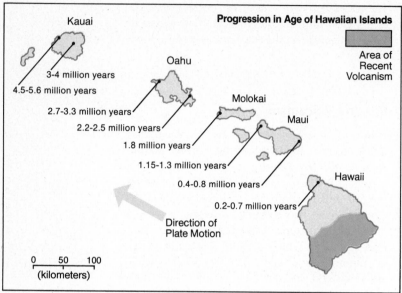

Figure 15.12 The ages of the Hawaiian Islands, as determined by the potassium/argon radiometric dating method, show a general progression from old to young moving southeastward along the chain. These observations were not explained until the development of the theory of plate tectonics, which attributes the formation of the Hawaiian Islands to a plume of volcanic activity fixed in the earth's mantle. As the Pacific plate drifts over the location of the plume, periods of volcanic activity cause the formation of islands of volcanic rock. (Andy Lucas after data by Ian McDougall, *Nature Physical Science*, Vol. 231, 1971)

types of basalt (Figure 15.14). The hottest flows issue in the form of very fluid *pahoehoe* (pāh-ho-ay-hó-ay) lava, which solidifies with a satin-smooth skin that is often conspicuously wrinkled by the continued flow of the lava underneath. Somewhat cooler basalt issues as slag-like rough-surfaced *aa* (ah-ah) lava. Pahoehoe lava is made fluid by the gases still dissolved in the lava. When these gases are lost by excessive agitation before eruption or within a flow, pahoehoe alters to aa lava.

In addition to producing enormous volumes of free-flowing basaltic lava, nonviolent eruptions of the type seen in Hawaii and Iceland are still forceful enough to hurl solid particles hundreds of meters into the air. These rain down to form deposits of *pyroclastic* ("fire-broken") debris. This includes solidified clots of lava and fragments of rock torn loose during the eruption. The larger particles, or *scoria*, (Figure 15.15) fall close to the vent and build up a *scoria cone* or *cinder cone*. Sometimes the magma is so gaseous that it rises to the surface as a froth, like the foam on beer. This solidifies as *pumice*, an extremely porous rock that is light enough to float on water.

Basaltic eruptions similar to the oceanic type also occur on the continents. Several land areas have been buried by enormous outpourings of very fluid basaltic lava, which has accumulated, flow upon flow, to thicknesses exceeding a thousand meters. Individual flows of these continental *flood basalts* have spread more than 160 km (100 miles) from their source, covering thousands of square kilometers to depths of tens of meters. The volumes of such flows measure in the tens to hundreds of cubic kilometers. A lava flood of this type occurred in Iceland in 1783, producing some 12 cu km (3 cu miles) of new basaltic rock.

Areas of continental flood basalts include the Columbia Plateau region of Washington and Oregon and the Snake River Plain of Idaho, as well as portions of India, Brazil, Patagonia, South Africa, and Antarctica. Some of the lava

Figure 15.13 Basaltic *shield volcanoes* such as Hawaii's Mauna Loa (horizon) have very gentle slopes, making their true sizes hard to appreciate. Mauna Loa rises almost 3,000 meters (9,800 ft) above the area in the foreground. In the middle distance is the crater of Kilauea volcano. The cinder cone and lava lake in the foreground formed during a volcanic eruption in 1959. (T. M. O.)

Figure 15.14 **(right)** Due to the gases dissolved in it, *pahoehoe lava*, shown here at Kilauea crater in Hawaii, is much more fluid than *aa* lava even though the chemical composition is the same. Note the thin layer of solidified crust on the molten rock. When the lava solidifies, the surface is left in the form of smooth billows or ropes. (Frank Sojka © Island Pictures, Honolulu, Hawaii)

(bottom) This advancing *aa* lava flow on Hawaii is characterized by chunky, angular blocks mixed with still-molten lava. When an *aa* flow solidifies, an almost impassable field of jagged lava is formed. (Charles A. Wood)

417

Figure 15.15   A *scoria cone*, produced by the fall of pyroclastic materials that have been hurled into the air by the force of escaping magmatic gases and steam during a volcanic eruption. The example here is Paricutín volcano, which rose from a Mexican cornfield in 1943 and erupted sporadically until 1952. (R. Segerstom/U.S. Geological Survey)

Figure 15.16   The Columbia Plateau in the state of Washington consists of superimposed layers of basaltic lava having an aggregate depth of about 1,000 meters (3,300 ft). The distinctive columnar jointing of the lava is evident wherever it is trenched by an erosional canyon, as in this view at Grand Coulee. Such masses of superimposed lava flows occur in areas of crustal tension and rifting. (T. M. O.)

floods on the land may have occurred during the breakup of continents as new oceans were being born. The lava flows of Triassic age that form the Hudson River Palisades and the Watchung Mountains of New Jersey probably originated in this way when North America began to separate from West Africa to initiate the present Atlantic Ocean, some 200 million years ago.

All areas of flood basalts on the land have been somewhat dissected by stream erosion. The resulting canyon walls expose the separate flows clearly, one atop the other, as can be seen along portions of the Columbia and Snake rivers in Washington and Idaho (Figure 15.16). Smaller basaltic lava fields with cinder cones and shield volcanoes are widely scattered in areas of recent tectonic activity, marking localities where fractures have penetrated through the earth's crust to the mantle beneath.

Andesitic Volcanism

Despite the enormous volumes of lava generated by basaltic volcanism on the continents and ocean floors, the eruptions are seldom violent or dangerous. But where crustal plates are converging and subduction zones are present, the volcanism is quite different. Subduction-related volcanism generates much smaller amounts of lava, but releases pent-up gases so explosively that vast areas are denuded, incinerated, or suffocated with tephra (ash). The volcanoes themselves are occasionally decapitated by the eruptive events.

The force of some volcanic eruptions is incredible. When Krakatau, an island volcano between Java and Sumatra, exploded in 1883, the sound carried 3,000 km (1,900 miles) to Australia. Krakatau itself vanished. Ash falls from a similar explosion at Santorin in the Mediterranean Sea may have destroyed the ancient Minoan civilization of Crete about 3,500 years ago. The explosive eruption of Mount St. Helens in the state of Washington in 1980 leveled 400 square km (150 sq miles) of forest, and removed 2 cu km (.75 cu miles) of the volcano itself. A similar event at El Chichón in Mexico in 1982 ejected so much tephra into the upper atmosphere that the climate of the northern hemi-

Figure 15.17   (a) *Strato-volcanoes*, such as Mount Hood in Oregon, are among the most impressive of all constructional landforms. Cones of this type are composed mainly of andesitic lava and pyroclastic material resulting from explosive eruptions.

(b) This volcanic plug dome in eastern California is less than 10,000 years old. Such domes, composed of siliceous dacite and rhyolite, are emplaced in association with extremely violent volcanic eruptions that produce voluminous falls of ash (tephra). The eruptions often include searing nuée ardente blasts that are lethal to all life in their paths. In the foreground is a "coulee" of the same highly viscous lava that forms the dome in the background.

(c) Oregon's Crater Lake is a caldera formed by the collapse of a large strato-volcano during an immense eruption that showered the Pacific Northwest with ash only 6,600 years ago. The resulting crater filled with a lake that is presently about 600 meters (2,000 ft) deep and 10 km (6 miles) across. Portions of the crater rim, such as the high point visible in the distance, rise 600 meters (2,000 ft) above the lake surface. Subsequent volcanic activity has built three new cones in the caldera, one of which protrudes above the lake surface as Wizard Island, visible at the right. (T. M. O.)

(a)

(b)

(c)

sphere seems to have been cooled measurably. Similar eruptions of other volcanoes in the past have lowered the average temperature of the northern hemisphere by about 0.3°C, with the effect lasting 3 to 5 years.

Andesitic volcanism occurs along continental edges where subduction is occurring, as in the Andes Mountains running the length of western South America; in areas of continental collision, as in the Mediterranean region and eastern Turkey; and in island arcs where one sea-floor plate is being subducted beneath another, as in the western Pacific region.

The first products of volcanic eruptions on the land are blankets of tephra and scoria cones like Paricutín, born in a Mexican cornfield in 1943 and built to a height of 400 meters (1,300 ft) in 8 months (Figure 15.15). Long-continued eruptions veneer scoria cones with lava, causing them to become *composite cones*. Andesitic lava flows usually consist of a stiff pasty interior covered by a rubble of hardened blocks, unlike either aa or pahoehoe lava. As lava coats the cone, it is also injected into cracks in the volcano, where it solidifies as lava *dikes*. These reinforce the cone's structure. As the cone grows higher, lava may issue from fissures opening in its sides rather than from the summit crater. Over hundreds of thousands of years giant steep-sided *strato-volcanoes*, composed of both scoria and lava, are constructed, such as Japan's Fujiyama, Washington's Mt. Rainier, and California's Mt. Shasta. Such volcanoes often rise 3,000 meters (10,000 ft) above the surrounding countryside (Figure 15.17). All the earth's large intact strato-volcanoes have been built within the last one million years—some within the last 10,000 years—and every one is potentially dangerous.

A peculiar type of volcano is the so-called *plug dome*. These are seldom of large size, but their eruptions are extremely hazardous. Plug domes are formed when a mass of unusually stiff and highly siliceous magma pushes to the surface. The pasty lava (dacite or rhyolite) congeals as it rises, jamming the surface vent so that enormous pressure builds beneath the volcano. This is released in catastrophic explosions that may hurl large blocks several kilometers. Plug domes can be recognized by the stubby masses of hardened lava that often ring their summits (Figure 15.17).

Associated with the plug dome is the *nuée ardente* (glowing cloud) type of eruption. A nuée ardente is an avalanche of glowing volcanic ash that moves down the flank of a volcano at speeds of more than 160 km (100 miles) per hour. Everything in the path of a nuée ardente is baked to a cinder or set afire instantly, and nearby life that is not incinerated may be asphyxiated by lack of oxygen or mummified by heat. In 1902, a nuée ardente issuing from Mt. Pelée, on the Caribbean island of Martinique, obliterated the town of St. Pierre, snuffing out almost 30,000 lives. Only two persons in St. Pierre survived—one in an underground dungeon.

Many giant strato-volcanoes have satellite plug domes, or have at times behaved like plug domes. They have erupted so violently that their summits have collapsed, producing vast craters known as *calderas*. This occurred in recent eruptions of both Mount St. Helens and El Chichón. If new plug domes do not refill them, these calderas usually hold circular lakes. Crater Lake in Oregon is a particularly beautiful example (Figure 15.17). It partly fills a caldera 10 km (6 miles) wide and initially 1,200 meters (4,000 ft) deep. Mt. Vesuvius, near Naples, was partially destroyed in its great eruption of 79 A.D., which buried the city of Pompeii in volcanic ash. In the last 1,900 years Vesuvius has rebuilt itself out of the caldera created by that eruption.

It is difficult to know whether a volcano is active, dormant (potentially active), or extinct. For example, Vesuvius had been inactive for centuries prior to its eruption in 79 A.D. Crater Lake was created by a violent eruption about 6,600 years ago, but subsequently much of the crater was filled by further volcanic activity. In fact, Wizard Island, at one side of the lake, is the summit of a volcanic cone built within the caldera. In the volcano-studded Cascade Range of Washington, Oregon, and northern California, only Mount St. Helens in Washington and Lassen Peak in California have produced violent eruptions in the last 200 years. Nevertheless, all of the major Cascade volcanoes have displayed some eruptive activity during this period.

The Cascade Range is part of the "Pacific Ring of Fire"—a nearly continuous chain of andesitic volcanoes circling from South and Central America through Alaska, Japan, the Philippines, and New Zealand (Figure 12.19). This gives clear evidence of crustal subduction around nearly the entire rim of the Pacific Ocean. The only interruptions in the volcanic ring are north and south of the Cascade Range. To the north in Canada, and to the south in California and northern Mexico, crustal plates seem to be sliding past one another rather than converging and being subducted.

All volcanoes eventually become inactive. When they do, erosion wears them away. All that is left in some areas are landforms resulting from the differential erosion of the lavas and ash composing the volcano. The most impressive of these forms are *volcanic necks* that rise as pinnacles composed of the lava left in the "throat" of the eroded volcano; *dike ridges* that form free-standing rock walls, often radiating from a volcanic neck; and *table mountains* consisting of winding, flat-topped ridges covered with lava that is more resistant to erosion than the rocks over which it once poured (Figure 15.18).

Volcanism and Humanity

Large volcanoes pose many kinds of hazards. Violent eruptions cause death by baking and suffocation. The force of the 1980 blast at Mount

Figure 15.18 **(top)** Flat-topped *table mountains*, such as this one in southern California, form when the land surrounding a lava flow is removed by erosion. This occurs when lava enters valleys cut in erodible rocks. The lava itself erodes very slowly since its great permeability allows rainwater to soak into it rather than running off over its surface. (T. M. O.)

(left) Shiprock, in northwest New Mexico, rises 430 meters (1,400 ft) and is a striking example of a *volcanic neck*. The original volcano has been completely removed by erosion, and all that remains is the lava that solidified in the subsurface conduit to the former cone. The volcano itself was active about 30 million years ago. Note the extensive volcanic dikes radiating from Shiprock. (John S. Shelton)

Figure 15.19 This lahar deposit, generated by the 1980 eruption of Mount St. Helens (background) in the Cascade Mountains of Washington, is as much as 200 meters (600 ft) in depth. The torrent of mud that created it was boiling hot and vaporized the water of the Toutle River, whose valley it invaded, causing steam eruptions to build pseudo-volcanoes of mud on the surface of the lahar deposit. (T. M. O.)

St. Helens stripped forested hillsides bare, removing even the soil to a depth of over a meter. Volcanic ash can be toxic to humans, animals, and plants, and its weight on roofs can collapse houses. Lava flows have been destructive to towns as well as to agricultural fields. Equally damaging are volcanic mudflows, known as *lahars* (Figure 15.19).These result from a number of causes: rainfall on fresh volcanic ash; lava eruptions under glaciers or snowfields; and slope failure on volcano summits weakened by eruptions, earthquakes, or acidic fumes vented from subsurface magma. Lahar deposits in the valleys surrounding some Cascade Range volcanoes are as much as 300 meters (900 ft) deep.

But there are also positive aspects to volcanism. Fresh mineral-rich volcanic ash and decomposed basaltic lava produce excellent agricultural soils. This is especially important in the wet tropics where leaching of nutrients leaves older soils infertile. The productivity of volcanic soils in Japan, the Philippines, Indonesia, and South America have, in fact, lured dense pop-ulations onto volcanoes, close to the hazards of explosive blasts, lava flows, ash falls, and lahars.

Another important aspect of volcanism is its role in the formation of metallic ores, such as those of iron, copper, lead, zinc, tin, tungsten, nickel, and chromium, as well as gold and silver. Metallic ores are deposited where hot richly mineralized solutions move upward from magma bodies through fissures in overlying older rock. Ore formation does not occur automatically, however; it depends on chemical reactions between the local rock and the invading magmatic solutions. Recent discoveries show that the volcanic vents at sea-floor spreading centers include veritable "sludges" of metallic minerals. It is assumed that these are incorporated into the sea-floor lithosphere and later remobilized into the new magma generated by melting in subduction zones. Thus metallic ores originating on the sea floors are eventually emplaced in the continental crust.

Pockets of subsurface magma can also be made to serve humanity. In a few areas of active, dormant, or extinct volcanism there are such natural phenomena as steam vents, geysers that periodically erupt boiling water, and sulphur springs with their smell of rotten eggs (hydrogen sulfide fumes). They all indicate that there is a subsurface *geothermal energy* source that humans can utilize (Figure 15.20).

To bring geothermal energy directly from the zone of fluid magma that is present everywhere beneath the earth's crust would require wells many tens of kilometers deep. These are costly and hazardous to drill. But the hot springs and steam jets in volcanic regions bring heat energy to the surface naturally. Hot springs occur where groundwater rises to the surface after being in contact with hot rock or steam vented from magma. The steam itself may remain trapped in the subsurface within reach of wells, or it may vent to the surface through rock fissures. Steam from subsurface sources is already used to heat water and generate electricity in Italy, Iceland, New Zealand, and California. As the global energy crisis deepens, people in those regions that have accessible geothermal energy supplies will be fortunate indeed.

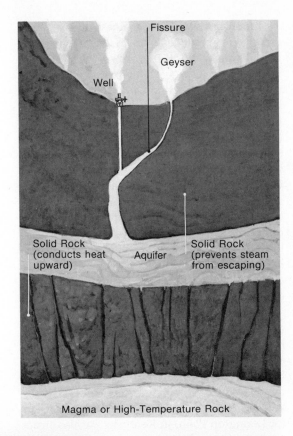

Figure 15.20 This schematic diagram shows the type of geologic structure in steam fields that can be exploited for power generation. A confining layer of rock covers an aquifer permeated with water. Water is heated by contact with hot subsurface rock, and steam is generated. The steam comes to the surface through fissures or is led to the surface through wells sunk into fissures or into the aquifer. The high-pressure steam may then be used to operate electric generators. Geothermal energy is a comparatively clean and inexpensive source of power, but its use is confined to regions having subterranean heat sources. (John Dawson)

The hot water issuing from Castle Geyser in Yellowstone National Park, Wyoming, is driven upward by steam generated from groundwater in fissured hot rock deep below the surface. Seismic studies and continuing measurable uplift in the Yellowstone area suggest a large body of magma unusually close to the surface. (David Miller)

CRUSTAL DEFORMATION AND LANDFORMS

Volcanism is a spectacular aspect of tectonic activity that creates impressive landforms on a variety of scales. However, less visible movements that are very slow but continue over millions of years have had much wider effects on the continental surfaces, having produced the structures that control the evolution of landscapes in many parts of the world. In Chapter 12 we outlined the major geologic structures produced by crustal tension, compression, and shearing due to lithospheric plate motions. The faults, folds, and broad warps of the lithosphere that result from plate motions are very important in the appearance of landscapes. It is now time to examine the consequences of these slow motions of the earth's crust.

Broad Crustal Warping

A geologic map of the eastern United States (Figure 15.21) shows many circular or arcuate patterns. These are outcrops of sedimentary rocks whose varying ages are indicated by the map colors. Strata of different ages outcrop at the surface because the strata are somewhat inclined by gentle deformation and the land surface cuts across their edges.

Figure 15.21 This portion of the geologic map of the United States displays the outcrop patterns of gently deformed sedimentary rocks from Michigan in the north to Tennessee in the south. The colors indicate the ages of the rocks exposed at the surface. Concentric circular patterns indicate structural domes and basins surrounded by cuestas.

Between Lakes Michigan and Huron is the Michigan Basin, which preserves relatively young (Jurassic) rocks. To the south is the Cincinnati Arch, the older (Ordovician) rocks of which have been eroded to form the Bluegrass Basin (Lexington Plain), seen in dark pink. Still farther south in Tennessee is the Nashville Dome, which has been unroofed by erosion to form the topographic feature known as the Nashville Basin. The dark blue area east of the Cincinnati Arch is a basin of coal-bearing Carboniferous rocks in southwestern Pennsylvania and West Virginia.

The linear patterns in the east represent the Appalachian fold belt, succeeded by the crystalline rocks of the Blue Ridge Mountains and Piedmont region of the Carolinas. In the extreme southeast are the young sedimentary rocks of the Atlantic Coastal Plain. (U.S. Geological Survey)

Taking a closer look at one of the arc patterns, the peninsula of Michigan may be thought of as the center of a saucer, part of whose outfacing edge arcs from eastern Wisconsin through northern Michigan and Ontario's Manitoulin Island, and then southward through the Tobermory (Bruce) Peninsula east of Lake Huron. Niagara Falls drops over this same edge after it suddenly turns eastward south of Lake Ontario. This conspicuous edge is a cuesta of resistant rock (in this case limestone), which, along with parallel lowlands on belts of more erodible rock, is the principal indicator of gentle crustal warps. The Paris Basin in France, crossed by the Seine River, is a similar structural basin surrounded by outfacing cuestas (including the Champaign region to the east).

In northern New York we find the opposite configuration—a structural dome surrounded by *infacing* cuestas. The center of this dome stands high in the form of the Adirondack Mountains, an upwarped portion of the continent's ancient crystalline basement. The Nashville Basin in Tennessee and the Bluegrass Basin (or Lexington Plain) in Kentucky are also structural domes surrounded by infacing cuestas; however, they are topographic lowlands because the sedimentary rocks exposed in their centers are more erodible than those forming their rims (Figure 15.22).

All of these humid-region cuestas are unspectacular forested slopes or well-dissected hill masses of low elevation. But farther westward in the semiarid and arid portions of the United States we encounter more sharply defined uplifts, with mountainous cores surrounded by infacing cliff-like cuestas, or even jagged hogback ridges (Figure 15.23). The classic example is the Black Hills Uplift of South Dakota—a partially unroofed granitic dome surrounded by a series of concentric homoclinal structures (Figure 15.22). To the west is the Bighorn Range, also an uplift with a crystalline rock core, and between the two is the Bighorn Basin, a structural depression rumpled into oil-bearing folds, but surrounded by outfacing cuestas. Other classically formed unroofed domes are the San Rafael Swell in central Utah (Figure 15.23), and the Kaibab Arch in Arizona, which is breached by the Grand Canyon of the Colorado River.

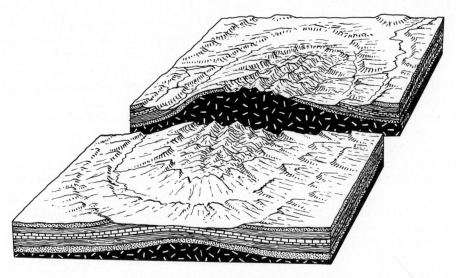

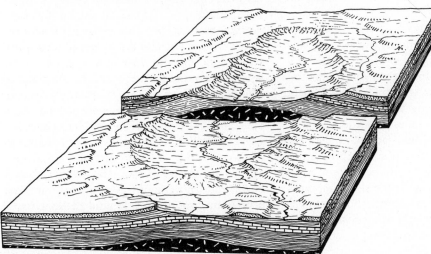

Figure 15.22 These drawings are diagrammatic representations of erosional forms developed in structural domes.

(top) Erosion of a structural dome exposing a core of resistant granitic rock. Encircling sedimentary rocks form hogbacks or cuestas with scarps facing the center of the uplift. Erodible sedimentary rocks form circular valleys between the homoclinal ridges. This diagram is representative of the Black Hills of South Dakota or the Adirondack Mountains of New York.

(bottom) Erosion of a dome structure in which only sedimentary rocks appear at the surface. As the rocks exposed in the center of this dome are relatively weak, a relief inversion results, with the structural high expressed in the landscape as a topographic low. The basin is surrounded by in-facing cuestas of resistant rock and encircling valleys carved in erodible rock. The Nashville Basin of Tennessee and the Blue Grass (Lexington) Basin of Kentucky are similar to this model. (T. M. Oberlander)

Figure 15.23 This LANDSAT false-color image taken from space depicts the structural dome in eastern Utah known as the San Rafael Swell. Clearly visible are the hogbacks and cuesta scarps that encircle the uplift, which exposes only sedimentary rock. The red areas at the left are irrigated cropland. (NASA and the Earth Satellite Corporation)

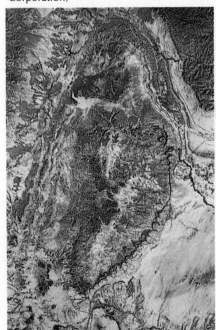

Mountain Building

Although volcanism has created some of our planet's most majestic individual mountains, all sizable mountain ranges, even those capped by volcanoes, have been formed by larger-scale motions of the earth's crust. Nearly all mountain systems began either as deep sediment-filled troughs in the sea floors, close to the margins of continents, or as immense embankments of sediment deposited along the continental margins at the base of the submarine continental slopes. In most instances, thicknesses of more than 10,000 meters (over 33,000 ft) of sediments accumulated along the edges of the continents before being crumpled into the complex structures seen in major mountain ranges.

Until the 1960s it was not agreed how or why these great sediment-filled oceanic troughs, called *geosynclines*, were formed, or why their deposits were eventually crushed together in the process that transformed them into mountains. During the 1960s, however, earth scientists came to understand the subduction process. It is now accepted that the descent of crustal plates in subduction zones first produces oceanic trenches that fill with sediment, and then crushes the sediment mass collected in the trenches and adjacent continental rise areas (Chapter 12). The present ocean trenches and continental rises are, in fact, living geosynclines—the birthplaces of future mountains. The blankets of sediment from which mountains are made also include the underwater continental shelves, though there the sediments are a thinner veneer over older rocks of the continent itself. Although every coast has a continental shelf and a continental rise, future mountain ranges of sedimentary rock will be built only where an ocean trench system is also present and lithospheric plate convergence occurs.

The crushing of geosynclinal sediments by horizontal movements is not the entire explanation for the building of mountain systems. Later, when the rate of compression slackens, strong vertical uplift of the rock that was squeezed in the geosyncline seems to occur. During the slow rise of California's Sierra Ne-

vada range, between 15 and 20 km (9 to 12 miles) of sedimentary rock has been removed by erosion, causing plutonic rocks to be widely exposed at the surface. The prior phase of subduction, folding, and plutonic injection ended some 70 million years ago.

Even greater uplift occurred in portions of the Rocky Mountains, which are exceptional in their mid-continent location, far from known lithospheric plate boundaries. The Middle and Southern Rockies, from Wyoming to New Mexico, appear to have formed almost entirely by vertical uplift, without the prior existence of a deep-water geosyncline.

The most recent period of world-wide mountain building began about 20 million years ago, with the first compression and uplift of the Alps, Himalayas, and coastal ranges of western North America. The geologic structures of the Rocky Mountains and Sierra Nevada are somewhat older, but the present ranges were lifted by the latest mountain-building movements. The structures of the Appalachian Mountains of the eastern United States were formed earlier still, about 250 million years ago, when seafloor spreading drove the African continent against the edge of North America, eliminating an ancient Atlantic Ocean. The present Atlantic began reopening about 200 million years ago, accelerating subduction around the margins of the Pacific Ocean.

It is these large-scale tectonic motions that produce the individual geologic structures we can see expressed in landforms in regions with variable relief. We shall now turn our attention to these local structures and the landforms they create.

Fault-Controlled Landforms

The Great Basin, which is centered in Nevada, has experienced a long and complex geologic history, involving early phases of compression and massive overthrusting (Chapter 12). But it stands now as the world's greatest example of crustal extension and collapse. The extension began some 30 million years ago and continues to the present.

The Great Basin is bordered on its east and

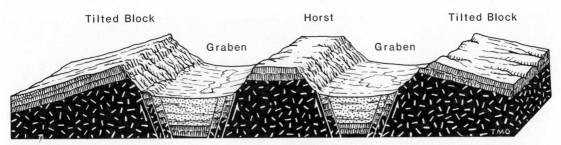

Tilted Block   Horst   Tilted Block

Graben   Graben

Figure 15.24 Structures developed by normal faulting in a zone of crustal extension. In the Great Basin, centered in Nevada, the relative movement of fault blocks is made evident by the tilting of formerly horizontal lavas and volcanic tuffs erupted just prior to crustal rupture. Depressions created by the subsidence of blocks between bounding faults are *grabens*; high-standing blocks bounded by faults on both sides are *horsts*; *tilted blocks* have rotated, sinking on one side and rising on the other to form fault scarps. Often the bounding fault systems are complex, with several separate fault planes present, as illustrated here. (T. M. O.)

west sides by two great tilted blocks, the Wasatch Range of Utah, and California's Sierra Nevada. Each of these presents a bold wall, or *fault scarp*, facing the center of the Great Basin. Between the two are a host of *tilted blocks* with fault scarps evident on only one side, upthrust *horsts* with fault scarps on both sides, and down-dropped *grabens* bordered by horsts or tilted blocks. The range-bounding faults are normal dip-slip faults in which the "hanging wall" resting on the inclined fault plane (Figure 15.24) has moved downward with respect to the "footwall". This type of motion not only creates vertical relief, but also causes horizontal extension of the crust, as indicated in Figure 15.24.

Fault scarps are recognized by the abrupt rise of mountain slopes from a linear base line. The base line of a fault block often cuts cleanly across both resistant and erodible rock. Where the base line is irregular, with spurs projecting different distances, no active fault is present, which is to say that there has been no movement along the fault in many thousands of years. Well-preserved fault scarps are dissected into *triangular facets* (Figure 15.25) that sometimes reveal polished or grooved *slickensided* surfaces caused by the friction of one fault block rubbing against another. Since faults extend downward many kilometers into the earth, they are often marked by hot springs and even volcanic cinder cones. Springs are common along all types of faults due to disruption of groundwater flow.

Thrust faults (Chapter 12) occur in areas of strong compression, and are often associated with folds. Where thrust faults (also known as reverse faults) occur, the hanging wall moves up and over the footwall, causing crustal shortening. Thrust faults can form linear mountain ranges, such as those in Idaho, west of Wyoming's Teton Range, but the scarps of thrust faults never retain the angle of the fault plane itself—to do so they would have to be overhanging.

Overthrusts, in which enormous slabs of rock are pushed horizontally up and over other rocks (Chapter 12), affect scenery mainly through the type of rock they carry into an area. Most have been folded after being displaced, so their landforms are those developed in folded rocks. Being associated with high mountains, their topography has often been retouched by glacial erosion (Chapter 17).

Strike-slip faults (Chapter 12) are those in which portions of the earth's crust slide past one another laterally rather than vertically, the motion being parallel to the fault strike rather than to the fault dip. The relation of strike-slip faulting to sea-floor spreading was noted in Chapter 12. Strike-slip faults are vertical and either straight or gently curving for distances of hundreds of kilometers. They create a distinctive set of landforms known as *rift topography*, which is very well displayed along California's San Andreas fault system. Three features are most evident: linear valleys eroded in crushed rock; elongated hills that are either compressed "welts" of pulverized rock or "slivers" of solid

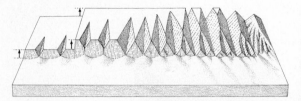

Figure 15.25 This diagram shows the evolution of an escarpment produced by faulting. Two periods of uplift are shown at the left. The forms are successively older toward the right. Note how the growth of valleys cutting back from the fault scarp gradually converts the scarp into a series of spurs ending in blunt *triangular facets*. These become smaller and less apparent with time. Alluvial fans issue from the valleys, becoming larger as the valleys reach farther into the uplifted block. Most real fault scarps resemble the forms on the far right. (T. M. Oberlander)

The east front of the Sierra Nevada is a fault scarp that has been dissected by erosion. Note the straight and regular appearance of the mountain range and its abrupt rise from the neighboring plain. Rising out of the lowland in front of the Sierran scarp is a second, much lower fault scarp. (T. M. Oberlander)

(**bottom**) Some of the terminology used to describe the motion on dip-slip faults comes from mining practices. Mine shafts often follow fault planes because magmatic solutions rise along the fractures, causing useful minerals to be deposited in them. Accordingly, the rocks of the roof of a mine shaft (fault *hanging wall*) may be different from those on the floor, or *footwall,* below the plane of the fault. In the diagram the hanging wall is composed of limestone and the footwall is basalt, with the mineralized fault zone appearing in a lighter color (inset).

Where the hanging wall has dropped relative to the footwall, the fault is a *normal fault* produced by extension of the crust. Where the hanging wall has pushed up relative to the footwall (as here), the fault is a *reverse fault* or *thrust fault* created by compression of the crust. (T. M. Oberlander)

rock dragged along in the fault zone; and closed depressions created by shifts in the ground. Small ponds form in the latter. When these so-called *sag ponds* can be seen forming a definite line, a strike-slip fault is clearly indicated.

One of the interesting features of strike-slip faulting is the horizontal offset of ridge lines and stream beds that cross the fault (Figure 15.26). Fences, orchard rows, streets, sidewalks, and buildings all may be similarly torn apart and offset. Some of this separation occurs in the slow process of *fault creep,* in which offsets of one or two centimeters a year may occur without any earthquake activity.

From the air, or on a topographic map, active strike-slip faults are relatively easy to detect despite the fact that they do not normally create major relief effects. Rift topography and the abrupt change in the pattern of stream dissection due to the juxtaposition of different rock types along the fault—all forming a conspicuously straight line—are unmistakable indicators of a zone of high earthquake risk. Unfortunately, many such areas were settled and developed long before the seismic hazard was fully appreciated, as in the case of many of California's coastal cities.

Fold Structures and Landforms

The terrain that develops in a region of folded sedimentary rock depends on many factors. These include the size and shape of the folds, the local stratigraphy (number, thicknesses, and relative resistances of rock strata present), the rate of tectonic uplift, and the climate. The landforms that develop at a particular point in such an area depend on the rock type, its thickness, and its dip at that location. Thus a great diversity of form is possible.

In real landscapes, anticlinal structures (Chapter 12) rarely produce simple ridges, and synclines are seldom seen as simple troughs. Generally, the anticlines have been "unroofed" by erosion, revealing layer upon layer of rock—some resistant, some weak (Figure 15.27). The scenery is dominated by edges of inclined rock strata that are resistant to erosion. These edges form *homoclinal ridges,* meaning that they are produced by rock layers that dip in one direction (rather than in two directions as along the axis, or center line, of an anticline or syncline). Often there are erosional basins in the centers of the anticlines, indicating that weak rock has been exposed there.

Synclines frequently form elongated plateaus or mountain peaks standing above anticlinal valleys. This is known as *relief inversion;* the structural highs form topographic lows, and the structural lows create topographic highs. Rock strata that have been folded are stretched and weakened by jointing at anticlinal crests, but are compressed in synclines. This makes

Figure 15.26   The San Andreas fault (arrows) is visible crossing this aerial photograph from the upper left to the lower right near the base of the hills. Note the offset ridges and valleys that mark the line of this right-lateral strike-slip fault, some 3 km (2 miles) of which can be seen here. This view is in the Carrizo Plain and Temblor ("earthquake") Range of central California. (U.S. Department of Agriculture)

the rock at the crests of anticlines less resistant to weathering and erosion than the same rock in synclines, thus facilitating relief inversion.

Most anticlines are elongated domes that die out eventually. Homoclinal ridges wrap around the plunging "noses" of folds, as shown in Figures 15.28 and 15.29. Where a series of folds press together, as in Figure 15.29, the homoclinal ridges loop back and forth, bending one way at the axis of an anticline and the other way at the axis of the neighboring syncline.

In most regions of fold structure there are two types of drainage. The smaller streams are controlled by geological structure and have a "'trellis" pattern. Their longer segments are parallel and lie in synclinal, anticlinal, and homoclinal valleys, with short connecting links across the grain of the topography. More often than not the main streams appear to disregard the geological structure. They cut through homoclines, anticlines, and synclines impartially, often crossing resistant layers in deep gorges. Near Harrisburg, Pennsylvania, the Susquehanna River cuts through five successive massive ridges of resistant sandstone (Figure

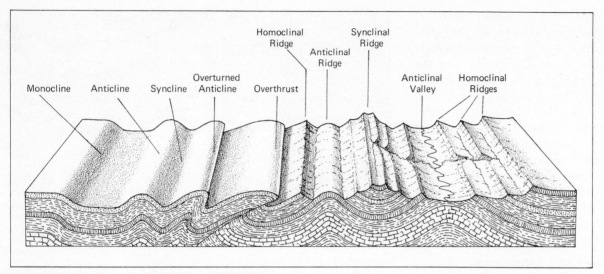

(a)

(b)

Figure 15.27 The left side of this diagram illustrates the geometry of folded rock strata and the terms used to describe it. A terracelike flexure in bedded rocks is a *monocline*. A downfold whose sides, or *limbs*, are rotated upward on either side is a *syncline*, and an arch whose limbs are rotated downward is an *anticline*. Where folding is extreme, the limbs of the fold may be asymmetrically inclined, and the strata on one limb of the fold may be oversteepened or even overturned. If the rock strata are subject to great pressure, an overturned fold may rupture along a zone of weakness, and one limb of the fold may be thrust over the other limb, forming an *overthrust*. Fold structures are rarely seen intact as in this half of the drawing.

The right side of this diagram illustrates the variety of landforms that can be produced by erosion of folded rock strata. The thinner layers are assumed to be more resistant than the thicker layers. Erosion of the weaker rock overlying hard rock on an anticline forms an *anticlinal ridge*. If a portion of the top layer of resistant rock over an anticline has been worn away, *homoclinal ridges* are formed by the exposed edges of rock strata, with an *anticlinal valley* between them. A *synclinal ridge* is formed where a remnant of resistant rock in a syncline acts as caprock to retard the erosion of the underlying layers. (T. M. Oberlander)

Figure 15.28 (a) The folded mountains of southwestern Iran have formed so recently that the uppermost layer of resistant limestone is only beginning to be attacked by erosion. Nevertheless even this young fold is crossed by a transverse stream that creates the gap in the center of the anticlinal arch. (Aerofilms and Aero Pictorial, Ltd.)

(b) More commonly, erosion of anticlines and synclines creates a topography of elliptical or winding hogback ridges, as in this view in central Wyoming. The ridges are formed by homoclines of resistant sedimentary rock that stand above erosional valleys in less resistant strata. (John S. Shelton)

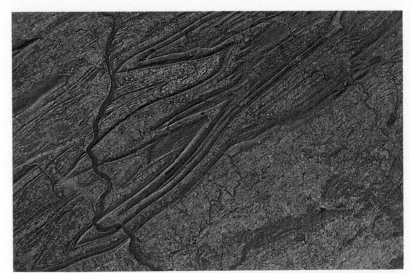

Figure 15.29 Folded strata form a distinctive ridge and valley landscape in southeastern Pennsylvania. This aerial view near Harrisburg was made from a computer-enhanced LANDSAT image (See Appendix II); the image spans a distance of approximately 50 miles. The ridges are resistant sandstone, and the valleys are eroded into shale or limestone. Note the nested hairpin ridges that change direction at the axes of anticlines and synclines. The water gaps through which the Susquehanna River flows are clearly visible in the image. (GEOPIC by Earth Satellite Corp.)

15.29). The Indus and Ganges rivers of Pakistan and India cleave through the folded overthrusts of the Himalaya Range in valleys more than 5 km (3 miles) deep.

Several hypotheses have been offered in explanation of streams that are transverse to geological structure. These include the possibilities that: (1) the streams were in place before the structures were formed and are therefore *antecedent streams;* (2) the streams developed on a blanket of younger unfolded sediments that buried older fold structures and are thus *superimposed streams* (Figure 15.30); and (3) the streams extended headward through points of structural weakness by a process of *progressive stream piracy* (Figure 15.30). Seldom is there any clear indication how transverse streams actually were established. In fact they constitute the principal mystery associated with landforms controlled by geological structure.

Figure 15.30 These diagrams illustrate one hypothesis explaining how streams such as the Susquehanna River in the Appalachian Mountains of Pennsylvania have cut watergaps through bold ridges of resistant rock. A region underlain by folded rock strata (a) becomes worn to a plain of low relief (b). The plain becomes covered with sediments upon which streams develop independently of the rock structures beneath the newer sediments. Renewed uplift (c) causes these streams to incise through the sediment blanket, superimposing themselves across the edges of the tilted rock layers beneath. As the major streams continue to incise downward erosion of the weaker rock (d) produces valleys between ridges of resistant rock. (Vantage Art, Inc.)

(a) Original eroding folds

(b) Region worn to plain and covered with sediments

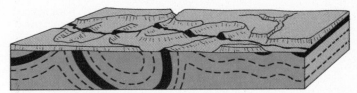

(c) Drainage incised in sediment cover to expose rock edges

(d) Covermass removed and watergaps cut across resistant edges

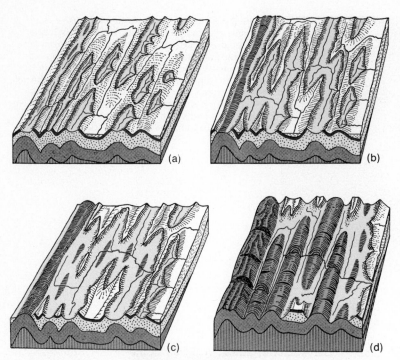

Figure 15.31 Drainage captures by progressive "stream piracy" in areas of weak rocks can produce large streams that are transverse to fold structures. This model is based on drainage evolution in the Zagros Mountains of Iran. Yellow represents easily eroded rock types; red indicates resistant rock. The original surface created by folding is left in white.

(a) The drainage follows the structures created by folding, and is therefore said to be *consequent* upon structure. Already the capping layer has been removed from the highest anticlines, exposing weak rocks (yellow) in which drainage extends rapidly by headward erosion.

(b) The expanding erosional basins in some adjacent anticlines have begun to merge, allowing drainage lines to cross the fold axes. Such drainage lines, which follow exposures of erodible material, are called *subsequent* streams.

(c) The expanding anticlinal basins have merged on a large scale, allowing a major subsequent stream to cross much of the fold belt before leaving it by a consequent path. However, a layer of resistant rock (red) is beginning to be exposed in the centers of the anticlinal basins, and the vigorous subsequent stream has been able to incise downward into this layer in several places.

(d) Gradual uplift of the area has caused the resistant layer (red) to be widely exposed, forming anticlinal mountains where basins were present in stage (c).

The powerful subsequent stream formed in (c) has incised downward into the resistant rock in a series of deep gorges. If no areas similar to (b) or (c) were preserved in the region, the origin of such drainage would be perplexing. (T. M. Oberlander, *Cambridge History of Iran, vol. 1, The Land.* Cambridge University Press, 1968)

SUMMARY

Much landscape diversity results from the nature of the bedrock and the large-scale arrangement of rock masses. These become expressed in the landscape through differential weathering and erosion, in which each body of rock responds in its own way to gradational processes. Differential weathering and erosion are most obvious in sedimentary rock, because of frequent variations in rock type, but they are also a factor in igneous and metamorphic rocks as a result of variations in chemistry and jointing. Shale is a weak rock that erodes rapidly. Sandstone and conglomerate are often resistant enough to form ledges, cuestas, mesas, buttes, and hogback ridges. Limestone is unique because of its solubility, which permits the development of karst phenomena, including surface sinkholes (dolines), disappearing streams, and large underground caverns. Limestone caverns form at or below the water table and are later entered by air when the water table sinks. It is then that caverns become decorated by speleothems.

Volcanic activity is a major source of new rock and an important influence on landscapes. Nonviolent basaltic volcanism occurs where lithospheric plates are separating, at hot spots related to rising magma plumes in the earth's mantle, and along fractures within plates. It creates the oceanic ridges and supplies the rock material of the sea floors. Such volcanism can also occur on the continents, especially where continents have broken up as new oceans were created. Eruptions of basaltic lava produce shield volcanoes and flood basalts. Explosive andesitic volcanism occurs above subduction zones, in which new silicic magma is formed by the melting of older rock. The violence of volcanic eruptions is a consequence of the explosive venting of gases separated from subsurface magma.

The first products of surface volcanic activity are scoria or cinder cones. In time these may develop into lava-veneered composite cones or towering strato-volcanoes. Plug domes are smaller than strato-volcanoes but extremely dangerous because of their violent behavior and release of nuées ardentes. Extremely explosive eruptions can cause the collapse of a volcano, producing a circular caldera. Basaltic plateaus, table mountains, dike ridges, and volcanic necks are erosional remnants of past volcanic activity.

Volcanic regions are hazardous, but soils derived from volcanic ash or basalt are usually very productive. Volcanism also is the source of metallic ores. Volcanic regions contain potentially valuable geothermal energy supplies. These have been developed for heating and electric power generation in a few areas and will be increasingly important in the future.

Broad warping of the earth's crust produces the geologic patterns seen in areas that have not experienced strong deformation. Structural basins in sedimentary rocks are surrounded by outfacing cuestas, with younger rocks toward the centers of the basins. Structural domes may have topographically high or low centers, depending upon the relative erodibility of the strata involved, and are rimmed by infacing cuestas or hogbacks. Major mountain systems are created by the compression and uplift of sediment-filled troughs, called geosynclines, which include the ocean trenches, continental rises, and continental shelves. The uplift of mountains occurs in pulses following the initial phase of compression.

Crustal deformation by compression and tension causes rock to fail by folding and faulting. Normal dip-slip faults create horsts, grabens, and tilted blocks. These are expressed in the landscape as elongated troughs and linear mountain ranges that rise very abruptly above flat-floored basins. Fault-produced scarps are commonly identified by the triangular facets that truncate mountain spurs. Thrust faults and overthrusts are important geologically but have no unique topographic expression. Strike-slip faults have their own distinctive set of landforms, known as rift topography, and are conspicuously straight.

Folds consist of anticlinal and synclinal structures in sedimentary rocks. Erosion in fold structures exploits strips of erodible rock, which become the locations of linear valleys, with resistant strata standing out in relief as homoclinal ridges. Trellis drainage patterns in fold regions feed into major streams that usually run transverse to the fold structures, and which rarely have a proven origin.

REVIEW QUESTIONS

1. Would you expect differential weathering and erosion to be more important in areas of igneous rocks or areas of sedimentary rocks? Why?
2. What is the relationship between buttes, mesas, and cuestas?
3. What sequence of events is necessary to produce a limestone cavern system that is decorated with stalactites and stalagmites?
4. What are some of the hydrological problems encountered in areas of karst topography?
5. Are any distinctive landforms associated with granitic rocks?
6. In what two types of locations do we find the majority of active oceanic volcanoes?
7. What phenomenon is exemplified by the Columbia Plateau of the state of Washington?
8. In what ways do andesitic volcanoes differ from basaltic volcanoes?
9. How are volcanic calderas formed?
10. What landforms are produced by long-continued differential erosion in volcanic regions after volcanism ceases to be active?
11. In what ways is volcanism important to humanity?
12. What is the modern conception of a "geosyncline"?
13. Describe the geological structure and related landforms of the state of Nevada.
14. What distinctive landforms are encountered in areas of strike-slip faulting?
15. What is the dominant landform in areas of anticlines and synclines? What is the usual drainage pattern in such areas?

APPLICATIONS

1. Draw north to south and east to west geological cross sections across your state, using information from state geological surveys, published articles, and maps.
2. How does geological structure affect the landscape in your area? If no specific structures can be seen nearby, where is the closest clear expression of a structural influence on the landscape?
3. In the library, consult the periodical *Geo-times* and look at the section titled "Geologic Events" for a period of twelve consecutive months. Make a catalogue of all of the geological events noted over this period.
4. What is the earthquake history of your area? How often have earthquakes occurred there, and what magnitudes have been experienced? What geological feature has caused the earthquakes?
5. Volcanism seems to occur both where crustal plates are pulling apart and where they are pushing together. Explain this apparent contradiction.
6. How do the individual mountains in regions of folded strata differ from those produced by faulting?
7. Obtain a copy of Erwin Raisz's physiographic map *Landforms of the United Sates* and divide the nation into physiographic regions of fairly uniform characteristics. Write an essay on the problems of such an undertaking.

FURTHER READING

Green, Jack, and **Nicholas M. Short, eds.** *Volcanic Landforms and Surface Features.* New York: Springer-Verlag (1971), 519 pp. This is a large collection of spectacular high-quality photographs covering every aspect of volcanism.

Harris, Stephen L. *Fire and Ice.* Seattle: The Mountaineers and Pacific Search Press (1980). This fascinating paperback presents the eruptive history of each of the volcanoes of the Cascade Range. Written with authority and enthusiasm.

Hunt, Charles B. *Natural Regions of the United States and Canada.* San Francisco: W. H. Freeman & Co. (1974), 725 pp. In Hunt's scheme, natural regions are primarily differentiated by geologic structure, which in North America creates the main contrasts in land surface configuration. This book presents good general descriptions of the physiography of each of the natural regions of North America, including information on the climate, vegetation, and human use of each region.

Iacopi, Robert. *Earthquake Country.* Menlo Park, Calif.: Lane Books (1964), 192 pp. Well-illustrated popular account of California's earthquake-prone fault landscapes.

Lobeck, A.K. *Geomorphology.* New York: McGraw-Hill (1939), 731 pp. Lobeck's textbook was constructed on the premise that a picture is worth a thousand words. This book includes about 200 pages on folding, faulting, and volcanism; these pages consist mainly of photographs and diagrams that have retained their value through the years. The material on folding is especially helpful.

Shelton, John S. *Geology Illustrated.* San Francisco: W. H. Freeman (1966), 434 pp. Nowhere is there published a better collection of photographs illustrating structurally-controlled landforms. The text is very readable and original in approach.

Shimer, John A. *Field Guide to Landforms in the United States.* New York: Macmillan (1971), 272 pp. The first portion of this handbook outlines the nature of the various structurally-determined physiographic regions of the United States, with portions of Erwin Raisz's detailed physiographic diagram of the United States used as index maps of each region. Following this is a catalogue of individual landforms, each illustrated by an excellent line drawing.

Simkin, T., L. Siebert, L. McClelland, D. Bridge, C. Newhall, and **J. Latter.** *Volcanoes of the World: A Regional Directory, Gazetteer and Chronology of Volcanism During the Last 10,000 Years.* Stroudsburg, Pa.: Hutchinson Ross (1981). The most complete record of volcanic activity available. Fascinating details and statistics on 1343 volcanoes and the 5564 known volcanic eruptions since the Ice Ages.

Sweeting, Marjorie M. *Karst Landforms.* New York: Columbia University Press (1972), 362 pp. This text by an English geomorphologist deals exclusively with the distinctive landforms that develop where limestone is the bedrock.

Twidale, C. R. *Analysis of Landforms.* New York: Wiley (1976), 572 pp. Twidale's massive geomorphology text contains many excellent illustrations of structurally-controlled landforms.

Wilcoxson, Kent H. *Chains of Fire.* New York: Chilton (1966), 235 pp. This stimulating book on volcanism, written for the general public, includes many eye-witness accounts of volcanic eruptions of various types. Excellent reading.

Williams, Howell, and **Alexander R. McBirney.** *Volcanology.* San Francisco: W. H. Freeman (1979), 397 pp. This is the most recent of a large number of authoritative texts on volcanic processes and resulting landforms.

CASE STUDY

The Cascades Awaken

On March 27, 1980, the volcanic Cascades mountain range, extending from northern California to southern British Columbia, awakened from a 60-year sleep. The signal was a towering cloud of ash and steam rising from snow-covered Mount St. Helens in southern Washington—the first such event since the eruptions at Mt. Lassen in California between 1914 and 1921. The renewed activity in 1980 culminated on May 18 in a gigantic blast that thoroughly altered the form of Mount St. Helens, replacing its conical glacier-clad summit with a crater 750 meters (2500 ft) deep.

In a single day, Mount St. Helens leveled a forest and killed everything in it, destroyed one lake and created fifteen others, destroyed one river and portions of several more, caused gigantic mud slides and floods, blanketed the Northwest with volcanic ash, paralyzed transportation facilities of every type, threatened the region's wheat crop, halted commerce on the Columbia River for weeks, and temporarily closed down the jobs of nearly half a million people.

Studies in the early 1970s had revealed that Mount St. Helens had a particularly violent past, having blasted away its summit 4,000 years ago and again only 500 years ago. Ash from explosive eruptions of the volcano reached as far east as Montana and Wyoming. These studies revealed that many artificial structures are far older than the mountain, much of which has been constructed in the past 500 years.

Through April and early May of 1980 the north flank of the volcano had been swelling outward some 2 meters per day. Early on the morning of May 18, a magnitude 5+ earthquake jarred the cone. The bulging north flank began to collapse, and as it did so, the volcano erupted in a terrific blast that removed 2.2 cu km (.55 cu mi) of the cone and flattened the forest over an area of almost 550 sq km (200 sq miles), stripping off the foliage and uprooting trees to a distance of almost 30 km (18 miles). Within 8 km (5 miles) of the gaping crater produced by the blast, the ground was stripped as naked as a lunar landscape, even the soil being ripped away. The loss of animal life is difficult to imagine. People died in their tents, in their vehicles, and in the open, as much as 32 km (20 miles) from the blast—scorched, choked by ash, and buried by mud, rock, and uprooted trees. As could be expected, melted ice and snow from the devastated summit sent a series of immense scalding mudflows pouring down the valley of the Toutle River, blocking the valleys of smaller streams and converting them into lakes.

Terrible destruction at the scene of a violent volcanic eruption is expected. But most Americans had never considered the wider effects of such an event. The first inkling of the problems in store for distant areas was a blackening of the skies, producing total darkness in midmorning. Many believed an incredible thunderstorm was in store, until the ash began sifting down.

Unforeseen effects of falling ash were soon apparent. Automobile carburetors became clogged, stalling vehicles on highways and back roads throughout central and eastern Washington. Aircraft engines were affected similarly, causing emergency landings by commercial jets and light planes. Airports and highways were closed and buses and trains were halted, stranding thousands of people. Clouds of fine ash whirled up by moving vehicles choked pedestrians in cities, and attempts to hose away the ash on the streets turned it into slippery mud, causing vehicles to slither out of control. Electrical failures were widespread as the ash accumulated on power lines and transformers, shorting them out. Where power was available, machinery broke down as it was penetrated by abrasive ash particles that were harder than crushed glass. Sewer systems became clogged with ash, and sewage treatment plants failed.

There was brief panic over the possible toxic effects of the ash on both humans and crops. Breathing was "like sticking your head in the fireplace and stirring up the ashes." Fortunately, the ash proved to be nontoxic.

In Pullman, Washington, 500 km (300 miles) from the eruption, the fallout of ash was as much as 8 tons/acre. The wheat crop in eastern Washington seemed to be threatened by a 10 to 15 cm (4 to 6 in.) blanket of ash; the orchard crops of the Yakima and Wenatchee areas also seemed endangered; and growers of potatoes, peas, and lentils in Washington and Idaho expressed alarm. Most of these fears proved groundless, as vegetation recovered when the ash was shaken off or blown away.

A greater problem was damage to streams, large and small. The eruption buried the upper Toutle River under 150 meters (500 ft) of mud and ash and sterilized the rest with steaming mud. The beds of other streams were paved by a concrete-like mass of ash. The loss of the upper 400 meters (1,300 ft) of the mountain made it too low to support the glaciers whose meltwater kept the streams flowing vigorously throughout the summer.

The debris flood down the Toutle River reached the Cowlitz River and then spilled into the Columbia River at Longview, Washington, shallowing the Columbia's channel from 12 meters (40 ft) to less than 5 meters (16 ft) over a distance of 6 km (3.7 miles). This stranded oceangoing ships upriver, loaded with grain for world markets, and blocked ships that had intended to move up the river to inland ports. The wheat crop of the Pacific Northwest was literally bottled up. Railroads and truckers diverted shipments of export cargoes intended for the blocked ports at Portland, Vancouver, and Kalama, increasing the costs of the goods and depriving the embargoed ports of jobs and income estimated at $5 million a week.

The gray wilderness of ash-covered tree trunks smashed down by the May 18 blast constituted some 800 million board feet of lumber, with a value estimated at between $250 million and $1 billion. Ironically, conservationists had spent years battling to save this very forest from the loggers' saws!

Meanwhile, the other young giants among the Cascade volcanoes—Baker, Rainier, Adams, Hood, Jefferson, the Three Sisters, and Shasta—tower over surrounding mountains, farmlands, hamlets, and cities—still asleep, as was shattered Mount St. Helens until March 27, 1980.

Mount. St. Helens following the great eruption of May 18, 1980. A gaping crater has replaced the previous symmetrical cone. Nearly 2 cu km of the pre-eruption cone slid to the north to form the vast debris flow seen in the foreground, and another 2 cu km of volcanic ash was blown into the sky to rain down over eastern Washington, Idaho, and Montana. The area's former lush forest was stripped off cleanly by the blast, with an estimated loss of life including 6,000 deer, 5,000 elk, 200 bears, some 50,000 small game animals, and about 60 persons killed or presumed dead. (T. M. O.)

Cliffs of the Upper Colorado River, Wyoming Territory by Thomas Moran, 1882. (National Museum of American Art, Smithsonian Institution Bequest of Henry Ward Ranger Through the National Academy of Design 1936–12–4)

The bold and colorful cliffs of the arid west of the United States are an expression of geological structure and a climate that discourages the formation of a softening blanket of soil. On the regional, local, and micro scales, climate determines the relative intensities of different geomorphic processes, and is a critical factor in the appearance of both regional landscapes and local landforms.

Looking at the paintings opening this chapter and Chapter 12 we see artists' portrayals of two especially striking landscapes. These landscapes actually exist and are not mere products of the artists' imaginations. Moreover, the paintings depict landforms that are "normal" in the areas where they are found. The rocks composing the two landscapes are not in themselves unusual. What gives the unique quality to both scenes is the combination of rock type, tectonic condition, and gradational processes active in the two areas. The first two factors may be regarded as direct, independent controls, but the nature and intensity of the gradational processes that operate anywhere are in large part consequences of local energy and moisture conditions, or in other words, the local climate. Thus climate is a crucial factor in the varying appearance of landforms from one part of the earth to another. Accordingly, this chapter discusses the ways in which variations in climate on different scales affect geomorphic process/form relationships. The chapter concludes with a survey of some distinctive landscapes that reflect the dominance of particular climatic types.

16
Climate and Landforms

Climatic Influences on Geomorphic Processes

Climate and Weathering
Vegetation and Erosional Intensity

Local and Regional Climate–Landform Relationships

Microrelief
Asymmetry of Slopes
Climamorphogenetic Regions

Geomorphic Processes in Different Climates

Landforms of Periglacial Climates
Landforms of Humid Midlatitude Climates
Landforms of Dry-Summer Subtropical Climates
Landforms of Arid Climates
Landforms of Tropical Climates
Other Climamorphogenetic Regions

CLIMATIC INFLUENCES ON GEOMORPHIC PROCESSES

Contrary to the ideas of early writers on landforms, there are no "normal" landscapes universally associated with particular rock types. This is because rock material weathers and erodes differently as the chemical and mechanical stresses applied to it differ in type or intensity (see Figure 13.4, p. 350). These stresses vary geographically in ways that reflect the wide range of global energy and moisture conditions discussed in Chapter 7. Moreover, at any locality climate-related geomorphic processes have changed with time as energy and moisture conditions have fluctuated or changed progressively.

It is obvious that geomorphic processes differ strikingly in both type and relative intensity from one climatic region to another. Processes

439

that dominate in some environments, such as the transport of sediment by wind in vegetation-free areas, or frost-heaving of the ground in cold-winter regions, may be insignificant or altogether absent elsewhere. Such geographic differences in geomorphic processes often lead to the development of distinctive weathering styles and erosional forms that are closely associated with particular climatic types.

Climate and Weathering

One of the fundamental ways that climate affects the development of landforms is through its control of weathering processes. Limestone blocks, quarried to provide facing stone for the great pyramids in the Egyptian desert, have weathered so little in 4,000 years that tool marks

Figure 16.1 (a) This schematic diagram suggests the relative importance of chemical weathering in various regimes of temperature and precipitation. Low temperature and high precipitation cannot occur together; thus this impossible combination is isolated by the diagonal line on the graph. Chemical weathering is most active where temperatures are high and moisture is abundant. The reactive H+ and OH− ions of rainwater are primarily responsible for chemical weathering. Even though chemical weathering is weakest in dry regions, its effects are nevertheless evident, and may be dominant in hot, dry regions where mechanical weathering processes are comparatively inactive.

(b) This diagram suggests the relative importance of mechanical frost weathering in various climatic regimes. Frost weathering occurs where free moisture is available to penetrate rock joints and pores, and where the temperature drops below freezing periodically or for long intervals of time. It is unimportant in regions that are perpetually warm. The diagram does not take into account processes of mechanical weathering that may be locally important, such as salt crystallization and the actions of burrowing animals and tree roots. (Modified from Peltier, 1950)

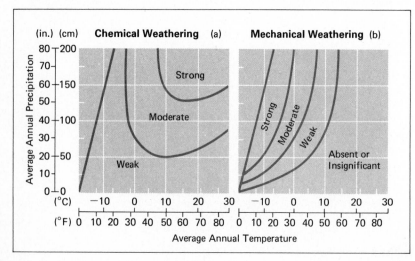

are still visible on them. On the other hand, limestone grave markers in moist New England have corroded to the point of illegibility in only 300 years. The obvious disparity in rates of weathering in moist and dry regions is significant, since weathering is the starting point in the erosional development of landscapes.

Weathering processes, both mechanical and chemical, are controlled by temperature and moisture availability. High temperatures and abundant moisture promote chemical reactions that cause rock decomposition. Nearly all chemical reactions require moisture because in solution ions are mobilized and interact vigorously, combining with and replacing one another. Additionally, oxygen and hydrogen ions in water combine with other compounds and displace weakly-bound ions, as noted in Chapter 11. High temperature speeds up chemical reactions because it causes the ions to move more rapidly and to collide more frequently. The rate of many chemical reactions roughly doubles for every 10°C (18°F) increase in temperature.

Low temperatures and water deficits favor mechanical weathering. Temperatures below the freezing point of water cause the growth of ice crystals in rock pores and more massive ice veins in rock joints. Both forms of crystallization exert expansive stresses leading to rock disintegration. In very dry climates, the evaporation of water that is laden with various salts causes the precipitation of saline crystals. The initial growth of the crystals and their later expansion and contraction with temperature and moisture fluctuations can cause the disintegration of rock—especially granular types like sandstone, tuff, and granite. Water is essential for chemical reactions to occur, and also for the process of mechanical disintegration due to changes of state (crystallization) and oscillations in the volume of substances, which apply pressure to confining materials.

Figure 16.1a illustrates the relative importance of chemical weathering under various combinations of temperature and rainfall. In hot, moist climates, the mantle of decomposed rock material may be tens of meters thick as a consequence of intense chemical weathering that has been minimally retarded by climatic changes. In the humid midlatitude regions, energy and moisture conditions have been subject to considerable fluctuation over the past few millions of years. Some areas in these regions have experienced large-scale removal of weathered mantles, due to periods of accelerated slope erosion as well as to glacial action. Chemical weathering is much less significant in dry climates, where little water is available to mobilize ions, and in cold climates, where chemical reactions are slowed and water is frozen much of the year.

Figure 16.1b suggests the relative importance of mechanical weathering in different climatic regions. Frost action, which roughens rock outcrops and generates coarse angular rock rubble, is unimportant in warm regions, but becomes increasingly significant in cooler regions that have a surplus of moisture. This figure does not show mechanical weathering by the crystallization of various salts, which is an important process in dry regions and along seacoasts wetted by salt spray. In general, "salt weathering" creates pitted surfaces and ground-level niches in granular rock, and it breaks pebbles along permeable fracture planes that are invaded by mineralized water.

Figure 16.2 summarizes the relative importance of the various weathering regimes under different temperature and moisture combinations. The diagram indicates that, in general, weathering processes are least active in dry climates, both hot and cold. Such a diagram should be interpreted only as a broad framework for thinking about the landscapes of different regions. It is obvious that many regions have quite different weathering regimes at different times of the year, and that most regions have experienced changing weathering regimes as a consequence of climatic fluctuations over the past few million years. Figure 16.2 omits the effect of seasonal variations in temperature and precipitation, which may have geomorphic significance, and does not consider the detailed processes involved in weathering. However, the general relationships depicted are valid.

Vegetation and Erosional Intensity

Climate affects the development of landforms both directly and also indirectly through its in-

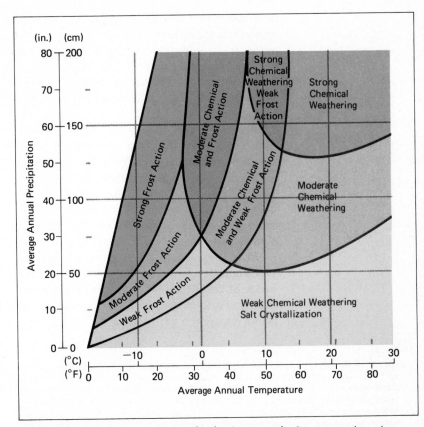

Figure 16.2 This diagram suggests the dominant weathering process in various climatic regimes. Chemical weathering and frost weathering cannot occur unless moisture is available; thus weathering is retarded in dry regions. However, even where weathering is slight, the continuous action of relatively weak processes can eventually produce landforms. Many parts of the earth's surface have been subjected to different weathering regimes with the passage of time because of changes in climate. The landforms in such regions have been formed partly by weathering processes that are no longer active there today. (Modified from Peltier, 1950)

fluence on the vegetative cover in a region. Plants influence weathering directly through their nutrient cycles, acid secretions, and decay products, all of which affect the chemistry of the water in contact with soil and subsurface rock. However, the most important influence of vegetation on landform development is exerted through its modification of the rates of erosion by wind and water.

A dense vegetative cover shields soil from the direct impact of rainfall and wind, and helps maintain a permeable surface. Because the presence of vegetation and an organic soil horizon tends to increase the rate of water infiltration into the soil, a vegetative cover significantly reduces surface runoff. This, in turn, reduces erosion. As indicated in Chapter 13, surface flow of water rarely occurs on well-vegetated slopes, where water moves downslope between soil particles as throughflow rather than on the surface as overland flow. On such slopes the number and aggregate length of stream channels per unit of land area is relatively small. Moreover, the deeply penetrating roots of trees help hold the soil and weathered mantle in place. This may contribute as much as 80 percent of the soil's resistance to shearing, which would otherwise take place in the form of slumps and earthflows.

In sparsely vegetated regions, the rate of precipitation often exceeds the infiltration capacity of the ground, so that surface runoff occurs, and evidence of erosion by running water is much more conspicuous than in densely veg-

etated areas. The drainage density is high—measured as the accumulated lengths of all stream channels per unit of area. Where hillslopes are not protected by a vegetative cover, loose material produced by weathering is flushed away almost as quickly as it forms. This prevents soil development and results in an angular rock landscape in which even minute variations in rock types and structure are apparent (Figure 13.4, p. 350, and Figure 13.18, p. 363). Thus the presence or absence of vegetative protection is a principal factor in landscape character.

If hillslopes in a humid area are cleared of vegetation, erosion soon begins to carry off soil, initiating a network of ever-deepening gullies, as in Figure 16.3. When forest is converted to grassland or brush, the eventual decay of tree roots may so greatly reduce the cohesion of hillslopes that slumps, slides, and earthflows begin to occur. Agricultural land usually remains free of plant cover for part of the year, and erosion accelerates during such periods, as in Figure 11.23, p. 300. Studies of lake sediments near ancient Roman towns indicate that the annual sediment yield from adjacent land increased ten-fold soon after the towns were founded and agricultural activity began. All around the Mediterranean basin, from Morocco to Turkey, phases of fluvial deposition that produced fills in erosional valleys tell a similar story.

Recent estimates of the effects of human activity on rates of erosion indicate that over the past 5,000 or so years the average erosion rate over all the earth's land areas has doubled or tripled. This is due to such activities as clearing land for agriculture, logging, and conversion of forest to grassland, all of which increase the amount of surface runoff. These rapid anthropogenic changes lead to immediate problems of erosion and sedimentation, but are interesting to geomorphologists because they compress in time the more gradual effects of natural climatic changes that influence the vegetative cover.

Figure 13.5 (p. 350) illustrates how the rate of surface lowering by erosion—the denudation rate—is affected by precipitation and vegetation. The example considered is a moderate (10°) slope in a hot climate where the average

Figure 16.3 This view illustrates landscape transformation in the humid climate of Tennessee due to removal of the natural vegetation cover. The area formerly supported a mixed hardwood and pine forest. Deforestation originally occurred to provide fuel for a copper smelter built in 1854. Toxic fumes from this and subsequent smelters prevented vegetative regeneration, keeping the land in a nearly barren condition for more than 100 years despite the region's abundant rainfall. An artificially established grass cover offers some protection, but it has been largely removed by gully expansion, as this view indicates. Although the locality receives about 140 cm (55 in.) of rainfall annually, the landforms resemble those seen in deserts. (T. M. O.)

monthly temperature is about 25°C (77°F). Potential evapotranspiration in such a climate is approximately 80 to 100 cm (30 to 40 in.); thus where the annual precipitation is less than 50 cm (20 in.) vegetation tends to be sparse. As the diagram shows, denudation rates are highest in semiarid areas, where the amount of moisture is insufficient to maintain a dense cover of vegetation but where rainfall occurs often enough to produce frequent periods of runoff. Denudation rates also increase rapidly when annual precipitation is greater than 150 cm (60 in.). Then precipitation considerably exceeds infiltration rates and potential evapotranspiration, generating a large moisture surplus. The resulting runoff erodes material mechanically or carries it away in dissolved form.

Figure 16.4 This mushroom rock, known locally as the Goblet of Venus, stood near Blanding, Utah, until it was toppled several years ago. Its narrow stem was created by the chemical action of trickling water and by salt crystallization, which is concentrated at the base of such rock projections.

LOCAL AND REGIONAL CLIMATE–LANDFORM RELATIONSHIPS

Varying climatic influences upon gradational processes can be seen in geomorphic features of all scales, from individual boulders to hillside slopes and mountain walls. The following discussion offers some examples.

Microrelief

The Goblet of Venus depicted in Figure 16.4 is a peculiar sandstone feature located in semiarid country near Blanding, Utah. It is common for the shapes of such "mushroom" rocks in arid regions to be erroneously attributed to the erosive power of the wind. In truth, however, aeolian abrasion is rarely capable of sculpturing such a form. The Goblet is, instead, a product of differential weathering controlled by microclimate.

Even in arid regions where the annual rainfall is less than 20 cm (8 in.), water is the principal agent of weathering. When infrequent rains fall on an upward-projecting mass of rock, some of the rain water runs down the sides of the mass and concentrates at its base. The shaded lower parts of the rock remain damp longer than the exposed upper parts. This allows minute physical and chemical changes to occur on the damper surfaces, weakening some minerals and partially dissolving the cementing substances that bind the grains of the rock together. As the water evaporates salt crystal formation also exerts pressure on confining mineral crystals. With enough repetition of the process, grains of rock will loosen and fall away. The lower portion of the original block becomes indented and is sometimes reduced to a narrow stem because it remains damp longer after rainfalls, allowing weathering processes to work for longer periods of times.

Asymmetry of Slopes

Single rocks in arid regions often illustrate the microclimatic effect on weathering. Microclimate also influences the form of larger features, such as valley walls. In the middle and higher latitudes, the angles of the slopes on opposite sides of east-west trending valleys often differ although there is no corresponding difference in rock type. In the middle latitudes of the northern hemisphere, slopes that face north are about 5° steeper on the average than south-facing slopes. Such differences do not exist in the low latitudes, where both the north and south sides of ridges experience similar exposure to solar radiation throughout the course of a year.

In the northern hemisphere, south-facing slopes receive more solar radiation than north-facing slopes, and thus south-facing slopes tend to be warmer, drier, less thickly vegetated, and covered by thinner soils than north-facing slopes. In Virginia, the drier south-facing slopes

Figure 16.5 The valleys dissecting Black Tail Butte, Wyoming, all show an asymmetry related to aspect and vegetative cover. North-facing slopes are thickly covered with coniferous forest and have steeper slopes, whereas the drier south-facing slopes are lower in angle and support only sparse clumps of shrubs and grasses. The nature of the vegetative cover influences the development of the slopes. Water runoff erodes the exposed south-facing slope more rapidly than the north-facing slope, which is protected from rainsplash by its vegetative cover and which produces less runoff because of its deeper and more permeable soil.

of some valleys are forested mainly by pine, while the steeper slopes on the opposite side of the valley support oaks and other hardwoods as well as pines. In central California, south-facing slopes are commonly covered by open grassland or chaparral, but the steeper north-facing slopes are woodlands of oak, laurel, and madrone.

In midlatitude regions, the slope with the more abundant vegetation and the deeper and more permeable soil tends to be steeper and less eroded than the facing dry slope, as in Figure 16.5. The complicated interplay of available moisture, vegetation, and soil affects the amount of runoff and the rate and type of erosion on opposite slopes. Surface erosion appears to reduce the angle of the drier, less permeable slope, and the stream in the valley between the opposing slopes may migrate laterally toward the more permeable slope, causing it to be undercut and kept relatively steep.

In cooler climates, the differences in snowmelt or in the thaw of frozen ground on opposite sides of a valley can induce asymmetry. The most striking asymmetry of all is seen in high mountain regions where glaciers form in the cool valley heads on the north and east sides of the crests, often causing mountain ridges to be deeply gouged by glacial erosion on one side while forms produced by fluvial erosion are dominant on the other side (see Chapter 17).

Climamorphogenetic Regions

In general, landforms are not as sensitive to variations in climate as are soils or vegetation. Although geomorphic processes may differ in intensity from one climatic region to another, the landforms shaped by the varying processes are often similar. Because we live in a time closely following a period of major climatic fluctuations, it is difficult to speak with great precision about the relationship between climate and landforms. The relationship is a complex one, and strong differences of opinion persist regarding the exact role of climate in landform development, present and past.

Nevertheless, certain entire landscapes may

Figure 16.6 (a) As this diagram suggests, each region on the earth can be characterized as a climamorphogenetic region possessing a certain combination of weathering and erosion processes that influence the development of landforms in the region. The boundaries in the diagram should be considered only as qualitative guides. The processes considered in classifying the morphogenetic regions include glacial erosion, mass movement, frost weathering, and landscape modification by water and wind (Figure 16.6b). Because of climatic change the morphogenetic character of a region may vary with time.

The dry summer subtropical and semiarid regions have similar annual temperature and precipitation averages, but very different seasonal distributions of energy and moisture conditions. The semiarid regions have hotter summers and colder winters, with their rainfall maximum in the summer rather than the winter. Also they may receive abundant snowfall, unlike the subtropical region.

(b) This chart summarizes the relative intensities of the dominant geomorphic processes in the most easily recognized climamorphogenetic regions. Other regions identified in part (a) (above) are transitional between those identified here. Of course, variations in geological structure, rock type, and tectonic activity will produce landform variations within each region, and some landforms are common to all morphogenetic regions.

The presence of two different process intensities in one morphogenetic region indicates that the process varies in effectiveness between upland areas (top half) and lowlands or open plains (bottom half). For example, linear erosion by streams is effective in humid tropical highlands, where abrasive bedload material is carried by streams, but is ineffectual in open lowlands, where stream loads consist of sand and silt. Channel formation refers to the tendency for a stream to develop a dense network of erosional channels in upland areas, and is the process-equivalent to drainage density and relief texture (process effects).

The most aggressive processes indicated here are frost action in periglacial regions; salt weathering in dry regions; the effects of wind in both tundra (periglacial) and desert areas; the vigor of slope wash resulting from both weak vegetative protection (arid, savanna, and subtropical dry summer areas) and heavy rainfall (humid tropics and summer-wet savannas); vigorous channel formation due to heavy rainfall in tropical highlands and to high proportions of runoff in desert uplands; strong linear erosion due to abundant bedload tools in tundra lowlands, midlatitude valleys, and tropical highlands; gravitational transfer by creep and solifluction in midlatitude and periglacial areas; rapid transfer by slumping and sliding in the wet tropics and areas of water-absorbing clays in the summer dry subtropics; and strong chemical weathering in the tropical forests and savannas.

Equally important are the areas of process ineffectiveness, such as the absences of frost action in the tropics and salt weathering and wind action in damp climates; the weakness of channel formation in the periglacial uplands that experience solifluction and on the permeable surfaces of the savannas; inhibited linear erosion owing to lack of abrasive bedload tools in areas of strong chemical weathering; and the near-absence of chemical weathering in the cold periglacial morphogenetic region.

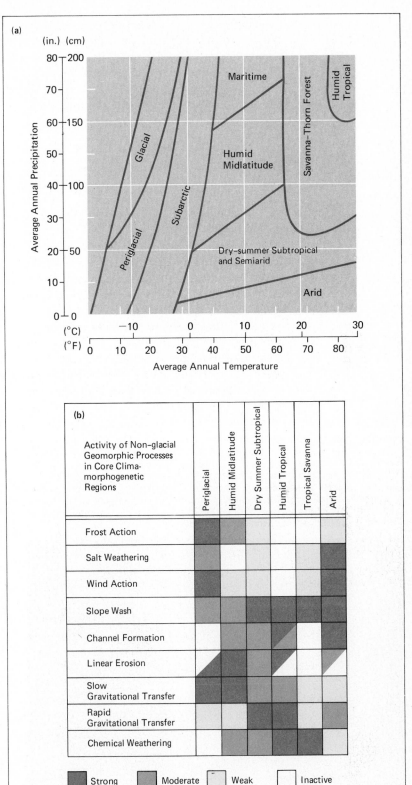

445

exhibit distinctive characteristics resulting from the geomorphic effects of particular climatic regimes. Figure 16.6a illustrates the use of the factors of temperature and precipitation to define *climamorphogenetic regions* that could be shown on a world map—that is, regions in which certain combinations of climatically-controlled geomorphic processes predominate and give a distinctive character to the landscape. The processes at work in the climamorphogenetic regions shown in the figure are listed in Figure 16.6b. Five of the most strongly contrasting regions are examined more closely later in the chapter.

If climate were the dominant factor controlling the appearance of landscapes, it would be possible to associate a distinctive set of landforms with each of the climamorphogenetic regions. However, no landscape owes its character entirely to climate, or to rock type, geologic structure, tectonic condition, or passage of time; the five factors are interwoven in a complex manner. Furthermore, external conditions may change more rapidly than landforms can respond. Consequently, the landforms in a re-

gion may reflect climates of the past rather than the present. The clearest example of this is the unusual scenery in regions that were glaciated 15,000 years ago, as seen in Chapter 17. The concept of climamorphogenetic regions focuses on the climatically-controlled processes that are presently active in a given region; it does not consider all the controls of landscape development and changes in climate throughout the history of the region.

Nevertheless, at the climatic extremes we find many landforms that are peculiar to but one climamorphogenetic region. It is impossible to imagine a thousand square kilometers of sand dunes developing anywhere but in a desert climate, where the absence of vegetation and abundance of loose small particles promotes active wind transport. Such distinctive forms are very important because we sometimes encounter them in regions where they could not form at present, as in the cases of the relict dune fields south of the Sahara near Timbuktu and Lake Chad, and also in the heart of Nebraska (Figure 16.7). This helps us to appreciate how much climates have changed with the passage of time.

Figure 16.7 Vast accumulations of sand in great dune fields are known as *ergs*. The largest ergs are found in the deserts of Arabia and North Africa. However, in Nebraska we can see an ancient erg, here lightly dusted with snow, that shows this region to have been a desert as recently as 15,000 years ago. The Nebraska Sand Hills cover an area of some 62,000 sq km (24,000 sq miles). The individual crescent-shaped mountains of sand in this view are about 1.6 km (1 mile) long, and as much as 100 meters (330 ft) high. Each of these mounds is a *draa* that is crossed by many smaller dune ridges.

GEOMORPHIC PROCESSES IN DIFFERENT CLIMATES

In the following pages we shall focus upon the five most easily recognizable climamorphogenetic regimes and the landforms that result from them. Throughout the discussion it should be recalled that around the core areas in which each of these regimes is clearly dominant there are zones of transition to neighboring regimes. Of course, geographic variations in the intensity of the different geomorphic process systems are as gradual as spatial variations in energy and moisture conditions, and climatically-influenced landform variations are as gradational as climates themselves.

Landforms of Periglacial Climates

Periglacial regions have near-glacial climates dominated by subfreezing temperatures throughout much of the year. Although the term suggests areas peripheral to ice sheets, it also includes much larger cold regions that are too dry to support ice sheets. More accurate terms used in the technical literature on landforms are *cryergic* (cold dominated) and *cryogenic* (cold generated). Thus there are cryergic processes and cryogenic landscapes.

Low-temperature regions at high latitudes and lofty elevations have a tundra vegetation cover of low growing, shallow-rooted shrubs and herbaceous plants. The surface soil freezes and thaws seasonally, and in many places is underlain by the permanently frozen subsoils known as *permafrost.* Mechanical weathering by the freezing of water is vigorous and shatters rock both at and below the land surface.

A particular feature of periglacial regions is the strong disturbance of soil by seasonal freezing and thawing as well as freeze-thaw cycles in the spring and fall. The resulting volume changes contribute to massive gravitational transfer on slopes. Added to this is the process of *solifluction*—the periodic movement of lobes of soil saturated with thawed water that is unable to percolate downward due to the permanently frozen subsoil (Figure 13.10, p. 355, and Figure 16.8). This churned surface layer, one or two meters deep, which freezes and thaws annually, is known as the *active layer.* Landscape features commonly developed in areas of periglacial climates are illustrated diagrammatically in Figure 16.21, p. 458–459.

Hillslopes in periglacial regions often are topped by a ragged cliff of rock that is shattered by frost action. Its upper surface may form a *periglacial block field*, or *felsenmeer* (German, *rock sea*), as in Figure 16.9. At the cliff base, a comparatively steep (30° to 40°) dry talus slope merges into a more gentle, wetter layer where solifluction commonly occurs. The solifluction slope is usually convex near its crest and concave toward its base, flattening to an angle of 10° or so at its lower end. Although the solifluction slope may be irregular in detail, the long-term effect of solifluction and associated mass wasting processes in the active layer is to smooth the landscape, producing hillcrests that are broadly convex and slopes that are concave, as in Figure 16.10.

Figure 16.8 This *solifluction lobe* in Canada's Yukon Territory is one of those on the slope depicted in Figure 13.10. The surface of the lobe is covered with hummocks created by freeze and thaw activity, and coarse rock rubble is conspicuous at the front of the lobe. Solifluction affects hillslopes on which silty, moisture-retentive soils overlie permafrost. (Larry W. Price)

Figure 16.9 This *felsenmeer*, or "sea of rocks" appears on a plateau-like upland at about 3,500 meters (11,600 feet) in California's Sierra Nevada. The granitic bedrock has been shattered by the growth of ice veins in rock joints, breaking loose the boulders visible here. In many other locations the rocks are larger, and no fine material is visible on the surface. (T. M. O.)

Mass soil transfer under periglacial conditions inhibits the development of rills and gullies, so although much water moves over the surface during the thaw season, runoff channels may be poorly defined and the drainage

Figure 16.10   This landscape in southern Ireland was smoothed by periglacial slope processes that were active in this region during Pleistocene time. Such landscapes rarely display linear erosion or rock outcrops, but their soils may be well supplied with angular rock fragments. Slopes of the type shown are common in both England and Ireland south of the limit of Late Pleistocene glaciation.

density (channels per unit of area) is uncommonly low. First-order valleys are often steeply sloping flat-floored *dells* that are aggraded with colluvium brought from adjacent slopes but which contain no stream channel at all. Because coarse rock waste supplied by frost shattering is carried down into lowlands by mass wasting processes, the larger streams in hilly periglacial regions are abundantly supplied with abrasive tools and have great cutting power during their annual floods resulting from the spring thaws. As their discharge declines, such streams deposit their coarse load in bars of sand and gravel. Thus many of the larger streams are intricately braided.

During the long winter freezes, moisture retentive silty soils in low-lying areas contract and crack, with veins of ice gradually filling the cracks. These ice veins grow into networks of *ice wedges,* each as much as a meter across and

(a)

Figure 16.11   (a) Ice wedge polygons seen obliquely from the air over the Mackenzie River delta of arctic Canada. The active ice wedges form a continuous network, and the tundra polygons they enclose can vary in diameter from a few to hundreds of meters. (J. Ross Mackay)

(b) Along the banks of large rivers, ice wedges are often exposed in cross section, as in this view. New ice is added annually as the ground cracks due to contraction at temperatures far below the freezing point. (Larry W. Price)

(b)

3 or 4 meters deep. Sometimes areas of many tens of square kilometers are covered with systems of *ice-wedge polygons* that can be fully appreciated only in an aerial view (Figure 16.11). Masses of clear ice formed in boggy lowlands may grow into lens-shaped ice blisters known as *pingos*, which dome up areas of ice-wedge polygons to form isolated circular mounds 10 or more meters high and as much as 100 meters (300 ft) across (Figure 16.21, p. 458). Conversely, areas of thawed permafrost subside to form depressions that hold *thaw lakes*, which are present by the thousands in some regions.

In areas of higher or more rolling topography, frost cracking of the ground combines with a sorting action resulting from repeated freezing and thawing to produce *sorted polygons*, in which ice wedges are replaced by concentrations of rock slabs that spill out at the surface to form *stone rings* (Figure 16.12); these are dragged downslope by mass wasting processes to form *stone garlands* and *stone stripes*. Many of these same landforms can be seen at present in high mountains above the treeline in the middle and low latitudes, where tundra vegetation dominates and permafrost is present in limited areas.

During the Pleistocene glacial periods, cryergic conditions were much more extensive than at present, and periglacial landforms were developed in many areas that now have more moderate climates. Some of these unusual landforms persist as relics in the mild climates of Western Europe and the eastern United States, and in many high-latitude regions; they are important evidence of the nature of climatic change over the past 10,000 years.

The periglacial climamorphogenetic region poses difficult problems for modern technological societies. Formerly the realm of seal hunters and reindeer herders, this region began to undergo increasing development in the twentieth century. For North America, change began with the building of military and communications facilities in the 1940s and 1950s, and the pace has been accelerated by the development of oil resources discovered in northern Alaska in the 1960s.

Construction of roads, railroads, bridges, pipelines, and buildings in the periglacial re-

Figure 16.12 The stone rings visible here cover a ridge crest at about 3,700 meters (12,000 feet) in the Rocky Mountains west of Estes Park, Colorado. Although the specific mechanism forming this type of patterned ground remains a question, such phenomena clearly are related to freeze and thaw processes, being seen in tundra landscapes at both high elevations and high latitudes. (T. M. O.)

gion is made extremely difficult by the nature of ground that is permanently frozen at depth (Figure 16.13). When vegetation is cleared or the ground is compacted by vehicle traffic, the insulating properties of the soil in the active layer above the permafrost are reduced, causing the upper portion of the permafrost to thaw. Heated buildings transmit heat downward, which also causes permafrost melting. Road and railroad fills increase insulation, causing a rise in the permafrost level, which heaves up the ground surface. Both heaving and thawing of permafrost produce differential ground movement. Local subsidence causes buildings to sink or twist off their foundations. Telephone

Figure 16.13 This heated roadhouse along the Richardson Highway in Alaska caused the underlying permafrost to thaw. The building has subsided vertically into the thawed ground, except for the porch at the left, which was not heated, and which remains close to its original level. (Troy L. Péwé)

poles, bridge abutments, and piers emplaced in the thawed layer during the warm season are frequently heaved out of the ground during hard winter freezes. In addition, solifluction movement on slopes sometimes damages roads and rail lines. Consequently, all construction in this zone must be very carefully engineered to minimize disturbance of the permafrost and to resist destruction by the flow of liquefied soil in the active layer. This, of course, enormously increases the cost of development in this environment.

Landforms of Humid Midlatitude Climates

The familiar landscapes of most of Europe and the United States east of the 96th meridian and south of the Missouri and Ohio rivers are examples of landforms developed in unglaciated midlatitude regions that were covered by forest under natural conditions. Slopes are soil-covered for the most part, and both chemical and mechanical weathering occur seasonally. In such regions slope erosion by running water is relatively ineffective because the mixed deciduous and coniferous forest minimizes rainsplash and slopewash (overland flow). However, the number of stream channels per unit of area, or *drainage density*, and the intricacy of stream dissection, or *relief texture*, are much

greater than in periglacial regions, though less than in the humid tropics (Figure 16.14). This reflects the fact that mass wasting, though present, is not as rapid as in regions of cold climate, and fluvial *linear erosion* assumes more general importance.

In humid midlatitude regions, hillcrests bearing permeable soils are broadly rounded; those with less permeable soils are narrower. The convexity of soil-covered hilltops and ridge crests is attributed to the process of soil creep. Even where resistant rock materials develop steep slopes and erodible materials make gentler slopes, the ubiquitous creep process nevertheless tends to blur distinctions between the different rock types. Thus, sharp slope breaks due to changes in lithology are not a normal feature of this morphogenetic region.

The more or less intermediate condition of the humid midlatitude morphogenetic region is evident in the fact that it grades into wetter and drier and colder and warmer regions on its various perimeters. All of the processes found in the more extreme regions also occur in the moist midlatitude region, each of them developed weakly or moderately rather than being a dominant factor. This region is the one to which William Morris Davis referred in his "cycle of erosion." However, it should not be considered the "normal" region just because it is the most familiar to the majority of Americans and Europeans. The earth's land areas are vast, and humid midlatitude regions occupy but a fraction of them. In fact, the landscapes considered normal by Davis have experienced the influences of several dissimilar climamorphogenetic regimes and contain many relict features that cannot be attributed to present geomorphic processes. Evidence of both the periglacial and humid tropical geomorphic systems is widespread in the humid midlatitude landscapes of Europe and North America, and the occurrence of both climatic types in these areas during the period of landscape formation is well established.

Landforms of Dry-Summer Subtropical Climates

We saw in Chapters 8 and 10 that the peculiar "Mediterranean" combination of rainless hot summers and mild wet winters occurs in widely separated subtropical west coast locations. We found also that this combination stimulates a distinctive vegetative response, consisting of drought-resistant evergreen shrubs and trees, and winter-active grasses. As this type of climate is associated with coasts poleward of coastal deserts, a conspicuous element of the scenery is itself coastal, being dominated by the erosional effects of waves and currents, which are discussed in detail in Chapter 18. Thus one aspect of many "Mediterranean" landscapes is the alternation of bold sea cliffs created by wave erosion, and sand and gravel beaches built by marine depositional processes.

Ocean beaches are fed by sand from diverse sources, as indicated earlier, and grow and shrink in rhythm with variations in wave energy. As "Mediterranean" winters are characterized by stronger winds and the passage of cyclonic storms, winter ocean waves are larger and more erosive than those of summer, so beach sand deposits tend to be reduced in size during the winter. The calmer atmospheric conditions prevailing during the summer droughts allow eroded beaches to be reconstructed. Annual "beach cycles" of this type are to be expected wherever storms are seasonal rather than spread throughout the year, but are especially marked in areas of "Mediterranean" climate.

Inland, the dry summer subtropical regions

Figure 16.14 This view in Tennessee is representative of landscapes produced by fluvial dissection in a humid midlatitude climate. Erosion by running water is more apparent than in periglacial regions but the drainage density is lower than in vegetation-free areas and regions of extremely high rainfall.

display other geomorphic tendencies that are more or less peculiar to this specific climatic type. Some of these are costly to the human occupants of these otherwise favored regions. The virtual absence of any moisture input during the long hot summers causes the vegetation to become thoroughly dried out and extremely flammable. Natural wildfires ignited by lightning, as well as artificial fires set accidentally or intentionally by humans, frequently rage for days through the chaparral and woodland vegetation (Figure 16.15). Heavy rains in subsequent winters fall on the fire-stripped hillslopes and produce sudden massive erosion, generating floods of soil, vegetation, and rock debris that pour through settled valleys, devastating buildings, roads, and cropland. The Pacific Coast Ranges of California, from the San Diego

Figure 16.15 (a) This hillside in southern California was formerly covered by chaparral that had burned in a destructive fire a few weeks earlier. Note the wide exposure of unprotected soil. Slopes of this type erode rapidly under the early winter rains of the dry summer subtropical (Mediterranean) climate, and often are the source of damaging debris flows in nearby communities.

(b) The bulldozer operator is attempting to clear the mud, boulders, and tree trunks brought down from the Santa Ynez Mountains in a series of debris flows that damaged the Santa Barbara area a few days earlier. The source area for these debris flows was denuded by a major fire in the preceding summer. (T. M. O.)

area to the vicinity of San Francisco, are especially prone to destructive fire-related debris floods.

Hillslope instability on a smaller scale is also a problem in areas of "Mediterranean" climate.

The combination of cool wet winters and dry hot summers generates Vertisol soils dominated by sodium- and magnesium-rich types of clay having an unusual capacity to absorb and retain water. These soils swell, increase greatly in weight, and lose cohesion when wet, and they shrink, harden, and crack when dry. Soils dominated by such clays can liquify and flow during the wet season. These clays are a product of the alkaline soil moisture associated with the unusual summer-dry subtropical climate. When clay soils on hillslopes become saturated with moisture and lose their internal cohesion, they liquify and collapse in slow, muddy slumps and earthflows (Figure 13.11, p. 356). The plane of detachment can be from a half meter to tens of meters below the surface, and may bite into the weathered rock beneath the soil itself.

Slumps and earthflows can be very destructive, blocking roads and causing the collapse of buildings. In the San Francisco area heavy rains in January 1982 triggered faster moving, more liquid flows that caused 26 deaths, nearly 500 injuries, and damage to more than 6,500 homes and 1,000 commercial structures, many of which were totally destroyed. In many other instances, slumps and earthflows have been caused by human activity in the form of excessive lawn watering, diversion of natural runoff by walls and fences, and the oversteepening of slopes during construction of roads, parking lots, and buildings of all types.

These unusual geomorphic processes create small-scale landforms that are visible in all regions of dry-summer subtropical climate. These include slump scars, earthflow tongues, valleys clogged with debris flow deposits, and chaotic hillside topography resulting from repeated slope failures.

Here as elsewhere on the fringes of the arid realms, fluctuating stream discharges and rates of slope denudation have caused the fluvial regime to oscillate between downward incision and upward aggradation, or, more simply, between scouring and filling. Often, the effects of tectonic activity, sea level changes, and human disturbance add to the complexity in fluvial behavior. Thus streams in regions of dry-summer subtropical climates are commonly bordered by terraces, and past cycles of gully cutting and filling are widely evident.

Landforms of Arid Climates

In arid regions, such as the southwestern United States, the vegetation cover consists of sparse grass or widely-spaced shrubs, and bare rock exposures or accumulations of rock fragments dominate the scenery (see Figure 14.24, p. 394). As there is little or no soil to smooth over variations in rock type, the slopes exhibit greater angularity than those of other regions. The absence of soil on slopes allows runoff to be unusually high during the occasional rainstorms. In these dry regions surface water erosion is more effective than creep, and the drainage density and immediate effect of erosion by flowing water are high. However, the duration and distance of transportation of rock debris by ephemeral flows of surface water are small. As a consequence, in desert landscapes depositional landforms are unusually conspicuous, taking the form of aggraded dry stream beds, large alluvial fans, talus accumulations, and extensive areas of sand dunes.

The surfaces of alluvial deposits more than about 10,000 years old are often armored by *desert pavement*, a continuous veneer of stones that fit closely together (Figure 16.16). Desert pavement seems to be produced mainly by water runoff that removes fine particles, eventually leaving the surface covered by a lag deposit of coarser, less mobile debris. The wind may assist initially, but the dark film of *desert varnish* usually seen on pavement rocks indicates that wind erosion has become inactive. Desert varnish is composed principally of clays and oxides of iron and manganese that have been superimposed upon the rock from exterior sources. The dark, manganese-rich variety is a product of bacterial oxidation of manganese in environments that are poor in organic nutrients.

Convex hillcrests in arid regions are rounded by rainsplash erosion and unchanneled soilwash rather than by creep. In the absence of a soil cover and the smoothing creep process, subtle variations in rock type result in conspicuous differences in slope form. Differential weathering and erosion in sedimentary rocks reach their maximum development in arid regions, producing sharply defined cuestas, mesas, buttes, and hogback ridges, with cliffs

Figure 16.16 This surface in Morocco is a *desert pavement* composed of closely packed gravel and larger rocks. It is a consequence of erosion of alluvial deposits by rare torrential flows of water that strip away all of the finer rock particles, peeling down the surface until it is armored by a continuous blanket of fragments too large to be flushed away. Thus the pavement is also called *desert armor*. There is some evidence of upward movement of rocks to the surface layer as a consequence of wetting and drying cycles that disturb the soil. Such surfaces are widespread in desert regions where there is no vegetative protection of the land surface. Formation of desert pavement requires a long period of time, in some cases at least 10,000 years. (T. M. O.)

of resistant rock jutting above talus slopes having angles of 30 to 35°. The talus is only a thin layer of caprock debris that veneers the more erodible sedimentary strata below the resistant caprock. Approximately equal quantities of material enter the talus slope segment from the cliff above and are flushed or blown away in finer form from the slope foot. The cuestas retreat laterally by the erosion of the weaker strata below the caprock, portions of which collapse from lack of support, producing the talus slope.

The lower reaches of slopes in dry regions often form an extensive gently inclined bedrock surface called a *pediment*. Pediments are erosion

Figure 16.17 Areas of granitic rocks in arid regions typically develop ramplike erosional *pediments* below sharply rising rock islands, or *inselbergs*. The pediment expands headward (upslope) as the mountain slopes erode back at a constant angle. The granitic pediment in this view in southern California's Mojave Desert is unusually well exposed due to erosion that has begun to dissect it. (T. M. O.)

surfaces left where hillslopes and cliffs have been worn laterally back by slope erosion. Pediments differ from the hypothetical peneplains of Davis (Chapter 13) in that they seem to be produced primarily by slope erosion by running water rather than by a process of stream dissection and lowering of the dissected landscape by mass wasting processes. Although many pediments are cut across easily eroded rock, others truncate the same resistant rock that creates bold hillforms that rise sharply above the pediment surface (Figures 16.17 and 16.24). Pediments of the latter type are most commonly developed on granitic or gneissic rock.

It is difficult to account for certain characteristic landforms of desert regions in terms of presently active geomorphic processes. Many of the landforms seen in deserts seem to be relict forms inherited from previous periods in which more humid climates prevailed. In areas of granitic bedrock, the presence of vast pediments that extend more than a kilometer outward from eroded hill masses seems to imply that major climatic changes have taken place. In such places, slope retreat and pedimentation appear to have occurred in a soil-covered semiarid landscape that was subsequently eroded down to the solid rock at the former lower limit of subsurface chemical weathering. The erosional acceleration occurred as a result of climatic changes that reduced the moisture supply

and greatly diminished vegetative protection of the land surface.

Too often deserts are assumed to be merely great seas of sand. This is an erroneous view, for desert landscapes consist of diversified surface types, whereas the formation of great sand seas requires special conditions, as outlined in Chapter 14. The vast areas of sand dunes, known as *ergs*, that are features of some deserts may themselves be products of climatic changes. To form an erg, or "sand sea", several things are required: an abundant supply of sand, a transport system that delivers the sand to an accumulation area, and a weakening of the transport system, causing massive deposition and erg growth. Climatic change may be a further requirement, for some desert sand accumulations are enormous, whereas production of sand by desert weathering processes is extremely slow. Thus erg formation may require a humid period of rock disintegration and stream transport of sand out of the supply region, followed by a climatic change toward aridity, so that the wind assumes the role of the sand transporting and concentrating agent. Major ergs have probably grown in several separate stages, including climatic changes back and forth between sand generating and erg-building geomorphic systems.

Landforms of Tropical Climates

Low-latitude regions have two characteristic climates and two characteristic landscapes. In the equatorial region we find vast lowlands covered with lush tropical rainforests, which occasionally extend up the slopes of bordering mountains. Here conditions are warm and moist all year around, with rainfall exceeding potential evapotranspiration in every month. Peripheral to the equatorial lowlands are the winter-dry tropical savannas and scrub forests consisting of vast tree-studded grasslands and dense tangles of thorny brush (Chapter 10). In the savanna and scrub forest regions, precipitation is less than evapotranspiration during those months of the year when the ITC has moved to the far side of the equator, but rainfall is heavy during the summer when the ITC dominates the region.

The almost daily rainfalls of the tropical rainforest regions generate enormous amounts of runoff, resulting in the world's greatest rivers. The average annual discharge of the Amazon River of Brazil is ten times that of the Mississippi and accounts for 15 percent of the world's freshwater runoff to the seas. The Zaire (Congo) River of equatorial Africa discharges only one-fourth as much water as the Amazon, but nevertheless has twice the volume of the next ranking stream system—the combined flows of the monsoon-fed Ganges and Brahmaputra rivers of India and Bangladesh. The tropical savannas, scrub forests, and monsoon forests likewise generate considerable seasonal runoff. But during the dry season, water tables in these regions sink far below the land surface, causing all but the largest stream channels to dry up.

Although the abundant rainforest vegetation of the humid tropical region is concentrated in the equatorial lowlands, in many regions it also extends upward onto steep mountain slopes, offering them some protection from severe gully erosion by surface water runoff. In areas of steep relief, the drainage density is extremely high and the landscape may become a maze of knife-edged, serrate ridges (Figure 16.18). Soilwash and removal of material by solution are important erosional processes, and the sliding of water-saturated soil on steep slopes is conspicuous where rainforests reach into highland zones.

Limestone plateaus in the wet tropics are commonly transformed into spectacular cockpit, cone, and tower karst forms (Chapter 15). Nowhere outside of the tropics do we encounter such features. In this environment, joint-controlled solution of limestone is accelerated by the organic acids derived from the abundant vegetation that flourishes throughout the year.

On level erosional surfaces in all humid

Figure 16.18 This view in the Andes Mountains of Peru shows a typical example of "feral relief"—literally "relief run wild"—in a wet tropical highland. The heavy rainfall in this area makes possible a complete vegetation cover, even on steep slopes, but also produces a high drainage density, strong stream incision, and frequent sliding of rotted rock on the precipitous slopes. (T. M. O.)

Figure 16.19   This flat-topped mesa in the arid south of Australia is capped by a laterite crust. The laterite is an isolated remnant of an ancient land surface that developed under the influence of a seasonally wet tropical climate. Climatic change toward aridity destroyed the former vegetative cover and triggered erosion that stripped away all but this remnant of the laterized soil. (G. R. Roberts, Nelson, N.Z.)

tropical climates, the weathered mantle is generally thick, attaining depths exceeding 100 meters (330 ft) in certain areas of granitic rocks. Chemical weathering is very aggressive, quickly decomposing rock and causing soluble bases and even silica to be leached away if not taken up quickly by vegetation and kept in the nutrient cycle. The remaining concentration of less soluble oxides and hydroxides of iron, aluminum, and manganese become part of subsoils that harden irreversibly when exposed to air. Thus deforested humid tropical landscapes and savannas develop a crust of rust-colored, iron-rich *laterite* overlying a white or mottled *pallid zone* of thoroughly decomposed rock ma-

terial. This is especially the case in savannas and scrub forests, where water table fluctuations seem to contribute to the process of iron concentration and laterite formation. The laterite crust may be several meters thick and often produces conspicuous ledges and mesas (Figure 16.19).

In many savanna areas, the large-scale relief features are abrupt escarpments and steep-sided domes composed of poorly jointed granitic or gneissic rock that has resisted subsurface chemical decay. These tropical granitic domes are known as *bornhardts*, after the German scientist who first described them (Figure 16.25). Waterfalls are common in the tropics, but deeply incised gorges are rare along large streams because thorough chemical decay of rock debris prevents the streams from acquiring cutting tools (Figure 16.20). European geomorphologists often refer to the ''paralysis'' of linear erosion in the humid tropics, where stream loads consist primarily of quartz sand and dissolved material.

Figure 16.20   The Jari River in Brazil flows southward into the Amazon River, and is typical of streams in the humid tropics in being interrupted by several waterfalls and cascades. The valley of this large stream is also poorly defined and the channel is often divided by forested islands. All of these features suggest that due to the intense chemical weathering affecting the rocks of the region, the stream lacks a coarse bed load and the abrasive tools necessary for vertical incision and the production of a smooth longitudinal profile cut through diverse materials. (Hilgard O'Reilly Sternberg)

Other Climamorphogenetic Regions

With the exception of the glacial morphogenetic region, in which the geomorphic processes are unique and important enough to merit separate treatment (Chapter 17), the remaining climamorphogenetic regions are transitional between various of the five major regions that we have described. Thus the *subarctic* region is the zone of transition between the periglacial and humid midlatitude regions; the *maritime* region is a still wetter mild-winter version of the moist midlatitude region; and the *semiarid* region is transitional between the arid region and either the humid tropical or humid midlatitude region.

To conclude this chapter, we present "A Gallery of Landscapes." The landscapes portrayed are products of some of the more extreme combinations of temperature and moisture conditions found on our planet. We have included more than one landscape in some of the morphogenetic regions to illustrate that although those landscapes differ because of geologic structure and rock type, they nevertheless show the dominance of certain climatically-controlled processes and are unlike landscapes developed in similar structures and rock types under different climates.

A GALLERY OF LANDSCAPES

Figure 16.21 Cold climate landscapes. Where temperatures remain below freezing the greater part of the year, landscapes take on a highly distinctive appearance. Below the top meter or so of the land surface, the water in all pore spaces in rock and soil remains frozen solid throughout the year. Permanently frozen rock and soil is known as *permafrost*. Permafrost reaches to depths exceeding 600 meters in northern Alaska and Siberia. Above the permafrost, the surficial blanket of soil thaws each summer, becoming saturated with water that cannot escape downward due to the frozen substrate. This water lubricates the soil here and there to the degree that the soil slips downslope as a mass, a few centimeters in a matter of a week or two. The water-saturated soil moves in lobes that are conspicuous on hillslopes, as the inset (bottom right) indicates. This type of movement is known as *solifluction*. Solifluction lobes are a few centimeters to a meter or so in height. Studies show that despite the appearance of great activity exhibited by these solifluction lobes, most are inactive, movement occurring in only a few places during each thaw. However, considered over a long perioid of time, the entire soil cover is draining downslope at a far more rapid pace than that produced in warmer regions by the creep process. The portion of the soil that freezes and thaws annually, expanding, contracting, and moving downslope, is known as the *active layer*. The long-term effect of the solifluction process is to smooth the landscape, filling preexisting depressions and peeling down projections. Small valleys are clogged with *colluvium* (slope deposits), forming flat-floored *dells* with poorly developed watercourses. Water seeps through the colluvial valley fills.

Bare rock within the active layer or projecting above it is subject to intense frost weathering during the long, cold winter. The expansion of water in passing from the liquid to the crystal state exerts enormous pressures on confining walls, causing rocks to split and, occasionally, to crumble. Thus projecting solid rock masses are rapidly reduced to rubble, which becomes incorporated into the solifluction lobes. High areas are thereby lowered effectively. Broad summits may be covered with angular rock rubble produced by the intensity of the freezing process in exposed sites. The result is a "sea of rocks," or *felsenmeer*. At the edges of rock exposures, large talus accumulations reflect the downslope movement of frost-riven blocks.

The annual freeze and thaw process produces a host of unusual minor landforms in addition to solifluction lobes. Repeated volume changes in the weathered mantle have the effect of sorting out coarse and fine materials. Due to the efficiency of freezing in springing loose slabs of rock, the weathered mantle on hills contains an abundance of coarse debris. This becomes shunted away from the fine material, which accumulates in masses surrounded by rings of slabs that are more or less "on end." The result is the stone rings shown in the inset (top center). On slopes, these rings are drawn out downhill into "garlands" and "stripes" by movements of the active layer. Another type of *patterned ground* de-

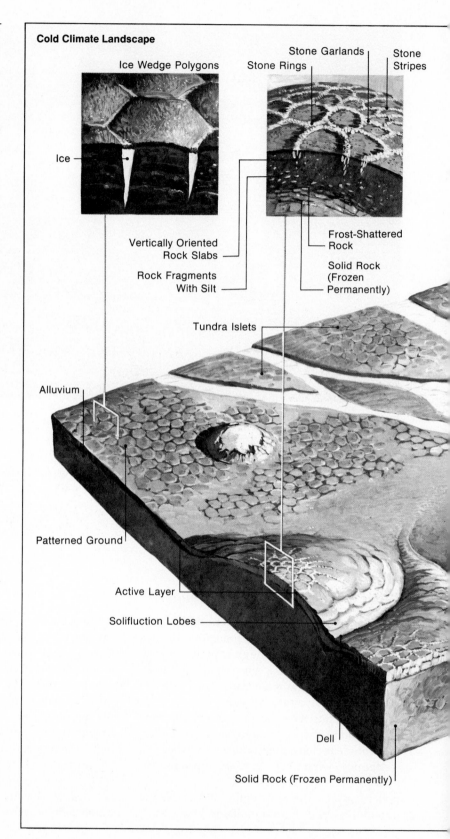

Cold Climate Landscape

Ice Wedge Polygons

Stone Garlands

Stone Rings

Stone Stripes

Ice

Vertically Oriented Rock Slabs

Rock Fragments With Silt

Frost-Shattered Rock

Solid Rock (Frozen Permanently)

Tundra Islets

Alluvium

Patterned Ground

Active Layer

Solifluction Lobes

Dell

Solid Rock (Frozen Permanently)

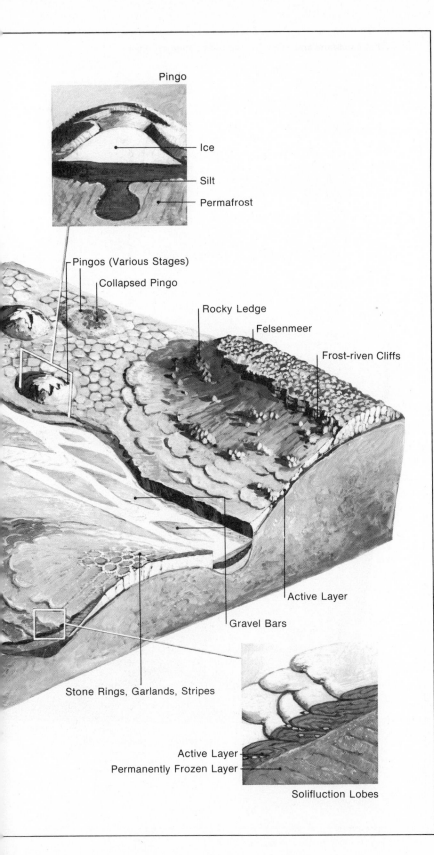

Pingo

Ice

Silt

Permafrost

Pingos (Various Stages)

Collapsed Pingo

Rocky Ledge

Felsenmeer

Frost-riven Cliffs

Active Layer

Gravel Bars

Stone Rings, Garlands, Stripes

Active Layer
Permanently Frozen Layer

Solifluction Lobes

velops in lowlands composed of silty alluvial deposits. During the freezing process, water-soaked silt first expands, but at prolonged low temperatures, eventually begins to contract, cracking in a network of polygonal fissures. Ice forms in these and eventually produces wedges, often a meter across and 5 meters deep (see inset at top left). These are *ice-wedge polygons*. Once formed they are permanent, continuing to grow very slowly.

Clear ice also forms in the ground in lens-like masses that heave up the overlying sod (see inset at top right). This phenomenon is known as a *pingo*. Pingos develop in old lake beds or marshes that are filling with sediment and vegetation. At a certain point, the encroachment of permafrost into the fill traps a lens of unfrozen water. Freezing of the water in turn draws moisture from the fill to produce a growing mass of ice that eventually domes up the recently developed sod above it. If a pingo is destroyed by thawing, it leaves a depression ringed by an earth rampart. A large pingo may be 100 meters across and 30 meters high.

Other oddities of landscapes formed by *cryergic* (low-temperature) processes are overhanging (frozen) riverbanks and unusually smooth stream profiles due to the abundance of fresh rock waste, which provides streams with abrasive tools.

The small, ice-cored domed hill (**below**) is a pingo near the delta of the Mackenzie River in the Northwest Territories, Canada. The surface of the ground exhibits the polygonal pattern often found in cold climate regions underlain by perennially frozen ground.

Figure 16.22   Desert landscape—relief by erosion.
This landscape is created by erosion by running water
in an area of very gently dipping sedimentary rocks of
varying resistance. The basic element is the *cuesta*, a
sloping plain that terminates on its high side in a bold
escarpment. The plain is the top of a layer of resistant
rock, usually sandstone or limestone, and the escarpment marks the edge of the layer. In such a landscape, escarpments wear back through time, always in
a down-dip direction. In the diagram, the dip is to the
right, and the main cliffs are being pushed back to
the right by erosion. Cliff retreat commences at structural highs, or domes, which are the first areas to be
attacked by erosion, producing "holes," which subsequently enlarge. Cliffs are also produced where
streams cut across the landscape, as in this view. The
main streams may flow obliquely to the structural
slope as a consequence of having been present here
before tilting commenced. Such streams are antecedent to the geological movement. If they flow across a
structural slope, along the strike, their canyons are
asymmetric: surface water will run down into them
on one side, producing tributary gullies, but not on
the other, leaving a straight cliff there. The Grand
Canyon of the Colorado in Arizona is a notable example. Commonly springs are present on the up-dip side
of the canyon as groundwater leaks through the inclined permeable sandstone or limestone layers.
Large springs create great cavernous alcoves, which
have been used by humans for shelter—whole villages were constructed within them by the Pueblo Indians of the southwestern United States (see inset at
top right).

The cliffs actually retreat by being undermined by
the more rapid erosion of less resistant rocks at their
base. The shale in this view is rapidly removed, causing the more resistant *caprock* (sandstone) to collapse
(see inset at top left). Collapse of rock faces occasionally produces an opening through a narrow wall of
rock, forming a *natural arch*. These should not be
confused with *natural bridges*, which are created
where two meander loops along a canyon stream
wear away the rock between them. Both arches and
bridges are rare except where the sandstone is very
thick and massive.

Mesas result where once-continuous layers of resistant rock have been fragmented by erosion into
widely separated flat-topped remnants. *Buttes* are
remnants too small to preserve a flat summit. Both
mesas and buttes commonly form erosional outliers
along the faces of retreating cuestas.

Where the less resistant rocks, usually shale or
marl, have been stripped off over a wide area, exposing the top of a little-eroded resistant layer, a *stripped
plain* results. Stripped plains commonly expose joint
sets, and these may be exploited by weathering and
erosion to form very rough rock badlands consisting
of humps, fins, and other strange bedrock forms.

Sandstone that collapses due to undermining often
shatters into loose sand. Streams in such areas also
convey a large quantity of sand, which may be blown
out of the dry stream beds between the rare water
flows. As a consequence, sand dunes are common in
these landscapes. In the extreme deserts, such as the
Sahara, the flow of sand driven by the prevailing
winds creates vast areas of sand accumulation, known
as *ergs*.

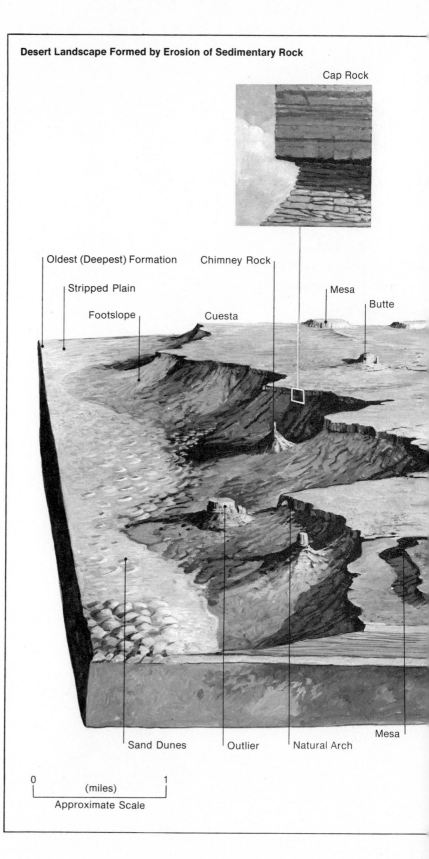

Desert Landscape Formed by Erosion of Sedimentary Rock

Cap Rock

Oldest (Deepest) Formation   Chimney Rock

Stripped Plain   Mesa   Butte

Footslope   Cuesta

Sand Dunes   Outlier   Natural Arch   Mesa

0   (miles)   1
Approximate Scale

Layered structures like those illustrated also are common in humid areas. However, the humid landscape result is different. Cliffs are less distinct due to a blanket of soil that creeps downslope, protected by a vegetative cover. Chemical decay affects nearly all rock types so that lithological variations are less distinct. Natural bridges and arches are rare due to weakness of rock resulting from deep chemical weathering. Dunes are absent because vegetation colonizes debris accumulations, preventing detachment of particles by wind. The often highly-colored bare rock that lends fascination to the desert scene is altogether obscured by a soil cover strongly anchored by vegetation.

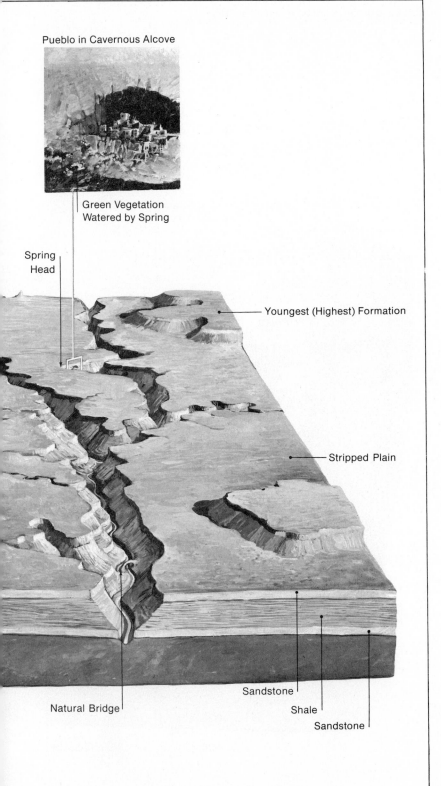

Pueblo in Cavernous Alcove

Green Vegetation
Watered by Spring

Spring
Head

Youngest (Highest) Formation

Stripped Plain

Natural Bridge

Sandstone

Shale

Sandstone

Figure 16.23 Desert landscape—relief by geologic movement. In this landscape, the difference in elevations of the highest mountain crests and the basin floors is a consequence of rupture of the land surface. The basin is an area that has dropped with respect to the adjacent high land. The line along which the rupture and displacement occur is a fault. The vertical displacement may attain thousands of meters. Such geologic structures may occur in any climate. What is distinctive in arid regions are the *playas, alluvial fans,* well-preserved evidences of *fault movement,* and *angularity* of form. In a humid area, such a basin would fill with water, eventually overflowing to produce an outlet channel to the sea. The lake would be eliminated by infilling with sediment to a point where a graded outlet channel would cross it, carrying off all runoff from the surrounding mountains.

In arid regions, water collects in the basin after rains, producing a temporary lake. This soon evaporates, leaving a dry lake bed, or playa. This may be a mud flat, or it may develop a crust of saline minerals precipitated during evaporation of the lake water or brought to the surface by capillary rise from a body of saline groundwater that is near the surface. Where a mud flat is present, the water table is very low and very large lakes have not been present recently (in the last several thousand years). The minerals of saline playas are zoned in order of solubility from carbonates at the edge, through sulphates, to chlorides (such as pure rock salt) in the center. These evaporite minerals are extremely valuable and contribute upwards of $100 million annually to the economy of California alone.

The alluvial fans reflect rapid torrential runoff from slopes lacking vegetative protection. Material flushed from slopes by cloudbursts moves into flooding channels, and is carried out of the mountains. Where the bedrock channel ends, the floodwater spreads and percolates into the gravelly debris at the canyon mouth. The resulting loss of discharge causes the coarse sediment load to settle out, adding to the volume of the fan. Two types of fan are present in this view. On the left, large fans drown the spurs of the down-tilting block, filling the canyons with alluvium. This is because filling of the basin creates a constantly rising base level. Across the basin the fans are small. They lie against the young fault scarp, and the valleys feeding them are actually "hanging" due to rapid uplift along the fault. The outer "toes" of these fans are drowned by alluvium due to subsidence along the right side of the basin block, and this partly accounts for their small size. Ordinarily, these fans are small, because their catchment areas, on the fault scarp, are smaller than catchments on the backslope of the block. The inset shows how fault uplift often creates a "wineglass" valley by rejuvenating erosion, causing slotlike gorges to be cut below older V-shaped valleys. The bowl of the wineglass is the open upper valley; the stem is the new erosional slot; the base is the alluvial fan below.

Fan surfaces in deserts are interesting due to their microtopography. Frequently, hardened streams of mud studded with sizable boulders (1 to 2 meters in diameter) are present, indicating that the fan is built largely by *debris flows*—mixtures of mud, gravel, and boulders that rumble out of the canyons during infrequent very heavy rains that fall on barren surfaces of

Desert Landscape in Areas of Block Faulting

Large Alluvial Fans Coalesce to Form a Bajada (Alluvial Apron)

Water Table   Fan Deposits

loose material. The inset (top right) shows a recent debris flow covering a slightly older one. Surfaces of different ages are apparent in the color they display because with the passage of time desert rocks tend to become coated with a patina of iron and manganese oxides called *desert varnish*. Thus new deposits are light in color, and successively older ones are increasingly darker. All sizable alluvial fans in deserts are a mosaic of tones as a consequence.

The prevailing angularity of form results from the absence of a smoothing cover of soil. Chemical weathering is greatly retarded by lack of moisture, and the absence of a protective vegetative cover precludes soil accumulation because loose material is flushed away by the infrequent rains as fast as it is formed by the weathering processes.

Fault scarps composed of both bedrock and alluvium are better defined than in humid areas because the limitation of weathering and absence of soil creep allow unusually long preservation of newly formed surfaces, even where they are merely banks of gravel and mud. Scarps commonly cut across the upper parts of alluvial fans, indicating recent uplift of the mountain block above them.

Alluvial fans **(below)** slope from bordering mountains toward the basin of Death Valley, California. This view shows the large fans on the west side of Death Valley. (John S. Shelton)

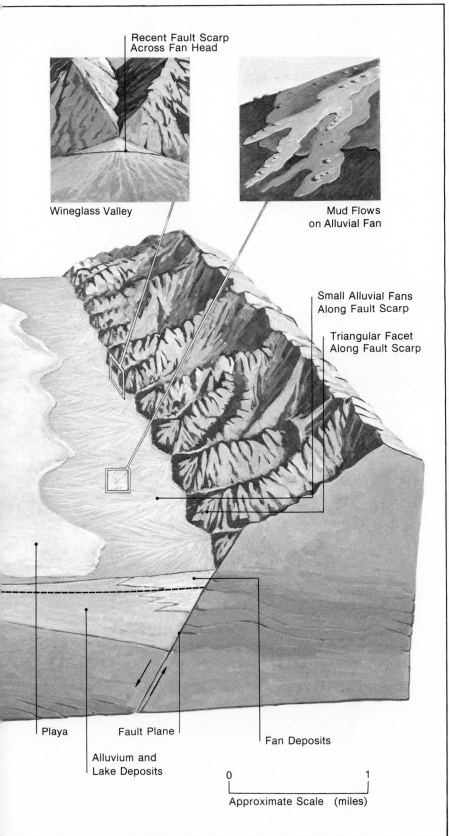

Recent Fault Scarp Across Fan Head

Wineglass Valley

Mud Flows on Alluvial Fan

Small Alluvial Fans Along Fault Scarp

Triangular Facet Along Fault Scarp

Playa

Fault Plane

Alluvium and Lake Deposits

Fan Deposits

0   1

Approximate Scale   (miles)

Figure 16.24 Desert landscape—relief by erosion in crystalline rocks. Where granitic or gneissic rocks appear in deserts, the landscape commonly consists of long, smooth ramps leading up to extremely rough bouldery slopes. The projecting relief forms rise very abruptly above the ramplike surfaces, resembling islands projecting above the sea; hence these summit relief forms are called *inselbergs* (island mountains). The lower portion of the ramp is an alluvial apron, but the upper part is an erosion surface, known as a *pediment*, that truncates solid granitic rock.

The summit relief clearly is a remnant of a long period of erosion, but unlike a monadnock on a Davisian peneplain, it is very distinct from the erosion surface below it, which it meets at a sharp angle in most cases. These steep-sided residuals (inselbergs) are thought to have been formed by the retreat of slopes at a constant angle, which is unlike the mode of slope development thought to characterize most nondesert landscapes, in which slope angle is imagined as diminishing through time. Thus in humid regions the landscape is thought to "wear down," whereas in arid regions it is believed to "wear back."

The crucial factor in the two types of development is thought to be the presence or absence of a mantle of soil subject to the gravitational creep process. Soil covers slopes in humid areas, where it is protected and anchored by vegetation, and slowly creeps downslope as a consequence of gravitational force. Alluvial material collects at the slope foot while removal continues on the higher part of the slope. Both on theoretical grounds and by actual measurement, creep causes slopes to decrease in angle as time passes, if the material delivered to the slope foot is not removed.

The absence of a protective vegetative cover in arid regions prevents soil formation, as weathered material is washed away as fast as it is formed. Slopes retreat through the erosive effect of running water rather than by soil creep; there is no accumulation at the slope foot; and this leads to parallel slope retreat, both theoretically and by measurement in the field.

In the landscape shown, the bouldery slopes are a consequence of the tendency of granitic rocks to be well jointed, forming a rigid mass of plane-faced blocks, which weathering tends to loosen and round at edges and corners. The steep slopes are regarded as the angle of repose of the loose blocks produced by the weathering process.

The lower blocks of the inselberg slopes are commonly perforated by cavities produced by desert weathering processes. The exteriors of the boulders may be very dark due to patination by desert varnish.

According to recent investigations of such landscapes, their development may have been misunderstood by geomorphologists. Evidence of the presence of a soil formerly covering them suggests that they are relict from a more humid period, having been altered to their present form by relatively recent erosion under desert conditions—the erosion being triggered by loss of the former vegetative cover. Nevertheless they evidence parallel slope retreat, even in the former soil-covered landscape, indicating efficient removal of material arriving at the slope foot. Thus the period of pediment formation seems to have been one of semiarid conditions, with erosion by surface runoff being more effective than the soil creep process.

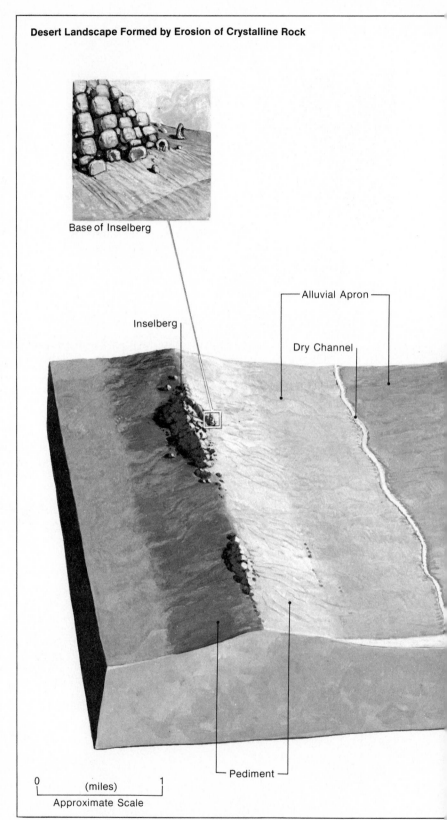

Desert Landscape Formed by Erosion of Crystalline Rock

Base of Inselberg

Inselberg

Alluvial Apron

Dry Channel

Pediment

0   (miles)   1
Approximate Scale

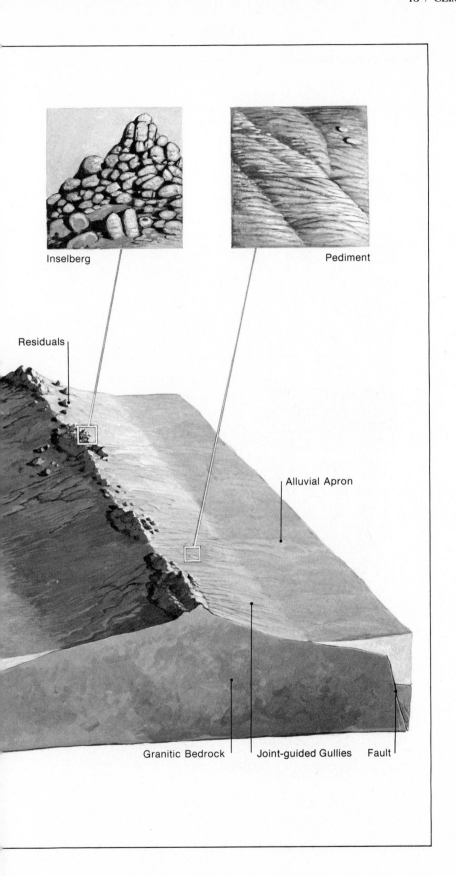

Inselberg

Pediment

Residuals

Alluvial Apron

Granitic Bedrock   Joint-guided Gullies   Fault

Figure 16.25 Tropical savanna landscape. This landscape is widespread in the seasonally wet tropics on stable continental areas wherever crystalline rocks are dominant, as in Brazil, much of Africa south of the Sahara, India, and Australia. The term "savanna" refers to the vegetation association consisting of drought- and fire-resisting tree species rising from a sea of grass. The important geomorphic features are very *flat plains*, into which only shallow saucer-shaped valleys are incised by ephemeral streams; *escarpments*, over which streams cascade spectacularly; and enormous rock domes, termed *bornhardts* (after the German scientist who brought those in Africa to the world's attention).

This peculiar landscape is the consequence of an aggressive climate and the peculiarities of granitic rock. The mode of landscape development has been characterized as *double planation*, as there are two separate interfaces along which geomorphic activity takes place. One is the land surface, subject to rainwash, gravitational transfer, and stream erosion. Erosional removal lowers the land surface and wears back projections upon it. The second interface is that at which decayed rock meets fresh, unaltered rock—the *weathering front*—which lies below the land surface.

The unusual aspect of this landscape is the aggressive chemical weathering under warm tropical conditions that decomposes rock in the subsurface as fast or faster than surface erosion can wear down the land. Boreholes in some locations in Africa indicate the weathering front to be 100 meters or more below the surface. They also show it to be highly irregular. Where the granite is closely jointed, the rock is deeply decayed, as water can penetrate it effectively. But where jointing is sparse, water is excluded and chemical decay cannot begin. Bornhardts, rising above the surface as much as 500 meters, are cores of monolithic granite that resisted chemical decay below the surface. They have been left behind as the land surface wears down and escarpments retreat. *Tors* are piles of boulders isolated by chemical decay below the land surface but exposed by erosion before they are completely decomposed.

The flat surrounding plains are armored with a *laterite* crust—a residue of oxides and hydroxides of iron and aluminum resulting from aggressive chemical alteration of the decomposed rock, such that all bases and even the silica are released and leached away. Mobilization of iron and aluminum and their fixation in the crust seem to require a water table that fluctuates in level seasonally, as it does in the savanna regions—the iron and aluminum being derived from the zone in which the water table fluctuation occurs. Because the water table varies in height most under hills and least under valleys, there is more chemical erosion under hills, so that their crowns literally sink as subsurface removal occurs. This probably contributes to the flatness of the laterized plains.

A last peculiarity of savannas is the presence of ungraded streams and the absence of stream canyons of any significant extent. Of course the two are associated. Canyons are cut by streams carrying abundant tools for abrasion, which permits irregularities in stream beds to be filed smooth. The absence of canyons and presence of ungraded profiles, including many waterfalls and cascades, are a consequence of a scarcity of abrasional tools. Stream bed loads in many

Tropical Savanna Landscape

Bornhardt

Laterite Crust

Saucer-shaped Valley (Dry)

Decomposed Rock

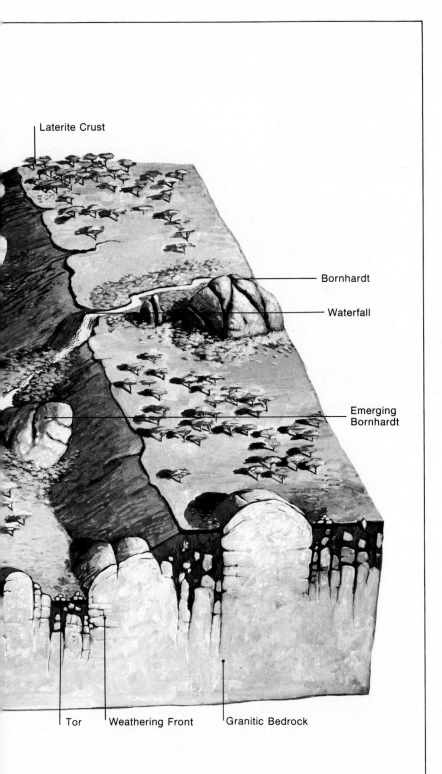

Laterite Crust

Bornhardt

Waterfall

Emerging Bornhardt

Tor   Weathering Front   Granitic Bedrock

tropical savanna regions lack cobbles because thorough chemical decay destroys coarse granitic rock waste before it reaches stream channels, which therefore transport only sand and silt. Rocks other than granite and related types do not decay as massively in tropical climates, although they may decay equally rapidly. Whereas chemical weathering eventually causes a granitic block to collapse into sand by permeating its structure, it will attack a basaltic block on the surface only, constantly reducing its size but permitting a durable kernel to persist in its interior. Thus basaltic outcrops are an excellent source of tools for abrasion by streams whatever the climate.

The massive granite dome of this bornhardt (**below**) in Rhodesia, Africa, rises abruptly from the flat plain of the surrounding savanna.

Figure 16.26   Tropical high island landscape—volcanic dome. The distinctive aspect of humid tropical landscapes is the effect of water in very copious amounts, combined with high temperatures. Chemical weathering is very active, producing soils even on near-vertical slopes. The soil is often saturated with moisture, leading to landsliding, and water runoff attains enormous volumes that produce highly intricate sculpture of steep slopes. Areas of granitic, metamorphic, and sedimentary rocks are generally maturely dissected, having a very high drainage density and slopes steeper than those in higher latitudes.

Certain geological configurations produce truly unusual landscapes in the tropics, such as those illustrated here, which occur widely. The figure shows the eroded flank of a volcanic dome, such as those composing the Hawaiian Islands. The landscape is peculiar in several ways relating to both the climate and the geological components present. The valley walls are as steep as 80°, yet covered with vegetation rooted in a shallow soil. Due to saturation with moisture, landslides are frequent and their scars are everywhere, exposing the bare basaltic lava composing the dome. The valley walls are fluted, each flute becoming a waterfall or cascade during the frequent rains. These flutes have been characterized as "vertical valleys." Where adjacent valleys have expanded so that their walls merge at the top, extremely jagged crestlines result. The main streams drop into the troughlike valleys over a series of waterfalls, each one terminating in a small *plunge basin*. The ephemeral waterfalls along the valley sides also drop from one plunge basin to another. Above the abrupt valley heads there may be an upland swamp; that on the Hawaiian island of Kauai may be the wettest place on earth (annual precipitation may aproach 1,500 cm, or 600 inches). Many of the valleys have flat floors resulting from subsidence of the islands and infilling by alluvium. It appears that the precipitous nature of the valleys dissecting these volcanic domes may be due to the low water table resulting from the permeability of the lava flows composing them—thus they are almost tantamount to gigantic spring alcoves, expanding due to sapping by spring action at their base.

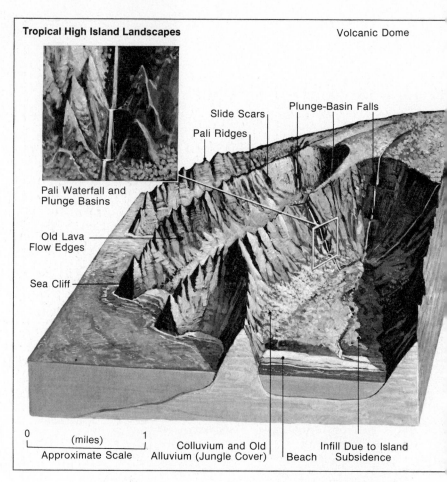

Tropical High Island Landscapes   Volcanic Dome

Pali Waterfall and Plunge Basins

Slide Scars

Plunge-Basin Falls

Pali Ridges

Old Lava Flow Edges

Sea Cliff

0   (miles)   1

Approximate Scale

Colluvium and Old Alluvium (Jungle Cover)

Beach

Infill Due to Island Subsidence

SUMMARY

Climate is a crucial factor in the processes of weathering and erosion that create most landforms. Climatic effects are visible on a variety of scales, from those seen on individual rocks to those imprinted over entire landscapes. Examples are undercut "mushroom" rocks in arid regions, asymmetric valley slopes related to differences in solar radiation and moisture availability, and the large-scale morphogenetic regions dominated by combinations of geomorphic processes that are characteristic of particularly aggressive climates.

Highly distinctive landforms are associated with climates that are dominated by extremes of cold, aridity, or heavy rainfall. These landforms often persist as relict features after the climate has altered to a less aggressive type. Although many individual landforms cannot be associated with particular climatic regions, the various landforms that collectively constitute landscapes are usually a fairly clear expression of the region's existing climate or prior climatic history. Thus it is possible to recognize a number of morphogenetic regions in which the

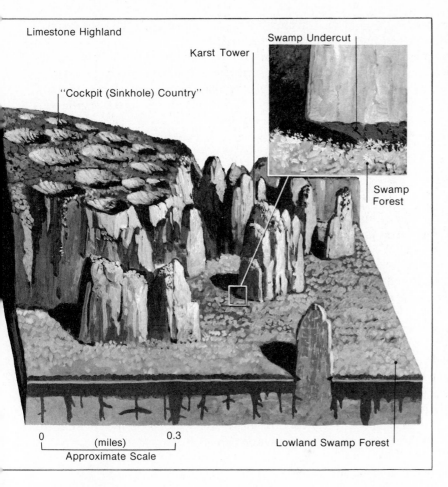

Limestone Highland

Karst Tower

Swamp Undercut

"Cockpit (Sinkhole) Country"

Swamp Forest

0   (miles)   0.3
Approximate Scale

Lowland Swamp Forest

Figure 16.27 Tropical limestone highland. Another highly distinctive landscape is found where massive limestones appear in moist tropical areas of strong relief. Plateau tops are pockmarked with large sinkholes, produced by the solution of limestone along intersecting joints. Due to its content of organic acids, groundwater in the tropics is particularly aggressive toward limestone, which it corrodes rapidly. Between the sinkholes is a maze of often sharp-crested limestone ridges. This terrain has been termed *cockpit karst*. Solution working deep into the joints at the plateau edge detaches enormous slabs and pinnacles, which become isolated as vertical-sided mountains hundreds of meters in height. These have been made famous in Oriental art. They are common in southwestern China, North Vietnam, Malaysia, and Indonesia. Once isolated, these *karst* towers are kept steepsided by undercutting and collapse. They rise from swampy lowlands, in which the water has been made acidic by contact with peat and other products of vegetative decay. This water corrodes a notch, sometimes 5 or 6 meters deep, around the karst towers. The overhangs collapse periodically, after which a new notch is formed, keeping the towers permanently "girdled." Tower karst develops only where limestone free of impurities is divided by well-defined but widely spaced joints. Where the limestone contains impurities of clay, marl, or chert, the towers are replaced by less imposing cone-shaped hills, which are common in Southeast Asia, Puerto Rico, and Cuba. Thus most tropical karst topography may be resolved into cockpit, tower, or cone karst, none of which is present in higher latitudes.

landscape reflects the domination of certain climatically controlled processes of weathering and erosion. Certain of these processes are clearly expressed, but many parts of the earth are transitional between them. Some regions have experienced the effects of different climates at different times; one such is the humid midlatitude morphogenetic region, which was the source of many of our early ideas concerning landform development. It presently experiences a wide range of geomorphic processes but is dominated by no single one.

A highly distinctive morphogenetic regime is that of cold "periglacial" climates, in which rock fragmentation by the freezing of water combines with surface volume changes and mass movements related to permafrost and an overlying seasonally thawed layer. A host of unique landforms is developed by this climatic regime. Dry-summer subtropical ("Mediterra-

nean") climates promote occasional summer fires followed later by winter debris floods, along with Vertisol soils that liquify and flow or slide when saturated with moisture.

In the arid realm the absence of vegetative protection and a soil cover results in angular rocky landscapes in which geological structure is strongly expressed. Depositional landforms are conspicuous in arid regions, owing to the lack of permanent streams and integrated outflowing drainage systems. The weakness of weathering in arid climates permits the preservation of relict forms such as pediments inherited from preceding semiarid to subhumid periods.

Tropical wet climates generate large rivers in forested lowlands, and intricate mazes of sharp-crested steep slopes in upland areas, where soil sliding repeatedly disturbs the forest cover. Massive limestone in the tropics is the basis for picturesque karst landscapes unlike those seen anywhere in the higher latitudes. In the summer-wet tropical savannas vast areas of plains are crusted with laterite over deeply rotted bedrock, and are frequently overlooked by bornhardt domes and retreating escarpments. In this climate, valley deepening seems to be in-hibited by a predominance of fine particles in stream bed-loads.

Areas of transitional climates, and those having experienced climatic changes, exhibit variations, combinations, and relics of the landforms developed by the more aggressive climatic regimes.

REVIEW QUESTIONS

1. Why are there no "normal" landscapes associated with particular rock types and geologic structures?
2. Outline the relationships between climate and weathering, and illustrate them with a diagram relating weathering to energy and moisture conditions.
3. What role does vegetation play in landform development?
4. How is vegetation usually related to slope asymmetry in the middle latitudes?
5. What is a "climamorphogenetic region"?
6. List five cryogenic landforms, and indicate how each one is thought to develop.
7. Discuss the problem of construction in the tundra environment.
8. What is distinctive about the geomorphic system in humid midlatitude regions?
9. Why is the dry-summer subtropical region recognized as a discrete morphogenetic entity?
10. Depositional features seem to be conspicuous in arid regions. Why?
11. What is a "pediment" and how is it thought to form?
12. Why does it seem that erg formation requires climatic changes?
13. Characterize the appearance of highlands in the wet tropical region.
14. Trace the sequence of events leading to a bornhardt.
15. Are mesas and cuestas restricted to arid environments?

APPLICATIONS

1. In what ways do specific landforms in your area reflect characteristics of the local climate?
2. Is there any geomorphic evidence of climatic change in your region? If not, what is your hypothesis about the landform modifications that would occur if the area became cooler, wetter, warmer, or drier than at present?
3. What sort of relict landforms would you expect to find in a desert? In a semiarid region? In a midlatitude forest region?
4. How do plate tectonic processes encourage the development of relict landforms?
5. Does a climamorphogenetic regime that has one exceptionally aggressive process usually "pay" for this by excessive weakness in some other important process of landscape formation?
6. How could the "Fall Line" discussed in Chapter 14 be explained in terms of "climatic geomorphology"?
7. How would you rank the morphogenetic regimes discussed in this chapter in terms of rates of landscape lowering by denudation?

FURTHER READING

Birot, Pierre. *The Cycle of Erosion in Different Climates.* Translated by C.I. Jackson and F.M. Clayton. Berkeley and Los Angeles: University of California Press (1968), 144 pp. This is a highly condensed outline of the erosional systems associated with different global climatic regimes—sparsely illustrated with diagrams.

Budel, Julius. *Climatic Geomorphology.* Translated by L. Fischer and D. Busche. Princeton, N.J.: Princeton University Press (1982), 443 pp. This volume offers the ideas of one of the leading proponents of climatic geomorphology as a field of research. Translated from the German edition of 1977. The perspective is entirely different from that of American geomorphologists.

Butzer, Karl W. *Geomorphology from the Earth.* New York: Harper & Row (1976), 463 pp. This textbook devotes about 100 pages to landform systems in different climatic regions. The treatment is up-to-date and more geographical in tone than that of most geomorphology texts, but is very concise.

Cooke, Ronald V., and **Andrew Warren.** *Geomorphology in Deserts.* Berkeley and Los Angeles: University of California Press (1973), 384 pp. This is a thorough and most readable presentation of research findings concerning present geomorphological processes in desert environments. The last third of the book, devoted to sand dunes, is rather technical.

Douglas, Ian. *Humid Landforms.* Cambridge, Mass.: M.I.T. Press (1977), 288 pp. For its size, a surprisingly thorough treatment of the landforms of humid regions, with emphasis on the wet tropics. Process-oriented.

Ferrians, O. J., R. Kachadoorian, and **G. W. Greene.** "Permafrost and Related Engineering Problems in Alaska." *U.S. Geological Survey Professional Paper No. 678* (1969), 37 pp. This treatise offers an interesting and well-illustrated account of the unusual problems encountered in the construction and maintenance of roads, railroads, bridges, and buildings in areas underlain by permafrost.

Garner, H. F. *The Origin of Landscapes.* New York: Oxford University Press (1974), 734 pp. Much of this large and well-illustrated volume is concerned with the geomorphic effects of varying climates, with considerable attention given to the specific role of climatic change in the formation of landscapes.

Mabbutt, J. A. *Desert Landforms.* Cambridge, Mass.: M.I.T. Press (1977), 340 pp. An interesting contrast to Cooke and Warren. A very compact but full treatment, with emphasis on different landscape types.

Stoddart, D. R. "Climatic Geomorphology: Review and Re-assessment." *Progress in Geography,* Vol. 1 (1969), pp. 161–222. Stoddart's 60-page article reviews the evidence for and against climatic geomorphology as an alternative to other schools of thought in landform studies. The main controversy is whether unique climates produce unique landscapes, or whether climatically induced morphological differences are more apparent than real. A wealth of excellent information is assembled and subjected to a critical eye.

Thomas, Michael F. *Tropical Geomorphology.* New York: Halsted Press (1974), 301 pp. This work focuses upon the bornhardt landscapes of savanna regions and is really a book on the behavior of granitic rocks in tropical wet and dry climates. The author is a leading investigator in this field, and the book mainly concerns his own findings.

Tricart, Jean. *The Landforms of the Humid Tropics, Forests, and Savannas.* Translated by C. J. Kiewiet de Jong. New York: St. Martin's Press (1973), 352 pp. This French text offers many interesting observations on the uniqueness of geomorphic processes in humid tropical and savanna regions.

——, and **A. Cailleax.** *Introduction to Climatic Geomorphology.* Translated by C. J. Kiewiet de Jong. New York: St. Martin's Press (1973), 295 pp. Tricart and Cailleux present the European view in favor of studies of geomorphic processes as conditioned by climate and the biosphere. The work overtly challenges the Davisian system, which the authors believe still impedes the progress of landform analysis. The treatment is interesting and nonquantitative.

Washburn, A. L. *Geocryology: A Survey of Periglacial Processes and Environments,* 2nd ed. New York: Wiley (1980), 406 pp. This is the most thorough current analysis of geomorphic processes and problems in periglacial environments, by a leading contributor and field investigator. Photographic illustrations are outstanding.

CASE STUDY

Climatic Change and Landforms

We have noted that the sawtooth peaks of mountains like the Rockies in Colorado, Wyoming, and Montana are an obvious indication of climatic change. As we shall see in Chapter 17, there is no mistaking the meaning of such glacially produced landforms in areas that today have only tiny glaciers or none at all. Equally obvious indicators of climatic change are the extensive systems of dry canyons in the deserts of North Africa and Arabia. These so-called *fossil wadis* seldom carry water and are clogged with dune sand and talus. However, they were obviously formed by stream erosion in the past. They too are a relict of the Ice Ages, which produced more rainfall and runoff in the subtropical regions. Less apparent effects of climatic changes have been imprinted upon virtually all land surfaces, though it may require an informed eye to detect them. Certain widespread types of scenery could only have been produced by climatic change, which is to say, by two dissimilar systems of geomorphic development acting at different times.

How did the major long-term climatic changes recorded by landforms come about? Several possibilities exist. We have seen that a change in solar output or the earth's relationship to the sun is thought by many to be the cause of the Ice Ages. The slow rise of mountain ranges, or their lateral migration by strike-slip faulting, gradually produces rainshadows extending far from the mountains, creating semiarid or desert areas. The changing elevations of the continents themselves may affect climates. Ocean currents are a strong climatic influence and must change as continents and ocean basins change shape and position due to sea-floor spreading. Of course, sea-floor spreading has caused the continents themselves to be moved across the planetary surface, from the poles to the tropics and from the tropics to the poles. Fossils of tropical plants are common in places like Spitzbergen, Greenland, and Antarctica, and ancient glacial deposits can be viewed in Brazil, South Africa, Australia, and the Sahara Desert.

French and German scholars were the first to comment, in the 1950s, on the red soils preserved under lavas in the central Sahara. Close examination of these soils in the midst of the world's largest desert revealed them to have been formed under summer-wet tropical savanna conditions. They indicate clearly that the Sahara has not been a permanent feature of North Africa. In fact, the erosional forms in granitic rocks in the central Sahara are very much like those seen in the summer-wet tropics, requiring intense chemical weathering in a moist climate some 20 to 50 million years in the past. Some French scientists have expressed the opinion that certain granitic domes and spires in the Sahara may be more than 50 million years old. Clearly these forms could not have originated in the present Saharan climate.

Large areas of the arid heart of Australia are dominated by flat-topped tablelands standing above a rough rocky relief typical of deserts. The level upper surface consists of a massive layer of iron- and aluminum-rich laterite produced by the soil-forming process that dominates in tropical savanna regions. Where the soil parent material lacked iron and aluminum, the capping layer is cemented by silica, being known as *silcrete*. It is impossible for either laterite or silcrete to form in a desert. Thus desert conditions must have become dominant after these soils had formed under a seasonally wet tropical climate. And long before the tropical phase, the same area, like the Sahara, was covered by an ice sheet.

Wherever deserts have been investigated, evidence has been found indicating a long earlier period of non-desert climate. In the Mojave Desert of southern California, as in the Sahara, a red-colored weathered mantle is preserved beneath lava flows having potassium/argon ages of 6 to 10 million years. This suggests a long predesert phase of deep bedrock weathering, with adequate vegetation to prevent immediate erosion of the weathered material. This is very significant in landform interpretation. Among the most common landforms in many deserts are erosional *pediments*—smooth ramplike rock surfaces that slope outward from the foot of rocky hillslopes. Pediments often extend 3 or more kilometers (2 or more miles) from associated hills and signify the trimming away of enormous volumes of rock, including whole mountain ranges in some areas. Because they are so obvious in deserts, pediments were long regarded as forms left by the wearing back of erosional hillslopes under arid conditions.

But pediments, though best seen in deserts, appear to have originated under non-desert conditions. In the California desert, lavas as much as 10 million years old rest on the pediments, indicating that the pediments

Island-like masses of granite rising above the bedrock erosional surfaces are known as pediments. *This view is in California's Mojave Desert. Pediments are widely distributed in arid regions and were produced by erosion during a period of greater moisture availability millions of years ago. (T. M. O.)*

are quite ancient. But there was no desert here 10 million years ago! The Mojave is largely a rainshadow desert, produced by mountains that did not rise strongly until about 6 million years ago. Thus the pediments are of predesert origin, as indicated by the fragments of red soils preserved on them. This is reasonable, since the wearing back of hillslopes required to produce large pediments requires a far more efficient erosion system than that existing in deserts. Since all desert regions show evidence of a lengthy predesert history, all so-called "desert pe-

diments" are most easily interpreted as relict landforms inherited from an earlier, more humid period. Pediments are well exposed in deserts because the loss of their original vegetation cover has allowed erosion to remove the soil that once covered them.

The "normal" landforms of unglaciated, humid midlatitude areas may also be relict features. Characteristic hillslopes in the midlatitudes are smooth and ungullied, convex at the top and concave at the foot. Many are beginning to be gullied today, changing the slope form. Most Euro-

pean investigators interpret this form association as evidence of slope smoothing by periglacial solifluction during the Ice Ages, followed by renewed dissection by running water in the post-glacial period. The evidence for this is strong in Europe but less so in North America; however, in North America little research has been done to resolve the question. Further investigation may reveal that all of the earth's landscapes bear the imprint of changing climates and changing geomorphic processes.

473

Maligne Lake, Jasper Park by Lawren S. Harris. (The National Gallery of Canada, Ottawa)

Many of the earth's most spectacular landforms were created by glaciers. Flowing ice is one of the most forceful agents of landscape change, and many regions owe their physical characteristics to glacial erosion or deposition.

We live perilously close to an Ice Age today, both in time and in terms of global climatic conditions. Only 15,000 years ago the sites of Chicago, Detroit, Minneapolis, Seattle, Vancouver, Toronto, and Montreal were buried under a thousand or more meters of ice. The Mississippi River was then an enormous braided stream that carried milky glacial meltwater down to a sea whose level was a hundred meters lower than at present. If there had been no Ice Age, there would be no Matterhorn, no Denmark, no Great Lakes, and no incredibly productive agricultural heartland in Saskatchewan, Manitoba, the Dakotas, Iowa, Illinois, Indiana, and Ohio. The Norwegian fjords, Niagara Falls, Chesapeake Bay, and Long Island would be missing from the map. All these were produced directly or indirectly by the formation and slow advance of both confined streams and continent-sized blankets of ice known as *glaciers.*

Ice Age conditions are likely to be repeated within the next 20,000 years. Indeed, the earth's average temperature now is about 2°C cooler than it was 6,000 years ago. A moderate increase in winter precipitation along with a decrease of merely 5° to 7°C (9° to 13°F) in the earth's average temperature would be sufficient to cause ice sheets to invade the northern hemisphere continents once again.

Conditions on earth today are so close to those of a glacial age that three-fourths of the earth's nonsaline water supplies are presently stored as ice, which covers 10 percent of the land surface. This is an exceptional condition that has occurred only occasionally throughout geologic time. The world's existing glaciers hold an amount of water equal to 5,000 years' flow of the earth's greatest river (the Amazon), or 60 years of rain and snow over the whole planet. By comparison, the storage of water in lakes, rivers, swamps, and artificial reservoirs is trivial.

Most of this frozen water resides in the immense Antarctic and Greenland ice sheets—85 percent in the Antarctic ice sheet alone (Figure 17.1). During much of the preceding 2 million years, two additional continent-sized ice sheets existed, covering all of northern Europe and

17
Glaciers and Glacial Landforms

Figure 17.1 The Antarctic continent is blanketed by some 29 million cubic kilometers of glacial ice that constitutes the earth's principal storehouse of non-saline water. As shown here, the ice, which elsewhere forms vast featureless expanses, also extends up the slopes of high mountains, but leaves a few small land areas uncovered. This view in Victoria Land includes 3,300-meter Mt. Herschel. (Warren Hamilton, U.S.G.S.)

the melting of enormous masses of debris-laden ice has completely submerged the previous landscape under a blanket of boulders, gravel, sand, silt, and clay. This has created rolling or flat plains where once there were hills and valleys produced by fluvial erosion. In some areas the effects of glaciers were detrimental to later human use of the land; in others the effects were very beneficial. This chapter explores these effects and the processes responsible for them.

GLACIATION PRESENT AND PAST

Glaciers form in the cool climates of high latitudes and lofty elevations where more snow falls each winter than can melt or evaporate during the succeeding warm season. In such locations the snow accumulates from year to year and the basal layers are gradually recrystallized into solid ice. Its increasing weight and low internal strength cause the ice to deform. A glacier is a mass of ice that is "flowing" under gravitational stress.

Glacier Types

Glaciers exist in several different forms (Figure 17.2). The broadest distinction is between confined and unconfined glaciers, which produce quite different erosional and depositional ef-

Canada, as well as 2.6 million sq km (1 million sq miles) of the United States. Smaller ice fields blanketed portions of the world's great mountain ranges. Strangely enough, most of Siberia, widely regarded as almost glacial today, was too dry to sustain large glaciers even at the peak of the Ice Age.

Glaciers of the past are of interest because they have significantly altered the environment we inhabit. Over vast areas in the higher latitudes, as well as at high altitudes, slowly moving currents of ice have scraped away all the soil and much of the weathered mantle that formerly covered the land surface. Elsewhere

Figure 17.2 Glaciers form where the annual accumulation of snow exceeds annual losses due to melting and evaporation. Glaciers may be unconfined, as *highland icecaps*, which submerge the older erosional topography in mountain regions, or confined, as *cirque* and *valley glaciers*. All glacial ice drains to lower elevations under gravitational stress. The valley glacier descending from the unconfined icecap is an *outlet glacier*. In some regions, the ice from one or more valley glaciers spreads out over a plain and forms a *piedmont glacier*. (John Dawson)

The surface of this valley glacier in the French Alps is broken by numerous crevasses. Cirques are visible in the far distance. (Anita Kolenkow)

The Malaspina Glacier in southern Alaska forms an extensive piedmont ice sheet that terminates at the coastline of the Gulf of Alaska. The ice flow is from right to left in this picture. The dark bands are deformed layers of rock debris carried within the ice. (Austin Post/U.S.G.S.)

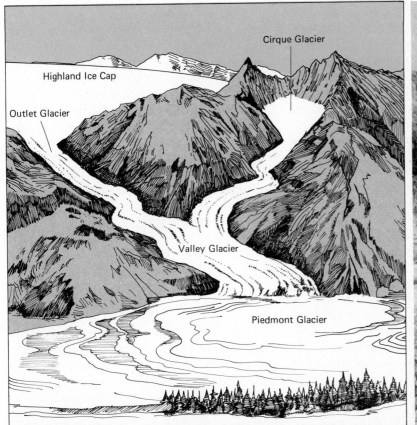

fects. Dwarfing all other types are the great *continental ice sheets*. These are unconfined blankets of glacial ice that submerge the land surface over areas of millions of square kilometers. Two exist today—the Antarctic ice sheet, which covers 12.5 million sq km (4.8 million sq miles), and the Greenland ice sheet, which has an area of about 1.7 million sq km (650,000 sq miles). Somewhat similar, but far smaller, are unconfined *highland ice caps* that blanket hundreds to thousands of square kilometers of mountainous terrain in the higher latitudes. Highland ice caps are conspicuous in Iceland, the Canadian Arctic Islands, and the Canadian Rockies. Often they find outlets to lower elevations through valley systems, resulting in confined *outlet glaciers* (Figures 17.2 and 17.20).

More common in high mountains are still smaller *alpine glaciers*. This term includes both confined ice fields, which occupy depressions below high mountain crest lines, and streams of ice that are hemmed in by valley walls as they drain from mountain crest lines to lower elevations. Alpine glaciers originate in high altitude rock-walled ice reservoirs called *cirques* (French, from Latin *circus*, or circle). Cirques are steep-walled rock basins having a distinctive appearance resulting from frost action and glacial erosion. In many mountain areas the only glaciers present are *cirque glaciers*—masses of ice that are restricted to cirques and do not enter valleys. Cirque glaciers range in area from less than one to five or more square kilometers (2 sq miles).

Where glacial ice does spill from cirques into the valleys below, it moves slowly in channeled streams called *valley glaciers*. These may have lengths of several tens of kilometers and depths of 1,000 meters or more (over 3,000 ft). Both valley glaciers and the outlet glaciers draining highland ice caps occasionally flow into open areas where they spread out in unconfined pools of ice known as *piedmont glaciers* (Figure 17.2). Thus alpine glaciers can take three forms: cirque glaciers, valley glaciers, and piedmont glaciers. Alpine glaciers are best developed in Alaska; the Canadian Rockies; the European Alps; the Caucasus of the Soviet Union; the Himalaya, Karakorum, and Tien Shan ranges of Asia; the Andes of South America; and the southern Alps of New Zealand.

More than 1,000 glaciers exist in the coterminous United States, most of them in Washington's Cascade Range. Nearly all are small, having a combined area of barely 500 sq km (200 sq miles). Nevertheless, they are important water sources, providing about 2.1 billion cu meters (1.7 million acre ft) of meltwater each year. Melting of glacial ice in the summer provides water for irrigation just when the demand is highest and when streams fed by rainfall and groundwater inflow are lowest. North America's greatest glaciers are in Alaska, where they cover more than 50,000 sq km (20,000 sq miles). The Malaspina glacier system of Alaska's St. Elias Range is itself more than 4,000 sq km (1,500 sq miles) in area (Figure 17.2).

Pleistocene Glaciation

We live in the *Holocene* epoch of the Cenozoic era (see Figure 1.8, p. 16)—the "post-glacial" period initiated with the rapid shrinkage of the great northern hemisphere ice sheets, beginning about 14,000 years ago. The last remnants of these ice sheets melted away some 6,000 years ago. The preceding *Pleistocene* epoch—the time of alternating cold (glacial) and warm (interglacial) periods—began about 1.8 million years ago, and is probably not truly ended. In other words, the Holocene epoch seems to be the latest in a series of interglacial interludes preceding glaciations yet to come. Most scientists attribute climatic cooling to astronomical phenomena such as the tilt of the earth's axis, the elongation of the earth's orbit, and the locations in the orbit at which the equinoxes and solstices occur (see Case Study following Chapter 8). On this basis it has been predicted that the earth should enter another Ice Age in approximately 23,000 years.

In the past 800,000 years there have been seven or more major episodes of glaciation, each lasting tens of thousands of years, separated by warmer interglacial periods when only Antarctica and Greenland maintained ice covers. There have been other Ice Ages in the earth's history, but they occurred hundreds of millions of years ago and have left no imprint on present landscapes.

At their maximum, Pleistocene glaciers

sprawled over some 44 million sq km (17 million sq miles)—almost one-third of the earth's present land area. The advances and retreats of these glaciers left clear evidence in the form of ice-scoured bedrock and vast blankets of glacially transported rock debris. The North American continental ice sheets, known as the Laurentide ice sheets, spread outward from the vicinity of Hudson Bay and at their maximum terminated close to the present line of the Missouri and Ohio rivers (Figure 17.3). Europe was invaded by the Fennoscandian ice sheet, which originated in northern Scandinavia and spread westward to an ice cap covering the British Isles, and southward to the uplands of Central Europe. In high mountains all around the world, erosion by confined cirque and valley glaciers completely transformed the scenery, giving it the picturesque "alpine" appearance.

The Pleistocene ice sheets and alpine glaciers left a magnificent legacy of landscapes especially suited for human recreation and enjoyment. The pinnacle of the Matterhorn in Switzerland, the fjords of Norway, the ski slopes of Tuckerman Ravine in New Hampshire, the deep gouge of California's Yosemite Valley, the sandy arm of Cape Cod in Massachusetts, and the ten thousand (and more) lakes of Minnesota—all were fashioned by Pleistocene glaciers.

The Indirect Effects of Pleistocene Glaciation

The effects of Pleistocene glaciation are not limited to the area actually covered by ice. The glaciers themselves, their deposits, and the cold climate accompanying them, had a variety of effects on landscapes. Wind erosion of loose, unvegetated glacial deposits coated the United States Midwest with loess, providing fertile parent material for soil development. Another productive loess belt having a similar origin lies along the southern margin of the North European plain from France to Poland, and is the most heavily populated portion of Europe. Meltwater streaming from the ice aggraded some valleys with sediment that was later carved into stream terraces that extend far beyond the boundaries of the glaciers. Drainage

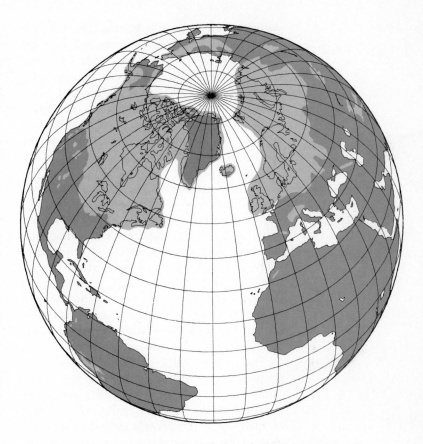

Figure 17.3 This map compares the existing northern hemisphere glaciers with the much larger Pleistocene continental ice sheets that attained their most recent maxima between 20,000 and 15,000 years ago. During the glacial stages of the Pleistocene, the Fennoscandian ice sheet covered much of northern Europe, while the Laurentide ice sheet blanketed nearly all of Canada and reached deep into the Midwest of the United States. Highland ice caps and alpine glaciers were also greatly expanded during the Pleistocene glacial stages, but most were too small to depict at this scale. (T. M. Oberlander, and Vantage Art, Inc.)

patterns were altered as glaciers blocked river systems and forced tributary streams to overflow into new paths. Portions of the Ohio and Missouri rivers were created as a series of spillways that carried water from one blocked drainage to another along the front of the Laurentide ice sheet. The landscape around the German city of Berlin is furrowed with a network of relict glacial spillways, which are now occupied by canals.

The cold climates of the Pleistocene intensified the breakup of rock by frost weathering. Over vast areas the ground became frozen solid, with only the surface soil thawing each summer. Cryogenic periglacial landscapes evolved, consisting of many distinctive elements, as seen in Figure 16.21, p. 458. Active examples of such landscapes exist today only in tundra regions and the cold margins of the coniferous forests, as in Arctic Canada, Alaska, Siberia, and certain highland areas.

The average thickness of the Pleistocene continental ice sheets is estimated to have been 1 to 2 km (0.6 to 1.2 miles), with a maximum thickness of about 4 km (2.5 miles). The weight of this ice—approximately 900,000 kg (1,000 tons) for each square meter of area—caused the earth's semiplastic crust to sink under the burden. The crust was depressed by as much as 1,200 meters (4,000 ft) under the thickest parts of the ice sheets. Ever since the retreat of the ice, the crust has been slowly returning to its preglacial level. Parts of Scandinavia and North America are continuing to rise, or "rebound," by as much as 2 cm per year; the area at the center of the former Laurentide ice sheet has risen 170 meters (560 ft) in the last 7,000 years.

For ice sheets to grow on the land, water must be removed from the seas. Consequently, a world-wide lowering of sea level has accompanied every period of glacial advance. The lowering of the sea causes rivers to deepen their valleys near their previous outlets. Several times during the Pleistocene epoch, sea level was lowered 100 to 150 meters (330 to 500 ft). Evidence for this is the fact that river valleys are cut into bedrock on the continental shelves far below present sea level.

The melting of the ice sheets restored the sea to its present level, drowning the deeply cut river valleys. This produced deep estuaries that now form some of the world's best harbors, such as those at Montreal, New York, Norfolk, San Francisco, and Seattle. If the Antarctic and Greenland ice sheets were to melt away completely, sea level would rise another 65 meters (210 ft), which would submerge most of the earth's largest cities and send arms of the sea far up the heavily populated valleys of the Mississippi, Ganges, Indus, Nile, Yangtze, and Yellow rivers, as well as into many other interior lowlands.

When sea level fell during the Pleistocene cold phases, the land area increased greatly, as shown in Figure 17.4. The Florida peninsula doubled in size, and land bridges joined Ireland, England, and France, as well as most of the islands of Indonesia and Malaysia. The exposed continental shelves allowed humans, animals, and plants to migrate freely between regions now separated by water. They also provided low-lying refuge areas with moderate maritime temperature regimes, enabling cold-sensitive plants to survive the rigors of Ice Age climates.

Origin of the Ice Ages

The mechanism that triggers and then terminates periods of glaciation is not entirely understood. An ice sheet is in a delicate balance between growth and retreat, and relatively small changes in climate can cause the balance to shift either way. Some of the hypotheses advanced to account for the Ice Ages include changes in the positions of the continents and in ocean circulation patterns as a result of sea-floor spreading; increased altitude of the land masses after a period of geologic upheaval; and variations in the amount of solar radiation the earth receives. The last could be caused by changes in the sun's energy output, in the earth's relationship to the sun, or in the atmosphere's content of carbon dioxide and volcanic dust.

Recent analyses seem to support the long-argued contention that cyclic changes in the inclination of the earth's axis, in the elongation of the earth's orbit around the sun, and in the orbital positions of the equinoxes and solstices account for the periodicity of warm and cold climates during the Pleistocene epoch. These astronomical cycles of varying lengths periodically reinforce and negate one another, resulting in intervals of above or below average solar energy input to the earth. Changes in oxygen isotope ratios in ocean sediments composed of microscopic marine organisms also reveal a periodicity. The isotope ratios are a reflection of land ice volumes that affect the chemistry of seawater. The oxygen isotope ratios have been found to fluctuate in rhythm with the astronomical cycles we have just discussed, making it appear that the latter have controlled the pulse of glaciation over the past 2 million years, as first proposed in 1912 by the Serbian scientist Milutin Milankovitch. Quite probably the movement of continents into arctic latitudes in late Cenozoic times was the major factor making continental ice sheets possible, with the astronomical phenomena subsequently producing cycles of ice sheet growth and decay.

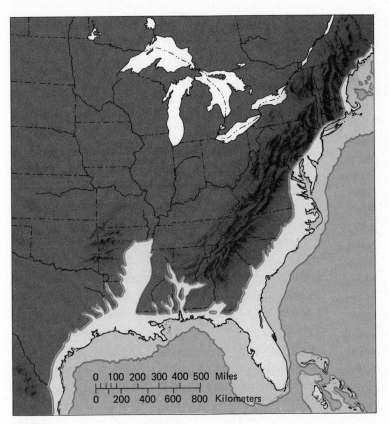

Figure 17.4 This map of the Atlantic and Gulf coasts of the United States shows the degree to which the storage of water as glacial ice affects the position of coastlines. The darkest blue tone represents ocean areas deeper than 130 meters (430 ft), which were unaffected by Pleistocene ice volumes. The middle blue tone shows presently submerged areas that were exposed as land during low stands of the sea that accompanied maximum advances of Pleistocene ice sheets. The lightest blue tone indicates land areas that would be submerged if the present continental ice sheets in Greenland and Antarctica were to melt, raising sea level by about 65 meters (210 ft). (T. M. Oberlander)

ANATOMY AND DYNAMICS OF GLACIERS

Glaciers are formed above the annual *snowline,* the elevation at which some winter snow survives the high temperatures of the following summer. Snowline elevation is thus a function of both the amount of winter snowfall and the temperature of the warm season. In polar areas the snowline is at sea level. With distance from the poles, the snowline rises progressively higher and reaches a maximum of 6,000 meters (20,000 ft) or more in the Peruvian and Chilean Andes, some 20 degrees north and south of the equator. Closer to the equator, precipitation increases along with cloudiness that lowers summer temperatures, permitting the snowline to descend almost 1,000 meters (3,300 ft) below its level in the subtropics (Figure 17.5).

In the Colorado Rockies and California's Sierra Nevada the snowline on slopes exposed to direct solar radiation would be at about 4,500 meters (15,000 ft), just above the highest peaks of both ranges. But on the shaded north- and

east-facing slopes the snowline descends to between 3,600 and 4,000 meters (12,000 to 13,000 ft). Thus snow can accumulate, forming glaciers in cool north- and east-facing depressions in the higher portions of both ranges. Farther north, the gradual descent of the snowline with increasing latitude permits even the south- and west-facing slopes of several volcanic peaks in Washington's Cascade Range to be glacier-clad, although their summits are lower than ice-free peaks in both the Rockies and Sierra Nevada.

Figure 17.5 (top) The summit of Licancabur, a volcano in the Andes Mountains of Chile, reaches 5,930 meters (19,450 ft). Because the volcano is at 23 degrees south latitude in a subtropical desert region, its summit does not reach the annual snowline and cannot support glaciers. This photo was taken early in the southern hemisphere winter and some snow dusts the summit. Peaks this high in the middle latitudes are heavily glaciated. (T. M. O.)

  (bottom) Huandoy in the Peruvian Andes rises to 6,356 meters (20,848 ft). It is only 9 degrees from the equator, where cloudiness and high precipitation cause the snowline to dip downward. Therefore the peak is covered by ice, with glaciers descending its flanks. (Servico Aerofotografico Nacional, Corporacion Peruana del Santo)

and heavy, the transformation of snow crystals to ice granules requires only a few days. But in the extreme cold of Antarctica the snow remains fluffy for years, and the transformation to ice is extremely slow.

Newly fallen snow contains much air space and has a specific gravity (weight relative to water) between 0.05 and 0.15. With time, and under the pressure of overlying snow, the ice granules formed from snowflakes gradually pack closer together. In midlatitude areas, the granular material begins to coalesce after one summer, and attains a specific gravity of about 0.55. Such material is called *firn*. In tens or hundreds of years, depending on summer temperatures, the pore spaces in the lower portions of the firn gradually disappear until the basal firn has become solid ice with a specific gravity of about 0.85. Then, as air bubbles gradually disappear, the specific gravity of the ice increases to about 0.9. The result is true glacier ice. Such ice usually forms at a depth of some 50 meters (150 ft) beneath the firn at the surface.

The Making of Glacial Ice

Snowfall is the raw material for glaciers, but glaciers themselves are solid ice. Snow usually arrives at the surface in the form of lacy hexagonal ice crystals. Eventually, compaction, vaporization, and local melting and refreezing convert the initial delicate snowflakes to rounded granules of ice. When the snow is wet

Glacier Mass Balance

Glaciers are dynamic ice flow systems—expanding or contracting, thickening in some sections and thinning in others, but always feeding ice forward to replace ice lost by melting or evaporation below the snowline. Whether a glacier advances or recedes depends on the glacier's *mass balance*—the balance between the in-

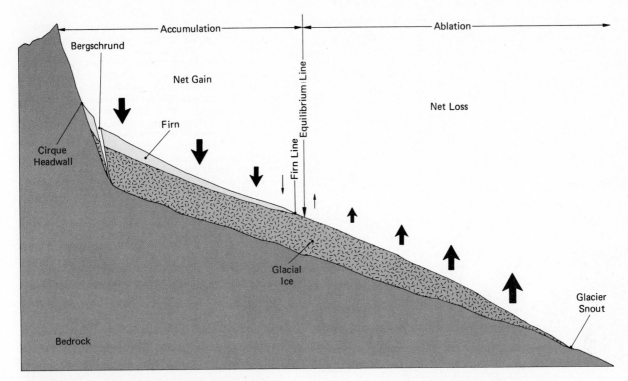

Figure 17.6 The dynamics of a cirque glacier. Arrows show relative gains and losses due to accumulation and ablation. The *firn line* marks the boundary between the annual accumulation of firn and glacial ice; the glacier's surface is white upslope from the firn line and bluish on the downslope side. Alpine glaciers exhibit a deep crevasse, called a *bergschrund*, at their heads. It has been suggested that glacial erosion of the rock headwall occurs because water from melting snow and ice seeps into the crevasse and helps to pry rocks loose when it freezes. (Vantage Art, Inc.)

put of snow above the snowline, and the loss, or *ablation*, of glacial ice by melting or evaporation below the snowline.

All glaciers consist of two portions, an upper *accumulation zone*, in which the annual input of snow exceeds the loss to ablation, and a lower *ablation zone*, in which the annual ablation loss exceeds the direct input by snowfall. Glacial flow transports the excess mass from the accumulation zone to the area of deficit in the ablation zone (Figure 17.6). The *equilibrium line* separates the zones of net gain and net loss, and more or less coincides with the *firn line*, which is the annual snowline on the glacier itself. During the winter, snow accumulation exceeds ablation on a glacier. In the summer, when

snowfalls are less frequent or absent altogether, ablation is dominant (Figure 17.7). Whether a glacier experiences visible growth or contraction depends on the annual mass balance between accumulation and ablation.

Ice is always moving forward to the front of an active glacier; this is true whether the glacier margin is advancing, retreating, or fixed in position. Where the forward ice margin (or "snout" in the case of a valley glacier) is stable in position from year to year, the volume of ice flowing to the margin is exactly balanced by that removed by melting and direct evaporation (sublimation). An increase in the rate of ice arrival or a decrease in the ablation rate will cause the glacier snout to advance. Such a glacier would have a *positive mass balance*. If the rate of ablation exceeds the rate of ice replacement, the snout retreats, and the glacier has a *negative mass balance*.

Glaciers in the Canadian Rockies, Alaska, and Scandinavia lose a depth of about 12 meters (40 ft) of ice to ablation each year. If they are to hold their positions, this loss must be replaced by inflow from above the equilibrium line. Most

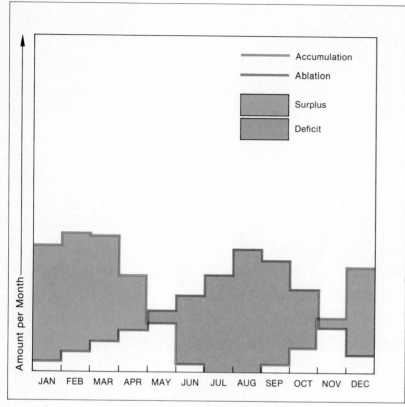

Amount per Month

	Accumulation
	Ablation
▨	Surplus
▨	Deficit

JAN FEB MAR APR MAY JUN JUL AUG SEP OCT NOV DEC

Figure 17.7 This diagram illustrates a typical glacier mass balance in the northern hemisphere. During the winter months, the accumulation of new snow on the glacier exceeds the loss of ice and snow by melting and evaporation. During the summer months, ablation exceeds accumulation, and the glacier experiences a deficit. Note that both accumulation and ablation occur during most months; the glacier may be losing ice near its snout by melting while gaining new snow in its upper, colder reaches. The mass of the glacier increases during periods of surplus and decreases during periods of deficit. If the net annual surplus exceeds the net annual deficit over a period of years, the glacier grows and advances. (Doug Armstrong)

Glaciers may also move by *basal slip,* in which the ice mass as a whole slides over its bed on a film of water. This water is produced by the melting of ice that is pressed against obstacles on the glacier bed (called *pressure melting*). Basal slip permits more rapid flow than does internal deformation alone. However, the total forward movement of ice in a midlatitude alpine glacier is rarely more than a meter a day.

The nature and rate of movement within a glacier is closely related to the temperature of the glacial ice. In *temperate glaciers,* meltwater is produced by pressure melting both within and at the base of the ice and by summer melting on the glacier surface. When the meltwater refreezes within the glacier, it liberates the latent heat of crystallization, which keeps the glacier "warm"—near 0°C (32°F), so that it can continue to melt when pressed against obstacles on its bed. Under such conditions, water-lubricated basal slip is a major component of ice flow.

In polar regions, however, summer melting is minimal and temperatures remain far below the freezing point throughout the depth of the ice. The result is *cold glaciers,* or *polar glaciers,* such as those in Antarctica. Cold glaciers are frozen to their beds, at least on their accessible margins. Because there is no basal slip and internal deformation is greatly reduced, glaciers of this type are extremely slow-moving. We shall see shortly that the distinction between cold and temperate glaciers is important in the development of landforms by glacial erosion.

In confined valley glaciers, the fastest ice

of these glaciers have been retreating since about 1900, indicating that negative mass balances have prevailed throughout the present century.

Movement of Glaciers

Because of its much greater viscosity (resistance to deformation), ice flows in a very different manner from water. When a mass of ice attains a thickness of about 50 meters (150 ft), it begins to spread and flow outward as a consequence of its own weight. It can move in two ways: by internal deformation and by slipping over its bed. Internal deformation is very gradual, amounting to a few centimeters per day, and depends on the fact that ice melts under pressure. The flow is generally laminar (Figure 17.8), with little internal mixing, and is the result of repeated partial melting and recrystallization of ice that is subjected to various stresses within the moving ice mass.

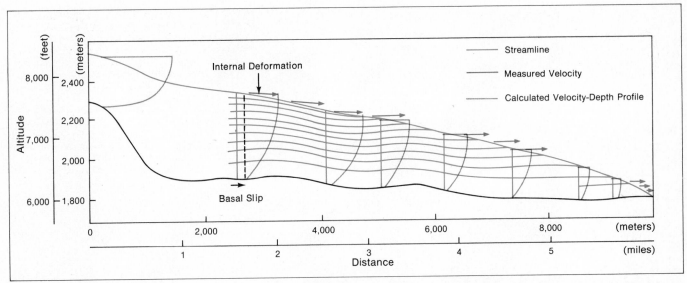

Figure 17.8 The movement of a temperate glacier is accomplished partly by basal slip over its bed (black arrow) and partly by internal deformation of the ice (red arrows). As the diagram indicates, basal slip causes the glacier to move as a whole, such that the distances traveled by the base and by the surface are the same. Due to internal deformation, however, the surface ice moves more rapidly than deeper portions of the glacier. The gray curves show forward displacement at various levels within the ice and are called *velocity-depth profiles*. Streamlines (blue) show paths of points within the glacier and indicate laminar flow of the ice.

This diagram, in which the vertical scale has been exaggerated, shows the ice velocity at various points in the Saskatchewan Glacier in the Canadian Rockies. Only velocities at the surface of a glacier can be measured directly; the internal velocities indicated in the diagram are inferred. The measured velocity of this glacier's surface in its thicker upper portion is about 400 meters (1,300 ft) per year. Near the snout, the velocity is approximately 100 meters (330 ft) per year. The streamlines, which show the trajectories of points within the ice, show that a point on the upstream portion of the glacier's surface tends to move slightly downward, whereas points on the downstream portion of the surface move slightly upward. Near the snout of the glacier, most of the movement is by water-lubricated basal slip. (Doug Armstrong after M. F. Meier, 1960)

currents are at the center of the ice surface, with the margins of the ice stream retarded by friction with the valley walls. The shearing that results within the ice stream causes its surface to split open in a series of parallel fissures called *crevasses* (Figures 17.2 and 17.9). Where steepening of the glacial bed causes a valley glacier to accelerate in speed, crevasses arc across the breadth of the ice surface. In some places glacial beds steepen so abruptly that ice streams collapse and break up in *icefalls* (Figure 17.9). Such areas are dangerous, since icefalls include enormous unstable blocks that frequently topple with crushing force.

Occasionally an alpine glacier moving by basal slip exhibits behavior called a *surge,* in which the ice front advances as much as 20 meters (65 ft) a day. Glacial surges may be caused in various ways. Some occur when large pockets of water form under the ice, or when lake water in tributary valleys blocked by the ice seeps out under the glacier. Other surges occur after periods of unusually high snow accumulation, which cause the glacier to thicken in the accumulation area. This produces a "bulge" or wave of ice that subsequently moves through the glacier at a speed that exceeds the normal ice velocity. When the bulge reaches the glacier snout, the ice front begins to push forward. This merely signifies that suddenly more ice is arriving than is melting away.

GLACIAL MODIFICATION OF LANDSCAPES

Flowing ice is a far more powerful agent of landscape change than running water. A thousand-meter-thick alpine glacier can excavate rock material from both the floor and walls of a valley, whereas a stream's direct action is confined to the valley floor. Furthermore, the special properties of glacial ice enable it to remove and transport debris by processes not available to running water. Like streams, glaciers have their greatest erosional potential when their depths and velocities are greatest, but unlike

Figure 17.9 These ice streams descending steeply from the summit of Mont Blanc in the French Alps are broken by crevasses. As the glaciers pass over convexities in their beds, they produce small icefalls. (T. M. O.)

streams they do not drop their loads of debris wherever their velocity decreases. In fact, large-scale glacial deposition occurs only where the ice terminates or where great masses of ice stagnate and melt with no forward motion at all.

Glacial Erosion

Glacial ice by itself has little destructive ability. Ice cannot generate enough pressure to break away unweathered rock, since ice that moves by basal slip is at the pressure melting point and yields by melting as it is pressed against solid obstacles. However, a glacier that is frozen onto its bed will tear out rock and soil material as it is pushed forward by the ice behind it. Ice that freezes onto rocks and pulls them loose

soon becomes armored with rock debris. It is then somewhat like coarse sandpaper, with the ice acting as the glue that holds the abrasive particles in place. A rock-studded glacial bottom, or *sole*, is an effective file. The faster the glacier moves, the greater the number of abrasive tools scraped over a given surface and the greater the potential for erosion.

All glacial erosion can be attributed to the processes of tearing out, known as *plucking* or *quarrying;* filing down, known as *abrasion;* and *crushing* of rock projections in the glacier's path. Plucking is best accomplished by cold glaciers in which the ice is frozen to the glacier bed (Figure 17.10). Abrasion is favored under temperate glaciers where basal slip is caused by water at the base of the ice. Crushing occurs where there is basal slip, an armored glacial

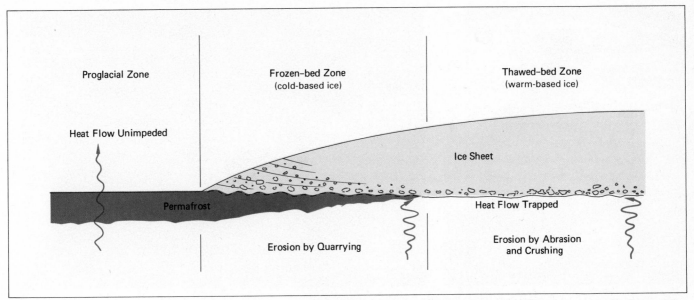

Proglacial Zone

Frozen–bed Zone
(cold-based ice)

Thawed–bed Zone
(warm-based ice)

Heat Flow Unimpeded

Ice Sheet

Permafrost

Heat Flow Trapped

Erosion by Quarrying

Erosion by Abrasion
and Crushing

Figure 17.10 This is a theoretical model of the process of glacial erosion by a continental ice sheet that is frozen to its bed along its periphery (frozen-bed zone) but moves by basal slip where the ice is thicker (thawed-bed zone). According to this hypothesis, this outward flow of heat from the earth's interior (about 4 calories/sq cm/year at the surface) is impeded by a thick insulating ice cover, causing the base of the glacier to warm and the ground under the ice to thaw. Where the ice is thinner, the earth's internal heat passes outward freely, the ground is not warmed, and the ice freezes onto the surface. In the frozen-bed zone, glacial erosion takes place by plucking or quarrying of frozen material. In the thawed-bed zone, glacial erosion is by crushing and abrasion. In the frozen-bed zone, debris can be lifted well up into the ice by thrust faulting within the ice. In the thawed-bed zone, debris remains concentrated in the lowest portion of the glacier. (After Clayton and Moran, 1974)

sole, and a highly irregular subglacial surface. In general, plucking roughens the surface under the ice and, along with crushing, provides the glacier with tools for abrasion. Glaciers probably remove far more material by plucking than by abrasion, particularly where the ice flow encounters unconsolidated (though frozen) soil, and where well-developed jointing divides rock into blocks of movable size. Abrasion tends to smooth rock surfaces and to gouge out grooves parallel to the direction of ice flow (Figure 17.11).

The effects of glacial erosion can be seen on many scales, ranging from an individual rock outcrop to an entire countryside. The first alteration a glacier makes when it invades a region is removal of the soil and the weathered mantle. This demonstrates that glacial erosion is more potent, at least initially, than the gradational processes that preceded it. The resulting exposure of the underlying bedrock enables the glacier to begin to remove unweathered rock, which would remain relatively untouched by normal processes of erosion.

The abrasion process causes bedrock to be scratched, chipped, gouged, and filed down (Figure 17.11). Crescent-shaped fractures and gouges called *chatter marks* or *friction cracks*, produced by the pressure of rock upon rock, are

characteristic small-scale effects of glacial abrasion on individual rock outcrops. Sometimes large expanses of rock are sculptured into smoothly contoured furrows and concavities. Smoothly sculptured forms may be created by either flowing meltwater under a glacier or a subglacial slush of sandy mud mixed with ice crystals.

The difference between abrasion and plucking can be seen on single rock outcrops, which often display *stoss and lee* topography. As Figure 17.12 indicates, abrasion tends to smooth and streamline the upglacier (stoss) face of rock knobs, while plucking and crushing steepens and roughens the downglacier (lee) side.

Severe glacial erosion is most evident where it has quarried out numerous rock basins that subsequently fill with water, as in Figure 17.13. Some rock troughs have been deepened more

Figure 17.11   The basal ice of a glacier is armored with rock fragments that abrade the bedrock under the glacier. Glacial abrasion has polished and striated the rock shown in the photograph. The rock also exhibits a prominent break caused by the crushing and plucking action of the glacier. (G. K. Gilbert/ U.S.G.S.)

Crescentic fractures (or "chattermarks") on glacially polished granitic rock. Scale is provided by a 9-cm-long jack-knife in the center of the view. The direction of the ice flow is indicated by the *dip* of the fracture, which is to the right here. The crescents may be either convex or concave in the downglacier direction, so their curvature is not a consistent indicator of ice movement. (Daniel O. Holmes)

Figure 17.12   In this diagram illustrating the characteristic action of glaciers on outcrops of jointed rock, the direction of glacier flow is from left to right. The upglacier (*stoss*) side of the rock is smoothed and polished by abrasion, and the downglacier (*lee*) side is roughened as the glacier crushes projections and plucks large rocks from the outcrop. (Vantage Art, Inc., after T. M. O.)

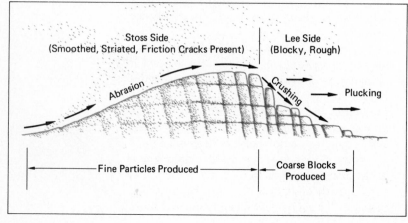

than 600 meters (2,000 ft) by glacial erosion. Examples are Norway's Sogne and Hardanger fjords and California's Yosemite Valley. Glacial erosion excavated some 400 meters (1,350 ft) of bedrock to form the basin now occupied by Lake Superior. The ice-scoured areas of Canada, New England, and Minnesota contain countless large and small lakes in basins produced by glacial quarrying.

The present rate of glacial erosion is not easy to measure because the area of active erosion is covered by ice. One method is to measure the rate at which sediment is carried away from the glacier by meltwater streams. Available data indicate that glacial erosion by abrasion alone is 20 to 50 times more rapid than the rate of erosion of nearby unglaciated areas. Glacial abrasion of marks chiselled into bedrock, or of objects artificially emplaced in the basal ice of glaciers themselves, indicates removal rates of 1 to 40 mm (0.04 to 1.5 in.) per year, the variation being related to differences in rock resistance to abrasion.

Glacial Deposition

In some ways glaciers resemble conveyor belts that steadily feed fragmented rock forward to the glacial margin. Most of the debris carried

Figure 17.13 Ice scouring in crystalline rocks produces multitudes of basins that later fill with water, as seen in this view of the Laurentian Shield north of Ottawa, in the Canadian province of Quebec. (Rolf and Mari Wesche)

by glaciers is concentrated in the lowest few meters of the ice, and along the ice margins in the case of alpine glaciers. Because flow in glaciers is laminar, there is little upward or lateral transport of debris into the body of a glacier except near its snout. Glacier snouts tend to develop thrust faults that push basal ice up over slower-moving ice near the glacier margin, thus thickening the debris zone.

At the glacier margin, transported debris is melted out and either overridden by advancing ice or piled into a ridge-like glacial dump along a stationary ice front. Even when the ice front is retreating, the "conveyor belt" remains in action, continuing to transport debris forward to the glacier snout. Forward transport stops only when the ice thins to the point where flow cannot be maintained over obstacles in the glacial bed. When this occurs the ice mass decays by melting in place.

The general term for all types of material deposited directly or indirectly from glaciers is *glacial drift*. The term "drift" dates from about 150 years ago, when it was supposed that the blanket of rock debris covering vast areas in Europe and North America was left by the Biblical flood associated with Noah. Glacial drift is quite variable in character, for it can be pro-

duced in many different ways. When ice advances it may become so laden with soil and rock detritus that it begins to release some of its load, which is plastered over the subglacial land surface in the form of *glacial till*. Till is a distinctive mixture of coarse and fine material, including fragments of all sizes, from boulders to clay; indeed, it was once called "boulder-clay." Till is most easily distinguished by the absence of any form of sorting or bedding (Figure 17.14 top). Usually some boulders within the till have facets that have been ground smooth by glacial abrasion. Till can be a problem for engineers because excavations in it are often delayed by unexpected encounters with large boulders or extremely tough clays.

When the ice front has pushed to its limit and finally begins to retreat, its marginal portions sometimes stagnate, melting in place, with great volumes of rock debris collecting amid the irregular topography of the decaying ice. When the ice finally disappears, a rather disorganized mass of hills and depressions is left behind. This is known as *ice stagnation topography*. Much of the material has been deposited in contact with the ice by glacial meltwater, and is somewhat sorted by size. Whether the ice stagnates in place or its margin retreats, enormous volumes of gravel, sand, and silt are carried away from it by braided meltwater streams. The coarser particles are rounded and deposited in vast aprons of *glacial outwash* extending away from the terminus of the ice, as in Figure 14.16 (p. 386) and Figure 17.14 bottom. Till, ice stagnation deposits, and outwash are all regarded as glacial drift.

Occasional components of glacial drift are enormous surface boulders known as *glacial erratics*. These can surpass the size of a house, and are composed of rock types not otherwise found in the local area. Some have ridden hundreds of kilometers with the ice, and their known sources help establish the flow patterns of continental ice sheets.

Specific landforms are associated with the various types of glacial drift. The general term for a deposit of drift lodged by glacial ice is *moraine*. Moraine includes both till and ice stagnation deposits, but not outwash deposits. *End moraines* are ridges of drift that mark positions

Figure 17.14   (top) At this location in California's Sierra Nevada, *glacial till* lies atop sheeted granitic bedrock. The till contains fragments of all sizes, from boulders to clay, and is distinctive by its lack of bedding or sorting.

  (bottom) In contrast to till is the deposit of outwash, in which discrete lenses of sand, gravel, and cobbles are visible. This deposit is in a recessional moraine in central New York State. (T. M. O.)

where the ice front remained stationary for a period of time, or where it pushed up a ridge of drift before retreating. End moraines may mark the culmination of an ice advance, or a pause in the retreat of a shrinking ice mass; thus end moraines are either *terminal* or *recessional moraines*. A morainic landscape that has

no conspicuous linear features is simply called *ground moraine*, a *till plain*, or a *drift plain*. Where outwash collects around the margin of a continental ice sheet the result is a flat *outwash plain*. Outwash plains that are confined by valley walls in areas of both alpine and continental glaciation are termed *valley trains*.

The depositional topography produced by glaciers includes a host of small-scale landforms made by advancing, retreating, and stagnating ice. As we shall see in the following pages, those produced by the channeled flows of alpine glaciers differ noticeably from those created by continental ice sheets.

LANDFORMS RESULTING FROM ALPINE GLACIATION

Alpine glaciation is responsible for some of our planet's most magnificent scenery—including the soaring mass of Mount Everest and the ice-clad pinnacles and knife-edged summits of Switzerland's Alps. Alpine glaciation either partially or totally remodels the upland topography of mountain regions, and transforms the valleys leading away from the high summits.

Erosional Forms

The basic element in alpine glacial sculpture is the *cirque*, the amphitheater-like rock basin that is the first product of mountain glaciation. When climatic changes lower the snowline below mountain summits, hillside depressions and the heads of stream valleys become occupied by firn. As this firn is transformed into glacial ice, the depressions and valley heads become deepened and expanded into steep-walled bowls with ice-scoured floors. In most glaciated mountain ranges there are areas in which glacial erosion never progressed beyond this point (Figure 17.15). Such areas are pockmarked with cirques that are separated by flat-topped plateaus.

In areas of more severe glacial erosion, cirque basins have expanded toward one another from the opposite sides of ridges, which

Figure 17.15 The steep-walled basins in this photograph of the Wind River Mountains, Wyoming, are *cirques* formed by the quarrying action of glacial ice. The steep walls of a cirque are believed to be produced by collapse when frost action at the bottom of a bergschrund undercuts the cliffs. Small lakes, or *tarns*, are present in the cirques and appear in the glacially scoured area in the foreground. This part of the Wind River Range preserves large areas of the rolling preglacial topography between the separate cirque basins. (Austin Post/ U.S.G.S.)

are subsequently reduced to jagged knife-edged *arêtes* (Figures 17.16 and 17.17). Where three or four cirques intersect, the dividing sharp-crested arêtes come together in a rock pinnacle known as a *glacial horn*. An extreme example is the Matterhorn in the Swiss Alps, shown in Figure 17.17. It is the growth of cirques that creates the saw-tooth peaks of alpine topography. Most of the rock faces forming such peaks are remnants of cirque headwalls.

Where snowfall is abundant, nourishing large glaciers, ice streams cascade from the cirques into the valleys below. The results are nearly as impressive as glacial modification of uplands. Valley glaciers deepen fluvial valleys and steepen their walls, converting them into *glacial troughs* (Figures 17.16 and 17.18). The interlocking spurs created by the irregular paths of deeply incised mountain streams are trimmed back. Valleys are expanded and simplified into open grooves as they are deepened. The result is a semicircular or U-shaped valley cross section.

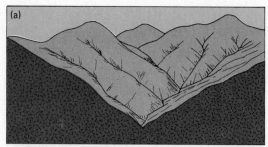

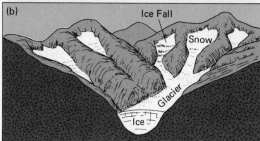

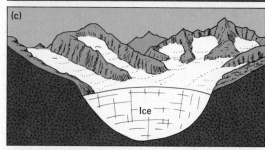

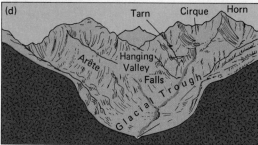

Figure 17.16 This series of diagrams shows the transformation of a fluvially dissected mountain region (a) into a glacially eroded landscape (d). At the highest levels, snowfields develop into *firn* basins that gradually produce glacial ice (b). Cirque growth by glacial scouring and frost action creates sharp crested *arêtes* that come together in horns (c). At lower elevations, ice streams convert sinuous fluvial valleys into open *glacial troughs*. Valleys are deepened and expanded, and their walls are made much steeper by glacial erosion (c). The greater modification of trunk valleys by large ice streams leaves smaller tributary valleys hanging along their margins, producing *hanging valley* waterfalls and cascades (d). Uneven glacial scouring of jointed bedrock excavates basins that later fill with *tarn lakes*. The result is alpine scenery (d). (Vantage Art, Inc. after P. E. James, *Geography of Man*, 3rd ed., 1966, Ginn and Company)

As a consequence of the greater volume of channeled ice, glacial deepening of major valleys is more rapid than deepening of smaller tributary valleys. Retreat of the ice leaves the tributary valleys hanging along the margins of the major glacial troughs. Streams occupying these valleys descend to the floor of the trough by picturesque cascades, or *hanging valley waterfalls* (Figure 17.17).

Because glaciers do not excavate weathered or jointed rock uniformly, erosion by valley glaciers produces irregular longitudinal valley profiles. The result may be a *glacial stairway* with high steps and waterfalls in the center of the trough itself. Lakes known as *tarns* occupy basins quarried out by the ice. Strings of tarns, such as those visible in Figure 17.15, are quite common in alpine topography, being called *paternoster lakes* because of their resemblance to the beads on a rosary.

Downward erosion by valley glaciers is best displayed in the *fjord* landscapes of Greenland, Norway, eastern and western Canada, Alaska, Chile, and New Zealand (Figure 17.18). Fjords are glacial troughs that have been cut far below present sea level, resulting in deeply penetrating arms of the sea hemmed in by rock walls 1,000 or more meters (over 3,300 feet) high. The glacial origin of fjords is demonstrated by their longitudinal profiles, which show deep basins that could not have been excavated by fluvial processes. The rock floors of some fjords are more than 1,000 meters below sea level. This is because a glacier with a density of 0.9 entering the sea, which has a density of about 1.0, continues to erode downward until it is nine-tenths submerged. Only then will it begin to float. Thus a 1,000-meter-thick ice stream in a fjord still has the potential to erode its bed in water that is more than 800 meters (2,700 ft) deep.

Depositional Forms

The depositional landforms produced by alpine glaciation are no match in scenic grandeur for alpine erosional forms, but they are nevertheless quite conspicuous. The largest depositional features produced by valley glaciers are *lateral moraines* built along the sides of the ice streams.

Figure 17.17 **(top left)** Severe frost weathering of the rock walls above cirque glaciers may erode the walls between adjacent cirques to the narrow, ragged rock ridges, or *arêtes*, seen here in the French Alps. (T. M. O.)

(bottom left) The deepening of the principal stream valleys in mountain regions by glacial erosion is more rapid than deepening of tributary valleys that channel smaller ice streams. After the glaciers retreat, the tributary valleys may be left as *hanging valleys*, which join the deep glacial trough high above the valley floor. The waterfall in the photograph is a stream falling from a hanging valley in Yosemite National Park, California. (T. M. O.)

(right) The Matterhorn in Switzerland is an extreme example of a *horn* formed by erosion where the headwalls of three or more cirques intersect. (John E. Kesseli)

Figure 17.18 **(left)** Valley glaciers follow preexisting fluvial valleys as they flow toward lower elevations. The glacier scours the valley floor and walls, removing irregularities and forming a smooth *glacial trough*. The glacial trough in the photograph is in California's Sierra Nevada. (T. M. O.)

(right) If a valley glacier reaches the sea, it can continue to scour out its channel below sea level until the ice is nine-tenths submerged, when it will finally float. When the glacier melts or retreats, it leaves a deep rock-walled ocean inlet, or *fjord*, such as Milford Sound, New Zealand. (Alvin Lynch)

Figure 17.19 Convict Lake, on the east face of the Sierra Nevada, California, lies in glaciated valley dammed by deposits of glacially-transported rock debris. The highest ridges are *lateral moraines*. Lower *end moraines* cross the valley and create the lake. The light-toned area shows the extent of a recent grass fire. (John S. Shelton)

Valley glaciers are also sources of glacial outwash, which aggrades stream valleys beyond the glacial margin (Figure 17.20). The deposits consist of sand and gravel, changing to silt at greater distances from the sediment source. In intervals between major glaciations, such as the present period, the decrease in stream loads increases available stream energy, causing the streams to incise the outwash deposits, converting them into terraces. Successions of glacial and interglacial phases may produce multiple terraces composed of nested or inset outwash deposits, as in Figure 14.23 (center), p. 392.

Figure 17.20 Outlet glaciers from a highland icecap on Baffin Island in the Canadian Arctic have built the outwash fan in the foreground. Note the discoloration of the sea produced by the arrival of silt-laden glacial meltwater. (Air Photo Division, Energy, Mines & Resources © Canadian Government)

These are ridges of till banked against valley walls or issuing from mountain canyons. They frequently rise more than 300 meters (1,000 ft) above surrounding lowlands. The waxing and waning of glaciers during the Pleistocene created many smaller end moraines that commonly make a succession of arcs across valleys, impounding lakes between the morainic crests (Figure 17.19). Many mountain campgrounds in the Sierra Nevada and Rocky Mountains are situated on these bouldery end moraines.

Climatic cooling since the post-glacial temperature maximum about 6,000 years ago produced glacial readvances in all the world's high mountains. The advances created Neoglacial (Holocene) moraines, visible mainly in cirques. These moraines are usually ridges or tongues of coarse rock rubble. However, present glaciers have retreated well behind their various Neoglacial moraines as a result of the warming trend of the present century.

LANDFORMS RESULTING FROM CONTINENTAL GLACIATION

The dual effects of continental glaciation are clearly illustrated in two very dissimilar North American landscapes: Canada's Laurentian Shield, an area dominated by the erosive action of successive ice sheets; and the Canadian Prairie Provinces and American Midwest, an area of deposition by the same ice sheets.

The Laurentian Shield is a vast expanse of ancient igneous and metamorphic rocks surrounding Hudson Bay and extending outward to the arc of large lakes—Great Bear, Great Slave, Athabaska, Winnipeg, Superior, Huron—and the St. Lawrence River. The Laurentide ice sheet stripped this area of its weathered mantle and subjected it to hundreds of thousands of years of glacial quarrying and abrasion. The resulting landscape is one of ice-scoured rock knobs and countless rock-bound lakes in quarried-out depressions (Figures 17.13 and 17.21). Stream patterns seem rambling and unsystematic, the old patterns having been deranged by ice erosion. In general, drainage lines are somewhat rectangular, following joints in the exposed bedrock. Within this glacially stripped region there are deposits of till, but they are thin and discontinuous. Most glacial drift here consists of ice stagnation deposits and *fluted moraine* that has been scoured by active ice (Figure 17.21).

Denuded of soil, and with a short growing season, the region has little agricultural potential. Most human activity here concerns mining and the logging of coniferous forest that thrives on nutrient-poor soils. But the forests, streams, and countless lakes of this wilderness give it enormous recreational value. Similar landscapes occur south of the St. Lawrence River in New England and northern New York. Here the relief is greater than it is over much of the Laurentian Shield, reflecting a more mountainous preglacial landscape. Scenery and water, both largely the result of glacial processes, have long made both of these latter areas lures for tourists.

A somewhat different erosional topography appears in central New York state, south of Lake Ontario, where the continental ice sheet

Figure 17.21 Vast areas of northern Canada have a surface of *fluted moraine* produced by active ice flow that leaves a thin layer of striated drift. Note the many lakes in this landscape. Some of these result from erosion, some from irregular deposition of glacial drift. (Air Photo Division, Energy, Mines & Resources © Canadian Government)

moved into a stream-dissected plateau composed of sedimentary rocks. In this region drainage divides were breached and valleys aligned with the ice currents were streamlined and greatly deepened. This produced New York's picturesque Finger Lakes. Even though they are 300 km (180 miles) inland, the bedrock floors of some of these glacial troughs are well below sea level. Glacial retreat from this landscape flooded the valleys with outwash, creating farmable flat land that was not a part of the preglacial landscape.

In contrast to the glacial effects in the Laurentian Shield area, much of the American Midwest and Canadian Prairie Provinces have benefited from glacial deposition on an immense scale. Lobes of the continental ice sheets spread

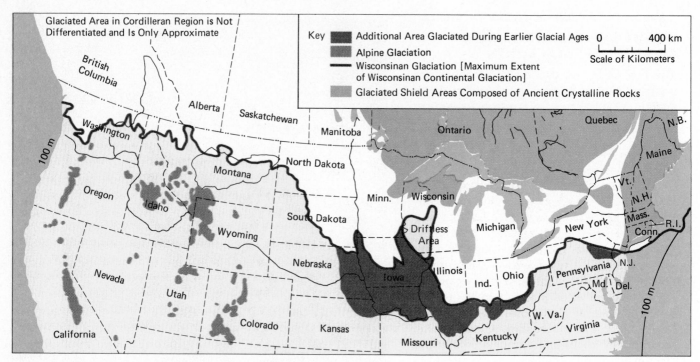

Figure 17.22 This map shows the southern limit of Pleistocene continental glaciation in North America, as well as the major areas of alpine glaciation. The limit of the latest Wisconsinan glaciation, which terminated about 10,000 years ago, is differentiated from the limit of earlier Pleistocene glaciations, the most recent of which ended more than 125,000 years ago. (Vantage Art, Inc. after R. F. Flint, *Glacial and Quaternary Geology*, 1971, John Wiley & Sons)

glacial drift over the North American Midwest from Alberta and Saskatchewan to eastern Ohio (Figure 17.22). Drift deposits in this area are deep enough to completely cover the preglacial erosional topography. An example of the hilly preglacial landscape is preserved in the so-called Driftless Area of Minnesota, Iowa, and Wisconsin—a triangle with Dubuque (Iowa), Winona (Minnesota), and Wassau (Wisconsin) at its apexes (Figures 17.22 and 17.23). This area remained unglaciated throughout the majority of the Pleistocene because of its slightly higher elevation, which caused the Lake Superior and Lake Michigan ice lobes to diverge around it.

The depth of the glacial deposits in some areas of Illinois can be measured in tens of meters, and surpasses 100 meters (over 300 ft) where preglacial stream valleys have been submerged in glacial drift. The old stream gravels in large buried valleys are important as aquifers, supplying well water for towns and cities in the region. From Alberta to Ohio vast areas are gently rolling till plains, and others are nearly flat outwash plains. Soils developed on the recent glacial deposits and overlying loess are fertile due to the short time that leaching has affected them. The largest relief forms are end moraines, which mark positions where the ice front remained stationary for long periods as debris was being conveyed forward to it (Figure 17.24). Both terminal and recessional moraines are present. They are arranged in concentric arcs, convex to the south, indicating that the Pleistocene ice front was lobate in form (Figure 17.22).

End moraines produced by continental ice sheets are usually rolling rather than hilly, and include many depressions that hold marshes and lakes. Most end moraines rise less than 100 meters (300 ft) and are 10 or more km (over 6 miles) wide. Thus end moraines produced by continental glaciation include huge volumes of drift but are much less conspicuous than the sharply defined lateral moraines built by alpine glaciers.

The rise in sea level at the end of the Pleistocene has fully or partly submerged portions of major end moraines deposited on present

Figure 17.23 Castle Rock, in the Driftless Area of southwestern Wisconsin, would have been destroyed if the ice sheets of the late Pleistocene had invaded this region. The Driftless Area reveals the nature of the preglacial landscapes that were eroded by ice and buried by glacial drift during the Pleistocene. (Robert J. Kolenkow)

Figure 17.24 End moraines in the midcontinent region of North America. Wisconsinan age deposits are shown in green, Illinoian in pink, Kansan in buff, Nebraskan in brown. Blue areas were submerged by lakes during ice retreat. Yellow indicates glacial outwash. Red indicates ice stagnation deposits such as kames and eskers. Each moraine, shown in the dark tone, marks a position at which the front of the continental ice sheet remained stationary long enough to build up an exceptionally large mass of glacial drift. Some of these end moraines are *terminal*, marking the farthest forward advance of the ice at a certain period; many are *recessional*, indicating long pauses in the retreat of the ice front. The lobate form of the ice front is conspicuous in the pattern of end moraines. (Geological Society of America)

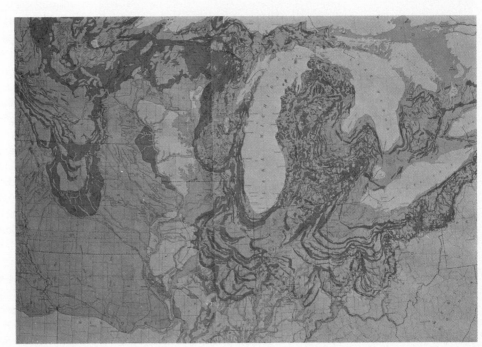

continental shelf areas, converting them into peninsulas or islands. Both Denmark and New York's Long Island are partly submerged terminal moraines and outwash plains built by the last major glacial advance, and Nantucket Island, Martha's Vineyard, and much of the Cape Cod peninsula are recessional moraines.

Drift plains include a variety of landforms smaller than the great end moraines. Especially distinctive are *drumlins* (Figure 17.25), which are

Figure 17.25   (top) This hill, shaped like the reverse side of a spoon, is a *drumlin*, photographed in western Ireland. The blunt end of the drumlin faced toward the ice currents that formed it. Drumlins are composed of drift that has been shaped by advancing glacial ice. (T. M. O.)

(bottom) The long sinuous ridge in the center of the photograph is an *esker* in northern Canada near Great Bear Lake. This deposit marks the course of a subglacial stream channel and is composed principally of sand and gravel. (Canada Department of Mines and Resources)

elongated streamlined hills composed of every type of drift. Drumlins occur in great swarms, with the long axis of each one parallel to the direction of ice flow. Clearly they have been molded by advancing ice, but the relative roles of erosion and deposition in drumlin formation remain a question. Much of the thin drift cover of the Laurentian Shield has been "drumlinized"—converted into parallel streamlined forms. The greatest drumlin fields in the United States are located in central New York and eastern Wisconsin, but the best known individual drumlins are Bunker Hill and Breed's Hill in Boston, Massachusetts, where British and Colonial troops clashed in 1775 in the first real battle of the American Revolutionary War.

Advancing ice also creates "rippled" till surfaces in many areas, known in North America as *washboard moraines*. Good examples are seen in North Dakota and the Canadian prairies. The ridges, which are transverse to the ice flow, are only a few meters high and tens to hundreds of meters apart. They have been interpreted in various ways and may be complex in origin.

Areas of ice stagnation frequently covered hundreds of square kilometers. In such areas the irregular hills of sand and gravel are called *kames*, and marshy or lake-filled depressions are known as *kettles* (Figure 17.26). Kettles mark places where ice masses have melted after being wholly or partially buried by drift or outwash. Many end moraines are *kame moraines* composed of *kame and kettle topography*. Where large ice masses were stranded on outwash plains and partially buried, they eventually melted, producing *pitted outwash plains*. Kame and kettle topography and pitted outwash plains are widespread in the American Midwest—especially in Michigan, Illinois, and Indiana—and in the lowlands of New England. To convert such topography to cropland requires large expenditures for artificial drainage of the land. Hilly kame and kettle complexes are ordinarily used as pasture land for dairy or beef cattle.

Among the most unusual of glacial landforms are *eskers*, which are sinuous ridges of well-sorted sand and gravel that resemble meandering railroad embankments (Figure 17.27). Eskers are stream deposits formed in tunnels under or within stagnant glacial ice.

Figure 17.26 The left half of this view, in the state of Washington, shows the irregular topography produced by the stagnation of the ice border in this region. The low hills, or *kames*, are composed of material deposited between masses of "dead ice," which subsequently melted, forming the depressions, known as *kettles*. (John S. Shelton)

Eskers many tens of kilometers in length and 3 to 30 meters (10 to 100 ft) high are common in New England and eastern Canada, and many also can be seen in Michigan, Minnesota, and Wisconsin. They once provided horse and wagon routes through marshy glaciated regions. Today they are extremely valuable as sources of sand and gravel for construction use.

Only glacial deposits formed within the last 20,000 years preserve the original depositional topography (Figure 17.25), including multitudes of small depressions that hold lakes or marshes. Because the latest deposits—having ages up to 70,000 years, but mostly less than 20,000 years—are well developed in Wisconsin, they are termed *Wisconsinan* in age. Drift older than 100,000 years has traditionally been subdivided into deposits of *Illinoian, Kansan,* and

Nebraskan age, each successively more ancient and named after the state in which it is best exposed. This subdivision is undoubtedly a gross oversimplification and does not accord with the oxygen isotope record from the sea floors. All pre-Wisconsinan drift is older than the last interglacial high stand of the sea about 125,000 years ago, and has been dissected by stream erosion to form undulating, well-drained topography.

Recognition of the ages of the various pre-Wisconsinan deposits is based on their degree of weathering and soil development. Agricultural use of soils on the older glacial deposits does not involve the expense of artificial drainage, but erosion, leaching, and eluvation make the older soils less fertile than soils on deposits of Wisconsinan age.

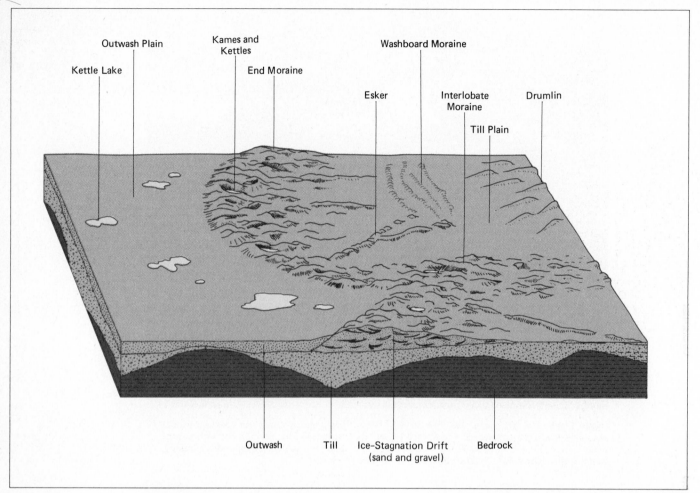

Kettle Lake

Outwash Plain

Kames and Kettles

End Moraine

Washboard Moraine

Esker

Interlobate Moraine

Drumlin

Till Plain

Outwash   Till   Ice-Stagnation Drift (sand and gravel)   Bedrock

Figure 17.27 This diagram represents the typical landforms encountered on drift plains in the midwestern United States and Canada, with an ice sheet advancing from right to left and then retreating back to the right. The length of this block would be on the order of 100 km (60 miles). The sizes of features are exaggerated for the sake of legibility (drumlins are only 1 to 3 km long). The drift covers an irregular bedrock topography. The largest features are the *end moraines* and *interlobate moraines* composed of drift delivered to the ice margin during a long period in which the ice front is stationary. The topography of the end moraines is often of the *kame and kettle* type. In front of the end moraine is commonly encountered a flat *outwash plain* with a scattering of *kettle lakes*, indicating positions of melted ice masses that were partly or fully engulfed by sand and gravel outwash carried away from the ice front by meltwater streams. Behind the end moraine may be an area of ice stagnation topography, including *eskers* built of sand and gravel deposited by streams flowing under the melting glacier. The otherwise regular surfaces of *till plains* may be diversified by *washboard moraines*, consisting of inconspicuous low ridges transverse to the ice flow, and streamlined *drumlins* aligned with the ice flow and composed of glacial drift that has been scoured by advancing ice. In front of this block (to the left), the entire sequence might be repeated, indicating an older position of the ice margin. (Vantage Art, Inc. after T. M. Oberlander)

SUMMARY

Our present global climate is only slightly warmer than the climates that brought on the Ice Ages, as can be seen by the fact that the majority of the earth's fresh water supplies are stored as glacial ice. In the present Holocene epoch, glaciers are found only in high mountains or in the higher latitudes, but during cold intervals within the Pleistocene epoch, glacial ice covered vast areas of North America and Europe, and also expanded in high mountains on all the continents. While glacial erosion has stripped vast areas of their soil, the subsequent deposition of glacial till and outwash has created extremely favorable conditions for agriculture in warmer areas.

The major glacier types are continental ice sheets, highland ice caps, outlet glaciers, cirque glaciers, and valley glaciers. Glaciers form at high altitudes and high latitudes where cool summers permit some winter snow to persist through the following warm season. As snow accumulates it is transformed into firn, which, in turn, is gradually metamorphosed into glacier ice. Gravitational force causes glaciers to move both by internal deformation and by slip over their beds. The glacier mass balance, including both net annual accumulation and ablation, determines whether a glacier margin advances, retreats, or is stationary. Glacier retreat occurs by means of downward melting of the ice surface at a faster rate than the ice loss is replaced by forward ice flow.

Pleistocene glaciers left a distinctive imprint on the land, being responsible for such features as North America's Great Lakes, the fjords of Norway, and the drift plains of the North American heartland. The indirect effects of Pleistocene glaciation include loess deposits, landforms produced by glacial meltwater, intensified frost action and mass wasting, local depression of the earth's crust, and sea level changes.

Glaciers erode the land surface by the processes of abrasion, crushing, and quarrying. Temperate glaciers that can move by basal slip are effective in abrading the surface and crushing bedrock projections, while cold-based glaciers that are frozen to their beds are more effective at quarrying rock and frozen soil. The debris carried by glaciers is deposited as glacial drift, which includes till, water-worked ice stagnation deposits, and outwash. Deposits of till and ice stagnation drift are called moraine.

The landforms generated by alpine glaciers and continental ice sheets vary considerably. Alpine glaciers create cirques, arêtes, horn peaks, glacial troughs, hanging valleys, tarns, fjords, lateral moraines, and outwash valley trains. Continental ice sheets produce ice-scoured plains, drift plains, fluted moraines, end moraines, washboard moraines, kames, kettles, drumlins, eskers, and outwash plains. Only the deposits of Wisconsinan age preserve such landforms. However, deposits of continental glaciation of every age have helped to smooth the topography of the North American heartland, making it more productive agriculturally than the stream-eroded unglaciated areas on its periphery.

REVIEW QUESTIONS

1. Approximately how large a temperature decrease, averaged globally, would be necessary to cause the return of an Ice Age?
2. Indicate three different ways of classifying glaciers. Name the glacier types within each classification.
3. What is the Holocene epoch? How does it relate to the Pleistocene epoch? What are the approximate beginning and ending dates for the Wisconsinan stage of the Pleistocene epoch?
4. How many major Pleistocene glaciations can be distinguished, based on oxygen isotope records from the deep seas?
5. What were the effects of Ice Age climates in unglaciated areas?

6. How is the "snowline" relevant to glacier formation? How does the snowline vary with latitude from pole to pole?

7. Define glacier "mass balance." How is the glacier mass balance reflected at the glacier "snout"?

8. How do "temperate" and "cold" glaciers differ? How are the differences reflected in glacial movement and related glacial erosional processes?

9. What is "stoss and lee" topography?

10. Under what conditions does glacier ice stagnation occur?

11. What types of deposits are included in the term "glacial drift"?

12. What landforms are created by the enlargement of adjacent cirques?

13. How could you distinguish between a valley that has been glaciated and one that has not been glaciated?

14. How do the moraines produced by continental and alpine glaciation differ?

15. What are the dominant landforms of the Laurentian Shield? of the areas of glacial deposition in the Midwest of the United States and in New England?

16. In what two ways are New York's Finger Lakes similar to fjords?

17. What are the beneficial results of glaciation in North America?

APPLICATIONS

1. It is usually possible to determine the limit of the major late Pleistocene glacial advance by merely looking at a map of moderate scale. Why?

2. If both drumlins and eskers are seen in the same area, which must have formed first?

3. How might the stones found in glacial till differ in appearance from those deposited by stream or wave action? If your campus is in a glaciated region, collect or photograph the most distinctively formed till stone you can find (often in or washed out of a road cut through a moraine). Compare your find with those of your classmates.

4. Obtain a topographic map of the closest area of unglaciated mountains. On such a map relief is shown by topographic contours. Redraw the bold index contours to show the area after transformation by alpine glaciation. Now redraw the contours again to represent the area after continental glaciation.

5. If your campus is in or close to an area of alpine or continental glaciation, make a catalogue of all the glacial landforms you can identify, indicating the specific location of each.

6. Obtain the *Atlas of American Agriculture* from your campus library. The atlas graphically displays agricultural statistics for the nation, using counties as data units. On how many of these economic maps can you identify the effects of continental glaciation? Wherever the glaciated areas stand out, attempt to explain the specific effect of glaciation on the particular economic phenomenon.

FURTHER READING

Andrews, John T. *Glacial Systems: An Approach to Glaciers and Their Environments.* North Scituate, Mass.: Duxbury Press (1975), 191 pp. This treatment of glaciers, glacial landforms, and glacio-eustatic and glacio-isostatic effects is concise and advanced; it is for students with an understanding of mathematics.

Coates, Donald R., ed. *Glacial Geomorphology.* Binghampton: State University of New York (1974), 398 pp. This symposium volume includes several outstanding papers on glacial models and glacial erosional and depositional landscapes. Many of the concepts presented were new and original.

Embleton, Clifford, ed. *Glaciers and Glacial Erosion.* New York: Crane, Russak, and Co. (1972), 287 pp. This collection of previously published articles contains many highly readable "classics" in the literature of glacial geomorphology.

Embleton, Clifford, and **C. A. M. King.** *Glacial Geomorphology.* New York: Wiley (1975), 573 pp. This is a very complete treatment of all aspects of glacier structure and dynamics, processes of glacial erosion and deposition, and the landforms resulting from each.

Flint, R. F. *Glacial and Quaternary Geology.* New York: Wiley (1971), 892 pp. Flint's book is the standard American source on all aspects of glaciation and the Pleistocene. The treatment is broader but less detailed than that of Embleton and King.

Imbrie, John and **Katherine P. Imbrie.** *Ice Ages: Solving the Mystery.* Short Hills, N.J.: Enslow (1979), 224 pp. The history of the astronomical theory of the Ice Ages, with anecdotes about the scientists involved. Easily digested and highly informative.

John, Brian, ed. *The Winters of the World.* New York: Wiley and Halsted Press (1979), 256 pp. Explores all aspects of the Ice Ages and the interglacials, including climatic change, glaciers, sea level fluctuations, and life.

Matthes, Francois. *Geologic History of the Yosemite Valley.* U.S. Geological Survey Professional Paper 160 (1930), 137 pp. A classic account of one of America's most spectacular glaciated landscapes. Many photographs and diagrams.

Paterson, W. S. B. *The Physics of Glaciers.* Elmsford, N.Y.: Pergamon Press (1969), 250 pp. Paterson's concise text concentrates on the structure and dynamics of glaciers. The treatment is advanced.

Post, Austin, and **E. R. LaChapelle.** *Glacier Ice.* Seattle: The Mountaineers (1971), 110 pp. The highlight of Post and LaChapelle's book is its excellent large-format photographs of glaciers. The text, dealing with the various types and structures of glaciers, is also excellent.

Sharp, Robert P. *Glaciers.* Condon Lectures, University of Oregon, Eugene, Oregon (1960), 78 pp. This small paperback is a good nontechnical introduction to the principles of glacial flow.

Sugden, David E., and **Brian S. John.** *Glaciers and Landscape: A Geomorphological Approach.* London: E. Arnold (1976), 376 pp. Although the subjects covered in this book are generally similar to those treated by both Flint and Embleton and King, Sugden and John's more selective treatment is highly original and thought-provoking.

CASE STUDY

Midwestern Glacial Deposits

It is hard to associate a windswept desolation of glacial ice with the wonderfully productive farmlands of Iowa, Indiana, central Illinois, and western Ohio. Yet these lands, among the most ideally suited to agriculture of any on earth, are the great gift of North American continental glaciation. Nearby areas beyond the glacial limit tell us what was here before the ice sheets came—stream-dissected terrain, with soils that were thin and stoney or leached and underlain by a hardpan. Without glacial modification, this region would have been useful for pasture and the small-scale production of some commercial crops, but it could not have become the agricultural cornucopia that sends meat to the rest of the continent and grain to much of the world.

There, in the incredibly productive "corn belt" from Iowa to central Ohio, everything came together. In this area of warm summers and adequate moisture, a rough landscape was fertilized, smoothed, and made ideal for the use of agricultural machinery by the deposition of a blanket of nutrient-rich glacial drift and glacio-aeolian loess.

But even in this bountiful area, physical conditions and agricultural opportunities are not uniform. Depressions in the drift plains contain heavy clays and are typically waterlogged; to be useful, they must be drained. Beds of former ice-dammed lakes are flat, silty, and productive, except for gravelly beach ridges. Kames and outwash plains composed of sand and gravel are quite porous and cannot support water-demanding crops. But they, along with eskers, are valuable in

their own way—as sources of aggregate used in construction.

The texture and chemical qualities of glacial drift are related. Soils developed on sandy till or outwash are likely to be acidic, requiring repeated treatment with lime to be kept productive. Such are the better drained soils of much of Michigan, Wisconsin, and Minnesota, north of the corn belt. Here the till was produced by glacial erosion of the crystalline rocks of the nearby Canadian Shield. Acid soils in cool climates are best suited to the growth of pasture and fodder grasses; thus these are the great dairy states. Close to urban areas, soils developed on sandy till and outwash are heavily fertilized to support truck farms feeding city populations.

The soils with the greatest natural fertility developed on calcareous loess or till derived from glacial erosion of limestone or dolomite. Ironically, much of the parent material for the corn belt's calcareous soils was transported from limestone and dolomite outcrops to the north that are now blanketed by nutrient-poor sandy till. Here, as elsewhere, the ice sheets shunted surface materials along, carrying away one type and replacing it with another, often thoroughly changing the raw material for soil development as well as the topography.

The age of the drift exposed in areas of glacial deposition also affects agricultural productivity. In the areas of most recent deposition, poor drainage is a problem; many lakes and marshes are present and stream systems to evacuate surface water are poorly integrated. However, this is where we find the least

leached and eluviated soils. As a consequence, in Ohio, Indiana, Illinois, Iowa, Michigan, and Minnesota, some 50 million acres of land have been artificially drained. Where the surface is composed of older drift—as in eastern Nebraska, northeastern Kansas, and the southern portions of Iowa and Illinois—mild fluvial dissection has produced gently rolling, well-drained surfaces, but the older soils covering them have already been leached of some nutrients and have begun to develop unfavorable hardpan structures.

Surface topography seldom gives a good indication of what may be encountered at depth in a mass of glacial drift. Where the drift cover is thick, several different layers of till, stratified drift, loess, and outwash can be expected. Unpredictable variations in materials at depth create difficulties in the construction of roads, wells, canals, bridges, dams, reservoirs, and large buildings. Enormous erratic boulders or tough clays often lurk unsuspected in the way of intended excavations, slowing progress and raising costs. In the 1950s improper assessment of construction difficulties in glacial deposits on the route of the St. Lawrence Seaway amounted to more than $27 million.

Loose materials of different textures and water contents behave differently under loads; thus glacial deposits must be analyzed by deep boreholes before structures can be placed on them. Sand and gravel lenses within drift produce elevated, or "perched," water tables and aquifers that can be drawn upon agriculturally or even for urban water supplies, but which create difficulties for construction projects. Waste disposal is a particular problem on drift plains. Deep excavations risk intersecting aquifers and perched water tables, causing flooding of disposal sites. Liquid waste disposal is complicated by the danger that toxic fluids may find pathways into aquifers that feed wells.

Although subsurface deposits of sand or gravel may be valuable enough to warrant considerable excavation to expose them, the major significance of the subsurface sands and gravels continues to be their role as aquifers supplying well water for human use. The aquifers consist of glacial outwash deposited in preglacial bedrock valleys that are deeply buried in drift and also in interglacial valleys cut into old drift and subsequently filled with younger drift. The depth of glacial drift over the old valley floors is commonly 100 meters (300 ft) and sometimes as much as 200 meters (600 ft). Where the drift cover is deep, aquifers may be present at several levels. Cities drawing their water supplies from drift-filled valleys include Champaign-Urbana, Illinois; Canton and Dayton, Ohio; and Peru, Indiana. Many smaller communities depend upon this water supply, which at present remains underutilized. Thus a thick drift cover not only increases agricultural opportunities, it also provides the reservoirs of water that large populations require.

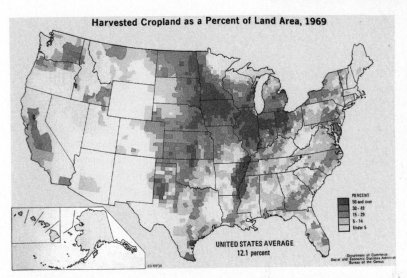

Harvested Cropland as a Percent of Land Area, 1969

UNITED STATES AVERAGE
12.1 percent

PERCENT
50 and over
30 - 49
15 - 29
5 - 14
Under 5

Department of Commerce
Social and Economic Statistics Administration
Bureau of the Census

On this map from the Atlas of American Agriculture the glacial limit in the United States is clearly evident in the high proportion of cropland indicated in the Midwest. Compare this map with Figure 17.22. The only nonglaciated area with such a high proportion of cropland is the flat alluvial plain of the Mississippi River in Mississippi and Arkansas. (U.S. Dept. of Commerce, Bureau of the Census)

Early Morning After a Storm at Sea by Winslow Homer. (The Cleveland Museum of Art; gift from J. H. Wade)

Water waves are a product of atmospheric energy, and convey that energy over vast distances, expending it against coastlines, molding them to accord with prevailing wave climates. The shoreline is a place of constant activity and change—one of the best locations to observe the dynamic relationship between landforms and the processes that create them.

Erosion and deposition by shoreline processes produce the most rapid large-scale changes in landforms observable in any geographic setting. Since the time of the Trojan Wars, described in detail by Homer in the *Iliad*, the coastlines of the Aegean Sea have changed so much that some portions no longer resemble those described by the ancient historian. The land has advanced miles in many locations and has been chewed back in others. The sites of towns have been carried away by wave action, and ancient harbors are now agricultural plains.

We need not look back 2,500 years to be aware of changes along coastlines. They often occur overnight. As this is being written, residents of the communities of Malibu in southern California and Aptos, Pacifica, and Stinson Beach, close to San Francisco, are digging beach sand out of their kitchens or picking through the splintered debris of their former beachfront homes, wondering whether to rebuild or relocate in a less picturesque but more stable environment. They are merely today's losers in the sea's relentless campaign to remake its shorelines.

The coastal zone is a particularly dynamic interface on our planet, a zone of continual energy conversion, where wind-initiated water waves and currents ceaselessly transform the edges of the land, molding shorelines to accord with motions of the sea surface. Two-thirds of the world's population resides near the edges of the land, close to the sea, constantly affected by its behavior and its many indirect influences. In this chapter we shall conclude our survey of the subject matter of physical geography by examining the nature of the interface between the land and the sea.

18
Marine Processes and Coastal Landforms

MOTIONS OF THE SEA SURFACE

Much of the sea's fascination as well as its power to shape coastlines (Figure 18.1) derives from the fact that it is never still. Since it is a liquid with little resistance to deformation, its surface is easily set in motion. The frictional shear stress caused by wind that blows over

Figure 18.1 Wave erosion of weak chalk formations along the coast of England has caused sea cliffs to retreat into a rolling, fluvially dissected landscape. In such settings, the rate of cliff retreat may be as much as 2 meters (6 ft) a year. (Eric Kay/Östman Agency)

water surfaces produces *waves* that roughen the water surface and *currents* that transport masses of water in slow-moving streams on both local and global scales (Chapter 4). The rotation of the earth/moon system and the gravitational attractions of the sun and moon result in the regular rise and fall of the sea surface known as *tides*. Waves, currents, and tides are daily phenomena along the earth's coasts. However, there have been much larger and longer-lasting changes in the level of the sea, which have significantly altered the sizes and shapes of the oceans themselves.

Long-Term Changes in Sea Level

Sea level appears to have risen to its present position approximately 6,000 years ago, following the melting of the Laurentide and Fenno-scandian continental ice sheets (Figure 18.2). We have seen that melting of the existing Antarctic and Greenland ice sheets would raise the level of the oceans another 65 meters (210 ft) or so, drowning the coastal plains and river valleys regarded as home by the majority of the earth's population.

How stable is the level of the sea today? Since fluctuations in climate have caused changes in the sizes of alpine glaciers even over the past 500 years, slight changes of sea level are to be expected. The current global trend is difficult to establish because many coastal regions are slowly rising, sinking, or tilting, as suggested in Figure 18.3. As a result, historic changes in sea level relative to the land have varied considerably from place to place. Thus, two phenomena can affect the level of the sea relative to the land: *eustatic* changes, which are world-wide changes in sea level caused by changes in the total volume of sea water or in the capacity of the ocean basins; and *tectonic* changes, meaning local uplift or depression of the land itself.

The Tides

More significant in the everyday lives of coastal dwellers are the smaller daily changes in sea level known as the *tides*. Tides most commonly cause two periods of rising water, or *flood tides*, and two periods of falling water, or *ebb tides*, in every span of 24 hours and 50 minutes. In the ideal case, the sea rises slowly for a little more than 6 hours, then falls for another 6 hours, and then repeats the cycle. The amount of tidal rise and fall, or *tidal range*, varies greatly from place to place, and also from month to month in any one location. The timing of flood and ebb tides is consistent and predictable at each locality, but varies from one locality to another.

The tides are actually created by two kinds of forces. One is the gravitational attraction of the moon and, to a lesser extent, that of the sun. The other is the centrifugal force generated as the earth and moon rotate about a common point at the center of gravity of the earth/moon system (Figure 18.4).

The point on the earth that is closest to the moon at any time is most strongly affected by the moon's gravitational force. Water on that side of the earth is pulled toward the moon, producing a *tidal bulge*. The moon's gravitational pull diminishes toward the earth's center as distance from the moon increases. Near the earth's center, lunar attraction becomes balanced by an opposing force that pulls in the opposite direction. This opposing force is the centrifugal effect due to rotation of the earth/ moon system.

While the earth spins on its axis, the earth and the moon revolve together about a common center of gravity that falls just inside the mass of the earth on the side facing the moon (Figure 18.4). Thus the earth component of the earth/

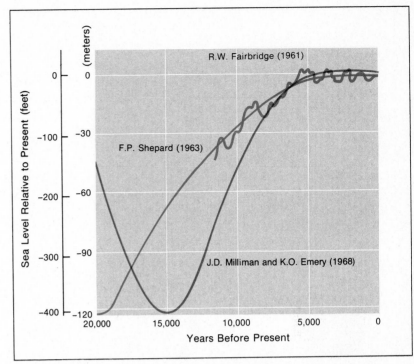

Figure 18.2   As this graph indicates, the level of the sea has risen by at least 120 meters (400 ft) since the late Pleistocene glacial maximum. Melting of the ice sheets returned large quantities of water to the oceans. The estimates of different scientists are shown. Francis Shepard assumed a fairly steady rise of sea level as the ice sheets melted. Rhodes Fairbridge suggested that sea level fluctuated during its rising trend. The curve of Milliman and Emory, based on the Atlantic continental shelf of the U.S., indicates a maximum lowering 15,000 years ago, which seems to agree with other data. Numerous coastal landforms that were exposed 15,000 to 20,000 years ago are now deep under water; many of the world's major harbors are located in what were once wide river valleys that were submerged by the rising seas. (Doug Armstrong after E. C. F. Bird, *Coasts*, 1969, by permission of the M.I.T. Press; and Milliman and Emory, *Science*, 1968.)

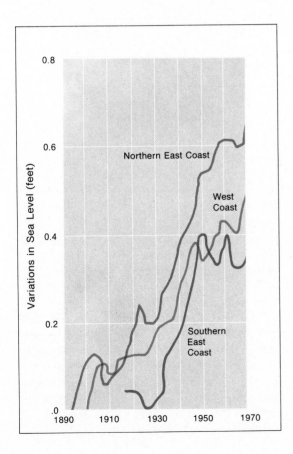

Figure 18.3   The level of the sea relative to the coastlines of the United States has been rising during this century. Part of the apparent increase in sea level is attributed to local lowering of the land. (Doug Armstrong after National Oceanic and Atmospheric Administration)

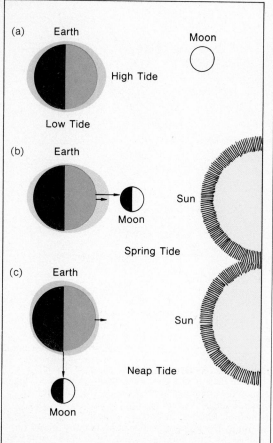

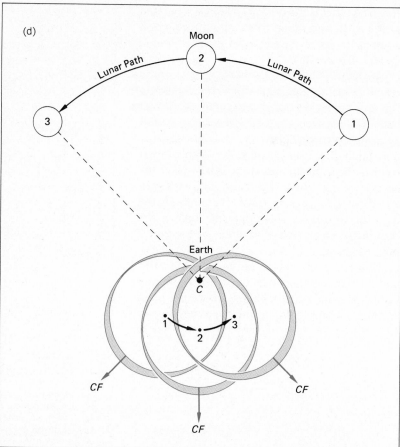

Figure 18.4 (a) Tidal bulges of water, here greatly exaggerated in height, are produced on the earth because of the gravitational effect of the moon and sun and the centrifugal force resulting from the rotation of the earth-moon system. As the earth rotates, two tidal bulges sweep across it, causing twice-daily high tides.

(b) The highest high tides and the lowest low tides, or *spring* tides, occur during new moon (shown here) and full moon, when the earth, moon, and sun are aligned.

(c) The lowest high tides, or *neap* tides, occur during the moon's first and last quarters, when the gravitational pulls of the moon and sun are at right angles, as shown. (Tom O'Mary)

(d) This diagram illustrates how the gravitational attraction of the moon gives rise to one tidal bulge on the side of the earth facing the moon, while the centrifugal force produced by rotation of the earth-moon system causes a tidal bulge on the side of the earth farthest from the moon. Three successive positions of the earth and moon are shown, numbered 1, 2, 3, with arrows indicating the path followed by the center of each. Note that the earth and moon rotate together about the center of gravity of the earth-moon system at *C*, which lies just below the earth's surface on the side closest to the moon. It is the centrifugal force *(CF)* created by the earth's rotation about this off-center axis that forces water toward the region on the earth that is farthest from the moon. (Vantage Art, Inc.)

moon system resembles an eccentric wheel with an off-center axis of rotation independent of its own rotational axis. As with a spinning off-center wheel, there is a strong component of outward centrifugal force on the "heavy" side, tending to fling it still farther outward. This centrifugal force directed *away* from the moon forces ocean water outward in a second tidal bulge on the side of the earth opposite the tidal bulge created by simple lunar gravity. Since the two tidal bulges draw water from the areas between them, at 90° from the moon, low tides tend to occur when the moon is lowest in the sky. Theoretically, the earth should rotate through both watery bulges every 24 hours and 50 minutes, or one every 12 hours and 25 minutes. This is seldom the actual case due to various complicating effects, including impediments to the movement of water imposed by the land masses themselves.

The sun too has an effect on the tides. But the sun's great distance from the earth limits its gravitational force to about half that of the much smaller but closer moon. The greatest tidal range in any region, known as a *spring tide*, occurs when the sun aligns with the earth and moon. This increases either the gravitational pull in the direction of the moon (at the time of new moon, when the moon is between the sun and earth), or the centrifugal effect (at the time of full moon, when the earth is between the sun and moon). Conversely, the tidal range is at a minimum (*neap tide*) when the positions of the sun and moon form a 90° angle with the earth, so that the sun's gravity draws off some of the water of both tidal bulges (Figure 18.4c).

The oceanic response to these tide-raising forces is affected by many factors, including ocean currents, variations in water density and atmospheric pressure, and the shapes of land masses. As a result, there are many local peculiarities in tidal characteristics, as can be seen in Figure 18.5. Tidal ranges are greatest where the flood tide is channeled into narrowing estuaries; there the mass of water is crowded to-gether so that it surges strongly upward. Where the tidal bulge pushes into the English Channel, its range increases from less than one-quarter of a meter in the open sea to 7 meters (23 ft) at the narrowest part of the channel. The Bay of Fundy in Nova Scotia, Canada, has the world's highest tidal range, the average spring tide range being 15.4 meters (50.5 ft). This is a consequence of rocky walls that confine the water in a channel that splits abruptly into two narrow arms that "shoal" (become shallow) rapidly.

When very strong onshore winds or storms occur at the time of high tide, a *wind tide* or *storm surge* may occur, raising the water as much as 3 meters (10 ft) above normal high tide. Storm surges can be tremendously destructive

Figure 18.5 The graphs show the tide heights relative to average sea level at Dover, England, and Victoria, British Columbia, for a period of 15 days. The tides at Dover exhibit the usual pattern of twice-daily high tides. Note that the heights of the daily tides vary through the period because the earth, moon, and sun change their relative positions. The tides at Victoria show essentially one high and one low tide per day. Once-daily tides can occur where certain of the twice-daily tides are unusually small or delayed because of special land configurations. If tides arrive at a location by way of two channels, with a relative delay of 6 hours between the separate flows, the twice-daily tides may be canceled. The tides at many locations show a mixture of twice-daily and once-daily tides. (Doug Armstrong after U.S. Naval Oceanographic Office)

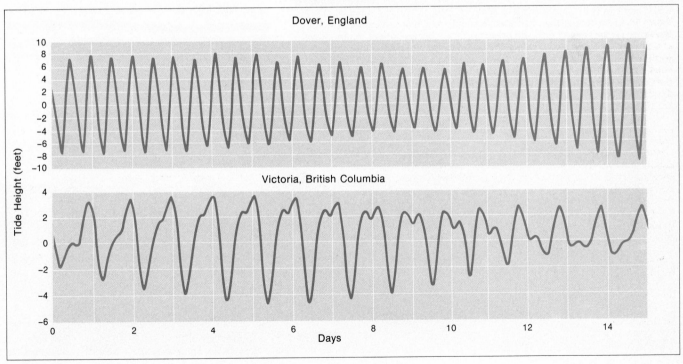

to life and property. America's greatest natural disaster occurred in 1900 when a hurricane-generated storm surge destroyed Galveston, Texas, and much of its population of 7,000. The severe property damage along the California coast in January and February of 1983 was the result of a combination of unusually high tides, storm-driven waves, and onshore winds.

Tides are extremely important in marine navigation. Certain harbors can be entered only at high or flood tides. In some places powerful tidal currents constitute a hazard to ships. Strong ebb tides occasionally draw pleasure boaters on San Francisco Bay out through the Golden Gate into the dangerous waters of the open sea. It was weather and tide conditions that determined the specific timing of the invasion of France by the Allies during World War II.

Waves in the Open Sea

The waves that make the sea surface so fascinating to watch are a clear example of an energy transformation. All waves, whether they are sound waves, light waves, or water waves, carry energy from one place to another. The energy that initiates wave motion is transmitted by the waves to some other point where it is ultimately transformed into heat and work of some kind.

The vast majority of water waves are generated by the kinetic energy of wind moving over a water surface. As the water surface moves under direct wind stress, waves are born. Subsequently, the wave disturbance is passed from particle to particle in such a way that the passage of a water wave merely involves a rise in the water surface, producing a *wave crest*, followed by a sinking of the surface, to produce a *wave trough*. The water surface elevation reflects the motion of the water molecules transmitting the waves. These move in circular orbits as shown in Figure 18.6. There is little forward displacement of the water itself, for what is transmitted is energy, not mass. This is obvious when a wave passes under a floating object, which is not carried forward, but merely rises and falls as the wave moves by it. The vertical distance between the wave crest and the trough is called the *wave height*. The horizontal distance between successive wave crests

Figure 18.6 The nearly circular motion executed by individual water particles is what transmits forward wave motion. When a wave crest passes, the surface water particles are at the top of their orbits. When a wave trough passes, the same surface water particles are at the bottom of their orbits. Orbital motion of water particles is negligibly small at depths greater than half a wave length. Waves in water deeper than half a wave length, which are called *deep-water waves*, are not influenced by the presence of the ocean bottom. As the wave enters shallow water near a shore, the ocean bottom begins to interact with the wave motion. The wave becomes steeper and higher and the wave length becomes shorter. The wave form eventually becomes unstable and the wave breaks. The wave dissipates its remaining energy as swash on the beach. (Doug Armstrong)

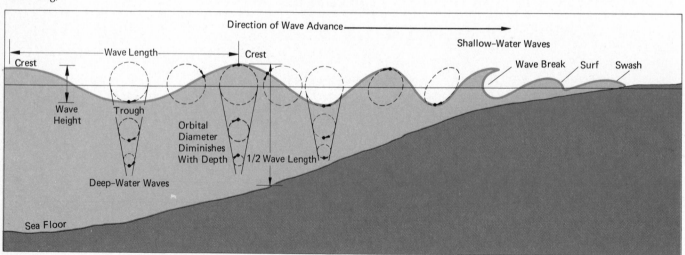

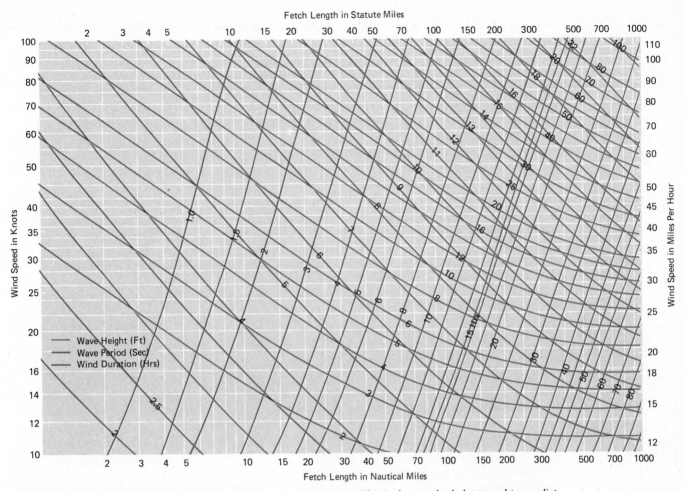

Figure 18.7 This is the standard chart used to predict the period and height of deep water forced waves on the basis of wind speed and either the wind duration or the fetch over which the wind blows. To use the chart, locate the appropriate wind speed at the right (miles per hour) or left (knots), and move horizontally to intersect either the fetch length (vertical lines) or the wind duration in hours (fine dashed lines). At the intersection point interpolate the wave period from the nearest dashed lines and the wave height from the nearest solid lines.

For example, a 30-knot (nautical miles per hour) wind blowing across a fetch of 400 nautical miles should produce waves close to 18 feet high, with a period of 12 seconds. So should a 30-knot wind blowing for a duration of 30 hours. To produce 100-foot waves—the largest observed on the open seas—would require 100-mile-per-hour (87-knot) winds sustained for 35 hours, or blowing across 1,000 nautical miles of open water. A nautical mile is an international unit equal to 1,852 meters (6,076.115 feet). (U.S. Army Coastal Engineering Center, Technical Report 4, 3rd ed., 1966)

is known as the *wave length*, and the time required for successive crests to pass a point is the *wave period*. In deep water, the greater the wave length, the more rapid the speed of the wave.

Water waves are of two types. The waves raised directly by wind stress on the water surface are *wind waves*, or sharp-peaked *forced waves*. Sizable forced waves are known to sailors as *sea*. The much lower, round-crested, linear waves traveling outward from the point of energy transfer are known as *swell*. The size of forced waves produced directly by the wind increases with wind velocity, wind duration, and the extent of water—known as the *fetch*—over which the wind can build waves (Figure 18.7). In storms on the open sea the height of forced waves is commonly about 6 meters (20 ft); however, in an unusually severe storm in the mid-Pacific in 1933, personnel on a U.S.

naval vessel measured a wave height of 34 meters (112 ft).

The vast majority of the waves that persistently roll in against our shores are swell, generated in some far-off place. When watching these waves come in, bear in mind that they very likely were born in storms raging in the seas off Japan, Alaska, Iceland, or even Antarctica. Typical swell with wave lengths of 100 meters (330 ft) moves at about 15 meters per second (30 mph). Swell of moderate size is capable of traveling half way around the world before all of its energy is dissipated. Nearly all moderate swell lives long enough to reach a coastline.

Seismic Sea Waves

Not all waves are initiated by wind stress on the water. The waves conveying the greatest amounts of energy are known popularly as "tidal waves," although they are quite unrelated to tidal action. They are also called *tsunamis*, the Japanese word for exceptionally large waves that strike the land. These greatest of all waves are actually shock waves in water—triggered by earthquakes, landslides, and volcanic eruptions on the ocean floor. Thus they are *seismic sea waves*.

Oceanic earthquakes are frequent, the largest originating in the oceanic trenches where subduction is occurring. Since the Pacific Ocean is ringed with these earthquake-producing oceanic trenches, the shores of the Pacific experience frequent seismic sea waves. Around the margins of the Atlantic, ocean trenches appear only in the area of the West Indies and east of Patagonia, so seismic sea waves are infrequent in the Atlantic basin. Most Atlantic coast tsunamis have been triggered by earthquakes or volcanic disturbances on the oceanic ridge system and associated transform faults.

A seismic sea wave is a peculiar phenomenon. In the open sea it is a barely detectable low swell, with a wave length exceeding 100 km (60 miles) and a height of less than a meter. But since the speed of deep-water waves is always proportional to the wave length, seismic sea waves have incredible velocities—as much

as 800 km (500 miles) per hour. All water waves decelerate in speed and bunch together when approaching the land, where the water shoals and wave motion begin to be retarded by friction with the sea floor (Figure 18.6). The law of energy conservation operates to make the wave height increase as wave speed decreases. In the case of a seismic sea wave, this causes a rapidly moving mountain of water to lift suddenly out of an ordinary sea, sometimes rising to heights of 10 to 30 meters (30 to 100 ft) as it strikes the land. The greatest height, or *runup*, occurs where the coast faces the approaching wave, and where the wave moves into a confined space, such as a bay or estuary.

In some cases, a great wave trough appears before the first crest, causing the sea surface to sink and recede far from shore. This happened at Hilo, Hawaii, in 1923. Many people ran out to pick up fish stranded on the newly exposed land and were engulfed by the gigantic wave crest that suddenly rose and drove the sea far inland. Because of their enormous wave lengths, the crests of successive seismic sea waves arrive at intervals of 10 or more minutes, and the third, fifth, or seventh wave in the series may be higher and more destructive than the first. Thus it is difficult to ascertain when the danger is past!

Historically, seismic sea waves have caused as much destruction and loss of life as have major earthquakes and volcanic eruptions on the land. Most of the 30,000 people killed as a consequence of the eruption of Krakatau in 1883 were drowned by the tsunamis generated by the volcanic blast, rather than by the eruption itself.

Hawaii lies in the path of tsunamis that come from three different sources: the Peru–Chile trench, the Aleutian trench, and the oceanic trenches of the western Pacific. Since tsunamis are triggered by earthquakes, warnings can be issued and loss of life minimized or averted. However, waterfront property damage can be prevented only by restricting building in the hazard area, as in Figure 18.8. Sadly enough, there are always a few people who go down to the beach to see the predicted "tidal wave," which then materializes too suddenly for them to escape.

Figure 18.8 This attractive park along the waterfront in Hilo, Hawaii, was developed after the tsunamis of 1946 and 1960 devastated the portion of the town formerly on the site. Development of shoreline parks in urban areas is a reasonable way of minimizing loss of life and property from tsunamis that are certain to strike such coasts from time to time. (T. M. O.)

WAVE ACTION AGAINST THE LAND

Analyses of coastal landforms in Greece reveal that in some areas wave erosion has driven back the shoreline some 800 meters (0.5 mile) in the past 6,000 years. The softer rocks of the Yorkshire coast of England have been eroded back even farther—some 3 km (2 miles) since Roman times. How can waves, which roll harmlessly past boats, swimmers, and children on inflated inner tubes, be so destructive to the land itself? In fact, waves are not destructive until they "break."

Wave Break

In the open sea, waves glide onward unaffected by the ocean floor, their velocities dependent on their wave length. But when waves move into shallow water and wave motion begins to be affected by friction with the sea floor, wave characteristics are modified. Waves then change from deep-water types, with velocities dependent on wave length, to shallow-water types, with velocities dependent on water depth.

When incoming waves begin to be retarded by friction, the waves decrease in length and increase in height. As shown in Figure 18.6, friction actually causes the orbit of the surface water particles to be deformed from a circle to a forward-leaning ellipse. This happens when the water depth decreases to about half the wave length. Three things can happen to this ellipse. Its base can collapse in a mass of foam, producing a *collapsing breaker;* its crest can slide downward, forming a *spilling breaker;* or its crest can shoot forward in a smooth curve to produce the elegant cylinder of a *plunging breaker* (Figure 18.9). Surfers attempt to ride the rising slope of water just ahead of the "curl" of a plunging breaker or the foam of a spilling breaker. A fourth possibility, the *surging breaker,* does not really break at all; a pulse of water merely surges onto the beach and then drains quietly back to the sea.

The way a wave breaks depends on the interplay of several factors, including the wave length, height, and steepness; the angle of the wave's approach to the shore; and the slope of the sea bottom, which can decelerate waves either gradually or abruptly. Steep waves, a direct approach, and rapidly shallowing water produce plunging breakers, whereas more gradual shallowing and an oblique approach favor spilling breakers. Collapsing breakers are produced by waves that are less steep than those that spill or plunge; nevertheless, they seem to be the most destructive of the various types of breakers, in terms of damage to seawalls and harbor facilities.

Wave Refraction

As waves enter shallow water, the lines of the waves are seldom exactly parallel to the shoreline; one portion of each wave starts to break before the rest. But even before this portion of the wave breaks, it is affected by friction with the sea floor. Its forward velocity decreases while the part of the wave that has not yet "felt bottom" moves on ahead. This local change in velocity causes the wave to pivot slightly at the point where it is first affected by friction, bending the line of the wave, as illustrated in Figure

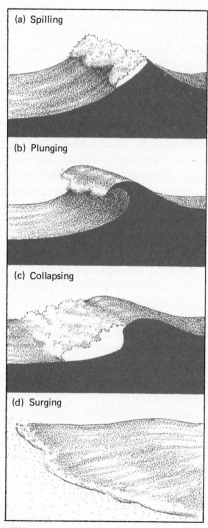

Figure 18.9 The manner in which waves break affects beach erosion as well as the stability of artificial structures such as breakwaters and seawalls. Breaker type is a response to wave height and steepness and beach slope.

Spilling breakers (a) result from the downward slumping of the crests of steep waves. In *plunging breakers* (b) steep wave crests curl over the front face of the wave and fall vertically. In *collapsing breakers* (c) the lower part of the wave front becomes vertical and the wave collapses in a mass of foam. When wave height is small, waves slide up the beach without collapse of the wave form, producing *surging breakers* (d). (T. M. O.)

18.10. This phenomenon is known as *wave refraction*.

Wave refraction is very important because it either focuses or spreads the energy conveyed by incoming waves. If a coast is irregular, waves are first slowed by friction as they approach a projecting headland. As Figure 18.10 shows, the resulting wave refraction causes waves to pivot toward the headland, concentrating their energy there. In bays between coastal promontories, refraction causes the lines of wave approach to diverge, spreading each previous unit of wave energy over a larger area. Given sufficient time and a constant sea level, the above-average wave energy at headlands and below-average wave energy in bays would smooth the coastline by wearing back the headlands and causing sediment to accumulate in the bays.

Coastal Erosion

In many places waves do not break on broad sand or gravel beaches, but impact against steeply rising land, which is cliffed and progressively worn back by wave attack. This erosion is accomplished in several ways. Some removal is by chemical corrosion of rock that is constantly wetted, and some is a result of the crystallization, thermal expansion, and hydration of salts in rock pores repeatedly dampened by salt spray and dried by the sun. There is also abrasion by rock debris hurled against cliffs by wave impact. Lighthouse windows 30 meters (100 ft) above sea level commonly have screens of heavy gauge wire to protect them from wave-tossed rock shrapnel. Part of the breakup of rock along coasts results from the enormous impact of heavy masses of water crashing into jointed rock during storms (Figure 18.11). One of the most effective processes for loosening and detaching rock is the compression of air within fissures in the rock when a wall of water smashes against it—followed by the explosive expansion of the air as the water falls away.

A coast's vulnerability to wave erosion depends on the local geological materials, the

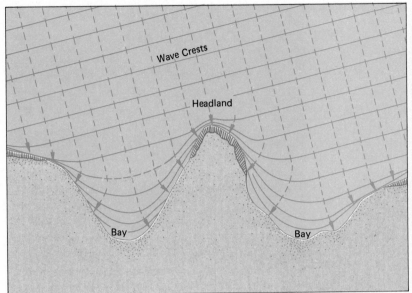

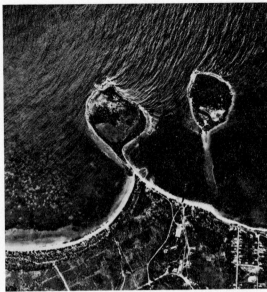

Figure 18.10 **(left)** Wave motion is altered in direction, or *refracted*, when it interacts with the sea floor. The diagram shows waves refracted at a headland, or coastal promontory. Solid lines represent wave crests; dashed lines show direction of wave energy transmission. Compartments bounded by wave crests and energy vectors contain equal amounts of wave energy. Refraction causes waves to concentrate their energy on headlands, accelerating erosion and causing the formation of such features as sea stacks and sea caves. In bays, wave energy is dissipated by divergence of wave trajectories. Therefore bays become areas of sediment deposition. (John Dawson)

(right) This photograph shows wave refraction along a portion of an irregular coast. (U.S. Navy)

fetch that controls the size of forced waves or storm surges, the pattern and magnitude of swell from distant storms, and the possibilities for wave refraction at the coast. In England the rate of coastal retreat under wave attack may be as much as 2 meters (6.5 ft) a year where the rocks are composed of poorly cemented sand or gravels, and less than a meter each thousand years where resistant rocks are present. In unusually intense storms, sea cliffs composed of glacial drift have been cut back more than 10 meters (33 ft) in a few hours.

Coastal erosion is concentrated in the zone of wave impact. This zone extends from the low-tide level to somewhat above the high-tide level. The effect of wave erosion on a sea cliff is to undercut it by notching the base at the *shoreline angle*, as illustrated in Figure 18.12. The unsupported upper portion of the cliff then retreats by collapse. The retreat of a sea cliff leaves behind a submarine rock platform called a *wave-cut platform* or *abrasion ramp*, whose seaward-sloping surface may be littered with de-

Figure 18.11 The power of ocean waves as they collide with the shore is evident in this scene at Point Lobos on the California coast. The rocks in the foreground were torn from granitic outcrops under wave attack, and subsequently were rounded by being tumbled about in the surf zone. (David Muench)

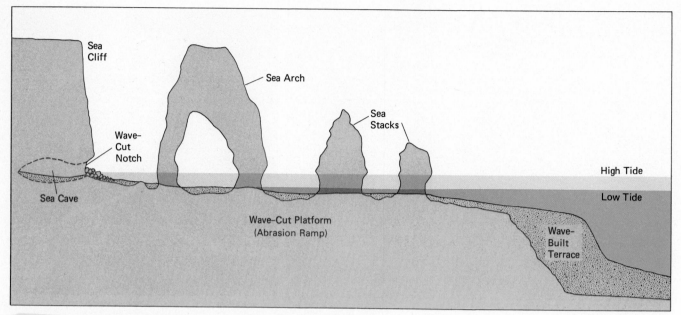

Figure 18.12 This diagram illustrates the principal features that may be seen along cliffed coasts. The major elements are *sea cliffs* that are kept steep by being undercut at the base along a *wave-cut notch*, producing an erosional *wave-cut platform* or *abrasion ramp* that extends seaward from the base of the sea cliff. The cliff and platform meet at the *shoreline angle*, which is usually at the high-tide level. Features occasionally seen are sea caves and erosional remnants such as *sea arches* and *sea stacks*. Depositional *wave-built terraces* may also be present. The existence and character of these subsidiary forms depend on the nature of the rock present. (Vantage Art, Inc. modified from Karl W. Butzer, *Geomorphology from the Earth*, 1976, Harper & Row)

bris removed from the land. Debris deposits may form a submarine embankment, or *wave-built terrace*, extending beyond the abrasion ramp.

As the width of a wave-cut platform increases, greater amounts of wave energy are lost by friction with the surface of the platform, until there is no further effective wave erosion at the shoreline. It is believed that the broad wave-cut platforms seen in some areas must have been produced during periods of slowly rising sea level, and that the zone of effective wave action and frictional energy loss must have remained narrow and close to the retreating shoreline.

Where coastlines are irregular, with projections and bays, wave refraction at the headlands concentrates wave attack on the projections from two sides. Selective wave erosion along lines of weakness in these projections may separate them into clusters of rock stubs

known as *sea stacks* (Figure 18.12). Undercutting along the flanks of rocky headlands often develops *sea caves*, which sometimes link together from the two sides to convert the promontory into a *sea arch* (Figure 18.13). Collapse of the arch leaves an isolated stack. Sea caves and stacks are common along the Pacific Coast in Washington, Oregon, and California, with sea arches occasionally present. These features are also well developed on both sides of the English Channel.

It is not unusual to find unmistakable wave-cut platforms (with former sea stacks) well above sea level, forming a terrace along the coast. Such *marine terraces* could indicate either a drop in sea level or uplift of the land relative to the sea. Since marine terraces are local phenomena and do not maintain constant elevations, uplift of the land is the more likely cause. Marine terraces are especially common along tectonically active coasts, such as the Pacific coast of the United States (Figure 18.13), where many areas have clearly risen relative to the sea. Some localities along the California coast display as many as six well-defined terraces, arrayed like a giant staircase. The highest terraces that are definitely of marine origin in California lie some 600 meters (2,000 ft) above sea level. Marine shells on the lowest well-defined ter-

(a)

(c)

(b)

Figure 18.13 (a) Wave impact and abrasion by wave-tossed rock shrapnel carve a *wave-cut notch* at the base of actively retreating sea cliffs. These cliffs are cut into basaltic rock on the west coast of the Hawaiian island of Kauai (T. M. O.)

(b) The chalk cliffs at Étretat on the coast of Normandy, France, were cut by waves to form sea arches and sea stacks. Note the vertical face of the cliffs; wave action at their base undermines the rock and causes the collapse of the upper portions. (Eric Kay/Ostman Agency)

(c) This Oregon coast illustrates the process of coastal straightening by wave action. Above the exposed surface of a wave-cut platform at low tide is a marine terrace that was an older wave-cut platform produced at sea level and subsequently tectonically uplifted. Meanwhile the low headland is worn by wave erosion and sediment from it is transferred into the bay in the background, where it accumulates in a broad beach. (T. M. O.)

races are more than 100,000 years old, indicating that these surfaces were abrasion ramps at sea level during the last interglacial stage, with varying amounts of subsequent uplift of the coastal region. Where the same terrraces descend well below current sea level, the land obviously has sunk relative to adjacent areas.

Beach Drifting and Longshore Drifting

While wave action gnaws at the edges of the land, it also causes a flow of sediment along the coast, nourishing beaches and associated marine depositional landforms.

Where beaches are present, wave break causes a mass of water to surge forward onto the beach as a turbulent sheet. This *swash*, or *uprush*, picks up fine sediment and pushes it up the beach slope. The speed of the swash diminishes due to friction and gravity, until forward momentum and gravitational pull down the beach slope are in balance. The uprush stops, much of the water sinks into the beach, and the remainder drains back down the beach slope as *backwash*, dragging fine sediment with it. When waves strike a beach obliquely, every grain of sediment that is driven onshore by uprush and dragged down by backwash is also shunted laterally along the shoreline in zig-zag fashion. This process, called *beach drifting* (Figure 18.14a), results in a flow of sand along the shore, which is necessary to maintain beaches and associated depositional landforms.

A second process is equally important in the flow of coastal sediment. In the offshore zone of wave break, the water is kept in an agitated condition and is often clouded with fine suspended sediment that is swirled this way and that. The sediment may be suspended only momentarily after each wave impact, but while it is in suspension it moves with the water. Be-

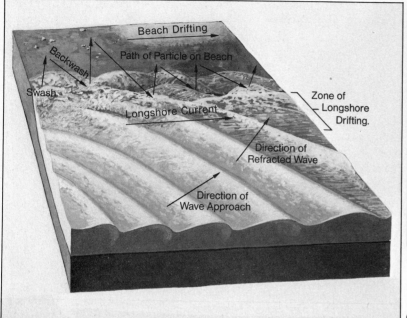

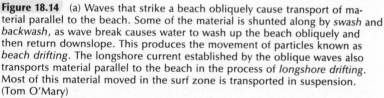

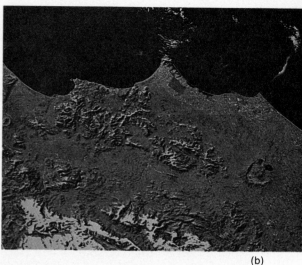

(b)

(a)

Figure 18.14 (a) Waves that strike a beach obliquely cause transport of material parallel to the beach. Some of the material is shunted along by *swash* and *backwash*, as wave break causes water to wash up the beach obliquely and then return downslope. This produces the movement of particles known as *beach drifting*. The longshore current established by the oblique waves also transports material parallel to the beach in the process of *longshore drifting*. Most of this material moved in the surf zone is transported in suspension. (Tom O'Mary)

(b) This LANDSAT false-color image of the western coast of Italy from Rome (gray spot at left) to near Naples (to the right of this view) illustrates repeated "half-heart" curvature of bays of varying sizes. This form, technically known as *zetaform* curvature, is created by longshore drifting of sediment, and reveals the refraction of swell by projecting headlands and deltas. The direction of longshore drifting is from left to right, with the dominant swell from the lower left. (NASA)

cause waves generally meet coastlines at a slight angle, they push water against the shore obliquely. This continual input of water cannot pile up vertically, so much of it drains off laterally, parallel to the coast in the general direction of wave advance. The result is a *longshore current*.

The oblique onshore piling of water by the arrival of one breaker after another causes the sediment-laden water in the zone of breaking waves to migrate slowly along the shore in the direction of the longshore current, resulting in *longshore drifting* of sediment. The direction may reverse from day to day, but most coasts have a preferred direction of longshore drifting during the season of maximum storminess and

most vigorous wave action. This is often visible in the configuration of the coast, as illustrated in Figure 18.14b.

The combined processes of beach and longshore drifting produce a constant flux of sediment along smooth coastlines. This flow waxes and wanes and even changes direction from time to time, depending on the state of the sea, but it creates and sustains all depositional landforms present in coastal regions. The net sediment flow may be evident in the configurations of depositional landforms and the shapes of bays. The latter frequently have a "half-heart" form, with decreasing curvature in the downdrift direction. The maximum curvature occurs in the lee of headlands where there is maximum refraction of oblique waves and minimum wave energy (Figure 18.14b).

Coastal Deposition

Sediment transport under marine conditions follows the same principles as sediment transport by running water and wind. Transport occurs by suspension, saltation, and surface creep. It is easy to see (and hear) sand and gravel moving in all these ways in the swash zone. Deposition occurs where wave energy decreases, permitting sediment to come to rest.

Every marine depositional form has a characteristic sediment budget in which gains must balance losses if the form is to persist. Many marine depositional forms periodically grow and shrink as a result of shifts in the balance between sediment input and outgo. Although waves strike coasts from varying directions from day to day, every coast has an average "wave climate" and an average direction of sediment flow. Except in severe storms, sediment normally arrives at one end of a depositional form and leaves at the other end. Any interference with the flow of sediment at either end of the system results in a change in the form of the deposit.

Artificial structures that interfere with the flow of sediment along coasts give us a measure of the amount of material passing by any point over a period of time in the same way that fluvial sediment trapped in reservoirs records the load carried by stream systems. As an example, the breakwater built in 1928 to protect the marina at Santa Barbara, California, traps an average of 215,000 cubic meters of sand per year. This has caused erosion of beaches for a distance of nearly 20 km (32 miles) in the normal direction of sand flow, accelerating the retreat of sea cliffs they formerly protected.

Figure 18.15 In many locations beaches are composed of "shingle," meaning pebbles or cobbles. This is often true where waves are eroding stony glacial drift or periglacial solifluction deposits, or along steep coasts where there are no rivers to transport sand to the sea. This shingle beach is on Anacapa Island, off the coast of southern California. The beach material is derived entirely from wave erosion of volcanic rock. (T. M. O.)

Beaches

Beaches are surfaces of sand or disk-shaped cobbles (known as "shingle") deposited by wave action between the low-tide line and the highest levels reached by storm waves (Figure 18.15). On the landward side of a beach is material of a different type or a surface of different character from the beach, such as sand dunes, permanent vegetation, or a sea cliff. Beaches form along both irregular and straight coastlines. On irregular coasts, wave refraction causes sand or shingle to collect in the embayments between projecting headlands, creating *bay head beaches* or smaller *pocket beaches*. Beaches also develop in continuous fringes along straight coasts where unresistant geological materials are present. Figure 18.16 indicates the main features of beach morphology.

Beaches are among the most sensitive indi-

Figure 18.16 A *beach* is the zone of transition between the land and sea. It extends from the low-tide level to the upper limit reached by the highest storm waves, which is the area subject to alternate erosion and deposition of sand. The actual profile of a beach is constantly changing. The diagram shows the principal geomorphic divisions of a beach. The *berm* is the nearly flat portion at the top of a beach; it is covered with material deposited by waves and constitutes the *backshore*. The *foreshore* extends from the edge of the berm to the low-tide line. Within this zone is the *beach face*, which is the area subject to swash and backwash. The *offshore*, which is permanently under water, contains *bars* and *troughs*. This is the zone of wave break and surf action. (Vantage Art, Inc. after Francis P. Shepard, *The Earth Beneath the Sea*, 2nd ed., 1967, Johns Hopkins Press)

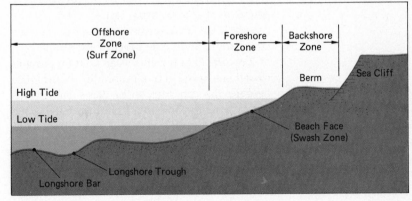

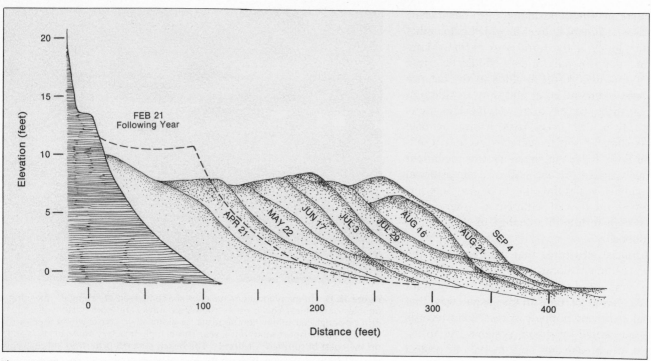

Figure 18.17   The growth of the berm on the beach at Carmel, California, is shown in this series of measured profiles. By February of the following year, most of the berm had been cut away again. The vertical scale is exaggerated 10 times. (John Dawson after "Beaches" by Willard Bascom, *Scientific American*. Copyright © 1960 by Scientific American, Inc. All rights reserved)

cators of changes in geomorphic systems. We usually enjoy beaches in the summer, when the relative absence of destructive storm waves allows them to be well filled with sand (Figure 18.17). The sediment may originate from wave erosion of headlands or from rivers carrying sediment from the land. Along a coast that has a prevailing longshore drift, sand eroded from one beach may be passed along to the next beach. On the southern beaches of Long Island, New York, sand produced by wave erosion of glacial drift moves from east to west. At the present time, loss seems to be outrunning accumulation, and some beaches are retreating 1 or 2 meters (3 to 6 ft) per year. Certain beaches elsewhere have been shrinking in recent years because the construction of dams on streams has reduced the amount of sand supplied to the coast, or because residential construction on coastal dunes has cut off part of the sand supply for the beach system.

One way to prevent beach erosion is to in-

tentionally interrupt longshore drifting with a wood, concrete, or rubble dam, or *groin,* that extends into the water at right angles to the shoreline (Figure 18.18). The longshore current is partially blocked at the groin and drops part of its sediment—just as a stream drops part of its sediment when its velocity decreases. Inevitably, the deposition caused by the groin robs beaches farther down the coast of incoming sediment. Those beaches erode all the more rapidly, to the anguish of resort owners. Miami Beach must import sand to balance erosional losses, despite efforts of individual hotels to maintain their beaches by groin construction.

Changes in the size of breakers, or a change from long wave length waves of low height to steeper short wave length waves of greater height, greatly accelerate the removal of the particles composing the beach. Large steep waves can peel a beach downward two meters or more in a day. The beach material removed is deposited as submerged *offshore bars* in the zone of wave break (Figure 18.16). In areas affected by seasonal storminess, beaches tend to be stripped downward and steepened by the erosive action of forced waves during the storm

Figure 18.18 The central portion of the New Jersey coast suffers from the presence of an excessive number of sand-trapping *groins* built perpendicular to the shoreline. Sand flow by beach drifting and longshore drift is impeded by the first groins, causing beaches in their lee to be starved and therefore narrowed or removed by wave erosion, as here at Deal, where waves break directly against a seawall, which will eventually be undermined and destroyed. (K. F. Nordstrom)

season—usually the winter in the middle latitudes. Some beaches periodically lose all their sand, retaining only large cobbles that are too heavy to be entrained even by the largest storm surges. During the season of calms, the sand stored offshore is slowly moved back toward the beaches by long wave length, low height swell, eventually refilling the beaches to their former level.

Spits

Beach drifting and longshore drifting frequently produce deposits of sediment that extend outward from initial shorelines (Figure 18.19). Linear sediment accumulations that are attached to the land at one or both ends are termed *spits*. The term *bar* is reserved for submerged depositional forms.

Some spits form where sediment is moving along a coastline that changes direction sharply, causing wave refraction and creating a low-energy environment in which sediment can accumulate. Where a coastline turns abruptly inland, as in a bay, the waves pivot and diverge, decreasing wave energy and sediment transporting ability. Sediment moved by beach and longshore drifting comes to rest where the incoming waves begin to pivot around the corner. The resulting deposit becomes the new shore-

line, which causes the point of wave pivoting to be shifted laterally. This, in turn, produces more deposition. In time a linear spit takes shape, usually with a curved end, as in Figure 18.20. At Cape Cod, Massachusetts, a large spit has been extended westward from the end of a wave-eroded glacial moraine.

The formation of spits from one or both sides of small bays may close off the bays, transforming them into *lagoons*. The lagoons eventually fill with fine sediment and are colonized by vegetation. This is one of the most important processes in the straightening of irregular coasts by marine action. Sand spits are common at river mouths, but river currents and floods normally maintain openings through such spits, preventing lagoon formation.

Sediment deposition also occurs in the low-energy wave environments in the lee of coastal islets or large sea stacks. Here wave refraction sweeps sediment behind the obstacle from two sides, producing a type of spit known as a *tombolo*, which ties the island to the land (Figure 18.19).

Spits are especially vulnerable to damage during hurricanes and storm surges associated with midlatitude cyclones. The topography of many spits reveals that they are compound features, having been rebuilt many times after partial destruction by storm waves.

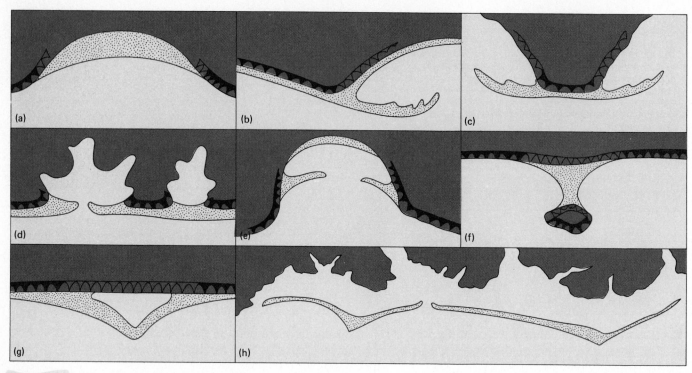

Figure 18.19 These diagrams illustrate the most common depositional forms produced by wave action. The dark sawtooth symbol indicates an active sea cliff; the open sawtooth symbol shows an active sea cliff. Land areas are dark. Depositional forms are yellow.

(a) *Bay head beach* with sediment deposition resulting from a low-energy wave environment.

(b) *Recurved spit* formed by sediment transport to the right, prolonging the previous line of the coast.

(c) *Winged headland* with sediment moved both ways from the eroding cliff as a consequence of changing directions of wave approach.

(d) *Bay mouth* or *barrier spits* straightening an initially irregular coast of submergence. Changes in wave approach produce spits extending from both sides of the bays, eventually closing them off and converting them into lagoons.

(e) *Mid-bay spits* with sediments accumulating before reaching the head of the bay.

(f) *Tombolo* formed by the deposition of sediment to the lee of an island and eventually linking the island to the mainland.

(g) *Cuspate spit* developed by reversals in the direction of longshore drifting along a straight coast.

(h) *Barrier island* developed along a coastline that is low-lying but irregular, suggesting recent submergence. The points on such barrier islands are called *cuspate forelands*. Cape Hatteras, North Carolina, is an outstanding example of such a form. (Vantage Art, Inc.)

Barrier Islands

Portions of the world's coastlines are paralleled offshore by narrow strips of sand dunes, beaches, and saltwater marshes, known as *barrier islands*, (Figures 18.19h and 18.21). Low, grassy dunes form the axes of the islands, with beaches facing the sea and salt marshes on the landward side. Barrier islands are more regular in form than the coastlines behind them, as can be seen around the Gulf of Mexico from Texas to Florida and along the Atlantic coast from Florida to New York.

There is increasing evidence that barrier islands have a complex history, probably originating as coastal dunes on the continental shelves during periods of low sea levels coinciding with Pleistocene glacial periods. Apparently, these sand deposits were moved landward by wave action during the post-glacial rise in sea level, and have greatly increased in size in the last 6,000 years, during which sea level has remained relatively stable. They continue to migrate landward, as much as 10 meters (33 ft) each year. This movement is caused by wave erosion of the seaward beaches and the deposition of sands in the salt marshes on their landward sides by storm waves that wash completely across them from time to time. Occasionally, adjacent barrier islands that are oriented differently, reflecting different wave sources or refraction patterns, become linked to form a prominent point, or *cusp*, as at Cape Hatteras and Cape Lookout, North Carolina (Figure 18.21).

Offshore barrier islands have great value as summer recreation areas, and they also create sheltered waterways between themselves and the mainland. In the United States an Intracoastal Waterway System protected by barrier islands is navigable all the way from Massachusetts to the Mexican border, with only a short interruption along the west coast of the Florida peninsula, where barrier islands have not formed.

The barrier islands along our Atlantic and Gulf coasts are valuable defenses for the mainland behind them. They receive the brunt of

Figure 18.20 Deposition by longshore currents has caused the gradual extension of this curved spit near Puerto Peñasco in Mexico. The bay behind the spit seen in this oblique aerial photograph may eventually be closed off, forming a *lagoon*. (John S. Shelton)

Figure 18.21 This view from the Apollo spacecraft shows the barrier island system off the irregular submerged coast of North Carolina. Cape Hatteras is near the center and Cape Lookout is at the bottom of the picture. Outgoing tides are visibly washing fine sediment seaward through tidal inlets, which are a normal feature of barrier island systems. However, the effect of storm waves is to erode the seaward side of the islands and to move sediment across to the landward side, causing the islands to migrate toward the mainland. (NASA, Pilot Rock, Inc., © 1976)

(center right) Each year some two million vacationers visit the barrier island of Assateague, off the coast of Virginia. Despite this, Assateague Island is still roved by wild ponies, and remains largely undeveloped due to the efforts of conservationists. Portions of the island are retreating toward the mainland as much as 20 meters each year. Assateague Island forms a marked contrast with the overbuilt barrier beach fringing the south shore of Long Island, New York **(bottom right)** where island migration is fiercely contested by shoreline engineers.

hurricane-driven storm surges whose full energy is seldom felt along the continental shoreline. But the conflict between their commercial and recreational development and their physical mobility poses difficult problems. How can roads be maintained along one of these shifting barrier islands—and parking lots, boat landings, motels, restaurants, and beach cottages? The only way is to stop the shifting. This can be done, at least for a while, by preventing overwash during storms. The technique is to heighten the dunes artificially, using sand-trapping fencing or dredged sand, and to plant them with stabilizing shrubs and dense grasses, creating a sort of seawall behind which roads and other facilities can be located.

The effect of this modification should have been foreseen. In the artificial landscape, the energy of storm waves is concentrated on the beach in front of the artificially elevated dunes instead of being exhausted in overwash across the full width of the barrier island. Thus beaches have become exposed to wave energies they have never before experienced. First, they lose their finer particles; the beach profile becomes steeper; and then intensified backwash drags away coarser particles. The process feeds upon itself, and the beaches become progressively narrower and steeper. Today the once broad beaches of many artificially "defended" barrier islands are strips only a few tens of meters wide in front of wave-steepened dunes. Migration of the islands has been halted for the moment, but their principal attraction, the beaches, are disappearing. If present processes continue, the dunes themselves will be undercut and washed away, and the islands will retreat at a faster rate than ever before.

Many solutions to this problem have been proposed, including elevated highways and other structures that would permit overwash, preserving the beaches, but this will not stop the inexorable migration of the island shorelines. The only real answer appears to be a more passive use of the barrier islands, maintaining them in their natural state as undeveloped beaches accessible by ferry from the mainland. This is the plan adopted for the Cape Lookout National Seashore Area, south of Cape Hatteras, North Carolina.

CORAL REEF COASTS

In some portions of the world, the coastline has been created by the activity of living organisms—corals and the marine animals associated with them. *Corals* are softbodied, tube-like marine animals that build a stony exterior skeletal structure composed of calcium carbonate. They live in colonies composed of many species, ranging from massive mushroom-like forms to delicate twig-like structures (Figure 18.22). Vital to coral growth are photosynthetic calcareous algae that live in symbiotic relationship with the coral polyps. The coralline algae extract CO_2 from seawater, providing the carbon necessary to form the corals' calcium carbonate exoskeletons. The calcareous algae also cement the reef structure together. As layers of coral die, new layers form on top of them, building up solid masses of organic limestone known as *coral reefs*.

Coral reef coasts, along with fjord coasts (Chapter 15), are the only coastal types that have a distribution controlled by climate. None of the reef-building corals can survive in water temperatures lower than 18°C (65°F), and corals truly flourish only where water temperatures are between 25°C and 30°C (77° to 86°F). Thus coral reefs are found only in tropical regions, seldom occurring poleward of the 30th parallel. Because the associated algae require light, the reef-building corals rarely grow at a depth exceeding 45 meters (150 ft). An additional requirement is clear water that is free of suspended sediment. The fact that corals cannot survive near the mouths of large sediment-carrying streams accounts for their preference for island rather than mainland locations. Coral reefs are seen in the West Indies, the Florida Keys, and as far north as Bermuda, which lies in the warm Gulf Stream, just north of the 32nd parallel. The greatest development of coral reefs is in the tropical Pacific Ocean, from the Hawaiian Islands to the Great Barrier Reef of Australia. Coral reefs are also widespread in the Indian Ocean.

Wherever they are seen, coral reefs take one of three forms: *fringing reefs,* which are built out laterally from the shore; *barrier reefs,* which are separated from an island or landmass by a la-

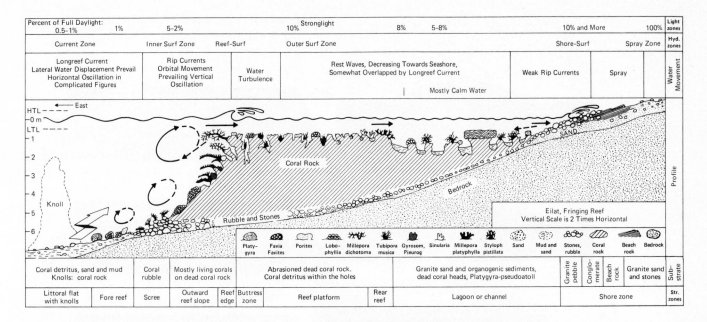

Percent of Full Daylight: 0.5-1%	1%	5-2%		10%	Stronglight	8%	5-8%		10% and More	100%	Light zones
Current Zone	Inner Surf Zone	Reef-Surf		Outer Surf Zone					Shore-Surf	Spray Zone	Hyd. zones
Longreef Current Lateral Water Displacement Prevail Horizontal Oscillation in Complicated Figures	Rip Currents Orbital Movement Prevailing Vertical Oscillation	Water Turbulence		Rest Waves, Decreasing Towards Seashore, Somewhat Overlapped by Longreef Current Mostly Calm Water					Weak Rip Currents	Spray	Water Movement

(diagram: HTL, 0 m, LTL, depth scale –1 to –6; labels East, Knoll, Coral Rock, Rubble and Stones, Bedrock, SAND; Eilat, Fringing Reef Vertical Scale is 2 Times Horizontal)

Symbol key: Platy-gyra, Favia Favites, Porites, Lobo-phyllia, Millepora dichotoma, Tubipora musica, Gyrosom, Pieurog, Sinularia, Millepora platyphylla, Styloph pistillata, Sand, Mud and sand, Stones, rubble, Coral rock, Beach rock, Bedrock

| Coral detritus, sand and mud Knolls: coral rock | Coral rubble | Mostly living corals on dead coral rock | Abrasioned dead coral rock. Coral detritus within the holes | | Granite sand and organogenic sediments, dead coral heads, Platygyra-pseudoatoll | | Granite pebble | Conglo-merate | Beach rock | Granite sand and stones | Sub-strate |
| Littoral flat with knolls | Fore reef | Scree | Outward reef slope | Reef edge | Buttress zone | Reef platform | Rear reef | Lagoon or channel | | Shore zone | Str. zones |

Figure 18.22 The diagrammatic cross section shows the details of a coral reef built atop granitic bedrock at the north end of the Red Sea near Eilat, Israel. In the photograph we see a reef face composed of the plantlike forms that are actually the exterior skeletons of lime-secreting marine animals. (Mergner, 1971, and John C. Hutchins/The Image Bank)

goon; and *atolls*, which enclose a lagoon, with no other land present (Figure 18.23 and Figure 18.24). Atolls are often thought of as circular, but they are normally irregular in plan. In all coral reefs, the outer (seaward) slope is very steep, and even overhanging, whereas the inner portion, though often quite rough, is close to horizontal. Barrier reefs and atolls are always pierced by several openings through which boats may pass from the interior lagoons to the open sea.

The formation of fringing reefs may be observed in countless locations where coral growth can be seen in reef platforms extending out from the shore. However, barrier reefs and atolls require somewhat more explanation. Although there have been several theories to account for them, modern investigations substantiate Charles Darwin's belief that barrier reefs and atolls indicate subsidence of islands initially bordered by fringing reefs. If subsidence of the central island is not too rapid, the living coral on the outer portion of fringing reefs can re-

main in the sunlit layer of the sea by growing upward. This produces barrier reefs. In the case of an atoll, coral has continued to grow upward, on the periphery of a subsiding island that has disappeared below sea level (Figure 18.23). Island subsidence is caused principally by the gradual cooling and sinking of lithospheric crust formed at sea-floor spreading centers. Some of the constructional activity in both barrier reefs and atolls seems to be a consequence of hurricanes and tsunamis, in which large waves tear loose masses of coral and toss them onto the reef or into the lagoon behind.

Atolls are picturesque but hazardous and meager environments, often lacking freshwater

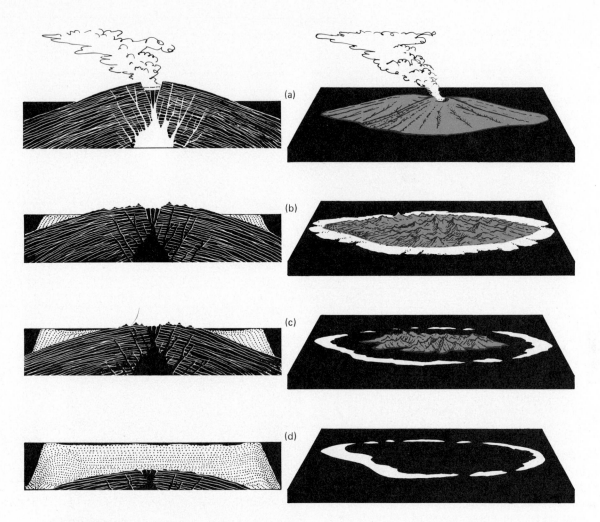

Figure 18.23 Deduced stages in the formation of a coral atoll based on observation of Pacific Ocean volcanic islands.

(a) An island is formed by volcanic eruptions that create a cone rising from the sea floor. Several volcanoes may be present. Molten lava is indicated within the volcanoes.

(b) The end of the volcanic phase is followed by erosion of the volcanoes, producing mountainous topography. A *fringing reef* grows outward from the shoreline.

(c) The beginning of subsidence of the island causes drowning of the erosional forms and upbuilding of coral to form a *barrier reef*. Storm waves toss up coral rubble to produce a rim higher than the lagoon behind it.

(d) Continued subsidence causes the volcanic island to become fully submerged and covered with coral. The result is a coral *atoll*. The ring surrounding the center lagoon is maintained by the action of storm waves. (T. M. O.)

supplies, having no resources other than coconut palms and marine life, and lying in the paths of hurricanes and typhoons. Nevertheless, many atolls are populated, demonstrating the adaptability of the human species.

CLASSIFICATION OF COASTS

Because of the many variations in the nature of coastal scenery, a number of schemes have evolved for the classification of coastal landforms. These classifications are a good indication of the complexity of coastal landforms and the processes that create them.

The broadest classification of coasts reflects their relationships to the motions of lithospheric plates. *Collision,* or *subduction, coasts* are

Figure 18.24 A portion of the Great Barrier Reef off the northeast coast of Australia. This reef is growing upward as well as outward around the small island to keep pace with subsidence of the east coast of Australia. The majority of the picture area consists of reef coral. Dark areas are the deepest water, except for coral "heads" surrounded by light "halos." (John Lewis Stage/The Image Bank)

These atolls photographed from the Apollo 7 spacecraft are in the Tuamotu Archipelago in the South Pacific. The view covers a distance of abuot 500 km (300 miles). The rings of coral have been able to grow upward fast enough to remain in the sunlit zone even though the volcanic islands forming their base have sunk far below sea level. Note that the atolls are irregular rather than circular in plan. White dots and "puffs" are cumulus clouds. (NASA)

the Pacific type, in which plate convergence and tectonic activity create strong relief and a steep coast with rocky promontories and bold sea cliffs. *Trailing-edge coasts* are low-lying because there is no strong tectonic activity where the continents and sea floors are moving together as a single unit—for example our Atlantic shores south of New England. Here we find estuaries, salt marshes, and coastal dunes. *Marginal seacoasts* are coastlines separated from the deep ocean basins by volcanic island arcs such as the Caribbean Antilles or the Japanese or Indonesian archipelagoes (island chains). The Gulf of Mexico and the coast of North China are examples of marginal seacoasts. Generally, marginal seacoasts have broad continental shelves and low relief similar to trailing-edge coasts.

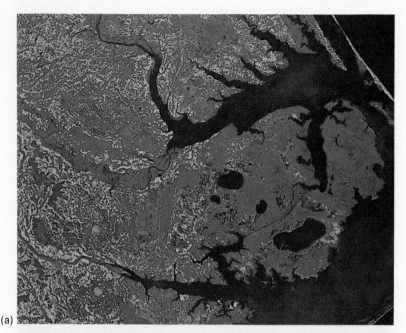

Figure 18.25   (a) This LANDSAT false-color image portrays the *ria* coast of North Carolina, created by Holocene submergence of river valleys incised to the lower sea level of the late Pleistocene. At the right is Pamlico Sound, protected by a line of barrier islands that link at Cape Hatteras (Figure 18.21). The drowned valley on the north is Albemarle Sound, which is the estuary of the Roanoke and Chowan rivers. Much of the drowned valley of the Roanoke River has been filled by Holocene and Recent alluvium. The Pamlico River creates the southern estuary. (NASA)

(b) This landscape was eroded by a continental ice sheet and then submerged by postglacial rise in sea level. The long, parallel projections of land and elongated islands that jut into the ocean off the coast of Maine were continuous ridges before being submerged. Submergence of the coast drowned the valleys and exposed the ridges to wave erosion. (John S. Shelton)

The most straightforward coastal classifications are purely descriptive, involving such categories as high-cliffed, low-cliffed, lagoonal, deltaic, estuarine, fjorded, barrier island, fringing reef, barrier reef, and so on. This type of classification is possible on any scale. The descriptive terms can be general enough to permit differentiation of the world's coastlines on a single map.

Other classifications differentiate coasts that have *prograded* (advanced seaward) by deposition of deltas, sand spits, and lagoonal fills; *retrograded* (retreated) by erosion; or developed their characteristics as a consequence of *submergence* or *emergence* of the land relative to the sea. Coastal emergence, or rise of the land relative to the sea, clearly is required for the creation of marine terraces, which are common on tectonically active coasts. Where tectonic activity is weak, emerged areas of sea floor are low-lying coastal plains. Along the Atlantic and Gulf coasts of the United States, there has been a long-term tendency toward emergence, with Pleistocene sea-level fluctuations superimposed on this general trend.

Two types of landforms have resulted from the world-wide coastal submergence caused by the melting of the Late Pleistocene northern hemisphere ice sheets. Very apparent on a map of the eastern United States are deeply-penetrating estuaries, called *rias*, where the sea has invaded river valleys deepened during Pleistocene low stands of the sea. The coast of North Carolina, shown in Figure 18.25, is a classic example, displaying the estuaries of several large rivers. Between rias we find cliffed headlands and sometimes sea arches and stacks. As the headlands are battered back, their debris helps produce spits that close off bays, transforming them into lagoons.

The form of submerged coasts varies according to the nature of the older subaerial (nonmarine) landforms and the orientation of relief features with respect to the coastline. The rocky islets off the coasts of New England, Labrador, British Columbia, Norway, and Chile result from submergence of glacially scoured landscapes (Figure 18.25). Other highly irregular coasts, such as those of Greece and western Turkey, are fringed by rugged islands that result from submergence of mountainous topog-

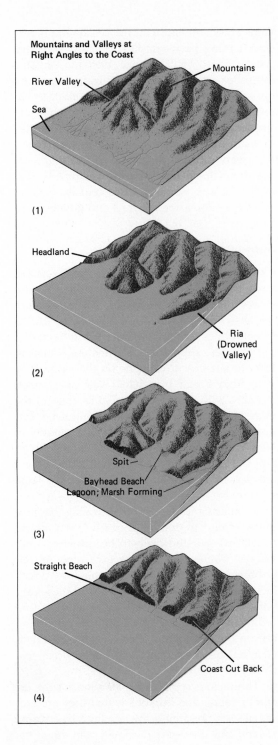

Mountains and Valleys at Right Angles to the Coast

Mountains

River Valley

Sea

(1)

Headland

Ria (Drowned Valley)

(2)

Spit

Bayhead Beach

Lagoon; Marsh Forming

(3)

Straight Beach

Coast Cut Back

(4)

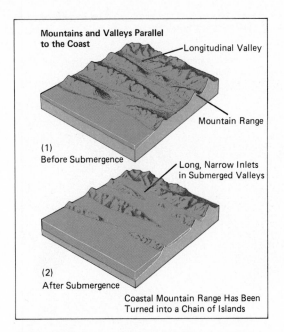

Mountains and Valleys Parallel to the Coast

Longitudinal Valley

Mountain Range

(1) Before Submergence

Long, Narrow Inlets in Submerged Valleys

(2) After Submergence

Coastal Mountain Range Has Been Turned into a Chain of Islands

Figure 18.26 The sequence of diagrams shows the typical evolution of a coastline formed where valleys and ridges transverse to the coast are submerged by a rise of sea level relative to the land.

(1) The initial coastline consists of valleys and ridges behind a narrow coastal plain.

(2) The relative rise of sea level forms drowned river valleys, or *rias*, and headlands jutting into the ocean.

(3) Wave erosion cuts back the headlands and forms vertical sea cliffs. Deposition of sediment by currents builds spits and beach areas; spit growth eventually closes off the bays, forming lagoons.

(4) Continued erosion wears the coast back to a straight line bordered by sea cliffs and a narrow beach. Stream erosion will reduce the elevation of the ridges and highlands of the land area.

(above) (1) The diagram shows a coastal region where ridges and valleys are parallel to the coast.

(2) Submergence of the coastline by elevation of sea level with respect to the land leaves numerous islands oriented parallel to the coast, such as those off the Dalmatian Coast of Yugoslavia.

raphy. In southern Greece and Turkey the line of the coast cuts across the trend of mountain ridges, resulting in a series of peninsulas. Where the line of a submergent coast parallels mountain ridges, the coastline is more regular, with linear islands parallel to the mainland, as along the coasts of Yugoslavia and southern California (Figure 18.26).

ENIGMAS OF THE OCEAN FLOOR

While the constant change observable at shorelines makes the coastal zone one of the earth's

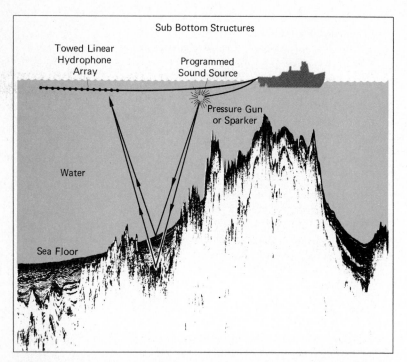

Figure 18.27 This figure illustrates the method by which information about the character of the sea floor is obtained. The oceanographic vessel tows a sound source that creates a loud noise by an electric impulse or an explosion of gas. This acoustic signal travels outward in all directions and is reflected back from the sea floor and, if strong enough, from separate layers of sediment on the sea floor as well as from the rock beneath the sediments. The reflected signals are sensed by a towed hydrophone array. Usually the signals are sent out every second. The reflected signals drive a recording device that makes a continuous trace of the type shown here in black. Such traces intentionally exaggerate the submarine relief to reveal minor details. Here we see irregular bedrock topography that is partially buried by marine sediments.

most fascinating areas of study, the ocean floors themselves, once a mystery, are proving to have quite diversified relief, including several features that have yet to be fully explained.

The use of deep-diving vehicles to observe the landforms of the ocean floor has barely begun. Most of our information about the form of the sea floors has been gained by indirect means. The most widely used method for charting the relief of the ocean floor is *echo sounding*, a technique that was devised to track enemy submarines during World War II. Sound waves are generated at the surface, and the time it takes for the echo to return from the bottom is recorded and automatically converted to a distance measure. In this way a continuous profile of the ocean floor is traced mechanically from echo data (Figure 18.27).

Coring and dredging devices lowered from

ships specially equipped for oceanographic research bring back samples of ocean floor material (Figure 18.28). Since 1968 the ongoing Deep Sea Drilling Project, utilizing the oceanographic research vessel *Glomar Challenger*, has retrieved cores from hundreds of boreholes in the deep ocean floor. Some of these were drilled in water depths exceeding 6,000 meters (20,000 ft), a very difficult procedure. While this work has done much to unveil the geologic history of the oceans, some puzzles remain. Two of these are submarine canyons and guyots.

Submarine Canyons

Canyons that rival any on the land surfaces gash the submarine continental shelves and slopes throughout the world. Their existence has been known for nearly a century. According to soundings and inspections using deep-diving vehicles, submarine canyons are similar in form to stream-cut canyons on the land. They vary from broad troughs to vast steep-sided trenches with dimensions equal to those of the largest erosional features on the continents (Figure 18.29). Canyons appear off the mouths of New York's Hudson River, the Columbia River, and the Zaire (Congo) River, but many others are not in the vicinity of any existing large stream.

Submarine canyons are puzzling because many extend to depths of nearly 5,000 meters (16,500 ft)—far too deep to be drowned river valleys. Some portions are excavated in soft sediments, but others are cut through hard rock in steep-walled gorges. The heads of these canyons come close enough to the land to trap sediment moving by longshore drift, diverting it from the continental shelves to the continental rises, ocean trenches, and deep ocean basins.

The most acceptable explanation for the origin of submarine canyons is that they are erosional forms created below sea level by streams of sand and silt that periodically pour down the continental slopes to the deep-sea floor. These so-called *turbidity currents* have been seen and photographed. The fact that turbidity currents occasionally rupture telegraph cables on the sea floor indicates a destructive potential that, given enough time, could account for rock-cut

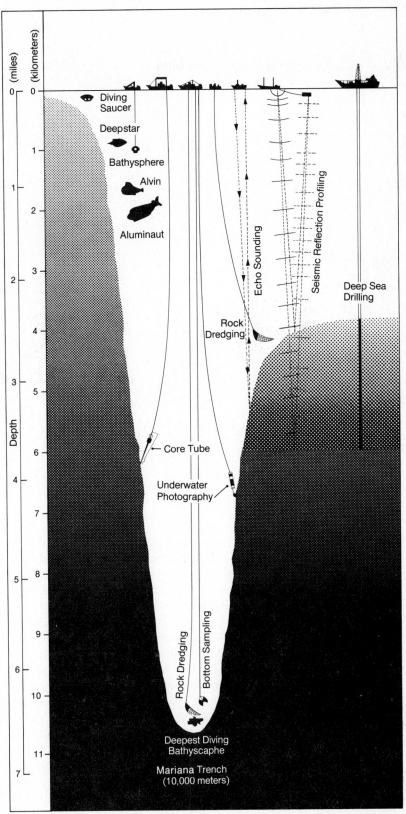

Figure 18.28 Detailed study of the ocean floor began with the voyage of the British research ship *HMS Challenger* (**top left**) in 1872. The *Challenger* made the first world-wide measurements, or soundings, of ocean depths. Some of its equipment (**above left**) included weights on lines used to measure ocean depths, and dredges to bring up samples from the ocean floor. Only about 500 deep-sea soundings were made during the 3-year voyage because of the primitive and laborious methods available.

(**right**) Modern methods of oceanographic study include echo sounding and seismic profiling for the determination of ocean floor topography; deep-sea drilling to retrieve sediment samples from the ocean bottom; and the use of a variety of deep-sea and surface craft, underwater cameras, and dredging tools. (Adapted from B. C. Heezen and C. D. Hollister, *The Face of the Deep.* © 1971 by Oxford University Press, Inc.)

submarine canyons. Thick masses of alternating sandstone and shale beds that are widely encountered on the continents are thought to be the deposits of turbidity currents in ancient oceans, and are known as "turbidites." The existence of turbidites suggests that submarine canyons have been present throughout most of the earth's history.

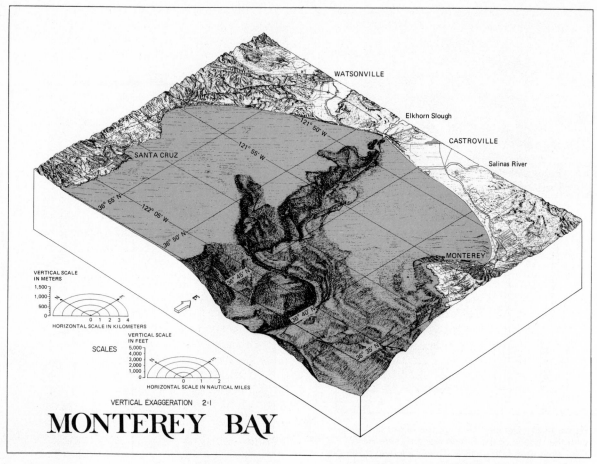

VERTICAL SCALE
IN METERS

HORIZONTAL SCALE IN KILOMETERS

VERTICAL SCALE
IN FEET

SCALES

HORIZONTAL SCALE IN NAUTICAL MILES

VERTICAL EXAGGERATION   2·1

MONTEREY BAY

Figure 18.29   This diagram shows the *submarine canyon* that cuts across the narrow continental shelf at Monterey Bay in California. Such canyons are cut into bedrock and may attain the dimensions of the largest fluvial canyons on the land. They descend much more rapidly than do fluvial valleys, however. They appear to be a principal means of channeling marine sediment down to the continental rises. The longshore drift process feeds sediment into the heads of the canyons, which lie quite close to the land. The canyons then funnel this sediment into the ocean depths as submarine *turbidity currents*. (Tau Rho Alpha, U.S. Geological Survey)

Guyots

Another puzzling feature of the sea floor is the *guyot*, a submarine mountain with a conspicuously flat top that is far below sea level. The form was discovered during World War II and has been named after a famous Swiss-American geographer of the last century. Hundreds of submerged volcanoes, or *seamounts*, are present in the world's oceans, particularly in the Pacific (Figure 12.10, p. 318). A large proportion of these are flat-topped guyots. Although guyots have the appearance of having been planed off by erosion at the surface of the ocean, their summits are often 1,200 meters (4,000 ft) or more below the sea surface—far too deep to be affected by wave action, even during times of lowered sea level.

It has been proposed that volcanic islands formed at oceanic ridges were planed off by wave attack, and then were carried laterally down the flanks of the ridges into the ocean depths as part of the sea-floor spreading process. This hypothesis of subsidence by lateral plate motion seems to be verified by age determinations on the volcanic rocks of the guyots. Most show increasing age with increasing depth and distance from the oceanic ridge system. It is probable that many coral atolls have been built up from summits of sinking guyots.

At present we have a more detailed knowledge of the surface of Mars than of the landforms of the earth's sea floors—our planet's last frontier. Future years are certain to turn up an ever-increasing list of mysteries of the deep, and more problems to engage the attention of physical geographers and their colleagues in the earth sciences.

SUMMARY

The action of waves, currents, and tides keeps the sea surface in constant motion. The sea also undergoes long-term fluctuations in level because of tectonic activity and changes in the storage of water in continental ice sheets. Most waves and currents are produced by wind stress on the water surface, whereas tides are created by the gravitational attractions of the moon and, to a lesser degree, the sun, and by the rotation of the earth/moon system.

Most water waves are of two types: forced waves and swell. The waves breaking against the shores of the land are usually swell traveling outward from distant storms. Forced waves vary in size as a function of wind strength, wind duration, and fetch. The largest waves are produced by intense storms, such as hurricanes, and by submarine earthquakes and volcanic eruptions that produce seismic sea waves known as tsunamis.

Waves transmit energy from areas of oceanic storms to far-off coastlines where the energy is expended in the work of erosion and sediment transport. Wave motion is transmitted by orbital motions of water particles. Where waves approach the land, they are slowed and refracted by friction with the sea floor. Along irregular coasts, wave refraction concentrates erosional energy on headlands and causes sediment to be shunted into bays. In this way the coast tends to become more regular. Coastal erosion by wave action creates sea cliffs, arches, stacks, and abrasion ramps. Where such features are found above sea level, uplift of the coastal region has occurred.

Beach drifting and longshore currents move sediment laterally along coastlines. Beaches, spits, and barrier islands constantly change in form according to shifts in wave climate and the balance of sediment arrivals and losses. Most of the sediment sustaining beaches is delivered by rivers that empty into the sea, rather than by coastal erosion. Barrier islands are in part relicts of Pleistocene low sea levels and are migrating landward by wave erosion and overwash during storms.

In warm tropical seas, particularly in island locations, coastlines may be composed of coral reefs. Fringing and barrier reefs encircle islands. Atolls are coral rings that surround lagoons above sunken islands. The subsidence of oceanic islands is probably a consequence of the sea-floor spreading process.

High-relief coasts are usually associated with converging oceanic and continental crustal plates, as around the margins of the Pacific Ocean. Low-relief coasts occur where the continent and the adjacent sea floor are part of the same lithospheric plate, as they are on both sides of the Atlantic Ocean. The world-wide rise in sea level at the end of the Pleistocene epoch has caused most coasts to be submerged, forming cliffed headlands, rias, and fringing islands, but this is a relatively recent development superimposed on a long-term emergent trend.

Two perplexing features of the sea floors are submarine canyons and guyots. Submarine canyons are erosional features on the continental shelves and slopes and are thought to be produced by turbidity currents. Guyots are probably former wave-bevelled volcanic islands formed along the oceanic ridge system and moved laterally into deeper portions of the oceans by sea-floor spreading. The ocean floors contain many unexplained phenomena, and are known principally by indirect means such as coring, dredging, and acoustical surveys.

REVIEW QUESTIONS

1. Describe the trend in sea level in the last 5,000 years. What was the trend in the preceding 5,000 years?
2. What two forces combine to produce the tides? What is the ideal tidal cycle?
3. Describe the paths of the water molecules that transmit the movement of waves in the open sea.
4. What are the relationships among wave length, wave period, and wave velocity in deep water? When does wave height begin to show a relationship to the preceding parameters?
5. Where are the principal sources of seismic sea waves? Why do seismic sea waves rise so high as they near the land?
6. What is the geomorphic importance of wave refraction?
7. What landforms are typical of irregular cliffed coasts?
8. Two things are required to create broad marine terraces. What are they?
9. Describe the mechanisms by which sediment is moved parallel to shorelines.
10. How can artificial groins affect coastal sediment budgets?
11. What is the difference between coastal spits and bars?
12. How are barrier islands important to the mainland areas behind them?
13. What are the habitat requirements of corals?
14. Distinguish between the different types of coral reefs, and explain the origin of each type.
15. What three different coastal configurations can result from coastal submergence?
16. How are submarine canyons formed, and what role do they play in submarine sediment movements?
17. How do guyots support the theory of sea-floor spreading?

APPLICATIONS

1. What large cities around the world would be submerged if the Antarctic and Greenland ice sheets melted completely? What inland cities would be transformed into ports if this happened?
2. How would it affect the tides if the center of rotation of the earth–moon system were closer to the center of the earth? if it were between the earth and the moon?
3. In what way is a beach similar to a glacier?
4. If there is a beach near your campus, note the relationship between the slope of the beach face and the size of the particles forming the beach. Both the beach slope and the associated particle size differ from point to point as well as from day to day. What explains the relationship? Is it a constant one? Can you see any relationship between breaker type and beach slope? between beach plan (map view) and beach slope or breaker type?
5. If your department or campus has a collection of topographic maps of the United States (or other areas), locate map examples of each of the coastal depositional forms illustrated in Figure 18.19. What seems to be the source of the sediment in each case: a river, or coastal erosion? Can you find a location in which the growth of a beach or spit has stopped wave erosion of a former sea cliff?
6. What accounts for the striking difference in the form of the Atlantic coast of North America north and south of Cape Cod, Massachusetts? Why is the coast of California so much less regular than the coasts of Oregon and Washington?

FURTHER READING

Adey, Walter H. "Coral Reef Morphogenesis: A Multidimensional Model," *Science*, 202 (1978), pp. 831–837. Advanced treatment of the controls of coral reef growth, emphasizing tectonics and sea-level change.

Bird, E. C. F. *Coasts*. Cambridge, Mass.: M.I.T. Press (1969), 246 pp. Bird's small book offers a concise and well-illustrated introduction to coastal processes and landforms, with many examples drawn from Australia.

Cronan, D. S. *Underwater Minerals*. New York: Academic Press (1980), 380 pp. Synthesis of current knowledge of the formation and possible exploitation of undersea minerals, with attention to environmental and legal pitfalls.

Davis, Richard A., ed. *Coastal Sedimentary Environments*. New York: Springer-Verlag (1978), 420 pp. Chapters on individual coastal environments: deltas, bays, estuaries, salt marshes, dunes, and beaches, by experts in the respective areas. Well-illustrated and readable.

Goreau, Thomas F., Nora I. Goreau, and **Thomas J. Goreau.** "Corals and Coral Reefs," *Scientific American*, 241:2 (1979), pp. 124–126. Excellent, well-illustrated article on all aspects of coral reefs, stressing the symbiosis between coral and photosynthetic algae that makes reef formation possible.

Heezen, Bruce C., and **Charles D. Hollister.** *The Face of the Deep*. New York: Oxford University Press (1971), 659 pp. This is a massive volume that deals in a thoughtful way with all aspects of ocean floors.

Inman, Douglas L., and **Birchard M. Brush.** "The Coastal Challenge," *Science*, 181 (July 6, 1973): pp. 20–31. This article details modern findings concerning the physical processes in coastal systems, and shows how human activities have changed coastal environments, often diminishing both their utility and aesthetic qualities.

King, Cuchlaine A. M. *Beaches and Coasts*, 2nd ed. New York: St. Martin's (1972), 570 pp. King combines theory, model experiments, and field observation in detailed analyses of coastal processes and landforms.

Masters, P. M., and **N. C. Flemming, eds.** *Quaternary Coastlines and Marine Archaeology*. New York: Academic Press (1982), 500 pp. Interdisciplinary inquiry into human movements between continents and islands during Pleistocene low stands of the seas. Techniques of underwater archaeology.

Menard, Henry W. "Insular Erosion, Isostacy, and Subsidence," *Science*, 220 (1983), pp. 913–918. Interesting advanced treatment of sea-floor spreading, erosion, and isostacy as they affect volcanic islands.

Russell, Richard J. *River Plains and Sea Coasts*. Berkeley and Los Angeles: University of California Press (1967), 173 pp. About half of this engagingly written book is devoted to Russell's personal observations about shoreline processes. The author concentrates upon low-latitude coasts.

Shepard Francis P. *The Earth Beneath the Sea*, 2nd ed. Baltimore: Johns Hopkins Press (1967), 242 pp. This very readable nontechnical treatment of the features of the sea floor is by a leading investigator of submarine canyons. As well as presenting facts and interpretations, Shepard indicates how data are collected beneath the sea.

—— and **Harold R. Wanless.** *Our Changing Coastlines*. New York: McGraw-Hill (1971), 571 pp. This large book is a complete inventory of the coastal morphology of the United States, including Alaska and Hawaii. The text is superbly illustrated with aerial photographs.

Strahler, Arthur H. *A Geologist's View of Cape Cod*. Garden City, N.Y.: Natural History Press (1966), 115 pp. This small, nicely written book details how glacial deposition and wind and wave action have fashioned the landforms of a popular tourist area. Very well illustrated and nontechnical.

CASE STUDY

New Resources from the Seas

Increased knowledge of the ocean floors may bring us both benefits and new problems. It is now known that the sea floors contain a host of valuable minerals, some of which are already being exploited. This has provoked important questions regarding the ownership of the sea floors and the effect of mineral extraction on oceanic life. The brief but costly war in 1982 over ownership of the Falkland Islands was in part motivated by the much more lucrative petroleum resources believed to exist on the adjacent continental shelf.

As noted much earlier, portions of the continental shelves contain vast reservoirs of petroleum. Offshore petroleum was first exploited from the land by drilling diagonally under the sea floor. Now, however, the technology exists to drill from immense platforms in the open seas (accompanying diagram) and operating offshore oilfields have proliferated in the Persian Gulf, the Gulf of Mexico, the Caribbean Sea, the North Sea, the Pacific coast of California, the Beaufort Sea, the Yellow Sea, the South China Sea, and the waters surrounding the Indonesian archipelago.

Other types of minerals of great potential value have accumulated on the sea floors close to the continents as *placer* deposits—materials carried down to the sea by rivers draining the land. Both gold- and diamond-rich marine placers are known, but mining them has generally been too costly to continue for long. The greatest lure in this category are the high grade diamond deposits off the coast of Namibia in southern Africa.

On the pioneering oceanographic voyage of the *H.M.S. Challenger* in 1872 (Figure 18.28) it was discovered that nodules and potato-size lumps of manganese oxides littered the deep-sea floors, far from land. Associated with the manganese are iron, copper, nickel, cobalt, molybdenum, lead, zinc, and vanadium: all commodities of great economic and strategic importance. Thus, scientists have urged that the popular term "manganese nodules" be replaced by the more accurate "polymetallic nodules." Phosphate nodules are also widespread on the ocean floors, and they too are a potentially valuable resource.

The sea-floor nodules of various types pose tantalizing problems: how to mine them economically? how to exploit them without endangering the biotic resources of the seas? how to extract the more valuable minerals from the extremely complex "ores"? how to determine who has the right to mine in specific areas far from land? whether to use the resources of the ocean floors to redress the economic imbalances between the technological societies and the less developed countries— even if the latter must count on the former to develop the means to harvest the new wealth? The nodules are not abundant everywhere; it is estimated that they occur on perhaps 15 percent of the bed of the sea, mainly in the Pacific Ocean, with especially rich deposits close to the Hawaiian Islands. The origin of the nodules and the factors that control their distribution remain matters of the purest speculation.

As recently as 1969 another oceanic phenomenon of possible economic value was discovered: richly mineralized muds associated with sea-floor spreading centers. The first discovery was in the Red Sea, where metal-saturated hot brines were observed venting from the central rift zone, presumably originating directly from molten magma beneath. Similar phenomena have subsequently been detected along the Mid-Atlantic Ridge and East-Pacific Rise. Thus the entire oceanic ridge system and all divergent plate boundaries have come under scrutiny as potential new sources of metallic minerals—to augment the dwindling reserves on the continents. As yet this resource is poorly known, and the ever-present problems of ownership and ecological consequences of exploitation loom large.

Finally, it must be noted that the seas are not only sources of food, fossil fuels, and perhaps metallic minerals, but they are also potentially inexhaustible sources of "clean" energy for human use. The sea has the potential to supply usable power in several ways, some of which are already operational. In estuaries the ebb and flow of the tides can be channeled through turbines to generate electricity directly, as is already being done at the mouth of the Rance, a French river emptying into the English Channel. Tidal energy can also be used indirectly to compress air that is held in underground storage, which can drive gas turbine generators even when the tide is not "running". Likewise, the kinetic energy of the up and down motions of the sea surface, produced by wave action, can be transformed into electrical energy. In addition, both thermal gradients and salinity gradients exist in the seas and each constitutes an immense

538

CASE STUDY

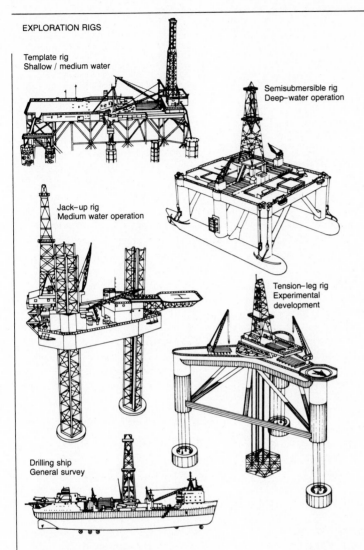

EXPLORATION RIGS

Template rig
Shallow / medium water

Semisubmersible rig
Deep–water operation

Jack–up rig
Medium water operation

Tension–leg rig
Experimental development

Drilling ship
General survey

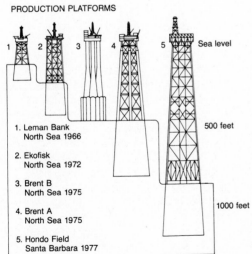

PRODUCTION PLATFORMS

1 2 3 4 5 Sea level

500 feet

1000 feet

1. Leman Bank
North Sea 1966

2. Ekofisk
North Sea 1972

3. Brent B
North Sea 1975

4. Brent A
North Sea 1975

5. Hondo Field
Santa Barbara 1977

reservoir of potential energy that can be converted to electrical power.

Techniques to utilize these oceanic energy sources have already been devised and pilot projects built, but costs remain too high to be competitive with conventional energy sources under current conditions. However, as fossil fuel supplies dwindle and the environmental problems associated with their extraction and use continue to mount, necessity may dictate that natural marine processes be utilized to yield safe, non-polluting, energy supplies that would be adequate for all time—in the coastal zone! This could lead to overdevelopment and environmental degradation along coastlines. No doubt unforeseen complications will arise in the utilization of such unfamiliar energy sources. Our task is to learn enough about the natural processes in this environment to avert the mistakes we have made elsewhere in similar attempts to utilize all of the earth's resources to the fullest measure.

These diagrams show various types of platforms used in exploration for offshore oilfields. In shallow- to medium-depth water the platform is anchored to the sea floor, but in deeper water the platforms are supported by huge partly submerged floats. Note the helicopter landing area on each type of platform. The drilling ships used in oil exploration are kept on the drill site by systems of propellers, visible under the hull. Nevertheless, deep drilling from a ship is an extremely difficult undertaking.

Production platforms are still larger and more complex, with multiple decks and crews of up to 100. Each production platform drills a large number of wells—as many as 27 in one of the North Sea fields—which angle downward and out to radial distances of as much as 3 km (2 miles) from the central borehole. North Sea platforms are designed to drill to depths of nearly 4 km (2.5 miles), and to withstand wave heights of 30 meters (100 ft) and winds of 200 kph (130 mph). (Rand McNally Atlas of the Oceans, copyright by Mitchell Beazley Ltd. 1977)

539

Appendices

APPENDIX I
Scientific Measurements

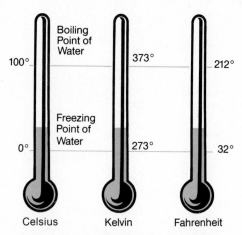

UNITS OF MEASURE

The length, volume, mass, or temperature of an object must be expressed according to a definite system of units if the measurement is to be meaningful to others. In the *metric system*, the system of units used in science, length is commonly expressed in meters or centimeters (cm) (1 meter = 100 cm). One meter is about 3 in. longer than a yard, and 1 cm is approximately ⅜ in. One kilometer (km) (1,000 meters) is approximately 0.6 mile. Mass in the metric system is expressed in kilograms (kg) or grams (kg or gm) (1 kg = 1,000 grams). One kg is equivalent to 2.2 pounds, and 1 gram is equal to about 0.035 ounce. A table of conversion factors is given in the next section.

Two scales are commonly used by scientists to measure temperature. On the *Celsius* (C) scale (formerly called the centigrade scale), the temperature at which pure water crystallizes, or "freezes" to ice, is taken as 0°C, and the temperature at which water "boils" into vapor is fixed at 100°C. The interval from freezing to boiling is divided into 100 equal degrees. Normal room temperature of 72° Fahrenheit (F) is equivalent to approximately 22°C, and normal body temperature is equivalent to 37°C.

The second temperature scale, called the *absolute*, or *Kelvin* (K) scale, has a different zero point from the Celsius scale. The temperature of a gas is a measure of the kinetic energy of its molecules; as a gas is cooled, the molecules move more slowly and have less energy. Theoretically, there is a temperature at which all motion would cease, and this point is taken as zero on the Kelvin scale. Zero on the Kelvin scale is equivalent to approximately −273°C (−459°F), so that the freezing point of water on the Kelvin scale is 273°K (Figure I.1). Many sci-

Figure I.1 The Celsius and Kelvin scales are used to report temperatures in scientific work. The freezing point of water corresponds to 0° on the Celsius scale, approximately 273° on the Kelvin scale, and 32° on the Fahrenheit scale. The temperature range between the freezing point of water and its boiling point is divided into 100 equal degrees on the Celsius and Kelvin scales, so that the boiling point of water can be expressed as 100°C or 373°K. The temperature of boiling water on the Fahrenheit scale is 212°F.

Since there are 100 degrees of temperature difference between the freezing and boiling points of water on the Celsius and Kelvin scales and 180 degrees of difference on the Fahrenheit scale, a temperature change of 1°C corresponds to a change of 1.8°F (9/5°F).

entific laws are simpler when expressed in absolute temperature because of its fundamental significance on a molecular level.

METRIC TO ENGLISH CONVERSIONS

Length

1 kilometer = 1,000 meters = 0.6214 mile = 3,281 feet
1 meter = 100 centimeters = 1.0936 yards = 3.281 feet = 39.37 inches
1 centimeter = 10 millimeters = 0.3937 inch
1 micron = 10^{-6} meter = 10^{-4} centimeter = 3.937×10^{-5} inch

Area

1 square kilometer = 10^6 square meters = 0.3861 square mile = 247.1 acres
1 square meter = 10^4 square centimeters = 1.196 square yards = 10.764 square feet = 1,550.0 square inches

Volume

1 cubic kilometer = 10^9 cubic meters = 0.2399 cubic mile

1 cubic meter = 10^6 cubic centimeters = 1.308 cubic yards = 35.31 cubic feet = 61,024 cubic inches

Mass

1 metric ton = 1,000 kilograms = 2,204.6 pounds

1 kilogram = 1,000 grams = 2.2046 pounds

Time

1 day = 86,400 seconds

1 year = 3.156×10^7 seconds

Speed

1 meter per second = 3.281 feet per second

1 meter per second = 3.6 kilometers per hour = 2.237 miles per hour

1 kilometer per hour = 0.62 mile per hour

1 knot = 1 nautical mile per hour = 1.151 miles per hour

Pressure

1 bar = 10^6 dyn/cm^2

1 atmosphere = 1.01325 bars = 1,013.2 millibars = 760 millimeters of mercury = 29.92 inches of mercury

Temperatures

°C = 5/9 (°F −32°)

°F = 9/5 °C + 32°

°K = °C − 273.15°

Energy

1 calorie = 4.186 joules = 3.968×10^{-3} British Thermal Unit

1 langley = 1 calorie per square centimeter

Power

1 calorie per second = 4.186 joules per second = 4.186 watts

1 calorie per minute = 0.07 watts

Figure I.2

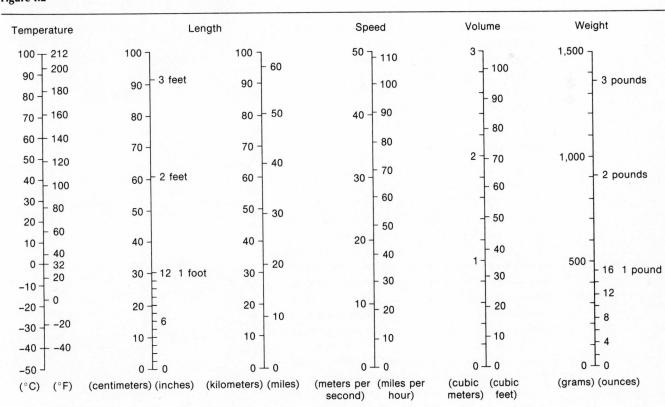

APPENDIX II
Tools of the Physical Geographer

MAPS: A REPRESENTATION OF THE EARTH'S SURFACE

Modern maps exist in great variety, from simple street maps to complex navigational charts for jet aircraft. A map can convey a large amount of information in a way that is easily assimilated: a map of climatic regions, for example, can make important climatic relationships much clearer than can lengthy written descriptions or tables of data.

A *map* can be defined formally as a two-dimensional graphic representation of the spatial distribution of selected phenomena. A map is *planimetric;* that is, it shows *horizontal* spatial relationships on the earth. In addition to specifying location, distribution, amount, distance, direction, sizes, and shapes, maps can also represent the form of the land surface or any statistical surface based on spatial data.

Scale and Distance

A map's scale gives the relationship between length measured on the map sheet and the corresponding distance on the earth's surface. There are several ways to express the scale of a map. It may be given in a simple *verbal statement,* such as "1 inch equals 3 miles." The scale of a map also may be indicated by a *graphic scale* marked off in units of distance on the earth, as shown in Figure II.2. One advantage of a graphic scale is that, unlike a verbal scale, it remains correct if the map is copied in a larger or smaller size.

Scale is often expressed as a fraction, called the *representative fraction.* A representative fraction of 1/5,000 (commonly written 1:5,000)

Figure II.1 (top) This Babylonian map from 500 B.C. is one of the earliest attempts to portray the world. (The British Museum)

(bottom) Claudius Ptolemy's map of the world, reconstructed from his descriptions written in the second century, is a comparatively accurate representation that takes into account the spherical shape of the earth. (*Atlas of the Universe,* p. 13, © 1971, Mitchell-Beazley, Ltd.)

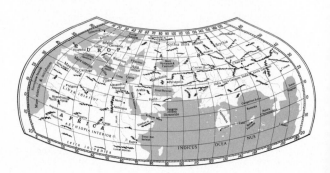

544

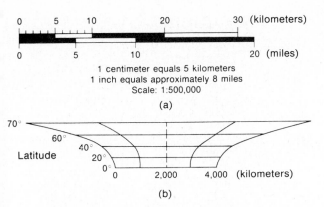

1 centimeter equals 5 kilometers
1 inch equals approximately 8 miles
Scale: 1:500,000

(a)

(b)

Figure II.2
(Andrea Lindberg)

means that a distance of 1 unit on the map represents 5,000 similar units on the earth. A representative fraction makes no reference to any particular system of units, because it represents simply the ratio between distance on the map and a corresponding distance on the earth.

A verbal statement of scale can be restated as a representative fraction by converting both members of the statement to the same units; thus a scale of 1 inch to the mile is equivalent to 1:63,360 because there are 63,360 inches in 1 mile. Conversely, a representative fraction can be expressed as a verbal statement of scale by assigning units and applying conversion factors as required. A scale of 1:100,000 can be stated as "1 cm equals 1 km."

When a portion of the spherical earth is represented on a flat map, distortions inevitably occur. Consequently the scale of a map cannot be constant for every portion of the map. However, scale does not vary greatly on a flat map of a small region, and even the coterminous United States can be mapped in such a way that the scale does not vary by more than a few percent. Significant variations of scale occur on all world maps, however. The scale on a Mercator map of the world, for instance, is several times larger at higher latitudes than at the equator (Figure II.2).

When the representative fraction is a small number, less than 1/1,000,000, a map is called a *small-scale map*. Small-scale maps are used when a large portion of the earth's surface, such

as a continent or an ocean, must be represented on a map of limited size. If a map has a representative fraction larger than 1/250,000 (1:250,000), it usually is called a *large-scale* map.

Large-scale maps of small areas are capable of showing greater detail than small-scale maps of large areas (Figure II.3).

Location

The principal method for specifying location on the earth's surface is by the system of latitude and longitude. *Latitude,* the position of a place north or south of the equator, is expressed in angular measure relative to the center of the earth. The angle of latitude varies from 0° at the equator to 90° at the poles. *Longitude,* the position of a place east or west of a selected prime meridian, is expressed in angular measure that varies from 0° to 180° east or west of the prime meridian. The most commonly accepted prime meridian is the one on which the Greenwich Observatory in England is located, but other prime meridians have been used in the past. The framework of lines representing parallels of latitude and meridians of longitude on a map is called the *graticule* or *grid* of the map. Depending on the method chosen to construct a map, the lines of the graticule may be straight or curved, and they may or may not intersect at right angles as they do on a globe.

A more easily computed location reference system uses a rectangular grid composed of straight lines that do intersect at right angles.

The first step in constructing such a grid system is to choose a standard map projection that meets the needs of the user. (The advantages and drawbacks of different map projections are discussed later in this appendix.) Once the map is chosen, a square grid is overlaid on the map and numerical coordinates are assigned to the reference lines of the grid. The coordinates are usually expressed in units of distance from a selected origin. The grid coordinates of any location can then be read from the map as illustrated in Figure II.4. By convention, the coordinate to the east, or the *easting,* is specified first. Then the coordinate to the north, or the *northing,* is specified. The rule is

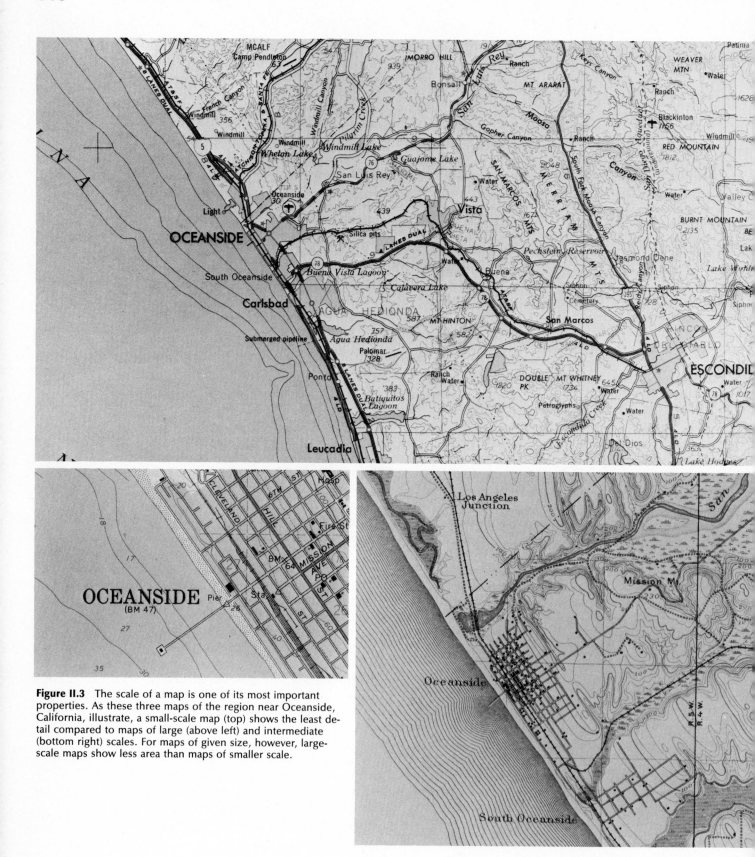

Figure II.3   The scale of a map is one of its most important properties. As these three maps of the region near Oceanside, California, illustrate, a small-scale map (top) shows the least detail compared to maps of large (above left) and intermediate (bottom right) scales. For maps of given size, however, large-scale maps show less area than maps of smaller scale.

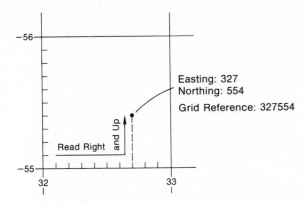

Figure II.4

to read toward the *right* and *up*, following the same order used for giving the x and y coordinates of a point on a graph. When using computers to construct maps or specify data points, rectangular coordinates may be recorded differently, as + or − values horizontally and vertically from a central point, or *right* and *down* from the top left.

The United States National Ocean Survey (formerly the United States Coast and Geodetic Survey) has designed a rectangular grid system for each of the states, called the *State Plane Coordinate System*. The basic grid square of the state coordinates is 10,000 ft on a side; eastings

and northings for the grid are listed in units of feet.

One of the grid systems most frequently encountered consists of the Universal Transverse Mercator (UTM) and the Universal Polar Stereographic (UPS) grids developed for military and civilian use. The UTM grids extend between latitudes 80°N and 80°S, and the UPS grids cover the regions poleward from latitude 80°.

For the purpose of setting up the UTM grids, the globe between latitudes 80°N and 80°S is divided into 60 sections, each 6° of longitude wide. A square grid with 100,000-meter spacing is then superimposed on a transverse Mercator map of each section. (The Mercator map projection will be discussed subsequently.) The central meridian of each section is assigned an easting of 500,000 meters; the equator is assigned a northing of 0 meters for locations in the northern hemisphere and a northing of 10,000,000 meters for locations in the southern hemisphere. Each 6° section is divided into twenty 6° by 8° quadrilaterals, and for purposes of zone identification, each quadrilateral is assigned an index consisting of a number from 1 through 60 and a letter from C through X (with I and O omitted). Each of the 100,000-meter squares is further assigned a two-letter identi-

Figure II.5 The UTM grid is subdivided successively into zones, quadrangles, and 10,000-meter grid squares.

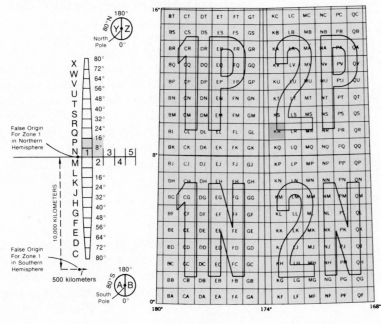

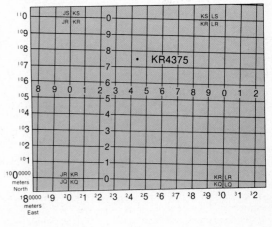

fication index (Figure II.5). The UPS grid zones are designated in a similar way.

Some maps contain references to one or more grid coordinate systems as well as to the conventional system of latitude and longitude. The topographic maps prepared by the United States Geological Survey, for example, bear tick marks around the margins that indicate UTM coordinates, state plane coordinates, and latitude or longitude. On the topographic maps published by the United States Army Topographic Command (TOPOCOM), formerly known as the Army Map Service, the UTM grid is drawn on the map, with latitude and longitude given as tick marks on the margins.

Aeronautical charts employ a system of coordinates that is equivalent to latitude and longitude but uses a different method of notation, known as the World Geographic Reference System (GEOREF). According to the GEOREF system, longitude is divided into twenty-four 15° zones and latitude is divided into twelve 15° zones. A two-letter index specifies each 15° by 15° quadrangle, and each quadrangle is divided into 1° by 1° sections (Figure II.6). Four letters are necessary to specify a given 1° by 1° section:

the first two provide the designation of the 15° quadrangle, and the last two, the designation of the 1° quadrangle. For accurate specification of location, each 1° interval is divided into 60 minutes of angular measure and subdivided into decimal fractions of a minute. Most aeronautical charts employ the GEOREF system rather thaan the UTM grid.

A modified grid system has been in use for many years in connection with the survey of public lands conducted by the Bureau of Land Management. The basic land unit of the survey, which was begun in the eighteenth century, is the *township*, a square plot 6 miles on a side. Townships are laid out with two sides along meridians and the other two sides along parallels of latitude. Because meridians converge toward the north, the north-south sides of the townships must jog eastward or westward every 24 miles to maintain the size of the 6-mile square.

Townships are laid out with respect to a north-south *principal meridian* and an east-west *base line*. Different land surveys established thirty-one sets of principal meridians and base lines for the conterminous United States and five sets for Alaska. The location of each township in a survey region is given with respect to the point at which the principal meridian and the base line intersect. The coordinates that specify a particular township are read as the number of townships north or south of the base line; the number of townships east or west of the principal meridian is called the *range*. The system for locating townships is an exception to the "right and up" rule of reading because northings are read before eastings. Townships are further subdivided into 36 squares, 1 mile on each side, which are called *sections*; sections are numbered 1 through 36 in a serpentine fashion, beginning in the upper right corner of the township (see Figure II.7).

Figure II.6 The 15° square of the GEOREF system is subdivided into 1° squares.

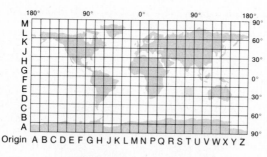

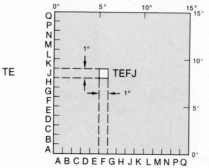

Direction

By definition, meridians of longitude are true north-south lines, and parallels of latitude are true east-west lines. Because of the distortions inherent in representing the surface of a sphere

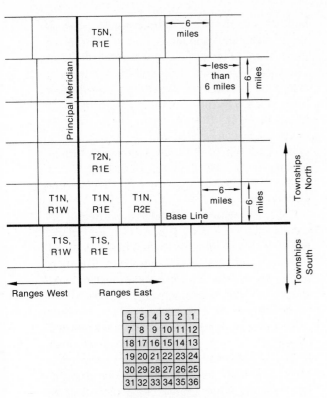

Figure II.7   (Andrea Lindberg after Army Field Manual, FM 21-26, Oct. 1960)

on a flat sheet of paper, meridians or parallels often vary in direction across a map. But for large-scale maps that cover small areas the distortions are barely visible and the map sheet can be aligned with respect to a single standard direction to establish a sense of orientation.

Many large-scale maps indicate the direction of *true north* by means of a star-tipped arrow or the symbol *TN* (Figure II.8). However, this direction is usually not the same as *magnetic north*, the direction in which a magnetic compass needle points. Large-scale maps usually indicate the direction of magnetic north by means of a half-headed arrow and he symbol *MN*. The earth's magnetic field is not uniform, and the magnetic poles do not coincide with the geographic poles, so the relation between magnetic north and true north must be specified separately for each region. The difference between magnetic north and true north is known as *magnetic declination* and is expressed in degrees east or west of the true meridian of a given location.

Across the conterminous United States the magnetic declination varies from 0° to as much as 25° east or west, and in polar regions a declination of 90° or more is possible. Furthermore, the direction of magnetic north at a given location varies with time, often by as much as 1° in 20 years. For precision map work, therefore, reference should be made to recent compilations of magnetic declinations, such as those prepared by the National Ocean Survey.

Direction on a map may also be specified as *grid north*, the northerly direction arbitrarily determined by a particular grid system, and symbolized on the map by *GN*. Grid north generally does not coincide with either magnetic north or true north. The grid north directions specified by two different grid coordinate systems usually differ from one another as well.

Directions other than north can be expressed in terms of *azimuth*, which is the angle of the desired direction measured clockwise from a chosen reference direction and expressed in degrees between 0° and 360°. According to the choice of true north, magnetic north, or grid north as a reference, the corresponding azimuths are termed *true azimuth, magnetic azimuth,* or *grid azimuth.*

MAP PROJECTIONS

A model globe is the only way to represent large portions of the earth's surface with accuracy, because only a globe correctly takes into account the spherical shape of the earth. A flat piece of paper cannot be fitted closely to a

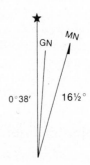

Figure II.8   (Andrea Lindberg)

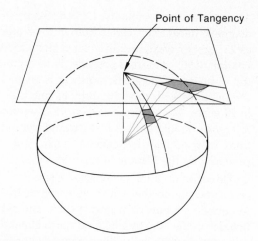

Figure II.9 Construction of the Gnomonic Projection (Andrea Lindberg)

sphere without wrinkling or tearing, so small-scale maps that represent regions hundreds or thousands of miles in extent inevitably introduce noticeable distortions.

The fundamental problem of map making is to find a method of transferring a spherical surface onto a flat sheet in a way that minimizes undesirable distortions. Any method of relating position on a globe to position on a flat map is called a *projection*, or *transformation*. Numerous projections have been devised, each with its characteristic advantages and distortions. Because no projection is free of distortion, the choice of a projection should be made with regard to its proposed application.

The principles of projection are illustrated in Figure II.9, using the example of the gnomonic projection. This projection can be constructed by tracing the rays of light from a light source at the center of a transparent globe onto a plane that touches the globe at one point, called the *point of tangency*. Each point on the portion of the globe that is projected onto the plane constitutes a point on the map. However, only a few projections, such as the gnomonic, can be visualized geometrically. Many projections can be expressed only as a mathematical rule that relates points on a globe to points on a flat sheet.

Projections and Distortion

Maps are commonly relied upon to show correct direction and distance from one location to another and the sizes and shapes of areas. A single flat map can depict one or another of these without distortion, but not all of them. The map user should realize where and to what extent inaccuracies are present in the projection being used. On the gnomonic projection illustrated in Figure II.9, for example, the scale is increasingly exaggerated with distance from the point of tangency, producing extreme distortion of shapes as well as sizes. Nevertheless, we shall see that this is an extremely valuable projection.

Scale distortion may lead to distortion of direction, shape, and size. On some projections, the scale in a small region is not the same in all directions, which necessarily leads to distortion of angles. Distortion of angles implies that shapes will not be geometrically accurate.

A number of projections, called *conformal* projections, have been devised so that azimuths from point to point are correct. Shapes of small areas are portrayed accurately. However, on a conformal map the scale necessarily changes from one region to another. Regions that span a large portion of the globe exhibit distortion when mapped on a conformal projection. The well-known Mercator projection is conformal, but it represents areas in the higher latitudes several times larger than areas of the same size near the equator.

Another important class of projections includes the *equal area*, or *equivalent*, projections. On such projections, point-to-point azimuths are incorrect, but the scale is designed to vary over the map in such a manner that the relative sizes of all areas are correct. Figure II.10 illustrates how shapes can be varied without altering their areas. Equal area projections should always be used when the geographic distributions of phenomena are displayed because regions are represented in their correct relative sizes. Many projections are neither conformal nor equivalent area but represent compromises to obtain adequate representation of shape without badly distorting size.

An impression of a projection's major properties can usually be obtained by observation of the lines of latitude and longitude. On a globe, parallels and meridians intersect at right angles. If a map shows them intersecting at right angles, the projection may or may not be confor-

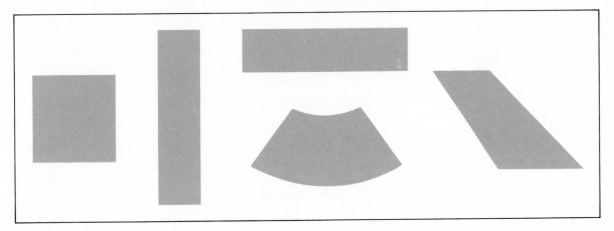

Figure II.10
(T. M. Oberlander)

mal, but if they are shown intersecting at other angles, distortion of shape is present and the projection is not conformal. Distortions of shape or size can be seen by comparing the map compartments bounded by parallels and meridians to those on a globe.

The globe may be projected onto a plane, cylinder, or cone, which are the only surfaces that can be spread flat. Projections are conventionally classified into families: *azimuthal*, or *zenithal*, projections onto a plane; *cylindric* projections onto a cylinder; and *conic* projections onto a cone. A fourth family of *geometrical projections* is usually used to portray the entire globe. The projections in a given family tend to have similar properties and similar distortion characteristics.

Azimuthal Projections

Azimuthal projections are projections of a globe onto a plane tangent to the globe at some point. The point of tangency is usually the north or south pole, but in principle any point on the globe may be used. Most azimuthal projections can depict only one hemisphere, or less, of the earth at one time.

The distortion of any azimuthal projection is least nearest the point of tangency and increases with increased distance away from the point of tangency. Figure II.11 exhibits some of the characteristics of standard azimuthal projections. Distortion patterns are depicted by degrees of yellow shading, with the deeper tints corresponding to greater distortion.

Cylindric Projections

A cylinder closely fitted to a globe makes contact with the globe along a *great circle*, called the *circle of tangency*. Cylindric projections are usually designed so that the circle of tangency is the equator or a meridian.

Most cylindric projections with the equator as the circle of tangency show parallels of latitude and meridians of longitude as sets of straight parallel lines intersecting at right angles. The spacing of the parallels can be manipulated mathematically to provide either conformality or equivalence. The best known such projection, the standard Mercator, is conformal; others are not. The standard Mercator projection is used for navigation charts because a straight line on this projection represents a line of constant azimuth, which facilitates navigation by compass direction.

Distortions on a cylindric projection are least nearest the circle of tangency and increase with increased distance from the circle of tangency. On a standard Mercator projection areas in the higher latitudes are grossly exaggerated. Conversely, on a cylindric equal area projection based on the equator, shapes at higher latitudes are badly distorted. Figure II.12 shows the properties of several cylindric projections.

Figure II.11

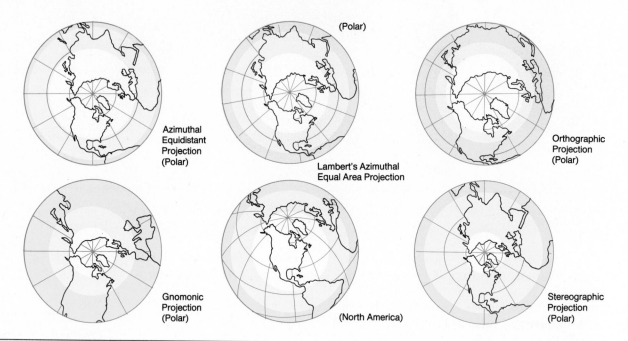

Azimuthal
Equidistant
Projection
(Polar)

(Polar)

Lambert's Azimuthal
Equal Area Projection

Orthographic
Projection
(Polar)

Gnomonic
Projection
(Polar)

(North America)

Stereographic
Projection
(Polar)

Table of Azimuthal Projections

Projection	Appearance	Properties	Distortion Pattern	Best Uses
Azimuthal Equidistant (Polar)	Meridians are straight lines radiating from the pole; parallels are equally spaced circles concentric about the pole.	Scale is constant and correct along meridians. Directions from central point are correct.	Distortion increases slowly with increased distance from the center. Shapes are represented comparatively well, but areas are distorted.	Directions and distance to the center are undistorted; useful for charts of radio propagation to a given location or for showing relative distance from a given location.
Gnomonic (Polar)	Meridians are straight lines radiating from the pole; parallels are circles concentric about the pole with rapidly increasing spacing outward.	Great circles anywhere on the map are represented by straight lines.	Distortion increases rapidly with increased distance from the center. Shapes become badly distorted.	Aircraft and ship navigation. Only projection on which shortest distance between any points on the earth plots as a straight line.
Lambert Equal Area (Polar)	Meridians are straight lines radiating from the pole; parallels are circles concentric about the pole with spacing decreasing outward.	The only azimuthal equal area projection. Directions from central point are correct.	Distortion increases moderately with increased distance from the center. Shapes are represented well.	Polar maps and maps of one hemisphere, especially where distributions are to be represented.
Orthographic (Polar)	Meridians are straight lines radiating from the pole; parallels are circles concentric about the pole with spacing decreasing outward.	Directions from central point are correct. Gives appearance of earth as seen from deep space.	Distortion increases moderately with increased distance from the center.	Used primarily for illustrations, shows how the earth looks from outer space.
Stereographic (Polar)	Meridians are straight lines radiating from the pole; parallels are circles concentric about the pole with spacing increasing outward.	The only azimuthal conformal projection. Directions from central point are correct.	Area distortion increases rapidly with increased distance from the center. Shapes are represented well.	Base map for the UPS grid system, poleward of latitude 80°.

(Figures II.11–II.14 by Andrea Lindberg after Arthur Robinson and Randall Sale, *Elements of Cartography*, 3rd ed., © 1953, 1960, 1969 by John Wiley & Sons, reprinted by permission)

Figure II.12

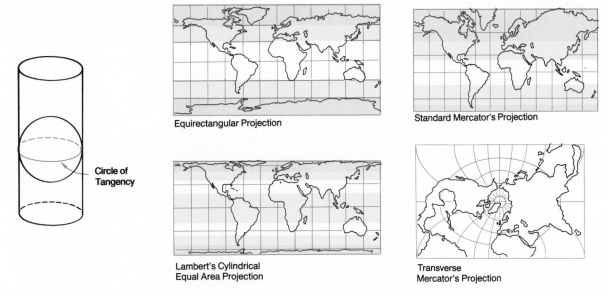

Equirectangular Projection

Standard Mercator's Projection

Lambert's Cylindrical
Equal Area Projection

Transverse
Mercator's Projection

Table of Cylindric Projections

Projection	Appearance	Properties	Distortion Pattern	Best Uses
Equirectangular	Parallels and meridians are equally spaced and form a square grid. Often a parallel is chosen to be the circle of tangency.	No major properties. Can depict entire earth. Neither conformal nor equivalent.	Scale is correct along the standard parallel and along meridians, but shape and area distortion increase with increased distance from the standard parallel.	Used only for large- or moderate-scale maps of limited areas in low latitudes.
Lambert Equal Area Cylindrical	Parallels and meridians form a rectangular grid. Employs two parallels equidistant from the equator as circles of tangency.	A global equal area projection excluding the polar regions.	Shape distortion increases with increased distance from the standard parallels. Shapes at high latitudes are severely compressed in the north-south direction.	Not widely used because of severe shape distortion at high latitudes, but satisfactory distributions in the lower latitudes.
Standard Mercator	Parallels and meridians form a rectangular grid, with the equator as the circle of tangency.	A conformal global projection. A straight line on the map is a line of constant azimuth on the earth. Excludes the polar regions.	Areas are magnified with increased distance from the equator. Shapes of small regions are represented well, but there are gross distortions of area at high latitudes.	Commonly misused for general purpose world maps. Only legitimate use is for navigation by compass headings after route is laid out on gnomonic projection.
Transverse Mercator	The circle of tangency is a meridian or portion of a meridian. Most meridians and parallels are curved.	A conformal projection. Most straight lines on the map are not lines of constant azimuth on the earth.	Areas are magnified with increased distance from the central meridian. Scale distortion is constant along lines parallel to the central meridian. Shapes of small regions are represented well.	Base map for UTM grid system, for some State Plane coordinate grids, and for the British Ordinance Survey maps.

Figure II.13

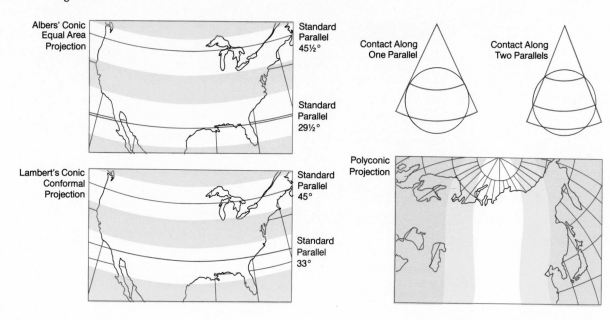

Table of Conic Projections

Projection	Appearance	Properties	Distortion Pattern	Best Uses
Albers Equal Area Conic	Meridians are converging straight lines, and parallels are arcs of concentric circles. Two standard parallels are used; for maps of the conterminous United States, they are 29½°N and 45½°N.	An equal area projection.	Distortion increases slowly with increased distance from the standard parallels. Scale along meridians is slightly too large between the standard parallels and somewhat too small outside them. Distortion is small in normal applications.	An excellent projection for midlatitude countries that extend primarily east-west. Usually applied over a restricted range of longitude. Distributions are shown without bias because it is an equal area projection. Chosen for the maps in the National Atlas of the United States.
Lambert Conformal Conic	Meridians are converging straight lines, and parallels are arcs of concentric circles. Two standard parallels are used; for maps of the conterminous United States, they are often 33°N and 45°N.	A conformal projection.	Distortion increases slowly with increased distance from the standard parallels. Scale along meridians is slightly too small between the standard parallels and somewhat too large outside them. Distortion is small in normal applications, but slightly greater than for Albers' equal area projection.	Used for countries that extend primarily in the east-west direction and for some air navigation charts because straight lines on the map represent great circles.
Polyconic	Meridians are curves converging toward the poles, except for one straight central meridian. Parallels are arcs of circles, except perhaps for one parallel, but each circle has its own center. Every parallel is a standard parallel for the projection.	An excellent compromise between a conformal projection and an equal area projection.	Distortion tends to increase with increased distance east and west of the central meridian. Distortion is small when small regions are mapped.	Base for the topographic maps of the United States Geological Survey. Satisfactory for maps of small regions, particularly regions extending primarily north and south.

Figure II.14

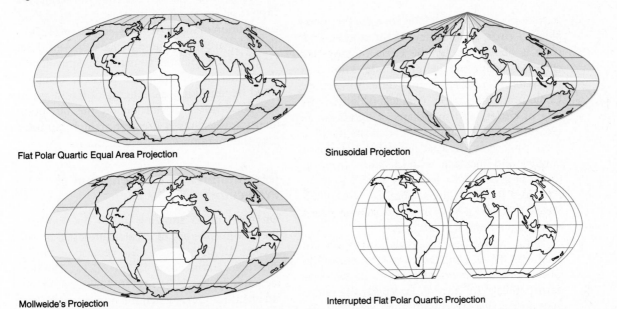

Flat Polar Quartic Equal Area Projection

Sinusoidal Projection

Mollweide's Projection

Interrupted Flat Polar Quartic Projection

Table of Geometrical Global Projections

Projection	Appearance	Properties	Distortion Pattern	Best Uses
Flat Polar Quartic Equal Area	Parallels are straight parallel lines. Meridians are curves in general but the central meridian is straight. The poles are represented by straight lines one-third the length of the equator. Meridians converge toward the poles and are equally spaced along each parallel. The spacing between parallels decreases slightly with increased latitude. The boundary of the map is a complex curve.	An equal area projection.	Distortion is least nearest the equator and central meridian and greatest at high latitudes near the boundaries.	Generally useful as a world map and for depicting global distributions.
Mollweide	Parallels are straight parallel lines. Meridians are elliptical in general, but the central meridian is a straight line half the length of the equator. The meridians converge to a point at each pole and are equally spaced along each parallel. The spacing between parallels decreases slightly with increased latitude. The boundary of the map is an ellipse.	An equal area projection.	Distortion is least in midlatitude regions near the central meridian and greatest at high latitudes near the boundaries.	Generally useful as a world map and for depicting global distributions.
Sinusoidal (Sanson-Flamsteed)	Parallels are straight parallel, equally spaced lines. Meridians and the boundary are sinusoidal curves. The central meridian is straight and half the length of the equator. Meridians converge to points at the poles. The length of each parallel is equal to its length on a globe of corresponding scale.	An equal area projection.	Distortion is least nearest the equator and the central meridian and greatest at high latitudes near the boundaries.	Somewhat inferior to other projections as a global map, but useful for maps of individual continents.

Conic Projections

A cone placed upon a globe contacts the globe along a circle of tangency. If the apex of the cone is above a pole, the circle of tangency coincides with a parallel of latitude, known as the *standard parallel* of the conic projection. Conic projections show parallels of latitude as arcs of circles, and meridians of longitude as radiating straight lines. The distortions of a conic projection are least nearest the standard parallel and increase with distance from it. Conic projections are therefore useful in mapping midlatitude regions, such as the United States, that have a considerable east-west extent and a limited range of latitude.

Conic projections of greater precision are produced by allowing the cone to intersect the globe along two standard parallels. If the standard parallels are not too far apart, the entire map area is displayed with good accuracy. Albers' equal area conic projection, which was chosen for the maps in *The National Atlas of the United States*, can show the conterminous United States with a linear scale distortion that does not exceed 2 percent. For the conterminous United States, the standard parallels of the Albers' conic projection are chosen to be 29½°N and 45½°N. For Alaska the standard parallels are 55°N and 65°N, and for Hawaii 8°N and 18°N. The properties of several conic projections are shown in Figure II.13.

Geometrical Global Projections

Equal area projections that show the entire earth are necessary to display global distributions of all types of phenomena. If the projection displays parallels of latitude as parallel straight lines, regions with the same latitude are aligned on the map, a useful property because many aspects of the physical environment as well as human activities are related to latitude.

On most of the commonly used global projections, the equator and the central meridian are shown as straight lines that intersect at right angles. The regions of greatest distortion lie near the outer margins. Distortion is least at the center of the projection. Figure II.14 shows the properties of several geometrical global projections.

The *flat polar quartic* projection, an equal area projection, is the basis for many of the global distribution maps in this text. In this projection, the poles are represented by lines one-third the length of the equator. To maintain equivalence, the stretching of the poles must be compensated by shrinking the lengths of meridians in the polar areas. Distortion on the flat polar quartic projection is greatest at high latitudes near the margins.

Interruption and Condensation of Projections

On most maps showing global distributions, the landmasses are of greater interest than the oceans. In such a case, the projection can be *interrupted* in the oceans (Figure II.14). The projection may then be reprojected to standard meridians through each major landmass so that no land area is far from a meridian. If ocean areas are of no interest in a particular application, portions of the Atlantic and Pacific oceans can be omitted entirely, producing a *condensed interrupted map*. Condensation permits the map to be at a larger scale without using more space. Conversely, maps concerned with oceanic phenomena on a global scale can be interrupted in continental areas.

RELIEF PORTRAYAL

Frequently it is useful or necessary to depict land surface form and vertical relief on maps. *Contour lines,* special kinds of *relief shading,* and *color tints* are some of the methods for indicating form and relief on maps. Relief and surface form are represented most accurately by contour lines.

A contour line on a map represents a line of constant altitude above or below a chosen reference level, called a *datum plane,* which is usually mean sea level. Consider the hilly island in Figure II.15. The figure shows horizontal planes at a regular vertical spacing, or *contour interval,* of 200 ft. Each horizontal plane cuts the surface of the ground in a circle that is the contour line for that altitude; every point on a given contour line is the same altitude above the datum plane.

Figure II.15

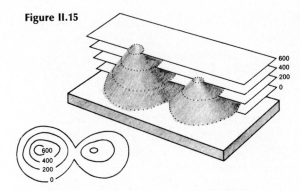

As shown in Figure II.16, contour lines can be used on a flat map to represent vertical relief. The contour interval should be chosen to be commensurate with the nature of the landscape being depicted; the contour interval will normally be larger for a mountainous region than for a gently sloping plain. The choice of contour interval also depends on the scale of the map. On a large-scale map the contour interval may be only a few feet, and the contour lines will show comparatively small changes in elevation. A small-scale map of the same region will employ a larger contour interval, and in addition the small kinks and bends of the contours will be smoothed and averaged, so that less detail will be represented. To make contour lines easier to read, index contours at regular intervals are thickened and labeled according to altitude.

Figure II.16 (Andrea Lindberg adapted from Whitwell Quadrangle, Tennessee, U.S. Geological Survey)

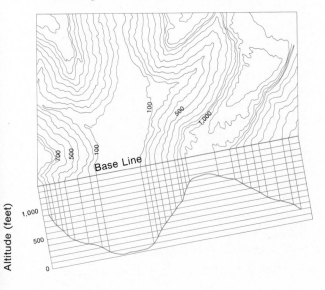

Figure II.17 The relation between contour lines and topography is illustrated in this landscape model. The model was cut from a block of plastic by a cutter set to trace contour lines from a map; the depth of cut was adjusted according to the elevation of each contour line. Although individual layers are distinguishable on the model, the closely spaced contours give a good impression of the topographic relief. (Gerald Ratto Photography)

The elevations of prominent hilltops and depressions are indicated numerically as spot heights.

As Figure II.16 indicates, the horizontal spacing of contour lines can be interpreted in terms of the local slope angle. Contour lines that are relatively close together represent steeper slopes than do contour lines that are farther apart. The convexity or concavity of a slope and the form of ridge crests and valley

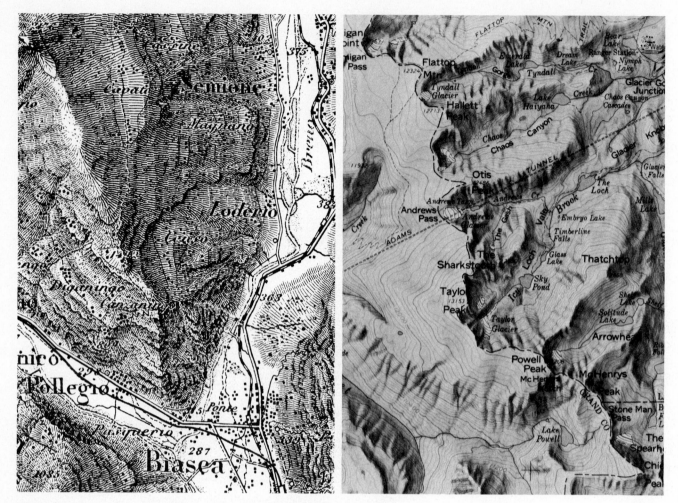

Figure II.18   Relief can be depicted on maps by hachuring **(left)** or by shaded relief **(right)**. (*Left*—Reprinted from Arthur Robinson and Randall Sale, *Elements of Cartography*, © 1953, 1960, 1969 by John Wiley & Sons, reprinted by permission; *right*—U.S. Geological Survey)

floors can also be inferred by inspecting the spacing and character of the contour lines.

For quantitative purposes, a *topographic profile* of the landscape along a given direction can be prepared from a contour map according to the method illustrated in Figure II.16. The vertical scale of a topographic profile is usually exaggerated in order to portray minor relief features more clearly.

In addition to contour lines and spot heights, a variety of qualitative or partly quantitative artistic techniques can be used on maps to indicate relief. In the nineteenth century, slopes were depicted by straight lines called

hachures drawn in the direction of slope descent, at right angles to contours. Hachures are drawn so that their width or spacing increases with increased slope angle, so that steep slopes appear darker than gentle slopes (Figure II.18 *left*). Expertly drawn hachures present a pleasing appearance, but they are seldom used today because they do not indicate the specific slope angle and because their construction is extremely tedious.

A modern method of symbolizing relief on a map is to add shading to give a three-dimensional impression of the terrain. Such *shaded relief* is normally drawn as though the area were illuminated from the upper left, or "northwest," corner of the map (Figure II.18 *right*). Shading can be combined with contour lines.

The use of color tints for successive ranges of altitudes, known as *altitude tinting*, is often employed on small-scale maps of large areas. Green is usually used for altitudes near sea level, with colors for higher altitudes progressing through yellow, orange, red, and brown. The green tint used for low altitudes should not be taken as indicative of vegetation cover. The progression of colors is based on human perception of "warm" colors (orange, red) as "advancing," or lifting above the page, while "cool" colors (green, blue) "recede" into the page. Hachures, shaded relief, and altitude tinting are sometimes used in combination on a contour map to depict the general character of a landscape while retaining the quantitative accuracy of contours.

Three-dimensional models with exaggerated vertical relief are mass-produced by molding thin sheets of plastic. On large-scale models, the quality of the molding is sufficiently good to represent the landscape accurately. Plastic "raised relief" maps are available for much of the United States at a scale of 1:250,000. Global relief maps are sometimes prepared from photographs of three-dimensional models, which are illuminated from the upper left to emphasize relief by light and shadow.

MAPPING IN THE UNITED STATES

Regional and city planning and the development of a country's resources are closely tied to the availability of specialized maps. In addition to maps of topography and those required for navigation, government agencies prepare maps of water, soil, geologic phenomena, timber and mineral resources, and other features. Maps are also employed to compile and report economic and demographic statistics. A map devoted to a single topic, such as the distribution of population or rainfall, is known as a *thematic* map. Figure II.19 gives examples of several kinds of maps.

The responsibility for mapping the United States is divided among many civil and military agencies, depending on the purpose of the map. Only a few examples are given here to indicate the extent of the mapping services carried on by the government.

The United States Geological Survey (USGS) publishes several series of topographic maps in different scales that cover the United States, Puerto Rico, and other territories. The USGS Topographic Maps are discussed in detail in the following section of this appendix because of their particular importance to physical geographers.

In addition to topographic maps, the Geological Survey publishes geologic and mineral investigation maps, and hydrologic maps for water-resource planning. Hydrologic maps prepared by the Geological Survey detail the expected flow rate from wells sunk into groundwater supplies and show flood risk for help in land-use planning. The Map Information Office of the Geological Survey is a useful source of information concerning the availability of maps and aerial photographs.

Maps related to navigation purposes are known as *charts*. The preparation of nautical and aeronautical charts is carried out by several agencies. Nautical charts of foreign waters are prepared by the Naval Oceanographic Office (formerly the Hydrographic Office of the Navy). This office also issues small-scale bathymetric charts that portray the topography of much of the ocean floor. The National Ocean Survey prepares nautical navigation charts of United States coastal and offshore waters, and some inland waterways, using the Mercator projection and a range of scales. The harbor charts prepared by the Ocean Survey may have scales as large as 1:18,000 to show accurate locations for important features such as main channels, marker buoys, and underwater cables. The Army Corps of Engineers issues nautical charts of the Great Lakes and navigation charts for numerous rivers and inland waterways.

Aeronautical charts are produced by several agencies, including the National Ocean Survey and the United States Air Force Aeronautical Chart and Information Center. The scales em-

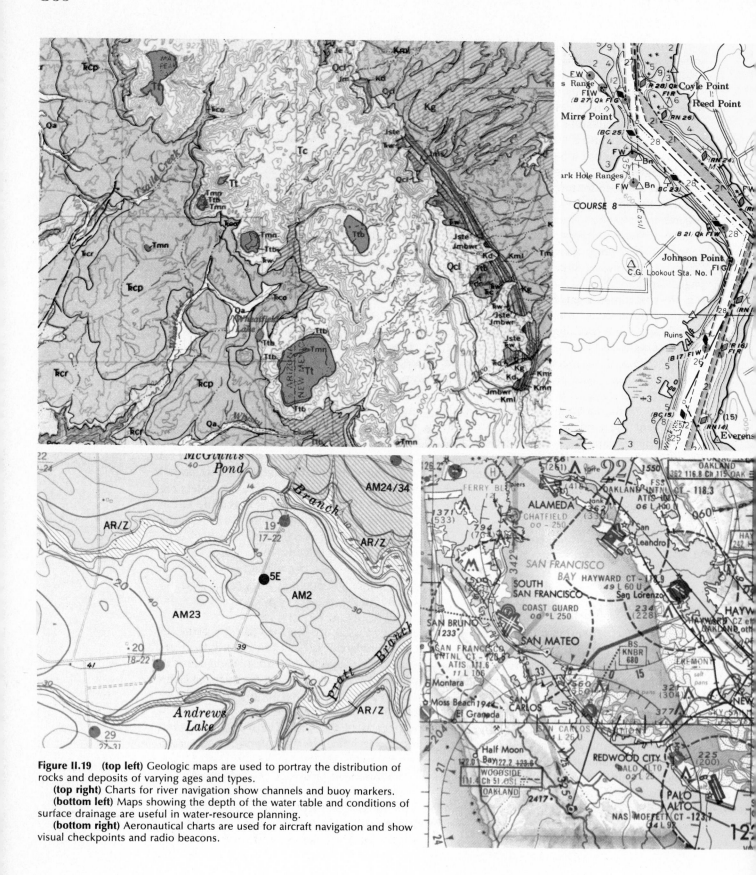

Figure II.19   (top left) Geologic maps are used to portray the distribution of rocks and deposits of varying ages and types.

(top right) Charts for river navigation show channels and buoy markers.

(bottom left) Maps showing the depth of the water table and conditions of surface drainage are useful in water-resource planning.

(bottom right) Aeronautical charts are used for aircraft navigation and show visual checkpoints and radio beacons.

ployed vary from 1:250,000 on charts for aircraft of moderate speed to 1:2,000,000 and smaller for the needs of global jet transport. Special features of aeronautical charts include identification of airports and air navigation radio beacons, and a simplified representation that emphasizes major relief features and landmarks easily visible from the air. Aeronautical charts are revised frequently from aerial photographs and can be helpful to geographers.

USGS TOPOGRAPHIC MAPS

A *topographic map* is a graphic representation of a portion of the earth's surface at a scale large enough to show human works such as individual buildings. Additionally, topographic maps employ contour lines to show the configuration of the land surface. The United States has maintained a topographic map program since 1879 under the direction of the Geological Survey. The Geological Survey publishes the National Topographic Map Series, an invaluable source of information on the physical characteristics and human activities present in nearly all parts of the national territory. The maps are compiled primarily from aerial photographs, but field surveys are used to establish highly precise networks of horizontal positions and vertical elevations, and to verify details of photo interpretation.

The Topographic Series is itself made up of maps at several different scales. Each of the maps is drawn so that its sides coincide with parallels of latitude and meridians of longitude, and a similar symbolism is used in all series (Figure II.20). Topographic map series covering most of the United States are published at scales of 1:24,000; 1:62,500; and 1:63,360 (Alaska). The entire nation is also covered by map series at scales of 1:250,000 and 1:500,000. In recent years, a number of maps using the metric system have been published at a scale of 1:100,000 (1 cm = 1 km). In addition, there are map series covering national parks, monuments, historic sites, major metropolitan areas, rivers and flood plains, Puerto Rico, other national territories, Antarctica, the moon, and Mars.

The 7½-minute quadrangle series (1:24,000) and 15-minute quadrangle series (1:62,500; 1:63,360) are large-scale maps suited to planning, engineering, and landform studies on a local scale. Such maps contain a vast amount of information.

Topographic maps employ more than 100 different symbols to depict natural and cultural features. The table of USGS topographic map symbols reproduced in Figure II.20 is not printed on the map sheets but is available separately. The colors on a topographic map are an integral part of the symbolization, with each color restricted to a particular class of features.

Black is used for all man-made objects other than major roads and urban areas. All names and labels are in black. Blue is used for all water features, including streams, lakes, marshes, canals, the oceans, and water depths indicated by contours and numbers. Brown is reserved for land surface contour lines, indications of surface type, such as loose sand or mine tailings, and certain elevation figures. Green is used for vegetation symbols: woodlands, orchards, vineyards, brush, and wooded marshland. Red is used for major roads and land survey lines. Pink indicates densely built-up urban areas in which only landmark buildings are indicated individually. Purple is used on interim maps to symbolize changes that have occurred since the previous edition.

FIELD WORK AND MAPPING

The physical geographer engaged in the study of streams, slopes, land use, or soils may find that the information required for the analysis of an area is not available from published sources. The important details may not be indicated on existing maps of the area of interest. The research may concern phenomena occurring during a short span of time, such as the erosion of stream banks during spring flood conditions, the analysis of which may require daily or weekly observation. In such cases, the researcher must be prepared to create new maps of the area. Because such maps are often the simplest and best way to record and display the

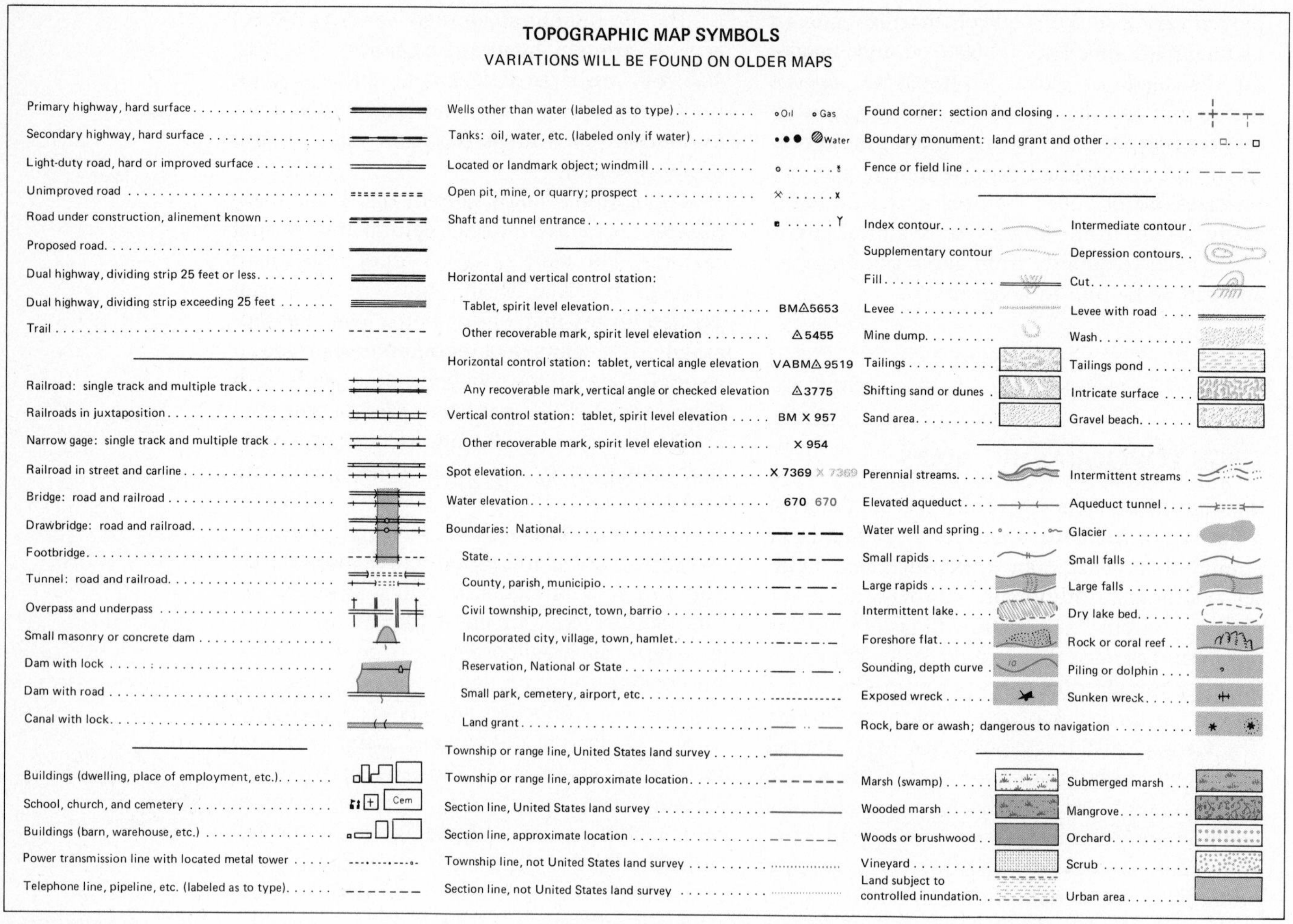

TOPOGRAPHIC MAP SYMBOLS

VARIATIONS WILL BE FOUND ON OLDER MAPS

Primary highway, hard surface . . . .

Secondary highway, hard surface . . . .

Light-duty road, hard or improved surface . . . .

Unimproved road . . . .

Road under construction, alinement known . . . .

Proposed road. . . . .

Dual highway, dividing strip 25 feet or less . . . .

Dual highway, dividing strip exceeding 25 feet . . . .

Trail . . . .

Railroad: single track and multiple track . . . .

Railroads in juxtaposition . . . .

Narrow gage: single track and multiple track . . . .

Railroad in street and carline . . . .

Bridge: road and railroad . . . .

Drawbridge: road and railroad . . . .

Footbridge . . . .

Tunnel: road and railroad . . . .

Overpass and underpass . . . .

Small masonry or concrete dam . . . .

Dam with lock . . . .

Dam with road . . . .

Canal with lock . . . .

Buildings (dwelling, place of employment, etc.) . . . . . .

School, church, and cemetery . . . .

Buildings (barn, warehouse, etc.) . . . .

Power transmission line with located metal tower . . . . .

Telephone line, pipeline, etc. (labeled as to type) . . . . . .

Wells other than water (labeled as to type) . . . . Oil Gas

Tanks: oil, water, etc. (labeled only if water) . . . . Water

Located or landmark object; windmill . . . .

Open pit, mine, or quarry; prospect . . . .

Shaft and tunnel entrance . . . .

Horizontal and vertical control station:

Tablet, spirit level elevation . . . . BM△5653

Other recoverable mark, spirit level elevation . . . . △5455

Horizontal control station: tablet, vertical angle elevation VABM△9519

Any recoverable mark, vertical angle or checked elevation △3775

Vertical control station: tablet, spirit level elevation . . . . . BM X 957

Other recoverable mark, spirit level elevation . . . . X 954

Spot elevation . . . . X 7369 X 7369

Water elevation 670 670

Boundaries: National . . . .

State . . . .

County, parish, municipio . . . .

Civil township, precinct, town, barrio . . . .

Incorporated city, village, town, hamlet . . . .

Reservation, National or State . . . .

Small park, cemetery, airport, etc. . . . .

Land grant . . . .

Township or range line, United States land survey . . . . . .

Township or range line, approximate location . . . .

Section line, United States land survey . . . .

Section line, approximate location . . . .

Township line, not United States land survey . . . .

Section line, not United States land survey . . . .

Found corner: section and closing . . . .

Boundary monument: land grant and other . . . .

Fence or field line . . . .

Index contour . . . .   Intermediate contour .

Supplementary contour   Depression contours . .

Fill . . . .   Cut . . . .

Levee . . . .   Levee with road . . . .

Mine dump . . . .   Wash . . . .

Tailings . . . .   Tailings pond . . . . . .

Shifting sand or dunes .   Intricate surface . . . .

Sand area . . . .   Gravel beach . . . . . .

Perennial streams . . . . .   Intermittent streams .

Elevated aqueduct . . . .   Aqueduct tunnel . . . .

Water well and spring . .   Glacier . . . .

Small rapids . . . .   Small falls . . . .

Large rapids . . . .   Large falls . . . .

Intermittent lake . . . . .   Dry lake bed . . . . . .

Foreshore flat . . . .   Rock or coral reef . . .

Sounding, depth curve .   Piling or dolphin . . . .

Exposed wreck . . . . . .   Sunken wreck . . . . . .

Rock, bare or awash; dangerous to navigation . . . .

Marsh (swamp) . . . . . .   Submerged marsh . .

Wooded marsh . . . . . .   Mangrove . . . .

Woods or brushwood . . .   Orchard . . . .

Vineyard . . . .   Scrub . . . .

Land subject to controlled inundation . . .   Urban area . . . .

Figure II.20
(U.S. Department of the Interior)

data of interest, geographers need to be familiar with the field study and mapping techniques suitable for their purposes. This section discusses some of the observational methods employed to study regions of small size using limited human resources and instrumentation.

An Approach to Field Work

It is essential to begin a field study with a definite plan in order to avoid wasteful, undirected effort. The first step in planning field work is to choose a suitable region for study. Maps and aerial photographs can be helpful in locating an appropriate location, but first-hand inspection of possible sites may also be needed. The size of the research area depends on the nature of the research problem and on the available resources. A study of stream erosion processes, for example, might involve an intensive study of only a few hundred feet of stream bank, whereas a study of land use changes might entail a broad survey covering several square miles.

Usually the geographer is interested in studying the spatial distribution or variation of some phenomenon or process. Proper execution of a map may be central to the success of the study. The scale of the map is determined by the degree of detail that must be represented. A study of erosion processes on a slope may require the mapping and description of every few square feet of ground, whereas the basic areal unit in a local soil survey might be several acres in extent. If the area to be studied covers several square miles, the data may be recorded directly on a map such as a USGS topographic map or on an aerial photograph of the region. However, the scale of USGS topographic maps is not large enough to represent smaller areas in detail, so that an original map must be created.

Mapping Techniques

Simple mapping techniques suitable for field use include field sketching, the compass traverse, and use of the plane table. A *field sketch* is a panoramic sketch of a local region; it can be used to show landforms and the relation of landforms to other features. For many purposes, a field sketch is superior to a photograph of the region because important features can be selected and emphasized in a sketch.

A field sketch need not be executed artistically to be of value in a field study; often a simple, uncluttered drawing is all that is required. The rule in sketching is to begin with the main outlines of a landscape, such as the skyline, and work toward smaller details. To show the correct relative locations of features, it is helpful to draw a horizontal line on the sketch pad to indicate the horizontal reference line, and a vertical line to indicate the central axis of the sketch. A ruler held at arm's length can then be used to estimate the relative distances of selected landmarks, as illustrated in Figure II.21. Once the major topographic features and a few landmarks have been sketched, the relative positions of other details can be located by eye.

A *compass traverse* is a simple map-making technique, which requires only a magnetic compass to measure directions. A *traverse* is a series

Figure II.21

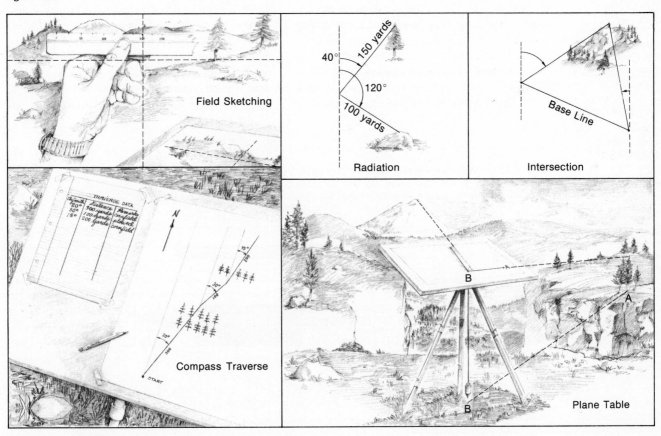

Field Sketching

Radiation

Intersection

Compass Traverse

Plane Table

of straight lines running from one point to the next. A *closed* traverse ends at the original starting point, and an *open* traverse does not. A closed traverse is used when features within a particular area are to be mapped. An open traverse is more suitable when features on either side of a central line are to be recorded and mapped.

Once the starting point and the approximate route of the traverse have been selected, a landmark along the first leg of the route is sighted from the starting point. The azimuth of the landmark is then measured using a compass, taking precautions to keep the compass away from magnetic materials that might influence its reading. When the azimuth has been measured and recorded in the data book, the investigator then walks toward the landmark while pacing off the distance. If the investigator makes observations along the traverse, the data and the location should be recorded. After reaching the end of the first leg, a "back-sight" to the starting point should be made with the compass as a check on the accuracy of the traverse. Then a sighting can be taken to a new landmark and the traverse can be extended.

The field record of a compass traverse consists of a list of azimuths, distances, and observations. To convert these data to a map, the legs of the traverse can be laid out on drawing paper to their correct scale and with their correct azimuths (see Figure II.21). Observational data can then be filled in.

Frequently it is necessary to determine the position of an object away from the line of the traverse. One method, known as *radiation,* is to sight on the object from a known point on the traverse, measure the azimuth by compass, and pace off the distance to the object. Steps can be saved by employing the technique of *intersection* instead. As Figure II.21 illustrates, the position of the object relative to the traverse is completely specified by measuring the azimuth of the object from two or more known points along the traverse. For best accuracy, two sighting points should be spaced so that the lines of intersection meet at the object with an angle of between 60° and 120°.

The methods of radiation and intersection are the basis for map making with a *plane table,* which is a horizontal drawing board supported by a stand or tripod. A map can be drawn in the field using a plane table, allowing the work to be checked until satisfactory accuracy is obtained.

When using a plane table, the first step is to establish a *base line* of known distance and azimuth between two sighting points that command good views of the area to be mapped. After the base line has been drawn to an ap-

Figure II.22 Remote sensing techniques utilize various portions of the electromagnetic spectrum. The visible and near-infrared regions are used for photography and television, and the infrared is used for scanning surface temperatures. Relief features are mapped by laser profiling and radar. (Calvin Woo)

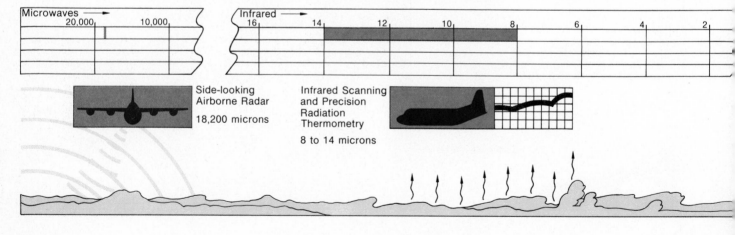

propriate scale on drawing paper fastened to the plane table, the table is placed over one of the sighting points and oriented so that the base line on the map sheet is exactly parallel to the the base line on the ground. A north arrow is drawn on the map, using the magnetic compass. Then a sighting device such as a triangular rule can be laid on the map sheet and aligned from the sighting point toward the object to be mapped (Figure II.21). In this way the line to the object can be drawn on the map at the correct azimuth without the need for compass measurements. The location can be fixed by employing the method of radiation, pacing off the distance from the sighting point to the object. Alternatively, the determination of azimuths from both sighting points fixes the location of the object by intersection. Normally the plane table must be moved and set up at several successive stations to complete the map. The position of each station must be precisely established on the map before the plane table is moved to it, and the orientation of the plane table at each station is checked both by compass and by sighting back toward the preceding station.

REMOTE SENSING

The space age has provided new techniques for obtaining information about the earth's surface. The new methods employ special sensing devices that can acquire data at a distance from the phenomena being studied; therefore the technique is known as *remote sensing*. The principal methods used are to sense and record electromagnetic radiation of sunlight, laser light, or microwave radar that is reflected from objects on the earth, or the infrared thermal radiation directly emitted by objects.

Remote sensing equipment is usually mounted in aircraft or in satellites in orbit around the earth. The advantages of remote sensing include the ability to cover vast areas in a very short time, the ability to provide data from very large to very small scale, the possibility of repetitive measurements to follow the progress of selected events, and the ability to penetrate areas inaccessible to ground survey. The techniques of remote sensing have been greatly extended recently in an effort to obtain the information needed to manage the earth's resources more efficiently, and much of the information being gathered is of direct interest to physical geographers. The capabilities of remote sensing technology include the detection

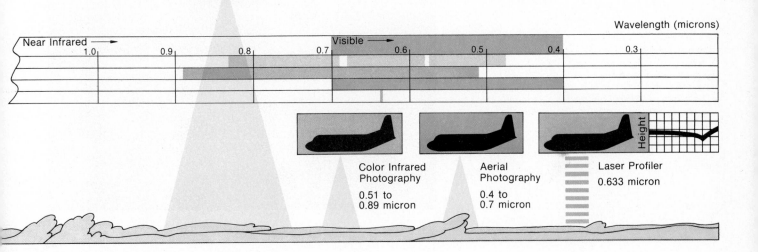

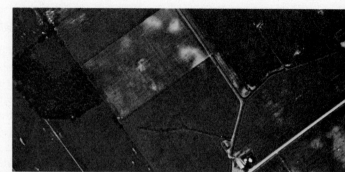

Figure II.23   (top, left and right) The aerial view on the right was photographed with ordinary color film. The same fields photographed with color infrared film are shown on the left; healthy vegetation appears bright red. (Courtesy of the Laboratory for Applications of Remote Sensing—LARS).

(**opposite, top**) Stereo pairs are used to make topographic relief apparent in aerial photographs. When viewed with a stereoscope, relief features in this view in eastern Utah stand out as a three-dimensional model. To see the stereo effect, the left-hand photograph should be viewed with the left eye, and the right-hand photograph with the right eye. A card held between the photographs helps to separate the images. (U.S. Geological Survey)

(**right**) Vegetation on irrigated land is clearly differentiated from dry desert in this color infrared photograph of the Imperial Valley of California taken from earth orbit. The Salton Sea appears as a large dark region at the upper left. The pattern of agricultural fields is clearly distinguishable; note in particular the border between the United States and Mexico that is evident toward the bottom of the photograph. The well-irrigated crops on the United States side show as bright red, compared to the bluish hues representing bare ground or desert shrubs on the Mexican side of the border. (NASA)

inal mapping. Conventional aerial surveys employ the visible light reflected from the earth and detected with cameras using black-and-white film. Optical techniques allow the distortions inherent in a photograph taken at low or moderate altitudes to be rectified so that distant areas in the photograph have the same scale as areas directly below the camera. Two exposures can be used to record a stereoscopic (three-dimensional) image from which contour lines can be drawn. A single photograph may cover an area of a few square km to hundreds of square

of vegetation types and their seasonal changes, the measurement of surface water distribution and soil moisture, the depiction of landforms and surface geologic structures, and the monitoring of human activities that affect the environment.

Aerial photography, the first remote sensing method developed, is now used for most orig-

Figure II.24   (bottom left) Monterey Bay, California, appears at the left of this ERTS-1 false-color composite image sensed from an altitude of 900 km (560 miles). The color print is made from a composite of three black-and-white images, each exposed to different spectral wavelengths; the image is similar to that produced by color infrared film. Healthy vegetation stands out in red. A patch of fog hides part of the coast south of the Monterey peninsula. (NASA)

(bottom right) Relief features, including linear features in the San Andreas fault zone, show crisply in this side-looking radar image of part of the San Francisco peninsula. The thin linear feature at the lower right is the 2-mile-long linear accelerator used in atomic research at Stanford University. (J. R. P. Lyon, Stanford University)

km, depending on the cameras and the altitude of the aircraft. Ground features as small as 1 cm in size can be detected on photographs made from aircraft flying at an altitude of a few kilometers, and satellite imagery can provide resolution of objects only a few meters in size from an altitude of several hundred kilometers. These advanced capabilities are gradually becoming available for civilian uses.

Aircraft equipped with detectors of longwave thermal infrared radiation can provide images of objects on the earth's surface with intensities proportional to their temperatures. Infrared imagery methods can be employed to detect cloud temperatures, ocean currents, or the mixing of warm water from industrial effluents with cooler river water.

Application of remote sensing methods to the study of the earth has been revolutionized by using remote sensors in conjunction with satellites in orbit around the earth. Surveillance of the earth's cloud cover from weather satellites has greatly improved the accuracy of weather predictions and the ability to track potentially damaging hurricanes and midlatitude storms. Weather satellites usually orbit at altitudes of several hundred kilometers.

Sunlight falling upon the earth consists of electromagnetic radiation in the visible region and in the near-visible infrared region of the electromagnetic spectrum. Different objects reflect different relative proportions of each wavelength, so that the radiation reflected from an object, when analyzed in terms of wavelength, forms a means of identification known as the *spectral signature* of the object. Vegetation, for example, usually reflects proportionately more short wavelength infrared radiation than does bare soil. Furthermore, different types of vegetation can usually be distinguished by their spectral signatures. Diseased plants or lack of soil moisture can also be detected.

Color infrared film is similar to ordinary color film, except that it depicts the previously invisible infrared radiation as red, the red colors as green, and the green colors as blue (Figure II.23). However, healthy vegetation appears red on a color infrared photograph. The intensity of the color is directly proportional to the amount of chlorophyll in the vegetation. Color infrared photographs of test fields and forests have proven that different trees and crops and their stages of growth can be distinguished by such means.

Another way to make use of spectral signatures is to employ several sensing devices, each fitted with a filter that allows only certain wavelengths of visible light to enter. LANDSAT, formerly Earth Resources Technology Satellite, ERTS, which views the earth from an altitude of nearly 1,000 km, employs four *multispectral scanners* and three television cameras, each sensing the same scene in a different range, or "band," of wavelengths. The resulting images can be assigned colors and combined to produce a "false-color" image (Figure II.23). From its orbit between the poles, LANDSAT can scan a given portion of the earth once every 18 days to allow the progress of crops to be followed through the growing season; to monitor streamflow, snow depth, and soil moisture; and to detect other short-term changes in the environment.

Many areas in the tropics are almost continuously covered by clouds, making normal photography and multispectral scanning unsuccessful. *Side-looking airborne radar,* which utilizes radio waves of short wavelength, can penetrate clouds and yield a high resolution picture of the surface. Radar waves emitted from the aircraft are partly reflected from the ground and detected by a receiving apparatus on the aircraft. Variations in the strength of the return signal are electronically translated into a picture of the terrain. Airborne radar has been used to map the Amazon Basin, Panama, and portions of Africa and New Guinea. Although radar methods do not differentiate between types of vegetation as clearly as multispectral methods, forest, range, and agricultural land can be separately identified on large-scale radar images.

APPENDIX III
Classification Systems

KÖPPEN SYSTEM OF CLIMATE CLASSIFICATION

The Köppen system of global climate classification recognizes five major climatic types, symbolized by *A, B, C, D,* and *E.* The *B* climates, which are the dry realms, are further divided into *BS* (steppe) and *BW* (desert) types; and the *E* region is divided into *ET* (tundra) and *EF* (perpetual frost) types.

Figure III.1 shows Köppen's criteria for establishing the boundaries between the *A, C,* and *D* climates and the *BS* and *BW* climates. The solid line on each graph marks the boundary between the humid and dry climates. Each of the three sets of criteria is based on average annual temperature and precipitation. Which criterion is used depends on whether precipitation occurs evenly throughout the year or primarily during the summer or the winter. Köppen considered a region to have a dry winter if at least 70 percent of the precipitation occurs during the 6 summer months, and to have a dry summer if at least 70 percent of the precipitation occurs during the 6 winter months. Regions not fitting either category are considered to have an even distribution of precipitation. Summer is interpreted as the season when the midday sun is high in the hemisphere being considered, and winter is interpreted as the low-sun season.

The classification of a place into *A, C,* or *D* on the one hand, or *BS* or *BW* on the other, can be determined graphically by plotting the average annual temperature and precipitation on the appropriate diagram. The equation for each of the boundary lines is also given in the diagrams. In each case, the upper equation is in degrees Celsius and centimeters, and the lower equation in degrees Fahrenheit and inches.

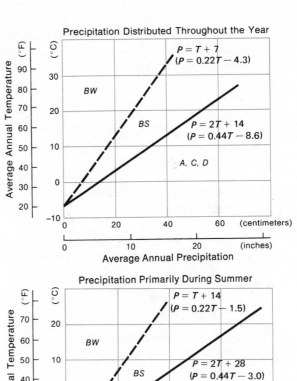

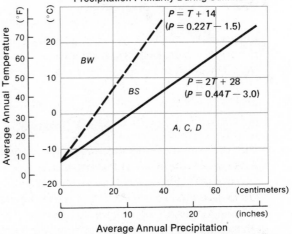

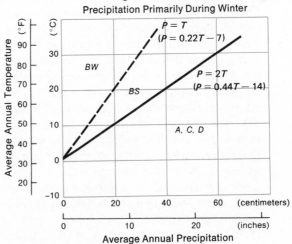

Figure III.1   (Figures III.1–III.5 by Andrea Lindberg)

569

The *A, C, D, ET,* and *EF* climates are distinguished from one another according to various criteria based on temperature, with the warmest being *A* and the coldest being *EF.* The criteria are as follows:

A: Average temperature of the coldest month exceeds 18°C (64.4°F).

C: Average temperature of the warmest month exceeds 10°C (50°F). Average temperature of the coldest month lies between 18°C (64.4°F) and −3°C (26.6°F).

D: Average temperature of the warmest month exceeds 10°C (50°F). Average temperature of the coldest month is below −3°C (26.6°F).

ET: Average temperature of the warmest month lies between 10°C (50°F) and 0°C (32°F).

EF: Average temperature of the warmest month is below 0°C (32°F).

H: Unclassified highland climates.

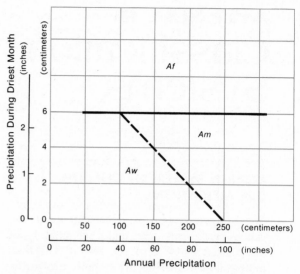

Figure III.2

Principal Subdivisions of *A* Climates

The *A* climates are subdivided into *Af, Am,* and *Aw* types on the basis of the amount and seasonality of precipitation. If precipitation in the driest month exceeds 6 cm (2.4 in.), the climate is classified as *Af,* as indicated in Figure III.2. *Af* regions, such as equatorial lowland rainforests, receive abundant moisture for plant growth throughout the year. If there is a winter dry period during which precipitation for the driest month is less than 6 cm, the climate is classified as *Aw* or as *Am.* The distinction between *Aw* and *Am* climates depends upon the relation between the amount of precipitation during the driest month and the annual precipitation, as Figure III.2 shows. Tropical wet and dry climates are classified as *Aw.* The *Am,* or monsoon, climate has enough annual precipitation that the moderate winter dry season does not exhaust soil moisture, and plant growth is not seriously affected.

A fourth subdivision, the *As* climate, would in principle be characterized by a summer dry season. However, *As* climates only occur locally near the equator, and the *As* classification is not significant on the global scale.

Principal Subdivisions of *BS* and *BW* Climates

The *BS* and *BW* regions are divided on a thermal basis, as specified by the additional notation *h, k,* or *k'.*

h (hot):   Average annual temperature exceeds 18°C (64.4°F).

k (cold winter):   Average annual temperature is less than 18°C (64.4°F) and average temperature of the warmest month exceeds 18°C.

k' (cold):   Average annual temperature is less than 18°C (64.4°F) and average temperature of the warmest month is less than 18°C.

The deserts of North Africa and central Australia are examples of *BWh* regions. *BWk* regions occur in the high plateaus of central Asia.

Principal Subdivisions of *C* and *D* Climates

The *C* and *D* climate regions are further differentiated according to the seasonality of precipitation. A second letter of notation, *s,* denotes regions with a dry summer (*Ds* seldom occurs); *w* denotes regions with dry winters; and *f* denotes regions with no marked dry season.

Cs, Ds: The driest month occurs during the warmest 6 months, and the amount of precipitation received during the driest month is less than one-third the amount received during the wettest month of the coldest 6 months. Also, precipitation during the driest month must be less than 4 cm (1.6 in.).

Cw, Dw: The driest month occurs during the coldest 6 months, and the amount of precipitation received during the driest month is less than one-tenth the amount received during the wettest month of the warmest 6 months.

Cf, Df: No marked dry season occurs.

The subdivisions *Cs, Ds, Cw, Dw, Cf,* and *Df* may be further differentiated according to the seasonality of temperature by adding a third letter of notation—*a, b, c,* or *d.*

a: Average temperature of the warmest month exceeds 22°C (71.6°F).

b: Average temperature of the warmest month is less than 22°C (71.6°F), but at least 4 months have an average temperature greater than 10°C (50°F).

c: Average temperature of the warmest month is less than 22°C (71.6°F), fewer than 4 months have an average temperature greater than 10°C (50°F), and the average temperature of the coldest month is greater than −38°C (−36.4°F).

d: Same as *c,* except that the temperature of the coldest months is less than −38°C (−36.4°F).

The Köppen system of climate classification possesses additional symbols to denote special features such as frequent fog or seasonal high humidity, but they are seldom used on the global scale. A global map of the Köppen climate classification is shown in Figure 8.6 on pp. 198–199.

THORNTHWAITE'S FORMULA FOR POTENTIAL EVAPOTRANSPIRATION

The tables and graphs presented in this section make it possible to calculate potential evapo-

transpiration for a given month at a given location from Thornthwaite's formula, if the average monthly temperatures are known. Thornthwaite's formula is designed to be used with temperatures expressed in degrees Celsius, so temperature data on the Fahrenheit scale should be first converted to Celsius before beginning the calculation of potential evapotranspiration.

The first step in applying Thornthwaite's method is to calculate a monthly heat index, *i*, for each of the 12 months. The monthly heat index is defined according to the equation

$$i = \left(\frac{T}{5}\right)^{1.514}$$

where *T* is the long-term average temperature of the month in °C. Approximate values of the monthly heat index can be read from the graph in Figure III.3.

The sum of the twelve monthly heat indexes is the annual heat index, *I*. The annual heat index is representative of climatic factors at a given location because it is based on long-term

Figure III.3

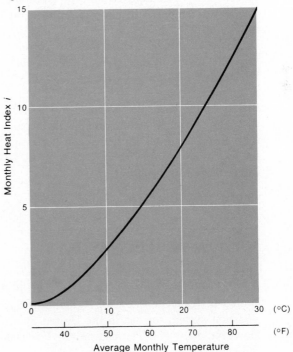

averages. Thornthwaite found an empirical formula that gives potential evapotranspiration, *PE*, for a given month of a particular year in terms of *I*. His equation for *PE*, unadjusted for duration of sunlight, is

$$\text{Unadjusted } PE \text{ (centimeters)} = 1.6 \left(\frac{10T}{I}\right)^m$$

where *T* is the average temperature in °C for the specific month being considered, and *m* is a number that depends on *I*. To a good approximation, *m* is given by the equation

$$m = (6.75 \times 10^{-7})I^3 - (7.71 \times 10^{-5})I^2 + (1.79 \times 10^{-2})I + 0.492$$

With these equations, unadjusted potential evapotranspiration can be calculated using tables of logarithms or a computer. Alternatively, approximate values of unadjusted potential evapotranspiration can be read from the nomogram in Figure III.4 with enough accuracy for most purposes, if the values for the average monthly temperature and the annual heat index are known.

If the average temperature of the month is below 0°C, potential evapotranspiration is taken to be 0. If the average monthly temperature exceeds 26.5°C, unadjusted *PE* is given directly in terms of temperature according to the graph in Figure III.5.

Values of unadjusted *PE* must be corrected for the duration of daylight in order to obtain the desired final values. Unadjusted *PE* for a given month at a given location should be multiplied by the correction factor listed in Table III.1. The correction factors for latitude 50°N are used for all latitudes farther to the north, and the factors for latitude 50°S are used for all latitudes farther to the south.

SOIL CLASSIFICATION SYSTEMS

The 7th Approximation soil classification system of 1960 has superseded the earlier 1938 United States Department of Agriculture classification system for use in the United States. However, many of the terms from the 1938 classification remain in common use.

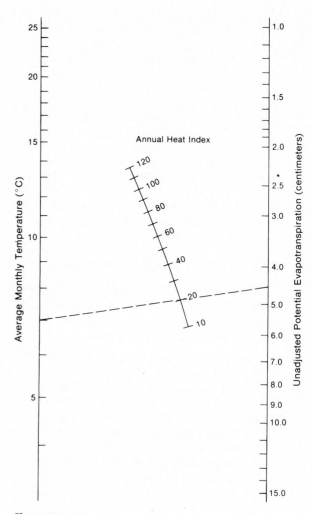

Figure III.4

Figure III.5

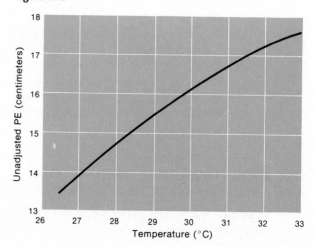

Table III.1 Daylength Correction Factors for Potential Evapotranspiration

Latitude	Jan	Feb	Mar	Apr	May	Jun	Jul	Aug	Sep	Oct	Nov	Dec
50°N	0.74	0.78	1.02	1.15	1.33	1.36	1.37	1.25	1.06	0.92	0.76	0.70
40°N	0.84	0.83	1.03	1.11	1.24	1.25	1.27	1.18	1.04	0.96	0.83	0.81
30°N	0.90	0.87	1.03	1.08	1.18	1.17	1.20	1.14	1.03	0.98	0.89	0.88
20°N	0.95	0.90	1.03	1.05	1.13	1.11	1.14	1.11	1.02	1.00	0.93	0.94
10°N	1.00	0.91	1.03	1.03	1.08	1.06	1.08	1.07	1.02	1.02	0.98	0.99
0°	1.04	0.94	1.04	1.01	1.04	1.01	1.04	1.04	1.01	1.04	1.01	1.04
10°S	1.08	0.97	1.05	0.99	1.01	0.96	1.00	1.01	1.00	1.06	1.05	1.10
20°S	1.14	1.00	1.05	0.97	0.96	0.91	0.95	0.99	1.00	1.08	1.09	1.15
30°S	1.20	1.03	1.06	0.95	0.92	0.85	0.90	0.96	1.00	1.12	1.14	1.21
40°S	1.27	1.06	1.07	0.93	0.86	0.78	0.84	0.92	1.00	1.15	1.20	1.29
50°S	1.37	1.12	1.08	0.89	0.77	0.67	0.74	0.88	0.99	1.19	1.29	1.41

Source: Thornthwaite, C. Warren, 1948. "An Approach Toward a Rational Classification of Climate." *Geographical Review.* 38: 55–94.

The 1938 Soil Classification System

The 1938 classification divides soils into three orders: *zonal, intrazonal,* and *azonal.* The orders are divided into suborders, which are in turn divided into the great soil groups. The listing given here is a modification of the 1938 classification prepared in 1949 by James Thorp and Guy Smith, soil correlators for the Division of Soil Survey of the United States Department of Agriculture. The great soil groups are listed under their appropriate suborder.

Zonal Soils

1. Soils of the cold zone
   Tundra soils

2. Light-colored soils of arid regions
   Desert soils
   Red desert soils
   Sierozem soils
   Brown soils
   Reddish brown soils

3. Dark-colored soils of semiarid, subhumid, and humid grasslands
   Chestnut soils
   Reddish chestnut soils
   Chernozem soils
   Prairie soils
   Reddish prairie soils

4. Soils of the forest-grassland transition
   Degraded chernozem
   Noncalcic brown soils

5. Light-colored podzolic soils of timbered regions
   Podzols
   Gray podzolic soils
   Brown podzolic soils
   Gray-brown podzolic soils
   Red-yellow podzolic soils

6. Lateritic soils of forested warm midlatitude and tropical regions
   Reddish brown latosols
   Yellowish brown latosols
   Latosols

Intrazonal Soils

1. Halomorphic (saline and alkali) soils of imperfectly drained arid regions
   Saline (solonchak) soils
   Solonetz soils
   Soloth soils

2. Hydromorphic soils of marshes, swamps, seep areas, and flats
   Humic-gley soils
   Alpine meadow soils
   Bog soils
   Half-bog soils
   Low-humic gley soils
   Planosols
   Groundwater podzolic soils
   Groundwater lateritic soils

3. Calcimorphic soils
   Brown forest soils
   Rendzina soils

Azonal Soils (no suborders)
   Lithosols
   Regosols (includes dry sands)
   Alluvial soils

The 7th Approximation Soil Classification System

The following list gives the orders and suborders of the 7th Approximation. The suborders are further divided into a large number of great groups, as tabulated in 1960, 1967, and 1975 publications of the United States Department of Agriculture Soil Conservation Service. The descriptions presented here are adapted from *The National Atlas of the United States of America.* Some indication of soil use is also given. Table III.2 gives the prefixes used to differentiate the great groups.

1. **ALFISOLS:** Moderate to high alkalinity, gray surface horizon, subsurface clay; usually moist, perhaps dry for part of warm season.

   Aqualfs: Seasonally wet; for general crops (drained) or woodland (undrained).
   Boralfs: In cool to cold regions; woodland and pasture.
   Udalfs: In midlatitude to tropical regions; usually moist, dry for short periods; row crops, small grain, and pasture.
   Ustalfs: In midlatitude to tropical regions; reddish brown; dry for long periods; range, small grain, and irrigated crops.
   Xeralfs: In regions with rainy winters and dry summers; dry for a long period; range, small grain, and irrigated crops.

2. **ARIDISOLS:** Semidesert and desert soils with definite horizons; low in organic matter; never moist for more than 3 consecutive months.

   Argids: Accumulation of clay in a horizon; range and some irrigated crops.
   Orthids: Accumulation of such salts as calcium carbonate or gypsum; range and some irrigated crops.

3. **ENTISOLS:** No definite horizons (generally azonal).

   Aquents: Permanently or seasonally wet; limited pasturage.
   Fluvents (Alluvial Soils): Organic content decreases irregularly with depth, formed in loam or clay alluvial deposits; range, irrigated crops (dry regions), and general farming (humid regions).

   Orthents: Loam or clay; organic content decreases regularly with depth; range, irrigated crops (dry regions), and general farming (humid regions).
   Psamments: Texture of loamy fine sand or coarser; range and wild hay (Alaska), woodland, small grains (warm, moist regions), pasture and citrus (Florida), and range and irrigated crops (warm, dry regions).

4. **HISTOSOLS:** Wet organic peat and muck soils, formed in swamps and marshes; truck crops where drained.

5. **INCEPTISOLS:** Weakly developed horizons; materials have been altered or removed but have not accumulated; usually moist except during part of warm season.

   Andepts: Formed in volcanic ash; woodland, range, and pasture.
   Aquepts: Seasonally wet, organic surface horizon; pasture and hay (Alaska) or woodland and row crops if drained (southeastern United States).
   Ochrepts: In crystalline clay minerals, light-colored surface horizons; woodland and range (Alaska, Northwest), pasture, wheat, and hay (Oklahoma and Kansas), and pasture, corn, and small grain (Northeast U.S.).
   Tropepts (Latosols): In tropical regions; pineapple and irrigated sugarcane (Hawaii).
   Umbrepts: In crystalline clay minerals, thick dark surface horizon; low in alkalis; woodland and range.

6. **MOLLISOLS:** In subhumid and semiarid warm to cold climates; dark surface horizon rich in organic matter, highly alkaline.

   Albolls: Seasonal perched water table, bleached subsurface horizon; small grain, peas, hay, pasture, and range.
   Aquolls: Seasonally wet; gray subsurface horizon; where drained, small grains, corn, and potatoes (north central United States), and rice and sugarcane (Texas).
   Borolls: In cool and cold regions; small grain, hay, and pasture (north central United States), and range and woodland (western United States).

Table III.2   Roots of Soil Terms in the 7th Approximation

Root	Derivation	Mnemonicon	Connotation
acr	Gr. *akros*, at the end.	acrolith	Extreme weathering
agr	L. *ager*, field.	agriculture	An agric horizon
alb	L. *albus*, white.	albino	An albic horizon
and	From *Ando*.	Ando	Ando-like (volcanic parent)
anthr	Gr. *anthropos*, man.	anthropology	An anthropic epipedon
aqu	L. *aqua*, water.	aquarium	Characteristic associated with wetness
arg	L. *argilla*, white clay.	argillite	An argillic horizon
calc	L. *calcis*, lime.	calcium	A calcic horizon
camb	L. *cambiare*, to exchange.	change	A cambic horizon
chrom	Gr. *chroma*, color.	chroma	High chroma
cry	Gr. *kryos*, coldness.	crystal	Cold
dur	L. *durus*, hard.	durable	A duripan
dystr	Gr. *dys*, ill; *dystrophic*, infertile.	dystrophic	Low base saturation
eutr	Gr. *eu*, good; *eutrophic*, fertile.	eutrophic	High base saturation
ferr	L. *ferrum*, iron.	ferric	Presence of iron
frag	L. *fragilis*, brittle.	fragile	Presence of fragipan
fragloss	Compound of *fra(g)* and *gloss*.		See the formative elements *frag* and *gloss*
gibbs	From *gibbsite*.	gibbsite	Presence of gibbsite (aluminum oxide)
gloss	Gr. *glossa*, tongue.	glossary	Tongued
hal	Gr. *hals*, salt.	halophyte	Salty
hapl	Gr. *haplous*, simple.	haploid	Minimum horizon
hum	L. *humus*, earth.	humus	Presence of humus
hydr	Gr. *hydro*, water.	hydrophobia	Presence of water
hyp	Gr. *hypnon*, moss.	hypnum	Presence of hypnum moss
luo, lu	Gr. *louo*, to wash.	ablution	Illuvial
moll	L. *mollis*, soft.	mollify	Presence of mollic epipedon
nadur	Compound of *na(tr)* and *dur*.		Presence of natric horizon
natr	Ar. *natrium*, sodium.		
ochr	Gr. *ochros*, pale.	ocher	Presence of ochric (light-colored) epipedon
pale	Gr. *paleos*, old.	paleosol	Old development
pell	Gr. *pellos*, dusky.		Low chroma
plac	Gr. *plax*, flat stone.		Presence of a thin pan
plag	Ger. *plaggen*, sod.		Presence of plaggen horizon
plinth	Gr. *plinthos*, brick.	plinthile	Presence of plinthite
quartz	Ger. *quarz*, quartz.	quartz	High quartz content
rend	From Rendzina.	Rendzina	Rendzina-like (calcareous)
rhod	Gr. *rhodon*, rose.	rhododendron	Dark-red colors
sal	L. *sal*, salt.	saline	Presence of salic horizon
sider	Gr. *sideros*, iron.	siderite	Presence of free iron oxides
sombr	Fr. *sombre*, dark.	somber	A dark horizon
sphagno	Gr. *sphagnos*, bog.	sphagnum moss	Presence of sphagnum moss
torr	L. *torridus*, hot and dry.	torrid	Usually dry
trop	Gr. *tropikos*, of the solstice.	tropical	Continually warm
ud	L. *udus*, humid.	udometer	Of humid climates
umbr	L. *umbra*, shade.	umbrella	Presence of umbric epipedon
ust	L. *ustus*, burnt.	combustion	Dry climate, usually hot in summer
verm	L. *vermes*, worm.	vermiform	Wormy, or mixed by animals
vitr	L. *vitrum*, glass.	vitreous	Presence of glass
xer	Gr. *xeros*, dry.	xerophyte	Annual dry season

Source: Modified from Steila, D. *The Geography of Soils.* Englewood Cliffs, N.J.: Prentice-Hall, 1976.

Rendolls (Rendzinas): Subsurface horizon with large amount of calcium carbonate; cotton, corn, small grains, and pasture.

Udolls: In midlatitude climates; usually moist, no accumulation of calcium carbonate or gypsum; corn, small grains, and soybeans.

Ustolls: Mostly in semiarid regions; intermittently dry for long periods; wheat and some irrigated crops.

Xerolls: In regions with rainy winters and dry summers; dry for long periods; wheat, range, and irrigated crops.

7. **OXISOLS:** In tropical or subtropical lowland regions; mixtures of kaolin (aluminum clay) and silicon dioxide, low in weatherable minerals.

Humox: Usually moist; high in organic content, low in alkalinity; sugarcane, pineapple, and pasture (Hawaii).

Orthox (Latosols): Usually moist; moderate to low in organic content, relatively low in alkalinity; sugarcane, pineapple, and pasture (Hawaii).

Ustox (Latosols): Partly dry for long periods; pineapple, irrigated sugarcane, and pasture (Hawaii).

8. **SPODOSOLS (PODZOLS):** In humid, mostly cool, midlatitude regions; acid, low supply of alkalis, with compounds of aluminum or iron.

Aquods: In seasonally wet regions; pasture, range, or woodland, and citrus and truck crops (Florida).

Orthods: Accumulation of organic matter and compounds of aluminum and iron in spodic (illuvial) horizon; woodland, hay, pasture, and fruit.

9. **ULTISOLS:** Usually moist but dry part of the year; low supply of alkalis, subsurface horizon of clay accumulation.

Aquults: Seasonally wet; woodland, limited pasture; hay, cotton, corn, and truck crops where drained.

Humults: In warm humid regions; high in organic content; small grain, truck, and seed crops (Oregon, Washington), and pineapple and irrigated sugarcane (Hawaii).

Udults: In continuously moist regions; relatively low in organic content; general farming, woodland, and pasture, and cotton and tobacco in some regions.

Xerults: In regions with rainy winters and dry summers; dry for long periods; low in organic content; range and woodland.

10. **VERTISOLS:** Clay soils; wide, deep cracks when dry.

Pelluderts (Grumusols): Black or dark gray surface horizon.

Torrerts (Grumusols): Usually dry; wide, deep cracks open most of the time; range and some irrigated crops.

Uderts: Usually moist; cracks not open for more than 2 or 3 months at a time; cotton, corn, small grains, pasture, and some rice.

Usterts: Wide, deep cracks open and close more than once each year, usually open for 3 months or more; general crops and range, and irrigated crops (Rio Grande valley).

Xererts: Wide, deep cracks open and close once each year, remain open for more than 2 months at a time; irrigated small grains, hay, and pasture.

Index

Index

Definitions of entries may be found on pages set **boldface**.

ABOUT THE AUTHORS

Robert A. Muller is Professor of Climatology in the Louisiana State University Department of Geography and Anthropology, and he also serves as the State Climatologist for Louisiana. Professor Muller received an undergraduate degree in geography from Rutgers University, and M.S. and Ph.D. degrees in Geography and Climatology from Syracuse University, where he also studied forestry. His first professional position was at the Pacific Southwest Forest and Range Experiment Station of the U.S. Forest Service, where he served as Forest Climatologist. Before coming to L.S.U., he taught at the University of California, Berkeley, and Rutgers University, with visiting appointments at the University of Delaware and the University of Wisconsin at Madison. Professor Muller's research interests and publications focus on applications of hydroclimatology, or the study of environmental and economic interactions and responses to the variability of weather and climate.

Theodore M. Oberlander, Professor of Geography at the University of California, Berkeley, received his doctorate from Syracuse University in 1963. His academic interests center on the ways climate and climatic changes affect landforms and are recorded in them. He is also much interested in the interrelationships between other physical phenomena in climatically differentiated natural regions. Professor Oberlander's published research has focused on the origins of desert landforms of various types. He has conducted field work in Iran, Morocco, Chile, and Hawaii, as well as closer to home in the western United States. He is an avid photographer, and an appreciator and occasional practitioner of most of the graphic arts.